GREEN BERETS Clan Militia & Blue Helmets

The Illustrated History of Army Special Forces in Somalia

Joseph D. Celeski
Colonel U.S. Army Retired

AUSA
ASSOCIATION OF THE UNITED STATES ARMY

UNG
UNIVERSITY of NORTH GEORGIA™
UNIVERSITY PRESS

Published by:
University of North Georgia Press
Dahlonega, Georgia

Cover and book design by Corey Parson.
Cover photographs courtesy of Joseph D. Celeski.

ISBN: 978-1-959203-12-4

For more information, please visit: http://ung.edu/university-press
Or e-mail: ungpress@ung.edu

UNG
UNIVERSITY of NORTH GEORGIA™
UNIVERSITY PRESS

AUSA
ASSOCIATION OF THE UNITED STATES ARMY

This book is dedicated to all the United States Army Special Forces Veterans of the Humanitarian Intervention in Somalia during the period 1992 – 1995, who served, fought, and sometimes lost their life to ensure humanitarian relief operations conducted by the UN were successful, saving the Somali people from disease and starvation.

It is also dedicated to those who also served from the Special Operations community and to the coalition partners who participated alongside them.

Table of Contents

Foreword

The United States' involvement in Somalia from 1992 – 1995 represented one of the last irregular wars fought by the United States military in the 20th Century and served as a harbinger of the complex military operations the nation would become embroiled in as a result of the Global War on Terrorism.

The very nature of operations in Somalia was a rehearsal for the irregular warfare spectrum outlined in recent Defense Reviews. To highlight the importance of irregular warfare for U.S. defense policy, the Secretary of Defense, the Honorable Chris Miller, implemented Congressional Directive 922 establishing the Irregular Warfare Technical Support Directorate within the office of the Assistant Secretary of Defense for Special Operations and Low Intensity Conflict. Major shifts of this kind strengthen the use of Special Operations power and increase its strategic utility. I can think of no better historical example of an irregular warfare case study than the Special Operations conducted during the armed intervention into Somalia.

When thinking about the American involvement in Somalia, ingrained into most Americans' psyche is the battle of Mogadishu by Task Force Ranger against Mohamed Farah Aideed's Habr Gedir militia after the downing of two Task Force Black Hawk helicopters, 3 – 4 October 1993. Since then, many remember Somalia from the vivid images of American soldiers being dragged dead through the streets of Mogadishu. Mark Bowden's excellent book, *Black Hawk Down* [Signet, 2000], followed by the stirring film, illustrated the savagery of Aideed's followers and the bravery and courage of the Rangers and Delta force operators, along with the pilots and crews of the 160th Special Operations Aviation Regiment, during their fight.

What this author found missing from the collective memory was the role of the U.S. Army Special Forces operators (the Green Berets) and their participation in the Somalia campaign. Little is known of the activities of the detachments during the period August 1992 through March of 1995. The purpose of this book was to capture their operations and experiences in the form of a case study before they are lost to history. In their exploits one can also find clashes with clan militia, unceasing work with coalition partners, and a doggedness to get the job done to further the diplomatic and military lines of operation.

A useful tool in crafting the story and conducting the analysis was David H. Ucko's and Thomas A. Marks' "Crafting Strategy for Irregular Warfare: A Framework for Analysis and Action," published by the National Defense University Press in July of 2020. The work guides the strategist towards a framework for a course of action in an irregular warfare conflict and is one of the better guides existing in the literature.

The Department of Defense defines irregular warfare as "a violent struggle among state and non-state actors for legitimacy and influence over the relevant populations." Somalia was a conflict involving armed, non-state actors using primitive, irregular, and unconventional warfare conducted by nontraditional military organizations. Unfortunately, the military lessons for irregular warfare were left behind with the sailing from Mogadishu of the last U.S. Navy vessel during the UNOSOM II extraction under Operation *United Shield* in March of 1995. Somalia would always be viewed as a failed 'humanitarian intervention' rather than the insurgency-like environment and people's war it actually resembled. Irregular warfare studies would not become firmly established in military academics and defense policy, along with how to use force in those environments, until after 9/11 and the war in Afghanistan.

The United States and the United Nations conducted the humanitarian intervention as a test case to resolve conditions resulting from a failed-state (ungoverned space) but refused to clearly see the conflict as an internal war and a civil war, based on ethnic and clan lines. Without question, the U.S.-led Unified Task Force (UNITAF) coalition partners and the United Nations Operation in Somalia (UNOSOM) military contingents were unskilled in irregular warfare and therefore seemed unwilling to tackle the counter-insurgency methods actually needed to gain victory in Somalia. The Army Special Operations Forces (ARSOF) deployed throughout the conflict were counted amongst the few organizations which would come to understand the "context" of the unconventional nature of the conflict – a clan versus sub-clan schism with various actors vying for power (not legitimacy).

While the United Nations struggled to implement a government focused on Western values, the belligerents struggled over a more lucrative form of government to control the resources of southern Somalia. The warlords of Somalia were not attempting to gain power to centrally govern in a democratic manner for the benefit of all Somalis; gaining power to benefit themselves and their clan drove their ambition. The major center of gravity was not the populace but rather clan support over who would rule Somalia. Mogadishu represented the center of military, informational, and economic power for the winner.

Somali resistance movements attempting to oust the dictator first developed as a result of the repressive actions of the Siad Barre government in the 1980's. Like all resistance movements with armed action as their strategy, the various movements were insurgent-like and adopted guerrilla warfare tactics in order to survive against the larger and better-equipped government forces. Many of the warlords united to cooperate against the central government and contributed their militias to the overall effort. With the collapse of the regime in 1990, pro-government forces fled the city of Mogadishu and moved out into the hinterlands to carry on a counter-revolution against the warlord factions.

Within the city, lines were drawn between the Abgal clan forces of Ali Mahdi (United Somali Congress) and General Aideed's Habr Gedir clan (Somali National Alliance) while each leader claimed the rightful title as the new ruler of Somalia. This uncompleted insurgency soon developed into a full-blown war between clan and tribe factions. As the fighting raged and grew more brutal, noncombatants were no longer protected and soon were forced to choose sides to support the various militias and guerrilla armies; whole villages and towns were uprooted in the clash between rivals, resulting in most food resources diverted to feed the burgeoning militias. Combined with severe drought in the region, and looting of the UN's and nongovernmental relief agencies' food supplies by the warlords, famine soon developed, resulting in the humanitarian crisis and the commensurate intervention by the UN.

Failure to analyze the context of the Somalia war led the UN and the United States to mistakenly believe that alleviating the causes of starvation and disarming the militias would assist in moving the Somalis to form a unity government. Like the later intervention in Bosnia, the lack of homogeneity of the clans in the region and the cultural differences as to forms of governance (Western vs. non-Western) would consistently hamper efforts to resolve the implementation of a central government. In the 'you break it, you fix it' model, the UN in essence became the government. The insurgency, civil war, counter-revolution, and then clan war would have to be addressed in consort with other nation-building measures if success was to be achieved. Surprisingly, the mandate for UNOSOM's peace-making (under Chapter VII of the UN charter) was very similar to lines of operation for counterinsurgency: legitimacy, development, and security (UN Security Council Resolution 814). This subtle fact was unfortunately missed.

Missing in the course of action to address Somalia's ills was a method to address the vulnerabilities of the clan insurgents while implementing the UN program. Lines of operation (or lines of effort) to address clan alliances, factional cohesion and unity, the lack of staying power and absence of any military form of shock system by the militias, and the illicit financing system were all vulnerable if the coalition's strategic analysis approached the problem from just seeing it as mere starvation.

The United States military sought different terms for their role during the three and a half years of in Somalia—Military Operations Other Than War (MOOTW), Low Intensity Conflict (LIC), and Humanitarian Activities (HA). Two names they would not call it were counterinsurgency or "nation-building." Although the United States supported Siad Barre in the 1980's with military aid and military trainers under various security assistance programs in his attempts to defeat the clan-based insurgents, there still was reluctance to recognize the effort as a foreign internal defense (FID) program to assist the Somali government in their own version of Internal Defense and Development (IDAD), the very doctrine designed to defeat insurgencies. Instead, America armed and supported the Barre regime to defend against the Soviet-backed Ethiopian military, part of the great power game in the Horn of Africa during the Cold War.

With the UNOSOM II declaration of hostilities against Aideed and his clan after the Habr Gedir ambush of Pakistani peacekeepers in June of 1993, the UN and United States entered the insurgency and clan war more aggressively and by default became counterinsurgents.

Somalia had a long history of clan-based, protracted, non-traditional warfare. While the United States and UN forces may have looked on in contempt at the military operations of the clans, their militias were in fact skilled irregular warfare adversaries. Both Generals Aideed and Morgan were trained militarily in the arts of guerrilla warfare and used their knowledge of non-conventional tactics to great extent against American and UN forces. General Aideed was well studied on the history of the "Mad Mullah" who had conducted a twenty-year resistance against British forces during the colonial period, using irregular forces.[1]

> In Mogadishu, Aideed and the other warlords adapted the traditional Somali way of fighting—small-unit guerrilla campaigns in which raiding parties fought hit-and-run wars—to the urban setting.[2]

It was also the era in the rise of terrorist movements. The Sunni Islamists of the al- Itihad al-Islamiya (AIAI) formed outside ties with Al Qaeda (who claimed they provided training to Aideed's militias in Mogadishu in order to defeat the Rangers during the October 3rd battle). Although terror was used as a tactic in Somalia, guerrilla warfare and primitive warfare were the predominant ways of war in many of the battles.

Just as terrorists and insurgents of today rely on nefarious criminal enterprises to finance and support their operations, Somalia also had its share of bandits, criminality, and drug trade. U.S. military forces were constantly embroiled in operations against this aspect of irregular warfare.

The Army Special Operations Forces deployed to Somalia were deeply involved in the insurgency, counter-revolution, and civil war, many times conducting operations on the front lines between various clan militias and often collocating with the various warlords. While Mogadishu seemed peaceful at times, Special Forces teams constantly deployed into active, front line engagements between ex-government forces and clan militias, and also the lines between warring clan factions. At times, Special Forces operators also clashed with the Islamic fundamentalists.

The experience gleaned in irregular, unconventional, and counterinsurgency arenas during their deployment to Somalia would contribute immensely to their success in the new wars of the 21st century. They had been through the rehearsal for the Global War on Terrorism.

The idea for this book was a result of my tour in Somalia, arriving shortly after the Battle of Mogadishu and leaving back for stateside after Christmas to attend joint schooling. I was keen to begin searching for what our Special Forces had been doing since they began operations in August of 1992, during Operation *Provide Relief.* Upon my return to the United States, I made it a point to begin collecting pictures and literature on the subject. I had been the 5th Special Forces Group (Airborne) S3 Operations Officer during 1992 and into 1993, so I only knew about our operator's deployments from the perspective of a faraway staff officer. In June of 1993, I began my joint tour of duty as one of the ground operation officers in SOCCENT's J3, placing me with their Joint Special Operations Task Force (as the J3) for the Somalia deployment in the fall of 1993. I later served as the team leader and one of the planners in the event of any pending noncombatant evacuation of American citizens in the summer of 1994. I was also on the staff to plan and launch SOCCENT's small JSOTF for Operation *United Shield*, the withdrawal of UN forces, from January to March 1995.

Although I had a good grasp of the activities performed by our operators, I was still missing information concerning the day-to-day missions, firefights, and long patrols executed by the SF teams. What were the decisions for the deployment of SF within the strategic approach and lines of effort? What missions did they conduct, and were they applied correctly for the situation? What was the campaign architecture and how did the use of Army Special Forces help to achieve those objectives? What were the command and control arrangements (as well as coordinating and liaison mechanisms)?

Next in the research was to comb the archives and existing literature for after action reports, operational reports, and identify the context of what drove decision-making, followed by numerous interviews from the Somalia Veterans themselves. Fortunately, it was a time of much picture taking on the part of U.S. military personnel. The picture collection from Special Forces operators and others who served in Somalia grew into the hundreds, with many photos yet to be seen by the public.

I desired to first retire from my thirty-year career with the Army before tackling my first book project. To prepare me for that venture, I was fortunate to contract with the U.S. Army Special Operations Command History office to put together a 200 – 300 page research paper and the beginnings of an archive collection a few years later, pursuant to a thorough written work on U.S. Army Special Operations (ARSOF) activities in Somalia, from 1992 – 1995 (PSYOP, Civil Affairs, Special Forces, Rangers, and the 112th Signal Battalion and the 528th Logistics Battalion). Of most importance in that process was the mentoring and training I received from all the historians on producing a well-written, accurate account of our involvement in Somalia.

In retirement, I chose to serve as a Senior Fellow at the Joint Special Operations University for the next ten years, conducting research and writing on relevant topics to the special operations community. It was a busy time and did not allow for the hours and months to work on a book of the scope I envisioned.

I was always intrigued on the lack of publication of comprehensive SOF official history, most of it left up to private authors. In 2014, I set my sights on a book about another period in Special Forces history of which little was known due to its secrecy, the War in Laos. By 2019, I completed two books on Laos (one about Green Berets and one about Air Commandos) and I turned my attention back to Somalia. With the passing of the years, much had been added to the Somalia research literature and I was fortunate to regain contact with Somalia Veterans who had helped earlier; a new version of Special Forces history in Somalia was warranted with all the new pictures and updated material.

The work is arranged as an open-source, illustrated history of U.S. Army Special Forces in Somalia; the story is told chronologically. The major events during the Special Forces deployments are easily divided into four main categories for the reader. First is a background to understand the country of Somalia and the effect of the clan system in every aspect of life. The background includes a study of clan and primitive warfare, important in understanding how the warlords and militias of the various clans fought. The background also includes events leading up to America's foray into the Horn of Africa, covering the Somalia Civil War, the humanitarian crisis, and our country's first humanitarian gesture to help alleviate starvation and suffering, Operation *Provide Relief.*

The next section covers the largest intervention ever accomplished by the UN, Operation *Restore Hope*. When efforts to provide more aid to starving people in southern Somalia were not enough, the United States then led the almost 30,000 peacekeeper force into Somalia to secure humanitarian relief sectors and stop the theft of food from various clan warlords. *Restore Hope* began in early December of 1992 and lasted until March of 1993. With its success, America departed and turned the operation back over to the UN to administer. A small, residual quick reaction force was left to assist the UN, along with a large logistical support system. The turnover to the United Nations Somalia II (UNOSOM II) Task Force was then named Operation *Continue Hope,* the third major section of the book.

During this period, operations began to hunt down Mohammed Farah Aideed, the criminal warlord whose men massacred a large body of Pakistani peacekeepers in early June 1993. The "Hunt for Aideed" resulted in the deployment of Army Rangers and other special mission units, culminating in the vicious Battle of Mogadishu, on October 3, 1993. In response, President Clinton pledged to withdraw from Somalia, but not without first deploying a Joint Task Force (JTF) and more troops to intimidate any clans believing they could nibble away at the UN's military forces until the American withdrawal by March 31, 1994.

The final part of the book covers a series of evacuations and withdrawals (a string of departures), culminating in the end of the United States' and UN's involvement in Somalia. The first action was to plan a noncombatant evacuation operation (NEO) of American citizens, potentially predicted to occur in the summer of 1994. If chaos resumed in Somalia, hundreds of U.S. citizens and a handful of State Department diplomats would be threatened. Although the NEO did not transpire, thankfully, the State Department did withdraw its

United States Liaison Office diplomats and staff from Mogadishu, along with the Marine platoon providing their security, in September of 1994.

With the United States no longer expressing interest in Somalia, or for that matter, any other foreign contingency resembling a Somalia-like scenario, other nations lost their desire to continue on the UN operation. The UN Security Council voted to withdraw its forces completely by the end of March 1995. The United States provided assistance, and deployed forces in January of 1995 to over-watch the evacuation from Mogadishu, named as Operation *United Shield*.

In each section, background is provided to understand the context of that period with respect to Special Forces' deployment of service. The work also identifies key decisions being made throughout the intervention, and why. Pictures, maps and organizational charts are included within the chapters to further clarify each of these major phases over three and a half years of involvement. Last, the appendices are used to explain clan organizations and Somali political affiliations (know your enemy), as well as a chronology to summarize major events.

"What do Green Berets do?" is an oft-asked question. The number one mission for Army Special Forces operators is the conduct of unconventional warfare. Unconventional Warfare (UW) is conducted 'Through, with, and by . . .' irregular forces – guerrillas and partisans, normally behind enemy lines or in non-permissive environments. The purpose of UW is to assist the populace to achieve its freedom and choose its own government. This mission was derived from the World War II role of special operations in the conduct of guerrilla warfare (GW) in the Philippines and working with resistance fighters throughout Europe under the Jedburgh program. In today's doctrine, this form of warfare is known as *Special Warfare*. In a sense, Special Warfare is one of America's answers to irregular warfare. Special Warfare is reinforced with psychological operations.

The Army Special Forces maneuver unit is the Operational A-detachment (ODA), often called the 'A-team,' consisting of twelve men with combat skills needed to teach and fight alongside up to a battalion of guerrillas or a group of partisans. So, Green Berets are also teachers, instructors and trainers of skills required to fight and survive an irregular warfare battlefield. They are neither Rangers nor Delta Force operatives, who practice more of a direct action role such as raids and hostage rescues, using specialized tactics and equipment.

Army Special Forces are fungible; they can also conduct special reconnaissance (SR), foreign internal defense (FID), and direct action missions (DA). All Special Warfare missions are conducted to obtain operational or strategic level objectives, through a combination of tactical operations. A strategic concept can center on the use of Army Special Forces employing Unconventional Warfare, such as the use of Green Berets with the Afghans in the opening phase of Operation *Enduring Freedom*. When special operations are in a supporting role to conventional forces, the job is to enhance or enable ground force maneuver. Special Forces operators often use irregular forces as an economy of force for a friendly ground commander, and create friction in enemy battle-space. However they are employed, Special Forces are an excellent source of information and intelligence for the command.

Enough time has passed that new sources on the Somalia conflict are being published each year. In the whole, the American experience in Somalia makes an excellent irregular warfare case study. While the bibliography lists relevant works, aficionados of irregular warfare history should also consider a few key works in the bibliography to understand the context of the policies adopted and the decision-making for what occurred in Somalia (to name a few):

Baumann, Robert F. and Lawrence A. Yates. *My Clan Against the World: U.S. and Coalition Forces in Somalia, 1992 – 1994*. Fort Leavenworth, KS, Combat Studies Institute, 2004.

Clarke, Walter and Jeffrey Herbst, eds. *Learning from Somalia: The Lessons of Armed Humanitarian Intervention*. Boulder, CO: Westview Press, 1997.

Hirsch, John L. and Robert B. Oakley. *Somalia and Operation Restore Hope: Reflections on Peacemaking and Peacekeeping.* Washington, DC: Institute of Peace Press, 1995.

Rutherford, Kenneth R. *Humanitarianism Under Fire: The U.S. and UN Intervention in Somalia.* Sterling, VA: Kumarian Press, 2008.

Stevenson, Jonathan. *Losing Mogadishu: Testing U.S. Policy in Somalia.* Annapolis, MD: Naval Institute Press, 1995.

Although Colonel Celeski completed the manuscript and had worked extensively on its revisions and historical documentation, he passed away before the book could be finalized for publication. This edition stands as a testament to his commitment to preserving the legacy of Special Forces service in Somalia and to advancing the study of irregular warfare.

Endnotes

1. Shultz, Richard H. Jr. and Andrea J. Dew. *Insurgents, Terrorists, and Militias: The Warriors of Contemporary Combat.* Columbia University Press, NY, 2006: pp. 86 – 100. Shultz and Dew provide an excellent overview of the clan-based, warrior traditional way of warfare and its implications to conventional forces. The study has a comprehensive chapter on U.S. and UN intervention in Somalia.
2. Ibid, 95.

Acknowledgments

First and foremost, this publication would not exist without the help and assistance of the Green Beret Veterans who served throughout southern Somalia as well as in Mogadishu and the Bossaso area in support of UN operations. It becomes important to fill this gap in special operations history before many more years pass and the knowledge becomes forgotten.

Very little detail exists of the experiences and combat of the U.S. Army Special Forces teams and their units while they served in the UN armed intervention in Somalia, from 1992 – 1995. This work serves as an unclassified, open-source story on their role in conducting *Special Warfare* to accomplish the mission, along with inclusion of rare photos never-before seen. As an illustrated history, it also includes maps, diagrams, and organizational charts to help the reader.

First, it was members of the Special Operations Command (Central) who taught me a lot while I served as the J3 of the JSOTF-Somalia in the fall of 1993, and later as a ground plans officer in the J3 at SOCCENT's headquarters in Tampa, Florida. Their mentoring and patience were instrumental to build my knowledge and experiences in Somalia and for teaching me how SOF should be applied correctly in an irregular warfare environment. Key among the staff for mention are: Tommy Smith, Gary Danley, Ed Townsend, Tim Hess (now deceased), Rich Stimer, Paul Holthaus, Dave Plumer, and Al Glover. They were also helpful in providing materials and pictures for this book. General Bill Tangney gave all of us a great opportunity and trusted our initiative.

Dr. Charles Briscoe, the USASOC Headquarters Command Historian at Fort Bragg, NC, along with Mike Krivdo, Eugene Piasecki, and Troy Sacquety (Command Historians) were instrumental in providing personal guidance and direction on the thesis, layout, and design parameters for the initial research, conducted 2006 – 2007. I would also like to thank their IT and documents staff who generously gave time to help and work on photos and documents.

I would also like to thank Ms. Roxanne Merritt at the Special Warfare Museum at Ft. Bragg, NC. The documents, pictures, and artifacts on the period of the Somalia contingency were invaluable.

It is also important to recognize the patience and assistance from the staffs of other museums and history archives which held artifacts and collections concerning the role of special operations during the UN peacemaking (under Chapter VII of the UN Charter) intervention in Somalia.

Four organizations were extremely helpful in providing materials: the Civil Affairs and PSYOP commands at Ft. Bragg, NC; the 16th SOS gunship squadron at Hurlburt Field, FL; the 5th Special Forces Group (Airborne) at Ft. Campbell, KY; and the Special Operations Command-Central Command (SOCCENT). Each organization provided a wealth of information as well as contact information to reach other Somalia Veterans.

The 10th Mountain Division historian was more than helpful to spend a day going through extant records on file as to their participation in Somalia. The materials were helpful to put their role in context, along with the use of the Special Forces from the 5th SFG(A) as a participant on some of the Division's operations.

Thanks to the Naval Special Warfare SEALs and their commanders, who served alongside or in support of our deployed Special Forces companies. Mike McGuire and Tom Bunce were of great assistance to help tell the story of the attached SEAL snipers who supported the SF from October 1993 through the U.S. withdrawal in March of 1994.

My thanks to the members of the U.S. Army PSYOPs and Civil Affairs communities for their initial research contributions, although this book is not specifically focused on all of their activities. Both of these organizations were an important enabler for the teams to get their job done in a socially complex, failed state. Bob Biller gave me a thorough understanding of how a Civil Affairs Detachment does their job in Somalia. Jim Treadwell and Charles Borchini were keen to ensure I had access to PSYOP records and to the files at the PSYOP battalion headquarters.

I would like to also thank our coalition partners from Canada. During one of my trips to lecture at the Canadian War College in Kingston, Ontario, the library and research staff was extremely helpful in retrieving what documents they held about the Canadian service during Operation *Restore Hope* (one of our SF companies served alongside them in Belet Weyne). If you are an aficionado of U.S. coalition operations in Somalia, you must obtain a copy of their excellent history, "*In the Line of Duty*" about the Canadian Joint Forces in Somalia during 1993, published by the Land Forces Command.

Notably, it was the Veterans of Somalia who contributed most to ensure a complete as possible story of this endeavor can be told. Archival materials, books, historical facts and figures provide the framework for military histories, but nothing can replace the actual experiences of those who participate and fight in irregular wars and conflicts. This kind of resolution and context is only found by listening to the Veterans themselves. In most cases the Veteran's recollections are the only existing accounts of what actually happened on the ground.

The stories and pictures in the work serve as the testimony of this time in their lives. However, some spent long hours and months during the project to ensure its completeness and to provide materials. Not all of the hundreds of Veterans, historians, and archivists can be mentioned in this small space, but significant is mention of those who readily spent personal time and effort to see the project come to fruition (many from the 5th Special Forces Group): Dave Asher, Jose Bailey, Pat Ballog, Ken Barriger, Joe Bovy, Ken Bowra, Lance Caffrey, Steve Cain, Lelon Carroll, Jon Concheff, Tom Daze, "Moe" Elmore, Bill Faistenhammer

Jr., Helen Fogarassy, Wendell Greene, Mark Hamilton, "Hawk" Holloway, Dave Jesmer, Tim Knigge, Kent Listoe, Frank McFadden, Kevin Murphy, Chip Paxton, Gary Ramsey, Bill Robinson, C.B. Smith, Dan Weber, Glenn Wharton, Bryan Whitman, Chaplain Wylie, and Rickie Young. There were many others, who I hope understand my appreciation of their mentoring and receipt of their materials throughout the project, even if space here precludes me from their mention.

Special thanks go to my wife Judy Celeski, who was extremely patient and supportive through this project. I would also like to thank Abbie Rindlisbacher and Ashley Rindlisbacher who both provided administrative assistance and help with artwork.

I would also like to thank the editor and staff of the University of North Georgia Press for their patience and input to make the work as complete and professional as possible.

Last but not least is acknowledgement of the professional job of various editors and colleagues to correct the work and make it the best possible story on this first complex irregular warfare conflict conducted by the Special Forces in consort with the United Nations. It is through their patience and diligence that the quality of the final work was ensured.

Joseph D. Celeski
COL (Ret.), USA
2021, Buford, Georgia

Introduction

If you liked Beirut, you'll love Mogadishu.

— Ambassador Smith Hempstone (U.S. Ambassador to Kenya)

With the victory over Saddam Hussein during Desert Storm in 1991 and the subsequent collapse of the Soviet Union, the "new world order" foreign affairs philosophy emerged to define a new era in international relations. The post-Cold War era began with America emerging as the remaining superpower, creating a unipolar world to replace the old bipolar world of the West versus the Soviet Union.

In a speech given on September 18, 1990, Charles Krauthammer delivered a lecture in Washington, which was later printed in the *Foreign Affairs* magazine, on the subject "The Unipolar Moment:"

> It has been assumed that the old bipolar world would beget a multipolar world . . . The immediate post-Cold War world is not multipolar. It is unipolar. The center of world power is an unchallenged superpower, the United States, attended by its Western Allies.[1]

This philosophy would soon be put to the test when the new world order met Somalia in 1992.

The term "new world order" has been used to describe those periods in history which upend existing political structures and the balance of power arrangements. Examples include the fall of the Roman Empire, the defeat of Napoleon in Europe, and the victory over Germany in World War I. What these epoch-changing scenarios have in common is when nation-states act together for the greater good, global problems can be solved rather than if attempted by one nation-state. This environment allows nation-states to focus on collective security or choose non-intervention as foreign policy tools, or acting in concert with other nation-states to tackle global humanitarian problems such as starvation, poverty, disease, and so on. The relative periods of peace in a new world order environment allow the time and space to use a country's elements of national power which may heretofore have been tied up in costly wars and conflicts.

International organizations, such as the United Nations, become useful as organizing mechanisms for nation-states to combine their efforts under one responsible authority. Such was the nature of the Gulf War when the UN led the effort to identify Saddam Hussein's invasion of Kuwait and provide the authority for President George H.W. Bush to lead a multi-national coalition to victory.

Seeing what could be accomplished, the new secretary general of the United Nations, Boutros Boutros-Ghali, began his office in January 1991 with the intent to use the UN in a more decisive way to solve problems along the North-South axis, now that the East-West tension was gone. Whether the UN was ready to deploy member states militarily, or if the UN even had the military expertise for a grand adventure, Boutros-Ghali chose to intervene in Somalia's civil war. Somalia, wracked with drought, starvation, and now declared a failed state, would serve as just the right challenge for the new world order.

He could have picked other challenges, but the newly emerging 24-hour media chose Somalia. With the United States emerging as a world leader from a bipolar world to what foreign policy analysts considered a unipolar world, the media emerged to cover humanitarian disasters and hopefully influence foreign policy decisions to help solve them. This has been called the "CNN effect" or the "media effect."

Select studies differ on what the effect is, exactly. Certainly, the media had a role in shifting non-interventionist and isolationist foreign policy positions to interventionist, primarily using humanitarian disasters as the means.[2] This position was no more apparent than in the constant news coverage of people starving and dying in Somalia. Out of all global problems, the Somali drought and starvation got the most media attention, framed as a moral issue for the world (Steven Livingston's "accelerant" effect; two other effects media can have on policy makers are to (1) reflect decisions policy-makers make then reinforce their decisions and (2) serve as an impediment to morale and operational security by exposing horrors and agendas–all of which occurred during the American involvement in Somalia from 1992 through the spring of 1995).[3]

Thus, intervening in Somalia was chosen over any other intractable problem on the North-South axis. Since America led in the Gulf War, Boutros felt the United Nations could lead this effort, evoking Chapter VII of the UN Charter "Peacemaking" rather than Chapter VI "Peacekeeping," since Somalia (mostly southern Somalia) was ungoverned. The road to Mogadishu began.

Somalia – The Failed State

There was a no more complex irregular warfare environment than that found in Somalia. Unfortunately, in the haste to intervene under humanitarian auspices, very few soldiers and marines on the ground were familiar with the intricacies of the land, geography, and human terrain. The clan system complexity alone boggled the minds of many commanders when issuing ultimatums or seeking negotiations. Where and when a peacekeeper served in Somalia differed from town to town, from one humanitarian relief sector to another, and from region to region, shaping each person's experience and recollection. Just as a peacekeeper felt they'd achieved a good grasp of the situation overall, it was time to leave. A larger context to all wars derives from the human dimension–the political struggle and tailored military operations shaping the daily lives of not only the inhabitants but also those who found themselves fighting an elusive enemy. Within this larger context, the Special Forces teams conducted their mission alongside their UN counterparts.

For the military, entering a contingency operation begins with the planners and commanders assessing the situation, known as strategic appreciation. This step provides the basis for understanding the application and use of force. If the situation is properly understood, the use of force when applied will match the conditions on the ground.

Geography, Terrain, and Weather

Somalia is among the largest land masses comprising the Horn of Africa. It appears in shape as the number seven, if viewed with its small tilt to the right. What can be hard to grasp for many is its size. If superimposed over the East Coast of the United States, Somalia would stretch from Albany, New York, through Norfolk, Virginia and down to Atlanta, Georgia; the northern portion of Somaliland would reach westward to Detroit, Michigan. In a more compact form, it roughly equals the area of Texas in size. Its southern coastline is along the Indian Ocean while its northern coastline is along the Gulf of Aden. The Straits of Hormuz separate Somalia's northern tip from Yemen. Somalia is bordered to the southwest by Kenya, to the northwest by Ethiopia, and to the north by Djibouti.

Looking from the shore inland, Somalia is formed by a set of flat, maritime plains and plateaus that then change into rolling hills. As one travels further inland, the land is arid and desert-like with mostly scrub vegetation; some mountain ranges begin the further north one travels. There are two major wet seasons–late spring and late fall–which only deliver about 500 millimeters of rain. Fortunately, it has two major river systems which have almost permanent flow: the Shebelle (named after the leopard), flowing north to south out of Ethiopia to outside Mogadishu, then paralleling the coast southwestward to end in conjunction with the second major river, which is the Jubba, near Kismayo (the Jubba also originates from Ethiopia). It has no major lakes. Almost all other water is seasonal and drawn from water wells or man-made ponds. In periods of no rain, Somalia experiences severe drought.

The Jubba and Shebelle produce the most fertile areas in Somalia, which include marshland, tall grasses, swamps, and lush thickets. These areas are also the most fertile centers for agriculture. Trees and small forests dot the landscape along the rivers. Large swaths of Somalia are used for farming, agrarian livestock, or pastoral purposes.

Somalia is hot and humid in nature. The temperatures can reach between eighty and one-hundred degrees Fahrenheit in the spring. Low temperatures range from the sixties to the eighties.

The harshness of the environment ensures living does not come easy. Food, fuel, water, and access to medical care in vast areas left inaccessible by a weak government meant most Somalis were very protective of the resources they had available. Without question, a clan or sub-clan would readily raid another to feed their families and have access to grazing grounds and water while protecting their own possessions.

Population and Culture

A country's physical place on the map, its geography, and terrain contribute to the shaping of its culture and locations where populations reside and fight. The modern Somalis descended from an original Nigeria/Kenyan-based Eastern-Cushitic ethnic group and became the "Samaal." The Samaal originated from the Omo-Tana sub-group and then its sub-group, the Sam. It is believed the Samaal began to inhabit the entire Horn of Africa around 100 A.D. Ancient Arab and Persian immigrants brought the Muslim religion to Somalia (Sunni) and called the Somalis Berberi. The Somalis were configured in patrilineal clans, mostly nomadic pastoralists, and further divided into sub-clans. In Somali society, clan affiliation began to define all parameters of life–politics, religion, and economic status. As described in the 1993 Country Area Study handbook on Somalia:

> Historically, Somalis have shown a fierce independence, an unwillingness to submit to authority, a strong clan consciousness, and conflict among clans and sub-clans despite their sharing a common language, religion, and pastoral customs.[4]

Most of the country is populated by four majority clan groups (called the noble clans, direct descendants of the Samaal). Following these coalitions, who make up the majority group, are the two minor clans, which are together regarded as the minority group. The four noble clans make up the nomadic pastoral group (70%): the Dir, Darod, Issaq, and Hawiye. The other two minor clans make up the agriculturalists: the Digil and the Rahanweyn (about 25% of the populace). They speak a common Somali language, with variations that developed in the coastal regions and riverine areas. Secondary languages include English and Italian. A common Somali script was introduced in early 1973.[5]

The Clan Structure and Demographics

Early Somalis were nomadic pastoralists and agriculturists. (The camel is a revered animal in Somali culture for all that it brings a family: food, milk, and transportation.) The products from both of these livelihoods were traded along ancient travel routes from Sudan, Ethiopia, and Kenya. For instance, both Kismayo and Mogadishu became trading ports along the Indian Ocean where Somalis brought bananas, livestock, and leather goods in exchange for spices, foodstuffs, gold, cloth, and other implements to improve their lives. Other major towns sprung up along the inland trade routes.

With scarce resources and grazing land, pastoralists became highly possessive of their stock and over access to grasslands and water wells. The Somali herdsmen were charged to protect family and clan wealth and were trained as *Waranle*, spearmen warriors (conversely, these skills were also useful for raids on another clan's livestock and territory). This responsibility made Somali men–toughened by their harsh lives and environment–experienced, natural, and nomadic guerrilla fighters. The clan system became competitive over power, land, and resources in a belief that "my loss is your gain."

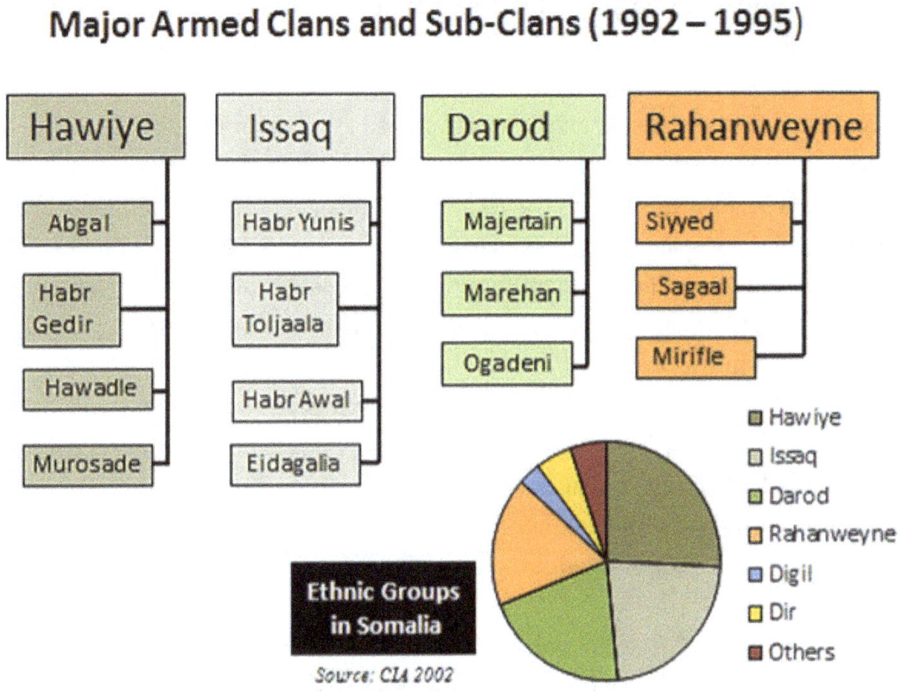

In the traditional social structure, clan and village elders ruled using the *xeer* system (customary law). Very senior clan leaders were called *soldans*, while local and village clan elders were called *oday*. Lacking any government, police, or judicial presence in many areas, the Somali people designed a set of cultural rules to handle transgressions within the community and bring swift punishment to perpetrators. No clan could afford to fight for long periods, given the lack of resources, and no clan could afford to lose men or livestock. Like other tribal cultures, Somalis designed a system to punish the worst of crimes (murder) yet spare life. The perpetrator, or his clan, had to offer a blood payment, known as the *Diya*. Thus, justice was meted out and no further life or resource was lost.

Some Somalis populated the coast, where a mercantile trade grew. The Chinese, Persians, and Arabs all plied their wares along coastal Somalia. Fishing also became a food source and provided goods for trade. A few coastal towns emerged that could be serviced with smaller boats or use lighterage from ship to shore.

Some of these towns included Hobyo, located in the northeastern coastal region of the country, and Merka, in the southern region bordering on the Indian Ocean. Mogadishu, a port city in southeastern Somalia, emerged as the preeminent trading center and grew to city size in a short span of time. It would eventually become the capital of Somalia.

In the 1980s, Somalia was a constitutional government with a prime minister and a council of ministers. Everyone 18 years and older had the right to vote for representatives in the national assembly–under Siad Barre, it was renamed the People's Assembly. Governance at the local and regional level was decentralized from whence it came, in the hands of respected elders or clan leaders. If one peeled away the veneer, all Somali politics was clearly clan based. The exclusion of major groups not part of the ruling government clan became the impetus for opposition, calls for secession, or regime overthrow. This disenfranchisement, combined with few resources for education, resulted in poor pay for the average Somali working to make a living.

Under colonial rule, the people of Somalia were poorly educated. There was an Islamic University in Mogadishu and there were some secondary schools for technical and vocational training, but no other educational facilities existed in the country at that time. In 1970, the Somali National University was founded and built (it was abandoned during the civil war and served as a base for American troops during their tours to Somalia).

Siad Barre, who rose to the rank of Major General in the Somali military, became President of Somalia in 1969 after the assassination of President Shermarke. Under Barre's vision of "scientific socialism," all media was controlled by the government through the Somali News Agency (for print) and the Somali Broadcasting Department, which ran Radio Mogadishu and various broadcasts on television that consisted of talk shows, news, and music.[6]

After independence, Somalia established a uniform legal and penal code based on the Italian jurisprudence system and English common law. This setup was created because, prior to independence, Somaliland was under British rule and at one point under a UN trust managed by the Italians. A system of lower courts allowed for limited application of Sharia law and clan customs. After the appellate courts, the government ran the higher court with judges, courtrooms, and prisons. Local and national police enforced the law.

Prior to the civil war, Somali maintained the largest military in the sub-Saharan region. In 1990, it boasted 65,000 personnel in the Somali Armed Forces (SAF): Navy, Air Force, Air Defense, and Army (land forces). The Somali National Army (SNA) mustered twelve divisions of infantry, mechanized infantry and armor, plus air defense, field artillery, and smaller elements of commandos and paratroopers. Able-bodied men mandatorily served in the Home Guard for a six-month period. They also had a small Navy. The Air Force was the most modern military arm, with contemporary Soviet and Italian aircraft. The squadrons were specialized into fighter, fighter-bomber, counterinsurgency, and helicopter squadrons. The Somali Armed Forces were supplied with arms, equipment, funding, and training from the Soviet Union, then the United States, China, and Italy (Under the Somali-Soviet Alliance, the Soviet Union loaned Somalia $55 million to update their army and to increase their military personnel). The SAF was equipped with Soviet, Italian, and U.S. gear.

The military was one of the few areas where a Somali could improve their economic condition and make something of life; Siad Barre favored the military over any other branch of government. To augment the SAF, Somalia also fielded a Border Guard, People's Militia, and some para-military police. Specialized units included a commando unit and a parachute regiment.

A military academy was located in Mogadishu, along with a war college. For Non-commissioned Officers (NCOs), General Daoud ran an academy in Kismayo. Other technical and specialty skills training occurred in a variety of foreign countries. Police training was primarily held in the Federal Republic of Germany.[7] The 1974 Treaty of Friendship and Cooperation was terminated by Siad Barre in 1977. At that time all training and military advice from Moscow ended. To increase a U.S. presence in the area, the U.S. agreed to provide military assistance and training to Somalia.

By 1992, Somalia had been wracked by another wave of severe droughts (1989–1991). This, along with clan fighting, transformed the country into a failed state. While these factors severely harmed the average Somali household in various regions of the country, the experiment with socialism, its centralized economic plans and heavy money borrowing to stay afloat, completed the ruin in the rest of the country. The military was slowly deteriorating and disintegrating. Much of the countryside's population had fled to towns and cities to escape the fighting; refugee camps abounded and exacerbated the government's problems. Approximately eight million Somalis were living throughout Somalia, mostly young in age. Militant Islam arose with the collapse of the state.

Few, if any, government services were available, so a subsistence economy arose. The state of health among the populace suffered and was primarily influenced by inadequate nutrition. Without sufficient doctors, hospitals, medical clinics, pesticides, and medicine, respiratory and intestinal parasites caused a rise in malaria cases. Many of the refugee camps also became breeding grounds for cholera.[8]

Mogadishu remained the principal port with an international airfield, although it was poorly managed. Telecommunications were almost nonexistent. Out of 21,000 kilometers of road within the country, only 2,600 kilometers were paved.[9]

History

Even though Somalis occupied the land with a common ethnic culture and language (very homogenous), prior to 1960 Somalis were not able to claim the country as free and independent. A succession of rulers from Oman, Yemen, Zanzibar, and the Ottoman Turks occupied and later ruled the country up until the colonial period, which ran from 1885 to 1960.

In 1885, the British occupied north-central Somalia, known as Somaliland. In short order, the French staked their claim to Djibouti. The British would be the first to suffer as outside occupiers when a resistance movement led by Mohammed Abdul Hassan, later dubbed the "Mad Mullah," used tribal guerrilla warfare against Ethiopians and the British in the Ogaden, a large Ethiopian territory bordering western Somalia. It is named for the Ogaden Clan that is part of the Darod Clan of Somalia. The first irregular war in the Horn of Africa persisted for 20 years.

The Italians arrived at southern Somalia in 1935. They captured British Somaliland at the beginning of World War II, but in 1941 Britain regained its colony. As for further punishing Italian transgression, the British also seized southern Somalia and the Ogaden region. During the war, a form of Somali nationalism began to rise, leading to the first political organization formed to seek independence–the Somali Youth League (SYL). After World War II, the newly created United Nations made southern Somalia a trust territory under Italian administration. The UN Security Council also returned the Ogaden region to Ethiopia. The Somalis' resentment and disdain for these actions made by outside powers on their land was palpable.

As most of the Western powers' colonial systems fell apart in the post-war years, Somalia gained its independence in 1960, becoming the Somali Republic. The hard part was deciding how to govern. The lack of clan unity was very frustrating to the early politicians. Since power was also a commodity for a clan, the ruling group's politicians ensured their clan reaped the benefits and rewards from their office and position, leading to corruption and cries of nepotism and favoritism. Those clans outside the ruling groups sought the Somali solution to rebalance the goods–President Abdirashid Ali Shermaarke was assassinated on October 15, 1969. In the confusion which followed, General Mohammed Siad Barre (a Marehan, a sub-clan of the Darod) seized control.

So far, so good, except it was an inauspicious moment in history for any of the Horn of Africa countries' rulers. The Horn became a Cold War-contested spot between the United States and the Soviet Union, threatening the free flow of oil, which was a major source of trade.

The Horn of Africa in the Cold War

Prior to World War II, the United States practiced a policy of isolationist views with respect to the South African region. Although Ethiopia and its Eritrean territory could claim to be a pro-Western leaning country with a Christian religious population, it did not figure into any grand geo-political strategy for the United States. This fact was apparent even during the invasion and seizure of Ethiopia by Mussolini early in the war. United States activities reflected concern but practiced benign neglect when the League of Nations condemned the act. The League of Nations had no muscle behind its assistance of Haile Selassie, the Emperor of Ethiopia. Superpower rivalries were non-existent during this period, and Somalia was ignored by the U.S. government.

After World War II, U.S. interests in the region were based on the Eisenhower Doctrine reacting to new superpower rivalries–which focused on replacing Soviet aid and support around the world with Western or U.S. support. As the British withdrew their forces from Ethiopia, fear arose that the Soviet Union would move into the region, particularly threatening the emerging oil markets and shipping routes in the Red Sea region. Some of this concern originated from a new ally in the region (based on the oil markets), Saudi Arabia, with their concern to protect this vital region from influences detrimental to the oil market. (They were also concerned with continuing stability on their eastern flank, which was threatened by radical Shia Islamic influences, and included Iran's activities. Thus, they, too supported stability in the Ethiopian government to counteract Islamic influences in Djibouti, Somalia, and, to some extent, Eritrea.)

Major General Mohamed Siad Barre led the opposition against the Somali President. Upon the assassination of President Abdirashid Ali Shermaarke, Barre then assumed the Presidency via a coup onOctober21,1969. Barre (a Marehan from the Darod clan), led a repressive, dictatorial regime to implement "scientific socialism" as a client-state for the Soviets. His policies led to a civil war and the country's disastrous ruin. 21 October Road in Mogadishu was named in honor of the coup (Courtesy of Tom Daze).

Pan-Arabism exploded in the region, capitalized upon by the Soviet Union. Communist-Marxist inspired movements, blended with socialism, characterized newly emerging Arab and nationalist governments like Egypt, Somalia, and, in some cases, Sudan. The Horn of Africa and northern African Islamic states took on a new American foreign policy descriptor, the "arc of crisis." In Somalia, this emerged as Pan-Somalism, or the desire to return to the "Greater Somalia" past. (The Somali flag is a five-pointed white star on a light blue background–the five points of the star represent Greater Somalia, the five lands of the Somali people: Djibouti, Ogaden, Northern Kenya, and the Italian and the British colonial holdings.)

The State Department reaction, aligned with the new strategy, resembled containment. Led by the U.S. Secretary of State, John Foster Dulles, the United States attempted to build a southern zone of stability in the Middle East, with the Horn of Africa as one of the foundations for the strategy. Ethiopia was a natural pick as one of the bulwarks against Soviet aggression. The Emperor readily joined, accepting U.S. aid to build up his military and granting the United States an intelligence eavesdropping post at Kagnew station in Asmara, Eritrea. This access alone drove U.S. policy through Ethiopia and heightened the Eritrean people's desire for independence. The updated strategy also favored Ethiopia over Somalia, affecting the dispute over the Ogaden region. When Siad Barre could not depend on the United States to build up his military to challenge the Ethiopians over the Ogaden, he turned to the only other superpower which would arm him with no strings attached–the Soviet Union.

The Soviets chose Somalia and Yemen as their proxies in the Horn. Socialism and Marxism were highly appealing ideologies to the downtrodden and "the wretched of the earth," promising a better way of life and the amelioration of grievances.[10] Whether Barre bought into the spin or not, he certainly had to act as if the ideology the Soviets were peddling was best for Somalia (if he wanted to receive aid). He soon patterned Somali life after a communist-like socialism–he declared his form of Marxism was "scientific socialism." To honor his communist sponsors, he renamed the country the Somali Democratic Republic. The government was renamed the Supreme Revolutionary Council (SRC). It evinced a complete lack of understanding about what the Somali clans and their Islamic faith could bear.

The Ethiopian annexation of Eritrea on November 14, 1962 came with the United States' tacit approval and acquiescence by the UN (even though, earlier, the UN Security Council adopted a favorable resolution on the notion of Eritrean independence via a referendum). The stinging words of John Foster Dulles struck the Eritreans during his speech in front of the UN Security Council in 1952:

> From the point of view of justice, the opinions of the Eritrean people must receive consideration. Nevertheless, the strategic interest of the United States in the Red Sea basin and considerations of security and world peace make it necessary that the country . . . be linked with our ally, Ethiopia.[11]

To exacerbate the problem in the Horn, the Somali irredentist factions blamed the Ethiopians over land disputes, resulting in Siad Barre overreaching to create Greater Somalia. His failed conflict in the Ogaden War (1977–1978) caused a geo-political shift; Ethiopia courted the Soviet Union for military aid the United States would not provide, concerned over a major war in the Horn between Ethiopia and Eritrea and between Somalia and Ethiopia over the Ogaden. United States policy then shifted to support the Somali dictator, once it was clear the Soviets would no longer support him militarily. America needed military basing in the Horn to counteract Soviet moves. After the United States shifted its support to Somalia when Ethiopia moved to a Marxist-socialist government (followed by USSR economic and military support), the Reagan administration shifted U.S. policy on the Horn from containment to one of détente. Reagan's policy would be to confront the Soviets anywhere they attempted to threaten U.S. security interests. The United States began a small series of military aid measures to Somalia, providing military equipment and U.S. Army military training teams (MTTs) from the 5th Special Forces Group (Airborne), beginning in 1982.

Thus, Army Special Operations Forces (ARSOF) were already familiar with Somalia prior to the United States and UN intervention in 1992. The United States government initiated a series of Security Assistance programs, beginning in 1980, to gain influence in the region and promote stability within the Somali regime. These measures included foreign military sales, IMET (International Military Education & Training) programs, and combined exercises with U.S. forces. As part of this agreement, Somalia gave the United States access to a variety of airfields and ports throughout the country, and Somalia even participated in Egypt's *Bright Star* exercise in the early 1980s. Special Forces Non-commissioned Officers (NCOs) who participated in this exercise would later run into a junior officer they met during that *Bright Star* iteration when they deployed for Operation Restore Hope–the notorious General Morgan of the Somali Patriotic Movement (SPM), an anti-Aideed faction operating in the Jubba River region around Kismayo.

Security assistance equipment included weapons, ammunition, anti-aircraft systems, light armor, and much more. Along with equipment came the requirement for security assistance training teams to aid the

Somalis in integrating the new systems into their defense organizations. These security assistance-equipping programs provided a variety of opportunities for Special Operations Forces (SOF) trainers and a formal TAFT (Technical Assistance Field Team), normally consisting of up to three or four Special Forces personnel, who were assigned to the Office of Military Cooperation (OMC) of the U.S. Embassy in Mogadishu. The members of the SF-comprised TAFT came from the 5th Special Forces Group (Airborne). The team helped implement a variety of programs for light infantry, anti-tank missile, and MOUT (Military Operations on Urban Terrain) security assistance programs involving small groups of SOF throughout the 1980s. TAFT members served for a one-year tour. They resided in Mogadishu and worked at a Somali field camp used for conscript training, located about 60 miles outside of the city. Combined exercises were conducted, some involving parachute operations with the Somali Commando Brigade.

Army Special Forces personnel from the 5th SFG(A) also participated in the six-month MTTs (Military Team for Training) to Somalia in 1985. The MTTs were comprised of a few SF soldiers training Somali Army and Somali Military Police.[12]

For its part, the U.S. Central Command (CENTCOM) conducted a series of combined exercises concurrent with various security assistance missions in support of attempts to promote stability in Somalia. The Special Operations Command for CENTCOM, SOCCENT, participated in these training exercises by deploying various joint SOF from its components. In September 1986, SOCCENT conducted a combined operations exercise focused on light infantry skills with a three-week airborne course running concurrent with deployment. The 2nd Battalion of the 5th SFG(A) (-) comprised the U.S. Army SOF element for the deployment and formed into a combination of two and a half Green Beret companies out of their three organic companies.

SOCCENT deployed a light headquarters under the leadership of LTC James Fletcher. The unit deployed to Bale Dogle on USAF aircraft and occupied an old Chinese camp of adobe buildings located south of Bale Dogle. Part of the camp was already inhabited by road construction employees. Prior to their arrival, a three-man Special Forces TAFT team arrived at the U.S. Embassy in Mogadishu to assist in implementing the security assistance program.

The 2/5th SFG(A) commander divided his forces into three training groups: one to conduct weapons training, one to conduct maneuver training, and one beefed up Operational Detachment–Alpha, ODA561, to conduct the airborne course at Bale Dogle. The SF weapons and maneuver cadres conducted light infantry skills classes for the Somali Army. The trainees were veterans of the Border War with Ethiopia (Ogaden War) and arrived on foot after walking miles to reach the training area. Due to ongoing tensions with Ethiopia, the entire command under SOCCENT conducted force protection drills throughout the deployment period.

The airborne school began its week one training after constructing ground-week (the first week of airborne training) platforms built out of wood. Some of the early Special Forces trainers and advisors included CPT Jerry Hill, SFC Bill Rambo, MSG Henry Beck, and SGT John Haines (pictured above).[13]

The USAF provided one C-130 aircraft for approximately 40 students. The T-10 parachute was used for Somali training. SGM Sloniger was on the training cadre and remembers aspects of the jump week:

> On the 1st day of Jump week, it was windy in the morning and the guys were getting drug through the thorns or landing on the PSP on the runway. Our SF team members were running all about to try to retrieve them. As a result of this first day's problems, we had our guys go back over the PLF

> and popping capewells class. On the second day, one of the Somali soldiers started yelling loudly on the DZ to 'pop capewells'. Unfortunately one of the students still in the air at about 90' took him literally and popped his, falling to the ground and dying. We stopped training because of the laws of the country concerning death and buried him. CSM Simms formed up the unit by the cemetery, conducted a memorial service, and had us all do the "gravestep march."[14]

The SF cadre fixed this problem and continued the training. A graduation for the airborne students was held at Bale Dogle, and many local Somali people came to watch.

SOCCENT conducted a similar exercise involving ARSOF in July 1989, once again at the Chinese camp location near Bale Dogle. The USAF were relying on a previous Combat Control Team (CCT) survey of the Bale Dogle airfield, indicating the runway and facilities were sufficiently constructed and in place to support air operations, but the airfield survey team deployed for this exercise ascertained that the runway would need to be hardened to support aircraft. From a USAF representative at the location, they hired 100 Somalis to do the manual labor and make the airfield suitable. They also found out the Chinese-made radar at the airfield did not work and would be of no help for air traffic control.

A SEAL team element (under LCDR Randy Goodman) deployed along with a Civil Affairs element to round out the force list. The Green Berets and SEALs focused on Military Operations on Urban Terrain training for members of the Somali Army, under the command of the 64th Area Commander, Lieutenant Colonel Farah, who remained in Mogadishu for most of the exercise. (MAJ Steve Sabarese was one of the Special Forces officers in charge of training.) The Civil Affairs element conducted well-digging projects in the local area. Again, the TAFT team made up of SF soldiers resided in the U.S. Embassy to facilitate the Security Assistance program implementation, working for the Office of Military Cooperation under Colonel Zarimba's direction.

Early in July, members of SOCCENT and its components were attending a U.S. Embassy dedication when, on the 9th, the Italian Bishop for Mogadishu, Salvatore Colombo, was assassinated. Large demonstrations by Somalis, mainly from clans not aligned with the President, erupted in the streets of Mogadishu; clan against clan fighting ensued, and the infamous "Technicals" came out on the streets. Siad Barre's son, the commander of the notorious Red Berets, moved his unit to the streets to quell the rioting, resulting in several deaths and the mysterious massacre of Somalis, mainly from the Issaq clan.[15]

During the increased tension, Captain (USN) Goodman later recalled, "During the fighting that month, the Somali Minister of Banking and a couple of Ambassadors came out to our camp at Bale Dogle for protection. We put all of this in a report to CENTCOM, and then conducted a no-notice recall."[16]

With growing instability, followed by regressive and repressive regime actions on the part of Ethiopia and Somalia against their opposition factions, the Carter administration began a slow withdrawal from the region in 1989, based on the president's human rights policy, which became an overwhelming influence on American foreign policy actions. The U.S. still harbored wounds from Vietnam and other interventionist failures, so no effort was made to support the insurgent and separatist movements in the Horn. Not readily noticed, but slowly and carefully, radical Arab and Islamic regimes took the opportunity to fill the vacuum and gain influence across the Horn, working with Muslim minority social and political movements.

With the now-apparent Soviet domination of the Horn, perceptions from those in charge regarding U.S. policy in the region did not change until the fall of Iran and the Soviet invasion of Afghanistan. The newly announced Carter doctrine, one of protecting the region against threats to United States national security interests (including oil-based economic resources), attempted to put some muscle against the now-recognized threat. However, the newly formed Rapid Deployment Force (RDF) failed, as did other missives of policy, and in effect no major change occurred in U.S. influence in the region. Benign neglect once again became the unstated policy direction for the Horn of Africa.

Disaffection from the Somali populace over the new socialist system strengthened. Somalia's failure to resolve the land disputes with Ethiopia in the Ogaden region added to the populace's grievances. Resentment grew over the government's handling of the affair. In response, Barre began a series of repressive measures, aimed mostly at the northern region of Somaliland. As this measure from the armed forces began and were combined with government corruption, political opposition parties formed under government elites and northern Somalis. These entities also became subject to Barre's oppression. The differing clans, under the new political parties, militarized and, for once, were unified about something–starting a civil war to overthrow the regime.

Political and Armed Opposition

The first major opposition to Barre grew from his repression of opposition clans. The Somali Salvation Front (SSF) was formed in 1978 in central Somalia from the dominant Issaq clan. The Majertain in central Somalia formed the Somali Salvation Democratic Front (SSDF), which later grew to include many Issaqs. Barre's response in the north to political opposition was soon perceived as a purge of the Issaq clan; he arrested multitudes of opponents, banned trade into Somaliland, and confiscated their transport and maritime assets. Civil war broke out in the northeast region.[17]

In 1981, a new opposition party arose in western Somaliland, branched off, and met in London–the Somali National Movement (SNM), a more militant, separatist party. In response, the regime doubled down on its repressive measures in the region. Government punitive actions began in Hargeisa and Burco, and ranged to Galkayo and Garoowe. To assist the Somali Army, Barre recruited other clans: the Ogadenis and the Dulbahantes (sub-clans of the Darod).[18]

In 1987, Barre felt strong enough to invade the Ogaden region in Ethiopia to return it to the fold and fulfill his vision of a "Greater Somalia." The Ethiopians, with Soviet assistance, lent him a handy defeat in 1988. Barre blamed elements of the Issaq clan and their officers for the lackluster performance. This criticism was his response to the soldiers' seeming lack of fighting skill and enthusiasm during the incursion and the opposition over how the war was managed.

The targeting and persecution of Issaqs led to unrest in the Somali military. The civil war's cost in the north was ruining the economy–Somaliland had already suffered an economic collapse due to the destruction wrought by the Somali Army. As if seeing the final chapter and collapse of the regime, many politicians and government officials tried to line their pockets and benefit their families and clans while they could. Corruption was rampant. When the military planned a joint exercise in Mogadishu in July 1989, riots broke out throughout the city.

As the country spiraled to its doom, more and more political opposition parties began to emerge. Among the Hawiye (south-central Somalia), the United Somali Congress (USC) formed in January 1989 after their delegation met in Italy. Following them, the Somali Patriotic Movement (SPM) was formed by COL Omar Jess, who deserted the Army with three-hundred followers in Hargesia and moved southwest near

Kismayo to establish the movement. His was among the first political movements in the Doble, Afmadow, and Kismayo region to be attacked by government forces.

The regime antagonist in the region, General Hersi Morgan (Siad Barre's son-in-law), moved south to the Kenya border region accompanied by Siad Barre and his fleeing followers, along with remnants of the Somali Army. Once secure in that region, Morgan established the Somali National Front (SNF).[19]

To defend the Jubba River Valley from Morgan, the Rahanweynes formed the Somali Democratic Movement (SDM). They became loosely aligned with the USC and SPM.

The armed groups of the three major anti-Barre movements consisted first of elements of the populace where the central government lost control. The first group was disaffected Somali Army and other military personnel, who deserted to join the anti-regime movements. The second element came from opportunists: gangs, criminals, and brigands. The third element came from arming the sub-clans who had suffered most under the regime's brutality. This clan and sub-clan alignment against the government during the civil war sowed the seeds for what was still to come, clan versus clan warfare. An accelerant to the war was the thousands of tons of weapons and ordnance lying around Somalia and the replenishment from illicit arms shipments.

By 1990 the Barre regime was in battle with the three armed fronts (mostly drawn along clan lines): the Issaq based Somali National Movement from the Northeast; the Hawiye clan-based United Somali Congress from central Somali (mostly to the north of Mogadishu); and the Somali Patriotic Movement from the southwest near the Jubba valley and Kismayo.[20]

The three factions' strategy was to aim their effort towards Mogadishu and remove Siad Barre and his loyalists. It would take the unified effort of the clans to access power and take back the state–no other path was readily available. After overthrowing the dictator, a new president for the country could be established. The struggle in Mogadishu to remove Barre lasted from December 29, 1990 to January 26, 1991. Much destruction and devastation was wrought on the city and countryside, resulting in thousands of people being displaced. The first major United Somali Congress (USC) armed faction to enter the city was Hawiye sub-clan militias belonging to the Somali National Alliance (SNA). The 1,500 militiamen were led by Mohammed Nur Galaa–fighters wore white headbands inscribed with USC in red letters along with "God is Great" armbands.

There were two major force commanders entering Mogadishu: Ali Mahdi of the Abgal sub-clan responsible for northern Mogadishu, and Mohamed Farah Aideed of the Habr Gedir sub-clan, taking over central and southern Mogadishu. It was Aideed's forces which ultimately ousted Siad Barre and his henchmen from the city. During the fighting, old clan enmities and *faidas* (tribal feuds) came to the fore as Marehans, Darods, and Ogadenis were targeted for removal, confiscation of property, or killing.

Thus, the USC declared its right to the capital and to forming a new administration. Of the two armed factions in the city the USC chose Ali Mahdi (from the Somali Salvation Alliance) as the provisional President of a transitional administration, effective January 28, 1991. Immediately, Aideed and other political factions rejected the USC's move to install Mahdi.[21]

After consolidating their victory in Mogadishu, the United Somali Congress turned to gaining control over the rest of southern Somalia through anti-Darod clan measures. Their initial move was an attack on the Somali Patriotic Movement militia in the town of Afgoye, just to the north of Mogadishu. (Perplexing, since the SPM was on Aideed's side–perhaps they were too close for comfort and the SNA needed the arms and ammunition stored in Afgoye.) The first strategic goal was clearing any opposition and regime remnants from central and south-central Somalia, down to Kismayo. From there it was imperative to clear remaining loyalist forces from the Jubba River line to the Kenyan border.

The former regime and Somali Army fighters gathered and formed a new defensive line near Kismayo and the Jubba River line, organized politically as the Somali National Front (SNF), with the goal to counterattack and retake Mogadishu.

By April 1991, the Somali Patriotic Movement and the Somali Salvation Democratic Front were aligned against the United Somali Congress. Militia of the Somali National Movement (SNM) in Somaliland who fought against the regime seceded from the transitional administration on May 18, 1991.

For all intents, the civil war was over and a new phase of war began—the "War of the Clans." The United Nations and the United States entered Somalia in this environment. Thinking the intervention in Somalia was for a noble humanitarian cause, they soon found themselves in an irregular warfare conflict.

Endnotes

1. Krauthammer, Charles. "The Unipolar Moment." *Foreign Affairs*, 69/5: (Winter 1990/91), 23.
2. Robinson, Pierce. "Media as a Driving Force in International Politics: The CNN Effect and Related Debates." E-International Relations, https//www.e-ir.info/2013/09/17/media-as-a-driving-force-in-international-politics-the-cnn-effect-and-related-debates/,ISSN 2053-8626, Sept. 17, 2013, 2.
3. Livingston, Steven. "Clarifying the CNN Effect: An Examination of Media Effects According to Type of Military Intervention." Boston: Harvard University JFK School of Government, The Joan Shorenstein Center, Research Paper R-18, June 1997, 2.
4. Metz, Helen C. *Somalia: A Country Study* (4th Edition. Washington, D.C.: Federal Research Division, Library of Congress, 1993, xxi.
5. Ibid, xiv–4.
6. Ibid, 158–171.
7. Ibid, 195–198.
8. Ibid, 110–111.
9. Ibid, 139.
10. The title of French psychiatrist Frantz Fanon's 1961 book which analyzed the dehumanizing effects of colonization upon individuals and nations.
11. Catlin, John D. "Ethiopia: A Case Study for National and Military Strategy in the New World Order." USAWC Military Studies Program Paper, 11 Aug.1994, USAWC, Carlisle Barracks, PA, 8.
12. Interview with CSM Jose Bailey conducted by COL (Ret.) Joseph D. Celeski 8 Feb.2006 at Ft. Bragg, NC.
13. Piasecki, Eugene G. "If you liked Beirut, you'll love Mogadishu. An Introduction to ARSOF in Somalia." *Veritas*, Vol. 3, No. 2, Fort Bragg, North Carolina, 21.

14. Interview between COL Joseph D. Celeski and SGM Sloniger 1/3rd SFG(A) during OEF, at Kabul Military training Facility, late spring of 2003.
15. ______________. Somaliland Forum Press Release 2005-01-11.www.somalilandforum.com, 1 Aug. 2006.
16. Conversation between COL Joseph D. Celeski and CAPT Randy Goodman, conducted at JSOU, Hurlburt Field, FL, summer of 2005 during JSOTF Training Conference.
17. Kapteijns, Lidwien. *Clan Cleansing in Somalia: The Ruinous Legacy of 1991.* Philadelphia, PA: University of Pennsylvania Press, 2013, 83.
18. Ibid, 85–86.
19. Harned, Glenn M. *Stability Operations in Somalia 1992–1995: A Case Study*. Carlisle Barracks, PA: Personal Monograph Series PKSOI (19950612 009) United States Army War College PKSOI: War College Press, July 2016, xi–8.
20. Kapteijns, 98–99.
21. Ibid, 125–133.

1

Irregular Warfare and the Somali Way of War

> But the conditions of small wars are so diversified, the enemy's mode of fighting is often so peculiar, and the theatres of operations present such singular features, that irregular warfare must generally be carried out on a method totally different from the stereotyped system. The art of war, as generally understood, must be modified to suit the circumstances of each particular case. The conduct of small wars is in fact in certain respects an art by itself, diverging widely from what is adapted to the condition of regular warfare, but not so widely that there are not in all its branches points which permit comparisons to be established.
>
> — COL C.E. Callwell's work, *Small Wars*

Somalia surprised peacekeepers—they weren't looking for war, they had been sent to conduct peace enforcement. But there was no peace to enforce. Peacekeepers were trained for humanitarian intervention and to deal with the types of violence inherent in a failed state lacking law and order: actions against bandits, outlaws, and thieves along with riot control measures to quell angry mobs of people at food sites. These strategies were not adequate training for the war they found themselves in. A belief among military decision-makers was if Somali warlords and militias wanted to take on American forces, they would be going up against a military which had just resoundingly won *Desert Storm* in Iraq with unmatched superiority in numbers and technology. (*Caveat*: Technology used in combat is a two-edged sword. Irregular warriors can employ it just as well as trained soldiers or adapt it in simple form for their use.)What was happening in Somalia, however, would be categorized as complex irregular warfare fought against nontraditional warriors who were using semi-modern military weapons and technology.

When Siad Barre was ousted from Mogadishu in January 1991, the Somali Civil War had changed goals to not only destroying the remaining Darod clan regime element through the opposing clans but also shifted to a new effort by the United Somali Congress (USC) to gain overall power. These events can be characterized as clan militia warfare–the Hawiye clan was now in power and took measures to ensure they removed any potential threats from other clans that might have eyes on power-sharing or were seeking to take power themselves. Clan warfare is a form of irregular warfare, elevated above a feud between clans. The object of removing other clans from the equation was to obtain additional power, territory, and resources.

Irregular Warfare

In 2020, the Department of Defense recognized irregular warfare as "a persistent and enduring operational reality." Irregular warfare is defined by the military as the "struggle among state and non-state actors to influence populations and affect legitimacy." The American military must then be prepared to fight both traditional wars, state on state conflicts with contending armies (and with the political objective achieved through military victory), and irregular wars (focused on the population).

If irregulars are fighting against a friendly government, they are often called insurgents or guerrillas. American support in this case comes in the form of Foreign Internal Defense (FID) and counterinsurgency (COIN). If irregulars are aligned with the government and are helping to fight a common foe, they are partisans, resistance forces, militia of some type, or also called guerrillas. Special Operations Forces (SOF) are often used to conduct Unconventional Warfare with these forces to support the conventional force maneuver. Terror is also a common tactic in this type of warfare, so often some form of counterterrorism (CT) is needed.

Because resistance forces are not formally trained and equipped, they must play their weaknesses against the strengths of the foe. This manner of tactics in this style of fighting is historically known as guerrilla warfare. In Somalia, there was no recognized government. Armed bodies of gunmen aligned with their clans and sub-clans. In this unsophisticated "gray stew," many irregular warfare characteristics were used when fighting. Overall, two different types of warfare clashed in Somalia. Western traditional warfare faced off against irregulars using nontraditional warfare. Local forces had both asymmetric and non-linear aspects in their strategic forms (nontraditional). Rarely in military history do the irregular forces adapt and change to conduct traditional warfare, so it falls to the traditional and formally trained military force to adapt to its nontraditional antagonist. The Western way of war evolved from the Clausewitzian model of war, adopting principles from Clausewitz's work, *On War*, which outlines state-on-state warfare to secure the strategic interests of contending powers. When diplomacy fails to solve a national security threat, then armies, air forces, and navies are unleashed to gain the political object of conflict. Traditional military forces are professionally raised, trained, and equipped by the state. They conduct combat using time-honored principles which, if orchestrated correctly, secure victory (proscribed by doctrine).

The Western way of warfare also has rules which prevent further horror than merely killing and destroying the opposition's military and resources. There are rules of war which limit just how far one may go to achieve victory and not fall into barbarism. Western armies are regulated by honor, tradition, laws of conflict, and rules of engagement to ensure their conduct is acceptable to international norms and that their armed responses are "just and proportionate."

Irregular warfare, on the other hand, is not based on theory, dogma, or doctrine nor bound by very many Western notions of rules. These types of wars are conducted as tribe versus tribe, clan versus clan, or one ethnic group against another. These conflicts may also arise as tribes, clans, or ethnic groups join against outsiders. The principles of war for nontraditional warriors are a result of experience and customs, not books, laws, or war colleges.

The objectives of primitive warfare are not strategic objectives or policy goals–primitive warfare is conducted for honor, revenge (vendettas), hatred, theft of territory and property, or any other emotionally-charged cause when one tribe or clan feels like they have been slighted by another. Many irregular wars emanate from religious differences in the populace or a desire to gain independence from colonial oppressors.

Primitive warfare is often unrestrained and unorganized violence (no Grotius being studied in the village). In most cases, the populace is not off limits. In its worst form, ethnic or clan hatred manifests as genocide or clan-cleansing.[1]

The two military types could not be more unalike. In traditional militaries, armed forces are made up of soldiers (or sailors, or airmen) under a regulated chain of command. They are well trained, disciplined, wear a formal uniform, and carry and employ standard equipment. They have a consistent order of battle, and they fight conventionally. Professionals in Western militaries abhor war, knowing the destructive toll it can take on their own forces but also the destruction it sows for other communities.

On the other hand, nontraditional forces are loosely organized, with no formal structure. The warriors or their leaders are rarely products of staff academies or war colleges. The uniform is non-standard, in an almost wear-what-you-want-or-can-procure fashion. The same goes for weapons–many of the weapons warriors use are handmade, procured off battlefield exploitation, or bought in a local black market (or from a smuggler). There are no existing support structures such as a logistical system, medical system, and so forth. A lack of resources precludes any such standing organizations. Improvisation and adaption are the hallmarks of irregular warriors.

Warriors organize and fight with their clans or tribes and often choose their own leaders, or a powerful clan or tribal leader takes charge of a set of followers who he then provides for with weapons and food, and, hopefully, pays them for their service. These groups almost always exist in a martial society. Nontraditional tactics are unconventional and asymmetric, evolved from tribal or clan rituals and customs. Irregular warfare is characterized by small actions, skirmishes, ambushes, and minor battles and is very often protracted in its execution.

Attributes of Irregular Warfare

In irregular warfare, the reason for fighting is not a victory for its end state; irregular warriors fight for personal honor and glory, lack of resources, or to solve grievances. Resolution is mostly centered around an emotional response. Warriors also fight as a means of advancement or to gain prestige and honor. Historically, most of the clan and tribal fighters have been men while the society's women remain servile; they can and do urge their menfolk to fight. Some of this encouragement is manifested during ceremonies or gatherings through ritual dances and self-mutilation. Primitive warriors are clannish and respond quickly to slights of honor or to losing face due to cultural norms. If from a religious society, there is also a sense of fatalism (why worry, what will be will be). Many ask why a nontraditional society would destroy its livelihood and infrastructure during fighting, as if not sensing it all will have to be rebuilt tomorrow.

Clan warriors live a subsistence life, which takes up most of their time. This situation provides little time for military training or any martial arts specialization. The result is a body of non-professionalized (or martially incompetent) gunmen or militias, all in small-sized units.

Some clans like Aideed's, who led the Somali National Alliance, were more disciplined and well-armed. They reaped the benefits of their conquest against the government and hauled in many modern weapons such as machine guns, tanks, rockets, and artillery. The SNA differed from other opposition militias–Aideed was well-educated, studied war, and received training in foreign military schools. He had his militia trained by ex-servicemen and outside non-state actors. Overall, he was also a good and charismatic leader.[2]

The clan served as the mobilization base for armed militias relying on kinship cultural norms. Thus, a clan-based system of warlords and their militias existed in Somalia. Since large armies maneuvering with large formations is not in the mix for nontraditional warfare, irregular warriors tend to gravitate toward terrorism, guerrilla warfare, and civil war as their form of armed action due to their lack of military resources, often resulting in an internal war versus another conflicting state.

Many irregular fighters are known for their lack of discipline in following orders, keeping cohesive maneuver, or any form of fire control, often choosing to "spray and pray" when firing automatic weapons versus using aimed fire. With this lack of discipline, clan warriors often retreat in the face of power.

As for why and when to go to war, every culture has the right to self-defense. Therefore, a certain segment of a tribe or clan must be prepared to defend their village if attacked. There were several root causes for the complex irregular war that was created and fought in Somalia. As mentioned earlier, the lack of resources resulting from the earlier civil war's devastation and the drought that destroyed food crops, animal herds, and livelihoods were all factors for the collapse of the state and descent into continual irregular warfare. There was also ethnic tension based on a discriminatory and exclusionary political and social system which kept Somali society divided and non-cohesive. Overall, one clan seeking power at the expense of another fueled the civil war, which then resulted in the clan militia war beginning in 1991.

In nontraditional societies, (and in some cases, nondemocratic regimes) power and control lies in the hands of a few (for example, Iraq, Russia, the Taliban). The select powerful make the decisions to go to war, not the clans, particularly when there is no state. Thus, nationalism is rarely the motive for nontraditional societies who go to war.

Authoritarian leaders at the local level also choose subordinate leaders and bestow awards, or punishment, on the fighters. Bravery and courage in battle is admired while cowardice is loathed. Although not governed by international law or rules of land warfare, most nontraditional armed groups consider the killing of women and children shameful, although a populace is not off limits if they are caught participating with the enemy.

Another motive for going to war is economics. Wealth for tribes and clans does not arrive as money but as commodities such as livestock, land, drugs, guns, and so on. In Somalia, a person's wealth could be counted in his camel or livestock herds. A warrior culture arose from the need to train herdsmen how to fight off predators and bandits. As pastoralists, Somali men were used to roaming far distances and having a sense of adventure. This made them tough men, used to deprivation and austere resources, perfect material to transform into a guerrilla or insurgent force.

Another factor for going to war is to resolve a grievance of some kind once frustration builds enough to take action. A feud is not considered a war-like activity and is below the threshold of what one would consider war. Feuds are fueled by hate and loathing and can take years to resolve.

Very few warriors become farmers; herding livestock provides mobility when it is necessary to move or flee. Food for crops can be taken and consumed on the go. Leaving food behind to an enemy was not uncommon as vegetables and fruits could be found another day, or one could return to the village and hopefully still find the crops and fields intact. However, leaving crops and fields behind could pose a dangerous risk when rival clans or militias use food as a weapon and destroy the crops, which did occur in Somalia during the civil war and drought years. Clan fighters were rarely mobilized from the agriculturalists who had a love for the land and were more sedentary. This sector of Somali society did not pose a threat to other clans or peacekeepers. They were considered weaker clans and less than warrior-like. This attitude among the warrior clans in Somalia often led to populace mistreatment in those areas and prisoner abuse by more war-like clans.

There exists an inherent desire within a tribal society for fertile grazing land, good water wells, and markets to sell or buy materials. In Somalia, a complex system of clan-ownership for land and water developed due to these needs. Any violation of the rules could spark a clash of clans.

Another means to increase wealth (or possessions) was through looting and banditry. Surprisingly, many Western humanitarian workers in Somalia observed that militias and gunmen did not destroy the vehicles and trucks they pilfered; the vehicles represented an upgraded form of mobility for their ragged armies and could be used to transport forces.

Special Warfare and Special Forces Utility

U.S. Army Special Forces experienced three distinct types of irregular warfare phases in Somalia: armed intervention, warfare against Aideed's clan and his SNA supporters, and counterterrorism activities. In each of these examples of irregular warfare, Green Berets were used to enhance conventional force maneuver or as extensions of American foreign policy objectives. When their application conformed with their doctrinal mission sets and was within their capabilities, the operators were successful. Irregular warfare is one of the environments where Special Operations Forces can reach their highest operational and strategic utility due to their unique skill sets.

Operation Provide Relief was not one of the three cases where Special Forces clashed with Somali irregulars. The United States had no national security interests at stake. Green Beret mounted teams were used in an unconventional manner to solve an unconventional problem: how could USAF aircraft operating into Somali be secured to deliver food relief. Although they faced daily risk during the flights, the Special Forces teams did not engage with Somali irregulars during any of the missions. The five months or so of operation did, however, help to increase environmental and clan system dynamics, which proved useful once Operation Restore Hope was launched.

Operation Restore Hope provided the opportunity for Army Special Forces to conduct almost every one of their doctrinal missions and perform Special Warfare (unconventional warfare and PSYOPs) tactics. The primary goal for American forces was to provide security for humanitarian relief, not feed the population. During this operation, Green Berets provided the most value to conventional forces through conducting Special Reconnaissance (SR) to find clan militias, identify trouble spots, and reconnoiter for weapons and mines in the hands of Somali irregulars. Most notably, Special Forces teams were used to contact various warlords and their militias, conduct negotiations with them, and assist UNITAF's efforts to gather heavy weapons systems into cantonments (a form of disarmament). No sides were chosen and none of the SF teams maneuvered with clan militias during this Unconventional Warfare (UW) mission. Other than destroying stockpiles of seized arms and munitions, the Direct Action (strikes and raids) mission set was not needed. Special Forces teams also conducted Foreign Internal Defense (FID) to train and advise foreign forces (coalition partners from more than twenty contributing countries included the Australians, the Belgians, the Batswana, the British, the Canadians, the Czechs, the Egyptians, the Finns, the French, the Italians, the Jordanians, the Moroccans, the New Zealanders, the Pakistanis, the Saudis and Zimbabweans to name a few).

These missions were not without risk or danger, however. Teams operating in the Kismayo, Belet Weyne, and Oddur Humanitarian Relief Sectors were all subject to firefights and minor clashes with Somali irregulars. The first Special Forces operator was killed in Somalia during *Restore Hope* in the Belet Weyne Humanitarian Relief Sector. Elsewhere, a Special Forces detachment clashed with al-Itihad al-Islamiya extremists near Luuq in the Bardera HRS, the first known clash with Islamic extremists in the country.

Operation Continue Hope was characterized by combat with Aideed's SNA forces, primarily in and around Mogadishu. After UNITAF (a humanitarian peacekeeping mission) successfully stopped starvation in the country, the mission was turned back over to UNOSOM II (a nation building mission). One Special Forces company was detailed to support the UN Task Force and the 10th Mountain Division QRF. Teams participated in the attack on the Aideed enclave in June 1993 and were involved in sniper operations against Somali urban irregulars up to the October 3rd Battle of Mogadishu. While supporting the mission in Mogadishu, the Special Forces company also assisted UNOSOM II to extend their reach into north-central and northern Somalia, furthering the UN's diplomatic mission and providing vital area assessments. These special reconnaissance (SR) missions were conducted by the mounted teams and, in some cases, took weeks to perform.

As a result of the October 3rd battle in 1993, American forces were reinforced with a Joint Task Force. With President Clinton's announcement of departing Somalia by March 31, 1994, the primary mission became force protection. Special Forces did their part to assist with calls for fire from AC-130 gunships and employed sniper teams around various U.S. and UNOSOM II garrisons.

When the United States reentered Somalia in 2014, Special Forces operators conducted Foreign Internal Defense as part of the mission to support AMISOM forces and train a counterterrorist and counter-guerrilla force for Somali Armed Forces (SAF). The *Danab* (Lightning) Brigade is now one of the most professional and well-trained units in the SAF and has seen significant combat against the al-Shabaab terrorist organization. During Danab Brigade direct action missions (DA) against al-Shabaab units and military infrastructure, SF operators helped plan and accompany the unit on combat operations to advise on tactics, provide intelligence, and assist during calls for fire support. This security assistance mission to "train and advise" continued up to January 2021; President Trump ordered all American forces out of Somalia prior to turning over his administration to President Biden. (However, American military observers from AFRICOM with Joint Task Force–Quartz were still allowed to participate in combined military exercises with the Danab and Somali Army in February 2021.)

In May 2022, President Biden authorized the return of 500 U.S. military personnel to support the mission in Somalia as part of achieving national security interests and continuing the campaign against terrorists in Somalia.

The Somali "Way of War"

The Somali way of warfare can be characterized by Clausewitz's nonlinear nature of social interactions in war. In his article for the *Joint Force Quarterly* [JFQ 96, 1st Quarter 2020], Brian Cole described this condition as a *Complex Adaptive System* as it relates to Clausewitz's Trinity by framing it within chaos theory. In Somalia, Clausewitz's "war is an extension of politics" did not exist due to a collapsed government. Thus, "primordial violence, hatred, and enmity" in Clausewitz's trinity replaced rationality and reason. This led to unpredictable behavior on the part of Somali agents (actors) in the "complex adaptive system."[3]

Cole writes that there are agents of the system who act on society, other agents, and the environment. This concept provides the framework for understanding warlord strategies in Somalia–a social-psychological endeavor (missing the political element). The agents of the system were the warlords and their subordinates, working as a network of agents acting in parallel; no activity could be fixed (predictable) as control in this environment is highly decentralized. At times a single agent dictated where battle would occur–other times, an aggregate of agents would act in a manner that was interdependent. Somalia military analysts became fixated on Aideed and activities in Mogadishu, but they were hard-pressed to see or find the interdependent-ness and interconnected nature of agents executing his strategy. [4]

Cole found that having various types of agents in a population gives rise to unpredictability. To prevent such chaos, he recommended that some sort of "moderating force" needed to be applied to the agents. If opposing agents become too close to one another, then there is a tendency to violence. If they are distant, then less violence occurs between them. Again, the challenge for military analysts and leaders was in determining just what that moderating influence could be to predict or anticipate where combat could occur.[5]

When peacekeepers arrived in 1992, Somalia was in the middle of an internal war between clans and sub-clans, protracted in nature and very much resembling the early phases of a Maoist-style guerrilla war (without the communism). It was also not a religiously inspired war. Somali clan wars were fought out of hatred and enmity of the "other." It was the 20th-century version of nontraditional warfare fought by a militaristic society which had adapted modern technology to their fighting style. Somalis were non-conforming when it came to any notion of acting like a modern, Westernized military. Conventional abstractions like rules and civilized legal norms for war were absent.

Militias were motivated by their loyalty to the clan and saw it as a sacred duty to kill those they hated outside the clan. In a sense, it was a war of feuds, without politics involved. Clans mobilized fighters and supported them once fighting began. The clan, not the state, represented the enduring supreme military body. The sub-clans were used as a command and control organizing mechanism.[6]

Assessment of Military Capabilities

Undoubtedly, there was a deficiency of thorough military assessments to provide deploying peacekeepers with accurate information on the Somali clans and how they fought. Although most militaries have a system for analyzing their enemy's military capabilities (like a net assessment), as the armed UN intervention into Somalia occurred, these assessments were based on patterns practiced by analysts during the Cold War.

A good military assessment will identify the enemy's strategy and operational approach (method) to fighting. In Somalia, it was cultural–hatred of others and outside intervention, layered with anti-colonial sentiment. Identifying the "style" of warfare can bring to light the ways clan militias conduct firepower and maneuvers.

Other questions for analysts include who the enemy is, where he operates, what his order of battle is (or force structure, that is, the size and purpose of his organizations), which doctrine and training is used, and how they fight (tactics, weapons employment, and so forth). These relate to the principles of war and functions practiced by a military body.

There is a myriad of other details that need to be addressed: Who is the leadership? What comprises the clan militias and their affiliations? What uniforms do they wear? What is their equipment, and where do they get it from? What are the beliefs and ideologies among the populace?[7] Answers to these questions will contribute to the overall military assessment. In addition, understanding the leadership within a military unit is a vital part of the analysis.

Leadership and Command and Control

An effective military body is always characterized by good leadership and a clear method to command and control a disciplined body of warriors. In Somalia, the clan warlords were first the leaders of their people, not the militias. A good commander maintained control over both. Depending on the clan or sub-clan family's culture and experiences, leadership could also be communal, where decisions could be gained through the elders' consensus in a meeting. The major warlords in Somalia were self-elected and had some level of military prestige they gained over their careers (i.e., General Aideed, General Warsame, Colonel Jess, General Morgan, and others).

In clan societies, community leadership is also attained through private and personal leadership, not by rank or appointment. Such was the case with Ali Mahdi, chosen by the United Somali Congress (USC) to head a new administration in Mogadishu. Each type of leader exerts a style based on their upbringing, and it is important for the military analyst to develop biographical backgrounds on all major contenders to help understand the types of decisions they may make, or better still, to use the information to develop schisms between them and their subordinates.

Clan leaders may rise to a level of lore and romanticism among the populace if they are involved in a "championship duel." Such was the case between Aideed versus UNOSOM and Admiral Howe. (The same situation occurred with Saddam Hussein taking on the UN coalition during *Desert Storm*; even though he lost, he was a hero in the eyes of his people for challenging the United States.) This trap should be avoided.

One thing is consistent, and was so in Somalia for nontraditional warriors, discipline was authoritative and sometimes brutal.

Due to the distances between groups and lack of modern communication sources, the command and control in Somali was decentralized. This condition should have been exploited by peacekeepers to isolate and contain pockets of irregular militia throughout the country.[8]

Planning and Campaigns (Operational Art)

Few tribes and clans go off to fight for strategic reasons. However, the Somali clans did demonstrate the planning required to conduct a campaign with the objective to oust the dictator. Achieving this goal required forming an alliance between the United Somali Congress, the Somali Patriotic Movement, and the Somali Salvation Democratic Front.

Aideed exhibited the acumen for developing a campaign plan for the Hawiye clan and Somali National Alliance (SNA), both during the clan wars and after the UN intervention. Aideed's strategy had the following objectives:

- Hawiye clan family domination of central and southern Somalia;
- Elimination of contenders;
- Rebuff all UN efforts for reconciliation and conduct a protracted campaign until they left Somalia;
- Adopt guerrilla warfare and adapt guerrilla warfare to the urban environment;
- Inflict unbearable casualties on outsiders (attrition);
- Control key centers of gravity (Mogadishu, Kismayo); and
- Conduct an aggressive PSYOP campaign.

On the other hand, ambushes and raids require little planning. With the clan warrior's individualistic nature, planning remained simple enough for all to understand. Due to the repetitive nature of raiding and ambuscades, militia leaders found very little reason to plan for what they knew worked from experience.

A plan for battle did not include much more than a straightforward march to battle, with fighters told to move forward, left, or right. Since most irregulars were not highly trained, standardized and simple maneuver was required. Some militia squads were designated to conduct enemy encirclement (surrounding), a holdover for how the clan hunted prey. There were few examples of Somali militia developing a battle line prior to offensive action–battles became chaotic melees.[9]

In Somalia, if there was advanced planning it took the form of ensuring secrecy and operational security prior to an attack. General Morgan, for example, showed a keen sense of planning when he infiltrated Kismayo at night to attack Colonel Jess's men in 1993.

Somali Principles of War

Somalis appeared to apply some of the principles of war in their own cultural way. They did exhibit the capability to achieve surprise before their attacks. They were also able to conduct fire and movement and seemed to have an understanding of their objective (the mission's purpose). They could achieve mass in the sense of gathering the clan or sub-clans and using "swarm" tactics. They did not mass in the manner of gathering several bodies of large military formations to conduct a combined arms operation. Mass could not be achieved through the gathering of disparate militia alliances.

The war in Somalia was not one of position. Such a war would require some level of expertise to conduct operational maneuver. The Somalis could execute tactical maneuver through movement, not operational maneuver. Somali militias rarely held territory if contested (Fabian tactics). In this manner, the Somali way of war appeared defensive in nature, not offensive. They did little in the way of shock to close with and destroy the enemy; shock was only delivered externally, if at all, through distant, indirect fires.

Nontraditional military forces prize intelligence and information just as much as their Western counterparts. Commanders and military forces require various types of information to know where the enemy is, provide for their own operational security (OPSEC), and, most usefully, achieve surprise. Having a system countering one's enemies from gaining information is also helpful.

Scouting and reconnaissance to ascertain the enemy's whereabouts is a timeless means to gather intelligence. Somalis were naturally adept at scouting based on their historical role as pastoralists and herders. They also developed techniques to procure game for food by learning how to stalk silently, using terrain and cover to their advantage, and observing their surroundings. Learning how to apply camouflage is inherent to achieve invisibility. (Assimilation into the populace is an easy way to hide among the enemy.) Scouting and reconnaissance techniques also assist in movement to battle and for silently approaching an enemy while remaining hidden. (In America's early colonial history, this form of movement was called "skulking," and was copied from the native Indians.)

Somali warlords utilized agents and spies to conduct reconnaissance, such as the myriad of Somali day workers roaming the U.S. Embassy compound taking note of distances between buildings, composition of forces, and other matters and then reporting back at night to the clans. If a day employee was not actively used as an agent, they were sure to be elicited at night by family members seeking information. The clan militia leaders also employed informants. Anyone residing near a coalition base became useful as a "hunter of information." The highest form of intelligence is to place a mole in the enemy's ranks (translators are usually preferred for this role–Americans counteracted this move by hiring vetted American-born Somalis).

The Somalis employed their own version of an early-warning system. Visually, if peacekeepers were seen deploying for a major operation, support from the populace included burning tires to send tall columns of black smoke into the air and notify everyone around. There was also a phone network which included manning from Somali youth on the streets.[10]

Somali militias sought to achieve attritional and punitive effects with their firepower. Even so, militia-like units are small and had few resources. In Somalia, it took an aggregated form of firepower to achieve overmatch against their intended target. They were not experts at combined arms, so they combined other skills to achieve an exponential effect. Somalis used asymmetric and indirect maneuver to approach their enemies. Stealth, covertness, secrecy, hiding, and confusion were all used to achieve surprise. During a firefight, misdirection, misinformation, and guile enhanced the effect of their firepower.

Psychological warfare was a component of Somali firepower and was used in every major battle. The objective of Somali PSYOP was to sow fear and confusion in enemy ranks and breakthe will of their foe (a cognitive form of firepower). Terror was very much a part of the Somali PSYOP agenda.

Civilians shaped the battlefield to assist in improving firepower delivery. They served as the intelligence network for early warning systems and as labor to help with defensive positions. They also provided militia fighters with food and arms. If needed, civilians were used as human shields so militias could deliver their firepower without fire being returned by coalition forces. Important roles for civilians also included evacuating, tending the wounded, and exploiting the battlefield to recover contraband, arms, or anything else not tied down. Many a Somali town and village were thoroughly gutted and destroyed by this primitive nihilistic rage. Somali civilians also provided the transport systems to move fighters and to run illicit markets for arms and ammunition.

Somali militias, on average, maneuvered with six to eight men. Civilians were used to confront armed peacekeepers, create friction, and sometimes act as human shields, indistinguishable from the militia, much to the ire of peacekeepers. This used UNOSOM's and American forces' Rules of Engagement (ROE) against them, as they were trained to not fire on civilians or create collateral damage in civilian areas or structures. Command and control was decentralized. For example, Aideed divided up his portion of Mogadishu into sectors, with a sub-sector commander in charge of each group.

The U.S. Army Special Forces are tailored for operations in irregular warfare. They are trained to operate in unconventional warfare and among foreign cultures. Their way of war is known as *Special Warfare*: the conduct of guerrilla warfare, psychological operations, escape and evasion, and special intelligence gathering activities. Small teams of Green Berets are infiltrated into austere, non-permissive environments, often behind enemy lines, to operate with and by friendly irregular fighters. Special Warfare principles (small units, decentralized control, non-standard tactics, unorthodox style, cultural rapport, guile, and relative superiority, to name a few) allow Army Special Forces to be successful in irregular warfare environments.

Military Functions

The Somali militia was for the most part a foot-borne, irregular infantry force. They were supported by heavy weapons fire and indirect fires. Militias performed both scouting and reconnaissance. A minimal logistics system ensured food and arms were supplied to the militias.

The strength of forces was harder to analyze. Agreed upon numbers include Aideed's Habr Gedir militia in Mogadishu at around 10,000 fighters. The amount of supporters and sympathizers also in his employ was unknown. Ali Mahdi's Abgal militia was estimated at around 5,000 fighters.

Somalia was awash in weaponry. Between the Soviet Union's military aid followed by the United States' military aid, plenty of small arms, heavy weapons, and aircraft were in the country when it collapsed. Much of it fell into the hands of opposition parties or was captured from the Somali Armed Forces during battle. A robust black market for arms existed, including support from diaspora Somalis living in Kenya. Other arms were smuggled from ex-Ethiopian Army stocks and from as far away as Serbia. The UN estimated over

400,000 mines were left over in the country as a result of the Cold War and the Somali Civil War. Somalis were also quick to place minefields and obstacles to defend their positions.

Innovative tactics included the use of Improvised Explosive Detonations (IEDs), some radio-controlled as well as booby-trapped. Later, with presumed al Qaeda training, the Habr Gedir SNA militia learned how to employ RPGs as anti-aircraft rockets, with devastating effect on Task Force Rangers during the October 3, 1993 Battle of Mogadishu. Many heavy weapons and automatic weapons were jury-rigged onto light and heavy trucks to provide mobile fire support (the Technicals).

Adept Somali militias leveraged special skills within their ranks during a firefight. They were known to employ snipers to pin down opposing forces. Another specialty included engineering work, using bulldozers to erect barricades and prevent coalition forces from encircling their position. Engineer work was crude; there was evidence of some prepared positions (anti-aircraft, artillery, and so forth) generally found in central and north Somalia. Trained mortar-men within the militia ranks seemed to do a better job than the few tankers or artillerymen, if present during a battle.

Tactics among the various clan militia varied, but there were some similarities used by all. Armed clashes with peacekeepers were generally raids and ambushes, with harassment fire. Rocket fires, mortar fires, and artillery shelling were used by all the warlords. (All were able to employ modern weapons, but the lack of formal military training led to indiscriminate damage wherever fighting occurred.) In some cases, if maintained and running, tanks and armored personnel carriers for fire support and maneuver were employed during a clash.

The Somali Clan Militia War

With the ousting of Siad Barre and his regime loyalists, the United Somali Congress (USC) turned its attention to consolidating the rest of the country under their control. Although this would not be possible in northern Somalia while it was under the Somali National Movement's (SNM) and the Somali Salvation Democratic Front's (SSDF) control, there was a good chance the Hawiye clan could rid itself of the Darods once and for all in central and southern Somalia. At first it was not apparent what the Somali National Alliance (SNA) under Aideed had in mind, while the other clans who had allied themselves with the SNA continued their support, thinking they were still on the move to expel Siad Barre and the Darod clan for good. They also desired to rid themselves of the loyalist militias in their area working for the regime-aligned warlords. With the civil war over, irregular warfare in Somalia changed its nature and became a war of clan militias.

Aideed's first move was to clear Afgoye of the Somalia Patriotic Movement forces (SPM under Colonel Jess) in February 1991. Afgoye was located only 30 kilometers northwest of Mogadishu and within easy range for SNA militia. Afgoye was also the site of a large weapons storage area–surface to air missiles, rockets, mines, and other munitions were piled to the ceilings in fortified bunkers. This cache was also an incentive to seize the area and get the arms and ammunition under Aideed's control. After a small pitched battle, the SPM retreated south toward Kismayo, chased as far south as Brava and beyond by motorized elements of Aideed's militia. The USC continued their attack almost 400 kilometers southward.

Aideed followed up with an attack that same month on rival clans in Galcaio (February 11, 1991) to remove Darod elements of the Somali National Front and SSDF from the Mudug region. The SNA launched their night attack with almost 1,000 well-trained fighters, supported by rocket and artillery fires. The result was indiscriminate collateral damage and a large number of civilian casualties. The Aideed fighters reportedly removed village elders and trucked them off, never to be seen again. Before the USC could consolidate its position, the SSDF regained control in April. The battle lines between the USC, SNF, and SSDF became a stagnant front line. It was the site of numerous skirmishes and clashes during the UN's intervention period.[11]

As a U.S. Army Special Forces Captain serving in the 5th Special Forces Group (Airborne) during *Restore Hope* in 1993, Charles B. Smith commanded Operational Detachment Alpha – 562, supporting the Canadian peacekeeper contingent assigned to the Belet Weyne Humanitarian Relief Sector (HRS). One of his A-team's major duties was conducting reconnaissance missions for the Canadian commander to ascertain the location of USC and SNF front lines and attempt meetings with those local commanders. On one such trip to Balenbale, he questioned the local leader, COL Mohammed, on the SNF tactics being employed against them:

> I wanted to get some insight into SNF tactics, in case we ended up having to fight them, so I asked Mohammed to show me how they operated. He misunderstood at first, showing me how the USC attacks. It was standard stuff. Lay down a base of fire and outflank the defensive position. FIATs and 37 mm technicals were priority targets. Dismounted infantry protected the flank and an assault group of dismounted infantry, led by someone with experience, would make a frontal assault. I didn't understand the point of that last part, but judging from the quality of marksmanship I had witnessed so far, they probably didn't run the risk of taking too many casualties.
>
> SNF tactics were basically the same. They would pair their vehicles up, habitually linking a FIAT with a 37 mm technical and a 23 mm technical with a jeep mounted 106 mm recoilless rifle. It was akin to our concept of armor/infantry teams, but it really just worked out as an automatic weapon protecting an antitank weapon. That, too, was standard military practice around the world. A small tactical force can only be organized in a limited number of ways without getting silly, so I hadn't expected any surprises. Without tanks, artillery, and aircraft to complicate command and control, fire and maneuver are all very basic. The Somalis saved their tanks, it seemed, for large, important attacks only. They were far too costly to run, and far too cumbersome to move, to use for raids.[12]

Aideed's next step in his strategy was to remove regime forces from the rest of central and south-central Somalia. His psychological ploy aimed at other clans was to demonize the Darod clan for all the ills of Somalia, labeling them the "others" and calling for the removal of any Darod influence or remaining power throughout Somalia. This systematic removal included denying them territory and population centers.

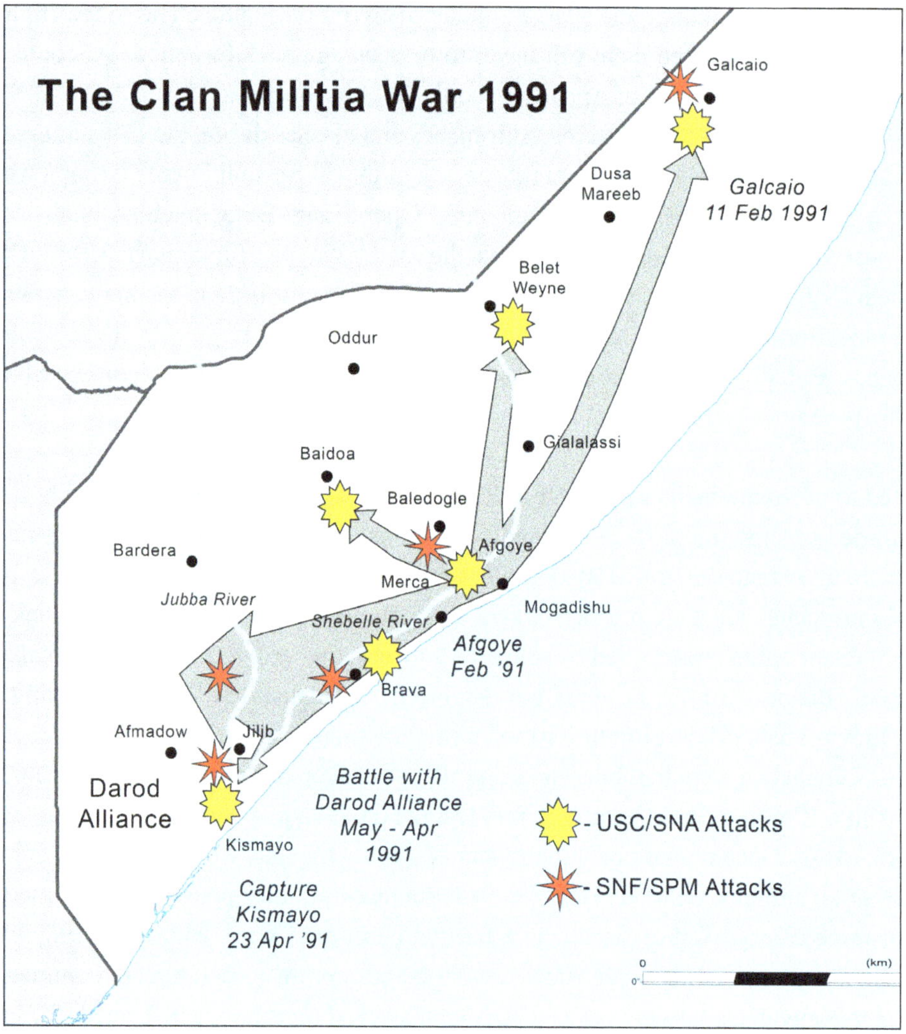

After Aideed's forces seized Afgoye and drove the SPM south to near Kismayo, the Darod-based Alliance forces attacked to seize Mogadishu in March 1991, reaching Afgoye and Brava. In April, General Aideed initiated a counterattack pushing Alliance Forces back to the Jubba River line, and then captured Kismayo. It was around this time Aideed also consolidated his SNA positions at Baidoa and Belet Weyne. In 1992, the Darod Alliance conducted a repeat performance and suffered the same results.

In this effort, the Rahanweyne of the Somali Democratic Movement (SDM) aligned themselves with Aideed. The SSDF in northern Somalia also aligned with Aideed, but as their displeasure grew about the increasing Hawiye influence across central Somalia, they joined the SPM in an anti-Aideed alliance in April 1991.

In March 1991, General Abdi Warsame (head of SNF and chairman of the Southern Somali National Movement–SSNM), General Ghanni, and General Morgan (SNF) counterattacked from their positions in southern Somalia towards Mogadishu, leading approximately 8,000 men of the Darod-based Alliance. The Alliance was formed by the Marehan, Dhulbhante, and Ogadeni sub-clan's militias. The Somali Patriotic

Movement (SPM) led by General Gardheere and Darod sub-clan elements in the SSDF, mostly Ogadenis led by COL Yusef, were also part of the Alliance.

The name Warsame means "bearer of good news," but a visit from his forces was anything but good. COL Abdi Warsame had been the head of the Somali Military Academy under the time of the regime. He was accused of conducting genocide in Hargesia during the civil war. When the regime was ousted by the USC in January 1991, he led the Somali National Front (SNF) in Gedo province and commanded the 2nd Tank Brigade. Oddly, when the rift occurred between General Aideed and Ali Mahdi over who was to be the new president of the country, Warsame sided with Ali Mahdi as his choice.

The SNF's objective was to defeat USC forces and retake Mogadishu, and then install Siad Barre back into power. The counterattack reached Afgoye on April 8, 1991. The southern wing of the attack reached the coastal town of Brava, where they clashed with Aideed's forces over nine times. The Alliance's movement towards Mogadishu created hundreds of thousands of refugees and internally displaced persons (IDPs), overwhelming Non-Governmental Organizations' (NGOs) abilities to feed and care for the populace. The diaspora poured into Kenya and Ethiopia.

In April, Aideed's forces were successful in pushing back General Morgan's forces in Brava and continued the attack south and southeastward to Kismayo, capturing the city on April 23rd. They were assisted by the Rahanweyne regional forces of the Somali Democratic Movement (SDM) from the Jubba River Valley, who helped liberate the city. On their right wing, they pushed COL Yusef's SSDF back across the Jubba River Line. It was during this time the USC also captured Baidoa. In time, the USC additionally moved to capture Belet Weyne.[13]

Due to the actions of the USC, a split then occurred in the Somali Patriotic Movement; COL Jess occupied Kismayo and aligned himself with Aideed. In May 1991, General Morgan and COL Gabiyu formed a new branch of the SPM (anti-Aideed faction) in Doble and became a second Darod-based front in the south, positioned northwest of Kismayo.[14]

Once the fronts stabilized in the south, the United Somali Congress then turned its attention to seizing and consolidating positions in Baidoa and BeletWeyne. A split was occurring due to the rivalry between the Hawiye sub-clans in Mogadishu, led by Ali Mahdi of the Abgals and Aideed of the HabrGedirs. The frequent clashes between the two forces led to the establishment of the "Green Line" in Mogadishu, with Ali Mahdi to the north and Aideed's forces in control of the city's southern portions.

As 1992 rolled around, Somalia was experiencing one of their worst droughts and starvation was rampant in central and southern Somalia. Refugees from the fighting, along with other displaced civilians from "clan-cleansing" began to reach enormous numbers. Between January and May 1992, the SPM (along with Siad Barre) once again moved on Mogadishu and retook Kismayo. Reports of large-scale clan-cleansing and atrocities began to reach the public.

Aideed counterattacked and was successful in driving the Darod Alliance forces back across the Jubba River, into Gedo province; his SNA retook Kismayo on May 14th.

The fighting between Aideed's forces and the SPM between January and May 1992, near the lower Jubba Valley region, purportedly killed many civilians and destroyed villages, exacerbating the famine.[15]

With the SPM's defeat, Siad Barre finally left Somalia in exile at the end of April. Once again, Colonel Jess held Kismayo. The feud-like fighting between the two sub-clans of Hawiye in Mogadishu continued into November 1992, with a loose cease-fire along the Green Line finally agreed upon. By then, large portions of central Mogadishu looked like the bombed-out city of Beirut during the Lebanese civil war.

The year ended with General Morgan attacking Kismayo in December and massacring civilians who had sided with Aideed and Jess. The United Task Force (UNITAF) coalition forces began Operation Restore Hope the very same month, with most units possessing very little understanding of the irregular warfare environment in which they were about to intervene. Richard H. Schultz Jr. and Andrea J. Dew noted in their book *Insurgents, Terrorists, and Militias* that "In Somalia in 1993 there could be no such thing as a purely humanitarian intervention. Rather, if the United States were to rescue the victims of that carnage it would have to tame the very clan militias that were responsible for dismembering Somalia. And simply put, that meant combat."[16]

Takeaways for Irregular Warfare

The Somali "way" of irregular warfare illustrates that irregular warriors, adapting modern technology in their fire and maneuver, can challenge modern military forces and create protracted and stalemated situations. Wars of this nature have lasted, on average, between eight years to over twenty years. Some successful lessons can be drawn from the Somalia experience which may be useful to strategists, planners, and practitioners of irregular warfare. Operational design for an irregular warfare campaign should take some of the main points from America's Somalia experiences into consideration when developing a strategy and course of action in future irregular wars. These points are:

1. In many irregular warfare cases, the "hearts and minds" of the populace can best be gained by providing security, law, and order. Without friendly forces accomplishing this task, irregulars capture "hearts and minds" through intimidation, fear, and terror while administering a sense of justice to criminals and marauding warlords as a method to gain initial acceptance. Such was the case in the popular support of the Islamic Courts Union and al-Shabaab in Somalia, as well as the populace's initial acceptance of the Taliban in Afghanistan. Ted Robert Gurr's work, *Why Men Rebel*, written in the 1970s, is still a highly useful manuscript for statesmen and the military during the conduct of a strategic appreciation before developing a strategy. A variety of grievances may spark an irregular war, and not all of them include politics.

   It will be just as important to build and equip police units, para-military police, border and customs police, coast guard, and navies, and place judges, prisons, courts, and jails to focus Security Assistance efforts on host nation military forces. A lack of arrests and war-crimes tribunals (no enforcement mechanism) feeds a major grievance of the populace.

2. The Powell doctrine works just as well in irregular warfare as it does in conventional warfare–if America puts boots on the ground there should be a clear objective–go big (strength and forcefulness), go fast (speed and violence of execution), and turn it over to competent authorities once the mission is complete.

3. Among the successes in irregular warfare by European and Western militaries, indirectness appears as one of the better options, given the mistrust of foreigners in many non-traditional societies.

Regional solutions from Frontline States (FS) should be preferred if there are no large national security interests for America and its allies presented during the conflict. Funding for Frontline States' militaries, equipment, and pay along with indirect military operations such as maritime interdiction and counterterrorism raids, combined with indirect fires from cruise missiles and unmanned aircraft and airstrikes have all worked well, as was the case in supporting the African Union and its AMISOM forces in Somalia. Instead of a "coalition of the willing," which makes everyone feel good diplomatically, only those military forces that are deemed professional and well-equipped should be let into any coalition and work for one commander.

4. Security assistance is an economically sound option to lessen the need for Americans to fight in foreign countries. If training a nation's fledgling military is required, it should be left to regional actors or contractors. In the case of training elite forces, consider U.S. military trainers operating out of another country and, if need be, insert them into the host country as a small footprint to address the problem where it exists. Paramilitary forces can be raised and used as proxies, but these types of units should be carefully controlled. In both cases, train the trainers then work Americans out of the job.

5. Propaganda and PSYOPs are fire support enhancers for irregulars. Terrorism is "propaganda of the deed." This area is not information operations. To counteract this capability from irregular warriors, a good counterpropaganda and psychological warfare program is needed to destroy the ideology or narrative of the irregular enemy. These can be especially effective tools to create divisions and schisms between factions and destroy their unity. Al-Shabaab has continually reemerged after each of its failures due to the intense unity exhibited by the organization, which should have been targeted. There were still an estimated 5,000 to 10,000 al-Shabab fighters within the organization as of 2020.

Special Operations Forces achieve their highest operational and strategic utility in the irregular warfare environment. It is in this "human and cognitive domain" that effective Special Operations power is achieved. SOF supports and enhances conventional force maneuver, conducts unconventional warfare, and performs high-risk, strategic missions. SOF operations build space for diplomacy to work, reassure allies, and help to prevent local forces failure. SOF operations produce a high level of "fog and friction" on irregular warriors and their leadership. SOF can also provide useful, human intelligence where no other intelligence systems can penetrate.

Endnotes

1. Schultz, Richard H. Jr. and Andrea J. Dew. Insurgents, Terrorists, and Militias: The Warriors of Contemporary Combat. New York: Columbia University Press, 2006, 4 – 6.
2. Schultz and Dew, 20–24.
3. Cole, Brian. "Clausewitz's Wondrous Yet Paradoxical Trinity – The Nature of War as a Complex Adaptive System." Joint Force Quarterly 96, 1st Quarter 2020, 42.
4. Ibid, 44.
5. Ibid, 48.
6. Turney-High, Harry Holbert. Primitive War: Its Practices and Concepts. Columbia, SC: University of South Carolina Press, 1991, 29–31.
7. Schultz and Dew, 20–24.
8. Turney-High, 63–72.
9. Ibid, 123 – 130.
10. Ibid, 107 – 122.
11. Kapteijns, Lidwien. Clan Cleansing in Somalia: The Ruinous Legacy of 1991. Philadelphia, PA: University of Pennsylvania Press, 26013, 162–163.
12. Stevens, Charles "C. B." Honor Bound: A Special Forces Detachment in Somalia. Houston, Texas: Unpublished Personal Draft, 1996, 394.
13. Kapteijns, 168.
14. Ibid, 170–176.
15. Ibid, 241 – 242.
16. Schultz and Dew, 3.

2

Operation Provide Relief

The road to hell is paved with good intentions.

— Proverb attributed to Saint Bernard of Clairvont, 12th Century

There was a new balance of power in the world leaving America on top, inheriting a new world order in international relations. With the Cold War gone, it was time to solve some of the more intractable global problems. President George H.W. Bush declared, "Now, we can see a new world order coming into view. A world in which there is the very real prospect of a new world order. . . . The Gulf War put this new world to its first head." The year 1992 began with some leftover Iraq business for President George H. W. Bush. The U.S. Central Command had the mission to create a safe zone in northern Iraq to protect the Kurdish population from Iraqi reprisals and provide humanitarian aid. This set of goals was called Operation Provide Comfort and was led by a joint special operations task force.

In Kuwait, a downsized Army Special Forces task force began 90-day rotations (from the 5th Special Forces Group) to provide training and assistance to the newly reorganizing Kuwait Army. The USAF still flew protective cover missions over southern Iraq, enforcing the no-fly zone, which required SOCCENT (CENTCOM's Special Operations component) to base aircraft and helicopter forces in Saudi Arabia in order to serve as Combat Search and Rescue (CSAR).

On January 1, 1992, Boutros Boutros-Ghali of Egypt became the UN Secretary General. He was a staunch advocate of the New World Order and freshly basking in the victory of Desert Storm. The victory boosted the UN's reputation to get things done in concert with the United States, now that there appeared to be a unipolar world with America left standing as the one superpower.

However, there was some left over business from the Gulf War; UN weapons inspectors began to comb Iraq in search of Saddam Hussein's weapons of mass destruction programs. Although news about Somalia's civil war and starvation from drought began to appear in the media, the level of coverage had not yet reached the "CNN effect" it would later achieve when American and coalition forces deployed (boots on the ground effect). The Secretary General would soon choose the quagmire of Somalia to prove the UN's worth in spreading peace and democracy.

The Cold War was continually thawing with the collapse of the Soviet Union and the concomitant dismantling of the Warsaw Pact. The Afghans overthrew their communist government. The country of Georgia seceded from the Soviet Union. In April, Bosnia-Herzegovina declared their independence from the former Yugoslavia, sparking a civil war between the Croats, Serbs, and Bosniacs; the city of Sarajevo became besieged. On May 30, the UN sanctioned Yugoslavia for indiscriminate shelling of the city and the additional forced removal of Bosniacs from villages. Boris Yeltsin and President George H. W. Bush were able to sign a "joint understanding" on the reduction of nuclear weapons (START II). The Presidential election season began that summer. The Democrats elected Al Gore and William J. Clinton for Vice President and President to run against President George H.W. Bush. Generally, the world's problems seemed manageable as the President entered his re-election year.

For the American military, Operation Desert Storm imbued them with a new pride in accomplishing the mission, using overwhelming force as prescribed by the Colin Powell doctrine (if you go, go big). Many in the military, including its think tanks and pundits, saw the Gulf War as a release from the entanglement of operations in low-intensity conflict and brush-fire contingencies (Military Operations Other Than War, MOOTW), pervasive since the Vietnam War.

It would come at a price. America was looking for a peace dividend now that the specter of war with the Soviet Union had disappeared. The so-called "Peace Dividend" was just over the horizon, and unfortunately for the military it began with about a twenty-five percent cut in strength, budgets, and modernization. It was an across the board cut in many organizations; even the Special Forces community suffered the loss of the 11th and 12th Special Forces Reserve Groups and began years of austere budgets (along with reduced operations tempo). The good news emanating from the Peace Dividend was balancing the federal budget toward the end of the decade.

It looked to be heady days for the "New World Order," with many international organizations, as well as the United States, looking to establish new relevancies. America was in the position of being the sole superpower standing, so, what to do next? UN operations around the world increased at an amazing pace. In his promotion of peace around the world, Boutros-Ghali intended to focus some of this new spirit on solving intractable problems in Africa, no surprise given his experiences in that region.[1]

In February, the European Union was formed under the Maastricht Treaty, creating a new organization to govern and regulate Europe's nation states. In Bosnia, the siege of Sarajevo began, initiating a civil war which would last for three years. European member states of the UN were involved in the Bosnia-Herzegovina to end the war. Participating member states deployed military assets to man the United Nations Protection Force (UNPROFOR) for the conduct of peacekeeping and to monitor humanitarian abuses perpetuated by the combatants.

The Ethiopian-Eritrean War ended in April with the UN recognizing Eritrean independence. A little breathing room for diplomacy opened to further peace and reconciliation efforts in the Horn of Africa.

Boris Yeltsin and President George H. W. Bush were able to sign a "joint understanding" on the reduction of nuclear weapons (START II), reducing tensions in strategic competition.

Somalia

The year 1992 remained chaotic in Somalia with the humanitarian relief efforts to mitigate the starvation crisis and increasing refugee flows to the Kenyan and Ethiopian borders. Clan factions continued to battle one another over territory and assets, destroying the country's capability to absorb food aid, conduct agricultural activities, and administer humanitarian medical assistance from the various non-governmental organizations (NGOs). Dire predictions of starvation emanated from Somali experts and pundits: over a million people had died to date from a lack of health services and food (some of these fatalities were also due to the conflict); estimates varied, but most settled around twenty-five percent of very young children had

Key Humanitarian Non-Governmental Organizations

ICRC - International Committee of the Red Cross
The Comite International Geneve was founded in 1863 to aid victims of armed conflict, particularly wounded soldiers and POWs. Its motto is, "Amidst War, Charity." They also conduct searches for missing people and help protect civilian populations in war zones, and can serve as a neutral intermediary when required. In Somalia, they partnered with the Muslim Charity, Red Crescent.

WFP - World Food Program
Founded in 1961, based in Rome, to fight global hunger due to conflict or natural disaster. Links donors to needs on the ground.

CARE - Cooperative for Assistance and Relief, Everywhere
CARE was founded in 1945 and began operations in Somalia in 1981. The NGO provides food, other nutrition, economic development, education needs, and disaster response.

MSF - Médicin Sans Frontières
Also called "Doctors without Borders." A medical-related NGO, established in 1971, Geneva. Provided doctors, nurses, and medical staff to victims in war-torn or disaster-stricken countries. Also provided sanitation and clean water services. Its logistic services provided medical supplies and helped establish clinics. Began operations in Somalia in 1992.

already died from malnutrition. Without some form of increased intervention, a million more Somalis of all ages could perish.

The international community looked for various ways to get increased humanitarian aid into Somalia, even though any breakthrough on cease-fire talks with the warring clans seemed unlikely. One method, among many proposed, was sending in aid to remote regions (outside Mogadishu) by aerial delivery. The situation became increasingly dire as even the possibility of escalating the maritime delivery of food was blocked by clan fighting in Mogadishu's and Kismaayo's ports (which made deploying deliveries and truck convoys to outlying regions impossible). Many of the NGOs had thrown up their hands and pulled their people and assets from the war-torn country, concerned over their workers' safety. As food became the new economic commodity (thus power), banditry, looting, and pilfering took its toll on the delivered aid.

President George H. W. Bush, now convinced America had to do something, announced on August 14, 1992 that America would participate in the aerial delivery portion of the donors' plan for increasing food deliveries. The Joint Chiefs of Staff alerted USCENTCOM to prepare for the mission on August 18.

The operation was named Provide Relief. Although not a very robust response, it soothed the American public, congressional lawmakers, and the JCS's concerns about getting too involved in Somalia, guaranteeing there would be no U.S. boots on the ground, yet demonstrating that the government was still doing something.

A Joint Task Force (JTF), commanded by Brigadier General Frank Libutti, deployed to Mombasa, Kenya after the President's announcement for the operation's conduct. Fourteen C-130 Hercules cargo aircraft and four C-141 Starlifters were initially positioned in Mombasa (at the Moi International Airport), working out of the U.S. Kenya Liaison Office's (USLO) aviation support warehouse facility.

To provide security for the USAF food flights when the aircraft were sitting on the ground, members of Army Special Forces teams, equipped with two armed M1026 weapons carrier HMMWVs, flew overhead on each delivery airfield, ready to land and respond to threats as needed. There were five delivery sites: the refugee site in Wajir, Kenya (on the border with Somalia) and four airfields inside Somalia which could not be reached by truck convoys to deliver food to the main NGO feeding stations–Belet Weyne, Baidoa, Bardera, and Oddur.

The Provide Relief period was characterized by four activities:

1. The humanitarian relief agencies' ongoing efforts to get back into Somalia and increase food aid delivery.

2. The implementation of UN Resolution 751 to create UNOSOM (United Nations Operation in Somalia) and place fifty cease-fire monitors in Mogadishu, along with 500 armed Pakistani peacekeepers.

3. The conduct of aerial food delivery via Operation Provide Relief alongside other international agencies' efforts; one U.S. Army Special Forces company assisted this effort for the first ninety days and was replaced with another SF company in November.

4. The transition of Operation Provide Relief to Operation Restore Hope in early December.

The Situation in 1992

In Somalia, the only "order" was found in clan-controlled territory, minus northern Somalia (the non-recognized breakaway Somaliland) which was under the control of the Somali National Movement (SNM) and formed from the dominant Issaq clan. Somaliland declared their independence from the rest of Somalia in 1960. The other predominant political faction in Somaliland was the Somali Salvation Democratic Front (SSDF), also Issaq controlled.

At the beginning of 1992, there was a palpable reluctance on the part of the UN and the Organization of African States to become involved in the chaos of Somalia, other than helping to feed starving Somalis. Even this measure was curtailed as the UN ceased assistance because of the ongoing clan fighting, affecting non-governmental organizations' abilities to operate in a safe environment.

This attitude began to change as horrific stories and pictures, along with thousands of deaths from the civil war and drought, entered the consciousness of the international community and American congressional lawmakers.[2]

In light of the deteriorating conditions in Somalia, the UN was moved to reassess their earlier position of abandoning the country in 1990. Boutras-Ghali, ever the peacemaker, charged a delegation to revisit Somalia and ascertain any hopes of once again seeking a peaceful solution to stop the warring clans (and sub-clans).

One of the few NGOs willing to brave the perdition-like state of Somalia was the International Committee of the Red Cross. Through their offices, the USAID was able to coordinate delivery of its donor's humanitarian supplies into the country. Other able and hearty NGOs involved in Somalia relief efforts included the World Food Program (WFP), CARE, and Doctors Without Borders.[3]

The intractable problem remained; both Ali Mahdi and Mohamed Farah Aideed contested one another for the position of President in Somalia. Clan factions and sub-clans aligned, generally, around this impasse and fought one another for the primacy of their "candidate." Further, Aideed was intransigent about any UN involvement or outside interference in his country's affairs (but did not mind the freely provided aid). Aideed was responsible for ousting the former tyrant, Siad Barre, and personally felt that action gave him the *bona fides* to be the next ruler. Ali Mahdi, a cosmopolitan businessman and Aideed's main competitor for candidacy, saw no problem with international assistance and development programs. These two contending views plagued Somalia up to the UN's final withdrawal in the spring of 1995.

Furthermore, the Somali people and main clan leaders had a huge distrust of the UN. Certainly some acrimony persisted in that Boutros-Ghali previously supported the Barre regime when he was serving in the defense cabinet in Egypt. Any initiatives forthcoming from the UN had to pass through this prism of Somali conspiracy theory.[4]

The UN did achieve a minor success–the two clan leaders arranged to meet in New York in February of 1992 and while there agreed upon a ceasefire to allow for more international humanitarian assistance into the country (surprisingly, this agreement lasted throughout most of Somalia for the remainder of the year). This agreement had one exception–the two clan leaders were not done fighting over Mogadishu. This conflict endangered the primary entry point for seaborne supply arrivals (as well as flights into the airport). The violence also hindered efforts to protect aid, once delivered, and continued to wreak havoc on the infrastructure needed to logistically support a renewed, massive aid effort.

In the February New York discussions, two questions emerged as part of the cease-fire–who would monitor it, and who would guard the aid delivered (again, contention centered on the main hub of Mogadishu and near the vicinity of the front lines).[5]

The answer came in Security Council Resolution 751. The Resolution allowed for the sending of fifty observers to monitor the ceasefire, armed and in uniform, to be followed by an additional 500 peacekeepers to protect the port and airfield. This operation was named United Nations Operation in Somalia (UNOSOM), but was not yet named UNOSOM I; at this time, no one foresaw UNOSOM II, which was created in April 1993. Mohamed Sahnoun became the UN special representative for the implementation of UNOSOM, residing in Mogadishu.

Sahnoun initially traveled to Mogadishu on a factfinding mission to prepare for the duty. While in the city, he met with all significant clan leaders to work out the details of the cease-fire implementation and the deployment of the observers and peacekeepers.

Aideed, typically, feared his loss of control over the seaport and airfield and viewed the oversight as an affront to his "power," especially considering that ceasefire monitors would be in uniform and bear arms. After all, Somalia was not at war with the UN or any other major power. Aideed was also wary when a UN-marked Russian cargo aircraft landed and delivered arms and supplies to Ali Mahdi earlier in the year, creating the impression that the UN was choosing sides (this moment was a major UN embarrassment; the aircraft had been recently chartered to fly in humanitarian supplies for the UN, but when it completed its contract, the crew forgot to remove the UN markings). Due to these points of concern and offense, Aideed refused to agree to the deployment of the observers and peacekeepers.

After a period of renegotiating and issuing diplomatic apologies, Resolution 751 was back in place, with one minor concession to Aideed–the ceasefire monitors would not bear weapons.[6] However, even with this UN measure, Somalia continued to slide into catastrophe. UNOSOM's mandate did not account for secure deliveries outside Mogadishu, or for the protection from the depredation on NGOs in the countryside. Famine still existed, with over a million deaths predicted for the rest of the year. The earlier 1989 – 1991 drought and cattle disease had created a perfect storm of human misery in the failed state. Refugee flow increased, along with thousands of displaced people as a result of the continuing civil war. Diseases in these conditions abounded, exacerbated by the collapse of any medical system to run clinics, hospitals, and provide medicines.

In the spring, experts in the Secretary General's report on Somalia predicted it would require at least 200,000 metric tons of aid to address these conditions; less than half of that figure was actually getting through (and still the target of theft and pilfering).[7]

Beforehand, the majority of clans were not focused on fighting one another as much as they were united against the former tyrant, Siad Barre. Clan fighting continued in 1992, especially in the south and near the Jubba River as the United Somali Congress-supported clans fought against remaining regime forces (primarily from the Darod clan). In some way, this effort united the clans. When Siad Barre fled into Nigeria for exile on May 17th, the clans now turned their aggression against one another, jockeying for territory, power, and control over humanitarian relief supplies. The fighting intensified, rather than lessening with the absence of Barre.[8]

The cries for help and outrage over the situation intensified, often reaching hyperbolic levels. CNN had become the go-to place for 24-hour coverage and was certainly making policy makers sensitive to the situation. The CNN effect began as an emotional appeal to the world and served as an accelerant to the decisions made during the summer of 1992 by President George H.W. Bush, notwithstanding the pressure put on him by a Democrat-controlled congress (and possibly the impending election).[9]The photos of starving children, rib-cages poking out and bodies covered with flies; dead livestock, bodies, and refugees in miserable conditions increased in intensity. Somalia shifted from being a humanitarian crisis to a humanitarian tragedy. As a result of the renewed fighting, delivery of relief supplies suffered and starvation was once again rising, with thousands dying daily. In late May, when tons of aerial-delivered supplies from two relief agencies were looted, the UN ceased all flights into Somalia.

The 5th Special Forces Group (Airborne)

Once the ceasefire from Desert Storm became effective, the 5th Special Forces Group (Airborne) began a series of battalion rotations to Kuwait, primarily to exploit the battlefield and help the Kuwaiti Armed Forces rebuild. Much of the operable Iraqi equipment gleaned from battlefield recovery was used to provide the weapons and armor for the new Kuwaiti armed forces. This effort included the slow clearing of each and every military compound and the airfield to ensure numerous mines planted by the Iraqis were at least, if not scheduled for clearing and removal, marked for future de-mining programs.

U.S. military forces were based outside Kuwait City inside a modern commercial and industrial area, located alongside the induction system for the country's water production facility. There were more than sufficient facilities and warehouses, parking areas, offices, and work trailers to house the force; the 5th's units were based in a large warehouse, which after cleaning dust and debris from its inside, was slowly partitioned off into offices and sleeping areas. The designated battalion kept its fleet of Desert Mobility Vehicles (DMV–a SOF-modified Army HMMWV) used in daily operations to reconnoiter and patrol the entire country, primarily focused on the areas around Kuwait City and north up to the Iraqi border. SF teams not assigned to daily reconnaissance began training and advisory missions to selected Kuwaiti units (Foreign Internal Defense–FID), reporting daily to their compounds.

The battalion rotations were three months in length and lasted into December, when the Gulf War was officially declared over. By 1992, the commitment to Kuwait's defense was reduced to a Special Forces company (mounted), again with rotations occurring every three months. A fleet of DMVs, along with modified Toyota trucks, was left as a pre-positioned package for use by the current SF unit on rotation.

No operations were intended for Somalia, even though many in the United States had heard and seen the devastation of the civil war and the starvation occurring due to the humanitarian crisis reflected on television and in the newspapers. In the 5th SFG(A), there was no notion that SF would become involved. However, by August, it was apparent the United States was considering some kind of contribution to the UN effort to stop the disastrous rate of starvation in Somalia.

The Summer of 1992

It took until June 23rd for the principal Somali clan factions to allow the deployment of UNOSOM contingents. The UNOSOM military commander was Brigadier General Imtiaz Shaheen of Pakistan (all the observers and peacekeepers were provided by Pakistan). The deployment was conducted in two phases, each one having to be arbitrated–the observers arrived in early July, followed by 500 peacekeepers in mid-September.

In early July, the Ambassador to Kenya, Smith Hempstone, shocked the Foggy Bottom elite with a paper titled "A Day in Hell," reflecting his thoughts after visiting Somali refugee camps on the Kenyan border. There were many attention-grabbing lines in the paper, but the one which resonated most was ". . . if you love Beirut, you'll love Mogadishu." President Bush and his National Security Council (NSC) read the paper.[10]

In the same month, President Bush asked his State Department for policy recommendations on what course to take with respect to the Somalia situation. When State queried the Pentagon, which was not convinced on the need for intervention, the reply was somewhat fueled by the glow of victory in Desert Storm–"we only do wars," and recommending the use of civilian charters.[11]

On July 30th, the Interagency Somalia Working Group's leader, Deputy Assistant Secretary of State Robert Houdek, requested the DOD provide an estimate for airdrop, airlift, and helicopter-lifted relief supply delivery. As discussions over Somalia responses increased throughout the American government and in the halls of Congress (and by now, the media assault was in full force), many defense experts and politicians remained wary of any involvement. After all, what were the vital interests to America? The U.S. could not have a peace dividend if America increased involvement in meaningless, benevolent foreign ventures around the globe.

Upon the return of a UN technical team which met with Sahnoun and assessed UNOSOM's effectiveness and status of the ceasefire (the visit occurred the 2nd–15th of August), the Secretary General proposed the establishment of four security zones (humanitarian feeding zones) for the outlying districts, in an attempt to now focus wider than just Mogadishu, UNOSOM's current mandate. In his proposals, he saw the need for about 750 troops for each zone (along with additional support and logistical units), to provide stability and security. Commensurate with his suggestion, the Security Council passed Resolution 775, the "all measures necessary" resolution, to create four additional UN zone headquarters and for the increase of troop strength in Somalia by an additional 3,000.[12]

It was this move by the UN, apparently uncontested by the United Somali Congress, and Ali Mahdi's willingness to go along with the decision which sparked Aideed's breakaway from the USC to form the Somali National Alliance–SNA. Ali Mahdi, not to be outdone, created his own Abgal-based SNA.

The first Course of Action (COA) response from the DOD was to provide three C-130 cargo aircraft and three CH-47 cargo lift helicopters, to be used for only sixty days. It was clear the DOD anticipated this pittance would dissuade those eager to involve the military. But, with Somalis continuing to die, and with Americans being reminded in the media and on TV almost daily, intervention voices became louder. On August 13th, President Bush ordered his administration to get something done.

In the August 14th Deputies Committee meeting, a better COA was hashed out; the military would fly aircraft to deliver food aid based out of Mombasa, Kenya. The COA designated the refugee camps at Wajir, Kenya as the first priority for flights and then the four airfields in Somalia designated by the ICRC. The operation would last for up to sixty days then be replaced with a commercial aircraft under contract by the State Department. Additionally, the United States would transport the hundreds of Pakistani peacekeepers to Mogadishu. President Bush agreed to the course of action and announced his decision on August 15th.[13]

The Joint Chiefs of Staff Chairman, General Colin Powell, issued an alert and execute order to the commander of USCENTCOM on August 15, 1992. In that order, the force list consisted of eight C-130s, four C-141s (the C-141s would service Wajir refugee camps on the border with Somalia, as some of the airfields in Somalia might have damage and some of them were too short). The operation was named Provide Relief.[14]

CENTCOM appointed Brigadier General Frank Libutti to head the task force. Brigadier General Libutti (USMC), a native of New York, graduated from the Citadel then entered the USMC Officer Candidate School to graduate as a new 2nd Lieutenant in October 1966. In 1967, he served in Vietnam with 1/9th Marines as an infantry platoon commander. He served in a variety of positions with the Marine Corps at Camp Lejeune, Quantico, and San Diego. After an overseas tour of duty in Naples, Italy, he served in the Washington D.C. area. Upon achieving the rank of Colonel, he was chosen as the commander of MAGTF 1-88 at Camp Pendleton in California, followed by command of the 11th MEU. He returned to Washington, D.C. and served in the Office of the Chairman, Joint Chiefs of Staff.

After his selection to Brigadier General, he moved to Tampa, Florida on assignment to USCENTCOM. He was only in his new job for about two weeks when he was selected for commanding the Joint Task Force–Provide Relief (General Libutti retired as a Lieutenant General at the end of his career, after serving as the commanding general of the 1st Marine Division and later the III Marine Expeditionary Force in Japan).

Within a week he was in Mombasa with his deputy, Colonel Kennedy (a CENTCOM assessment team visited Mombasa before his arrival to conduct a survey). Within the first days of his arrival, and before the military air contingent had arrived, BG Libutti took advantage of a flight with a civilian-chartered C-130 for the World Food Program and ICRC scheduled for delivery of aid to Baidoa. He was able to see first-hand the conditions NGOs were operating under, and he spoke with their representatives on the ground. He also spoke with the media (BBC) to explain Operation Provide Relief's mission. From that time on, he urged his staff to seek these types of flights to have a better understanding of the situation.[15]

Along with the task force's aircraft, BG Libutti was also assigned a Disaster Assistance Response Team (DART) to assist in the coordination of the JTF's efforts with UNOSOM and the NGOs operating in the region. For airborne security, a company of mounted Army Special Forces teams rounded out the organization (over 500 troops). The Office of Foreign Disaster and Assistance (OFDA) provided additional charter cargo aircraft (flown by USAF pilots and crews).

There was one sticky point–how would the safety of the aircraft and crews be secured while operating inside Somalia. The planners identified a need for a company of security personnel. Although the USAF Air Police and the U.S. Army's infantry were considered, Army Special Forces were the preferred choice due to their capabilities and knowledge of conducting rescue, survival, and escape and evasion with USAF pilots and crews, if the need arose.

As per the plan, the first flights by C-141 began delivery of aid to Wajir, Kenya (22 tons per load). On August 22nd, the first USAF C-130 flight landed in Baidoa, Somalia, to deliver its cargo (10–12 tons per load). The largest relief organization operating in Somalia was the ICRC. After discussions between the ICRC team and General Libutti and his DART concluded, it was agreed that all planes would bear the Red Cross symbol (national markings could remain on the planes), and no planes or crews operating on the ground would be armed (all of these issues were quietly resolved with the addition of the SF A-teams remaining airborne in a separate aircraft and "wink and a nod" approval from Ambassador Hempstone).

Map of various locations for PROVIDE RELIEF flights (*Provided courtesy of Mark Hamilton, a member of the JTF Provide Relief staff*).

The 2nd Battalion of the 5th Special Forces Group (Airborne)

SOCCENT chose the 5th SFG(A) as the Army Special Forces component to support Provide Relief (the 5th Group's mission area of responsibility and specialty was the Middle East and East Africa, thus their alignment with USCENTCOM and its special operations component, SOCCENT). The Group S3 received the order mid-August with a deployment date scheduled for August 20th. After the S3's consultations with the 5th SFG(A) Group commander, COL Kenneth R. Bowra, the 2nd Battalion, commanded by LTC William Faistenhammer, Jr., was selected for the mission. The 2/5th was in its GREEN contingency deployment cycle.

With the Group commander's guidance in hand, the S-3 personally notified Lieutenant Colonel Faistenhammer. Given the deployment date, the unit had about seventy-two hours to launch (this was a well-practiced procedure by all GREEN battalions). After a course of action analysis, Lieutenant Colonel Faistenhammer chose Company A, commanded by MAJ Kent Listoe, with a provision to provide five SF teams and four Desert Mobility Vehicles to conduct the mission (other companies had other deployment requirements; Company A could easily muster the four teams needed at the time).

On Saturday, Company A was enjoying a unit picnic when Colonel Bowra visited to speak with Major Listoe. The men wondered about the nature of the conversation; was something going on again in Kuwait? They were also aware of Somalia in the news; it was prevalent in all the media, but they thought it highly unlikely humanitarian relief would require a military intervention.

CPT Steve Moniz was new to the 5th SFG (A) and assigned as the Detachment Commander for Operational Detachment Alpha 542. It was a solid team, with experienced Noncommissioned Officers (NCOs), who recently participated as a Coalition Warfare Team (CWT) in Desert Storm. MSG Dan Kaiser was the Team Sergeant. ODA542 was designated as a mounted team, equipped with four modified HMMWVs, dubbed Desert Mobility Vehicles (DMVs), to conduct long-range operations in desert environments. Other experienced NCOs included SFC Wendell Greene, SFC Larry Foster, SSG Mike Clark and SFC Wade Bush.

Captain Moniz attended Major Listoe's mission briefing along with the other Detachment Commanders and Detachment Technicians (the SF Warrant Officer on the team served as Assistant Detachment Commander, like an XO). The requirement was vague, and no mission was assigned as of yet. The information about the mission was unclear; at best, most were getting their intelligence from CNN broadcasts (Captain Moniz could not ascertain the task and purpose for the mission at the time).In order to prepare the teams and identify the equipment needed, the toughest course of action was chosen as the baseline; a team would fly into Somalia, land, and then move out to help protect NGOs at various feeding locations.[16]

Major Listoe selected Operational Detachments 542, 543, 544, and 545 (ODAs 541 and 546 were on other missions), and was augmented with ODA556 (from Company B). There were also some operators from Company C augmenting the staff and SF teams, for instance, Chief Warrant Officer 3 Richard A. Detrick, from Company C, augmented Company A's staff. The A Company commander and his staff deployed as an Operational B-team (ODB540), per doctrine.

By the next day, ODA542 reported in to the barracks with the other teams of the company to begin their pre-deployment checklist. Major Listoe departed on August 16th to join the assessment team forming at CENTCOM (a thirty-seven man Humanitarian Assistance Survey Team–HAST) which then traveled on to Mombasa. This advance party would form the core of the JTF staff.

Colonel Bowra had requested the consideration of deploying a larger contingent, a battalion to serve as a Forward Operating Base (Light); his request was denied. Colonel Bowra, after discussion with the U.S. Army Special Forces Command (Airborne) Commander, urged Lieutenant Colonel Faistenhammer to consider personally deploying with a small staff to serve as the senior SF Army representative and provide a buffer between the company and the General Officer-led task force. Faistenhammer agreed with this idea.

For this portion of the mission, the battalion S-2, Captain Martinez, and Staff Sergeant Terriborne would serve as Lieutenant Colonel Faistenhammer's intelligence section. They were augmented with SFC Wendell Greene who served as the battalion intelligence collector but who was still allowed to participate on missions with his team, ODA542. Major Carroll was the battalion Executive Officer; this would give him on-the-ground experience before he assumed command of Company C and replace Major Listoe in November.

The company and battalion staff prepared for the mission and were put on ninety-day temporary duty orders (TDY), as they did not know if the operation would last the sixty days projected, or longer. All were to receive a full allowance of per diem while in Mombasa since the only place for lodging and food would be from a commercial source. In Mombasa, preliminary JTF staff planning identified no requirement for SF vehicles (only Major Listoe advocated for the unit to have vehicles). Once again, the U.S. Army Special Forces Command (Airborne) Commander and Colonel Bowra expressed their concerns to SOCCENT. Fortunately, the vehicles were approved along with Lieutenant Colonel Faistenhammer deploying with his light Forward Operational Base (SF battalion)–the FOB.

The HAST was also able to meet with major representatives of the humanitarian relief agencies and gained a little more in situational awareness; all information updates were passed back to the teams.

Lieutenant Colonel Faistenhammer sent his S-4 and Chief Warrant Officer 3, Clark, from the battalion to serve as the advance party in Mombasa. On Wednesday, their C-5A Galaxy cargo jet arrived. Seventy personnel deployed.

The deployment became delayed in Aviano, Italy, due to a misunderstanding of the mission (and lack of diplomatic coordination by the United States) by Kenya's President, Daniel arap Moi. With keen diplomatic finesse from Brigadier General Libutti and the Ambassador, he soon blessed the mission, asking that it be kept low key and the unit's arms kept out of sight.

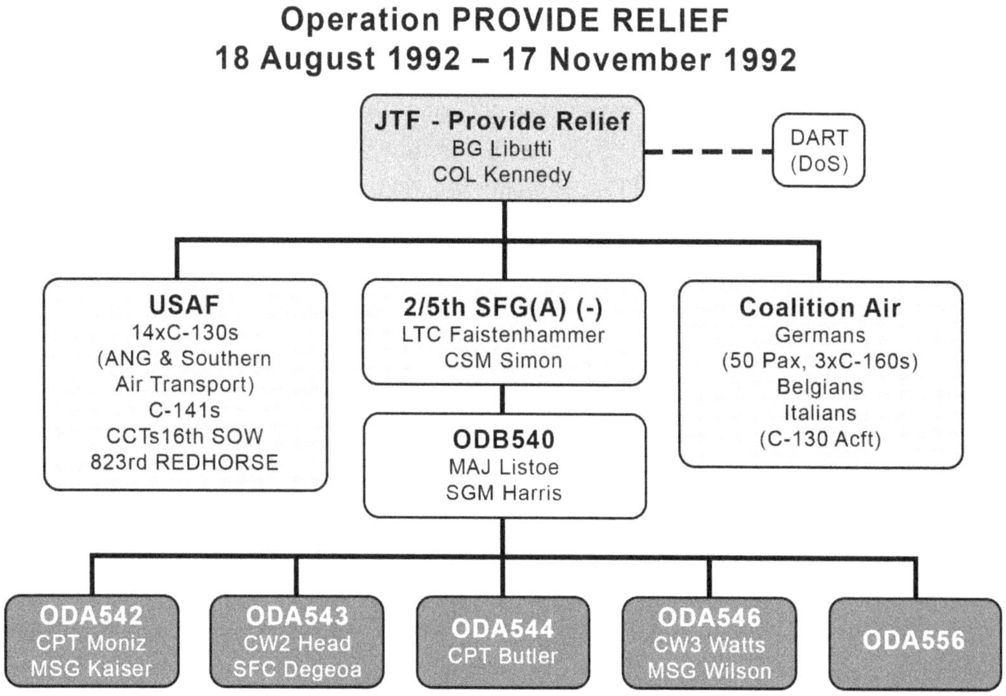

Major Listoe and the advance party met the company upon their arrival on the commercial side of the airport (Moi International) and then moved to the military side of the Mombasa airport. The unit was soon introduced to the equatorial heat and humidity of Africa. The Provide Relief JTF Headquarters was located in the U.S. Liaison Office's aerial support warehouses. These were two long parallel buildings, with office spaces, a cargo storage area between them, and large bays with roll up doors. The B-team established its operations in one of the buildings, with an operations room, Major Listoe's office, a team equipment area, and a parking area for the DMVs. The teams and attached Combat Control Teams (CCTs) modified their assigned area with the construction of plywood walls and steel framed gates to secure their gear and weapons.[17]

The unit lodged in the Golden Beach Hotel and the Dianee Reef Hotel in Mombasa. Later, Brigadier General Libutti changed their arrangements to the Intercontinental Hotel, where the rest of the Joint Task Force lodged, due to security concerns with the unit traveling in the early morning hours through some seedy portions of town. This move also eliminated the ferry ride; now all were on the same side of the estuary.

The first order of business was to provide Brigadier General Libutti with an equipment display and capabilities briefing. He seemed a bit curious about why they brought diving gear with their assigned SCUBA team and asked if they could perform underwater salvage operations (not knowing the mission parameters, the teams were directed to bring all of their equipment).[18]

After the briefing and further discussions among the leadership, it was clear that the mission would incorporate security for the aircraft and crews. Brigadier General Libutti initially saw the teams as a quick reaction force, with at least a Special Forces medic aboard the landing aircraft to accompany the CCT conducting air control. General Libutti saw the operation in three phases: first, get the aircraft bedded down in Mombasa; second, begin relief flights; and finally, downsize and close the operation.

As more details for the role of the SF Company became clear, ODB540 received the following mission–provide ground and airborne security for Provide Relief aircraft (ODB540); also, conduct airfield surveys and area assessments, if possible (gather intelligence). With further discussion between the JTF and SF leadership, the Airborne Security Augmentation Team concept (ASAT) was born.

The Threat

The threat at the airfields was primarily clan-based and under the control of a local warlord. The clan militia tended to operate in sizes ranging from 40 to 100 fighters. Their primary job was to serve as armed security guards for NGOs. The fighters included children as young as ten years old. They rode around in a variety of vehicles but primarily in Toyota trucks and land rovers. Military vehicles were rarely seen except for the ubiquitous Soviet jeep (GAZ), old U.S.-model 2-1/2 tons, and some Soviet trucks. On one occasion, a French-made *Panhard* truck was seen by SF team members at Belet Weyne.

The clan militia members were armed with a variety of firearms; Thompson submachine guns, Czech weapons, .30 caliber Brownings, Soviet light and medium small arms, and even up to an M1939 37-millimeter cannon were all seen at one time or another.

The most dangerous threat during these missions were "Technicals" moving in and around various locations in Belet Weyne. A gun jeep or gun truck hired by humanitarian organizations was labeled a "technical"–NGOs could not contract for military-related equipment while providing humanitarian assistance, so they often listed this service as a "technical service" on their invoices. Technicals normally were armed with DsHKs, and SG43 machine guns. Larger trucks were seen armed with Bofors 40-millimeter cannons. When the Technicals appeared to menace humanitarian NGOs at Belet Weyne, they were neutralized by the SF airfield security teams (this often meant just taking the breech blocks or firing pins from weapons).

Many of the former military airfields were littered with cannon shells (from MiG munitions), wrecked older MiG aircraft, and destroyed anti-aircraft artillery guns of the 57-millimeter variety. All of the radars were destroyed. The vast amount of scattered munitions provided the Somalis gunpowder, but after beating or extracting the powder out, the primers were left in place and still posed a risk. Minefields surrounded most airfields. Many of the militias extracted the larger anti-tank mines to use as part of their roadblocks.[19]

It was incumbent on the various NGOs to coordinate airfield security through the local warlords. While the International Committee of the Red Cross attempted oversight of this process, tensions often arose between competing "contractors" at the airfield sites. If the argument over who provided the service escalated, bullets would often fly between the factions, necessitating the shutdown of the food flights to that locale.

This issue created an operational dilemma for the CJTF staff. Earlier proposals by the SF battalion staff to place armed SF teams on the food flight aircraft were vetoed to maintain the neutrality of the humanitarian relief efforts–no weapons were allowed on any aircraft landing in Somalia, including concealed weapons. Having an orbiting AC-130 gunship as a protective measure was considered but not adopted.

The Operational Detachment–Alphas (ODAs), challenged by these restrictions, developed a course of action for an airborne security force remaining aloft during food flight deliveries, only landing in an emergency. This measure was acceptable to all and became known as the ASAT–Airborne Security Augmentation Team.[20]

The Airborne Security Augmentation Team Concept

The Airborne Security Augmentation Team (ASAT) typically comprised eight SF operators with two weapons carrier HMMWVs (M1026 variant). The team and vehicles rode in the cargo section of a C-130, designated the Airborne Battle Command, Control, and Coordination aircraft (ABCCC), circling above each C-130 food delivery aircraft on the ground. The ASAT served as a quick reaction force, landing in the event of threats or emergencies.

The first order of business was updating the current status and condition of the proposed landing sites. The first rehearsals to check the airfield sites were conducted between the 24th and 28th of August. All aircraft sorties were controlled by the JTF USAF component at the Mombasa air operations center, "Cricket." Communications were linked between Cricket and the ABCCC aircraft.

ODA556 had the honor of the flight on the 28th, assisted by ODA542. After ascertaining all was well, the teams began their full-up C-130 missions. On the 29th, four C-130s serviced Belet Weyne. General Libutti went in on the flight with ODA543 and ODA544; Sergeant First Class Schumer from the 2/5th Forward Operations Base served as his radio man (RTO). On the 31st, Major Listoe escorted the media for their first Provide Relief flight to Somalia (Belet Weyne).

A Day in the Life of a Special Forces ASAT Team

Each day the assigned ASATs arose around 0300 hours, ate breakfast, and boarded a minibus. Between their location and the USLO warehouse was a ferry (an old World War II amphibious landing craft–an LST!) to transport vehicles across an estuary. Upon arrival at the USLO Warehouse, teams drew their weapons and gear and packed and moved their vehicles to the fenced area between the buildings until called forward to the assigned aircraft. Team leaders took this time to get their assigned C-130 tail number, gather the latest information update on the destination airfield, and coordinate with the pilots and crew(along with the four designated airfields to service, on any given day an individual could be tasked for a separate flight to Mogadishu to conduct coordination or deliver materials to UNOSOM from the JTF). Locations requiring food delivery each day were provided by the ICRC or other NGOs (represented within the JTF).

This was a great benefit; the teams were able to talk directly to the NGO representatives before any mission, because the NGO representatives on the JTF staff were in constant radio contact with their people at the designated airfield. This input was essential to configure the appropriate security posture needed.

The Nongovernmental Organizations in Somalia had to coordinate each daily delivery through the local warlord (or clan militia leader). A tense situation could arise if competing groups providing the gunmen on the airfield began a dispute over their share of the haul, and how much pay they received (or could extort). These moments could easily break out with gunfire, with the NGO caught in the crossfire. The locals' most dangerous move was the appearance of a Technical not authorized to guard the delivery of humanitarian aid.[21]

To prepare for a food delivery, supplies for the designated airfield were palletized by USAF Loadmasters. The load generally consisted of 100-pound bags of beans, rice and cooking oil, and sometimes dehydrated and multi-mixed food (wheat, flour). Medical supplies were also included on the pallets. The pallets were then loaded aboard the C-130s until the entire cargo bay was full (three pallets high, four stacks deep).

During the loading phase aboard the ABCCC aircraft, any heavy weaponry was removed from atop the DMVs and kept out of sight. Personal weapons, ammunition, and load-bearing vests were kept in a footlocker during loading. Greene recalls:

> Each vehicle was set up for three day operations as far as ammunition, food, water and fuel–three days desert operations. Because of the distance to the airfields we were operating on, we figured it would take us three days to get back to Kenya from there, driving time. If we encountered any problems along the way, then we could use our discretion to take care of that problem. But also, that's why the GPS's–we had two GPS's [TRIMPACs] just in the vehicles with the MK-19s. They had all the airfields programmed in there, so if we had to land at one airfield, like if a plane had crashed at an airfield or something happened where we couldn't land, we would land at the closest airfield and go overland to the next airfield.[22]

At the twenty-minute mark in the flight, all members of the team procured their weapon, magazines, and loaded their vests. Personal load for the weapons was three magazines of 9mm for the pistol and seven magazines of 5.56 millimeter for the rifles. Ammunition for the crew-served weapons was opened from cans and laid atop the vehicle, with the rounds in the feed tray, but the weapons were not loaded or any ammunition chambered.[23]

In the event of threats from armored vehicles, AT-4 anti-tank rockets and the M-72 Light Anti-tank Weapon (LAW) were issued (this distribution did not occur until the second SF company rotation).

Now prepared for the mission, the six or eight team members spent the long flight time conducting hands-on drills, reading, or catching up on sleep.

Any given mission might require a response to threats on the ground. There were three plausible scenarios for this contingency. The first response was in the event there was a threat to a plane and crew on the ground. The ASAT team would immediately land. Two SF operators aboard would exit the parachute doors and work within a thirty-meter distance from the aircraft to assess the situation.

While outside, remaining SF operators and the CCT inside the aircraft armed with weapons and wearing type III-A Kevlar vests covered their movement with an M60 machine gun out of one of the right doors and a Squad Automatic Weapon (SAW) protruding from the left door. Stacked nearby were the weapons and vests of the two operators on the ground, to throw to them if needed. In this scenario, operators abided by the Rules of Engagement–use lethal force for self-defense if threatened with bodily harm or damage to the aircraft. If the threat level was serious, the machine gunners stepped out and joined the two operators outside the aircraft, bringing those operators their arms and individual equipment. The remainder of the team boarded the DMVs and removed the shackles on the vehicle for a rapid exit; the aircraft ramp was lowered. This scenario would also be used in the event of a lengthy time on the ground for the cargo aircraft, due to a maintenance malfunction. Once fixed, everyone boarded and took off; if not, the USAF crew was moved to the ABCCC aircraft and evacuated unless ordered to remain with the aircraft and provide security until assistance could be flown in.

In the most dangerous scenario, the DMVs exited the aircraft, drove over to pick up all who were outside, and manned the crew-served weapons. In some team drills, the DMV mounted with the .50 caliber swung down the left side of the airstrip once exiting and the vehicle with the MK-19 down the right side. Hopefully this imposing appearance thwarted any of those predisposed to attack a food flight.

Wendell Greene explained the alert for the circling aircraft, stating, "the code word TAILPIPE was used to get one of those ABCCC birds to come down and land. As it was, they just stayed in orbit over top of the airfields when the food was being delivered. If we were servicing just two airfields a day, each one would orbit an airfield; if we were using more than two airfields, then they would racetrack around all those airfields."[24]

The first two-ship formation departed Mombasa in the early dawn–the airborne security bird departed Mombasa first, followed by the food cargo aircraft. Two hours later, the second two-ship formation departed Mombasa. The first sortie arrived in Somali at daybreak, the second sortie in the afternoon or evening (it took about two hours from take-off to reach an airfield inside Somalia). The ABCCC security aircraft remained overhead, circling the designated airstrips for the day; with two ABCCC ships, this could last up to twelve hours (a normal off-load of supplies for the C-130 on the ground lasted an hour or so, but could take up to four hours). At the end of the day, the two aircraft from the second sortie returned to Mombasa by mid-evening.

While inbound and prior to any landing, the aircraft commander requested a situational update from relief agencies on the ground. All manner of things could occur between the time of take-off and the arrival to the designated airfield in Somalia. It could have been last minute disagreements between local hires and gunmen over the food allocation, or appearing with non-approved technicals and heavy weapons, or worse–mines thrown on the airfield. The NGOs kept radio contact with their representatives located at the JTF headquarters to apprise them of any changing conditions on the ground.[25]

As the C-130s landed, NGOs waited with prepositioned trucks and security guards; there was always a gathering of Somali civilians looking for immediate feeding or to rush out after the transaction was complete to gather any morsels of food spilled out during the process. No food deliveries were airdropped just for this reason (and to prevent anyone from being killed by a falling pallet); all were air-landed. The planes either parked on the end of the runway or, if available, in keyholes along the airstrip. Immediately upon the aircraft stopping, a truck (with laborers) backed up to the loading ramp and the process to load the food began; as each truck filled, another would take its place.

NGOs ran the security and logistics for humanitarian relief on each airfield. Local clans or factions, under the control of a warlord, were contracted by the responsible NGO and provided armed men in platoon strength for the immediate security of the air-delivered food distribution point. Upon arrival of the delivering aircraft, the Somali security forces typically fanned out on every corner of the airfield with about a 150 to 200 meter separation between armed fighters. This force was often augmented by the warlord through subcontracting. The competition to be a subcontractor could often become violent if clans or militias felt left out on a money-making deal.

While the aircraft sat on the ground, a small number of SF operators exited (two in civilian clothes–a medic and one operator, carrying a small duffle bag with communications, emergency gear, and a loaded pistol–some concealed their pistol in a waist pack). A control point was established for the Combat Controller (CCT) where he emplaced a beacon. The beacon was about the size of a briefcase. Air control was required to control the next set of aircraft, incoming. The SF operator had the duty to walk around and survey the airfield and designate a prominent feature as a rally point for all in the event of emergencies. This normally involved pacing down to the opposite end of the runway. Of interest was identifying and recording any minefields.

One or two members of the ground security detachment remained at the aircraft to protect the pilots and crew. If possible, SF medics traveled into the nearest town or village to assess medical conditions. The medics also spoke with NGO medical personnel at the airfield to gain further knowledge.

The assessment team took advantage of their time on the ground to count Somali forces, ascertain the condition of the runway, identify how many weapons and what type were at the airfield, what militia tactics were being used, and given time, assessed threat potential, and questioned relief workers and Somali civilians on current activities in the area. Meanwhile, the mounted SF security team (the ASAT) circled overhead as the ABCCC prepared to land in any emergency.

Major Listoe recalled the activities from the flights he personally flew on as a member of the security team into Belet Weyne and Gialalassi:

> Basically, when we would go into a new airfield that we had not been on before, the Air Force wanted to check the airfield out to make sure that it was capable of receiving the aircraft that they were in and sustaining extended operations. So, traditionally, what we'd do is we'd take a C130, we'd load up some of their combat control personnel, load up a few of my personnel, usually two, and if we were looking at opening up a new airfield, fly into that airfield, land, and make contact with the relief organization that was on the ground.
>
> Of course, that would have been done ahead of time to ensure that they knew we were coming and that they were able to accept the level of supplies, the volume of supplies that would be flown into that field . . . and essentially do a security assessment of that airfield by determining

> were there any heavy weapons? Was there a high level of military activity in the area? What were the people doing? Were they generally friendly; were they hostile? That sort of thing. And while we were doing that, the combat control team would check out the runway to make sure it met the Air Force requirements for landing in that area . . .
>
> You provided medical assistance if necessary to any Americans or the relief personnel that were there who required medical assistance. Spoke with the aid personnel. Spoke with the locals. Walked around to the extent that you could determine what military activity was going on. Get a feel for the level of hostility or the level of friendliness that was there that day. And, based upon what you determined, went into determining also whether you flew in there the next day or not.[26]

Additionally, the special operators could see the physical condition of the townspeople and refugees for themselves. It was also a chance to speak with tribal elders and gather their biographies (a priority intelligence requirement from the JTF–PIR). If possible, this inspection included riding with the NGOs into the nearest town. The team members were able to see the plight of many of these communities and the absolutely stark situations the populace had to endure first-hand.

The mission to Oddur gave Sergeant First Class Greene an opportunity to see "the wretched of the earth." Oddur had a dry riverbed as an unimproved landing strip. A few landings were a bit dicey, bouncing along when loaded down with the weight of the cargo. (One aircraft was lost around October due to structural failure after operating at Oddur. The Air Force cannibalized it.)

Greene described the toll so many deaths there were having on the populace:

> Oddur had established its own little police force and maintained a sort of military wing. So the Chief and his elders all dressed in military uniform and they maintained a pretty good military presence. Across the lake bed was a cemetery. Between plane runs, they'd cart little stick people bodies over there and bury them. The gravediggers worked from the time daylight came to when the sun went down.[27]

Captain Moniz secured a vehicle ride one day into a refugee camp during his mission to Baidoa.

> The team boarded a Land Rover provided by the local NGO and went into the refugee camp at Baidoa, where we saw dying people and bodies of dead people lying around who had attempted to get to the refugee center for food and water. The crowd was a sea of humanity and as we tried to get through, they suddenly parted. Looking up the street we saw a Technical with a machine gun aimed directly at us, but we continued to drive on and no incident occurred that day.[28]

Master Sergeant Bobby Phipps, A Company's Operations Sergeant, also remembered one of these experiences:

> There were times where we'd send individuals out, go to travel out and see, you know, the starving. And that put a lot in your mind seeing these people in the shape that they were in. The memory and the mission, the biggest thing is remembering the people that you had seen, the shape they were in.

> Basically that was the big thing about being there. Just remembering how the people were. That's the biggest thing about the mission in my memory.[29]

Once the humanitarian food unloading process was complete, the outside survey team called on the radio for the aircraft to taxi to their location, drop its ramp, and pick them up. The plane took off and flew about 200 feet above the deck for about five minutes in order to evade any ground fire.

All the information and photos were put into a report, saved to a disk, then sent to General Libutti's intelligence staff on the JTF. In this manner, the special reconnaissance mission for SF in a semi-permissive environment was carried out (the activities of the SF to gather this intelligence was very often the sole source of intelligence for Brigadier General Libutti). It was also briefed in the evening update meetings, held at 1600 hours. With updated information, the coordination could be made with the DART and NGOs on the next day's missions.

All personnel who deplaned to conduct assessments remained all day on the airfield until recovery by the second sortie.

Operational Challenges

The missions were not without peril. Some aircraft were shot at in the conduct of the missions and situations on the ground could become tense, with the team never knowing what level the Somalis wanted to escalate things. SSG Glen Wharton, the communications sergeant on ODA546, recounts his and Staff Sergeant Salinas' experience:

> Once we landed on the ground at Baidoa, we were off-loading the aircraft with all the equipment needed to establish an air traffic control type scenario for the Air Force guys and we were going to conduct our observer security type mission. Shortly after that, we noticed that the ICRC, or International Red Cross members, came across on the radio and told us to abort the mission, to go ahead and leave that particular airfield as soon as possible. Well, prior to that, the CCT guys had already departed the aircraft along with Sergeant Salinas, the detachment medic. We got word from the loadmaster, from the aircraft, and they told us to get back on the aircraft so we could depart the airfield. The reason, we later understood, was they did not have personnel to unload that particular air load of food that we had brought in with us.
>
> As I got off the aircraft to tell the other guys that we were going to abort the mission, and we were going to depart the area, immediately after that there were about four or five Technical vehicles that were approaching the aircraft at a high rate of speed . . . I gave him that message and I got back on the aircraft. I noticed these vehicles approaching, like I said, at a high rate of speed towards the aircraft. A Technical vehicle was on the runway at the 6 o'clock position with the aircraft; the aircraft ramp was open and all I could see was him approaching me with a .50 caliber machinegun.
>
> . . . But that was something that really made me think, you know, that particular situation. Kind of stuck out in my mind, close call right there."[30]

Captain Moniz remembered observing one such incident:

> Now, on one occasion, I'm sure it was at Belet Weyne, one of the subcontractors decided that he had enough guns to be the primary contractor. And, of course, they put a land mine on the runway, after one of our aircraft had taken off, which leaves us, the four observers or four-man element on the ground. After the aircraft had taken off, they put a land mine on the runway and decided to start labor negotiations at that time. This guy had just, I guess, just bought his new Technical. He had a .50 caliber on it, and he was quite proud of it. And he rolled this thing out. Well the primary contractor, not the primary contractor for nothing, after a few back and forths, he comes out with his deuce and a half that had twin Bofers, 40mm antiaircraft guns, mounted on. And they come out and they're doing the crew drill and they're swinging the guns around, and you know, knocking the guys off the truck. It was actually rather comical. And once the guy got a look at the old twin Bofers, he kind of backed down. Some negotiating, and they pulled the mine off the runway, and we continued the mission."[31]

The first C-130s flew into Belet Weyne on the 28th of August. Belet Weyne would soon earn the nickname "Bullet Weyne" after a rise in incidents from clan militia. As an example, one day a Southern Air Transport C-130 landed and was caught in the crossfire from two warring factions. ODAs 542 and 546 flew the Belet Weyne missions on the 8th through the 10th, and again on the 13th through the 18th. Greene easily remembered one of those "Bullet Weyne" missions captured later as an "incident" in message traffic:

> At Belet Weyne, Mr. Watts [CWO3 Bruce Watts, ODA545] was sitting there and they cracked some rounds across the airfield and hit the berm beside him just to see if we would jump like monkeys. They did the same thing at Baidoa. And the Air Force CCT guys just about came unglued and I told them to sit still, just wave at 'em, if we jumped like a bunch of monkeys every time we come down here,"hey, watch this," then they would shoot at us every day.
>
> Two aircraft were hit by ground fire; one while on the ground. The C130 was hit in the tail area of the cargo bay–it was probably a ricochet–it entered at an upward angle. It hit one of the 5-gallon thermos cans they had. The hydrostatic force blew the lid off and the round landed down by the Crew Chief's feet. So the crew was apprehensive and we evacuated the airfield. They had already unloaded their stuff and we were leaving for the day. We figured it was somebody high on khat just taking target practice.[32]

Reassessing the Mission

By the 16th of September, Major Listoe and Lieutenant Colonel Faistenhammer adjusted the mission based on these now predictable and recurring activities. It was clear this type of mission could be conducted by Military Police, and the chain-of-command advocated their replacement with MPs; General Libutti heard their concerns but preferred the maturity of the SF over any other form of security for the food flights. The battalion was still on the mission but did not require the force size initially deployed. Any team with more than eight personnel was cut.

Lieutenant Colonel Faistenhammer turned the mission exclusively over to Major Listoe, who now reported directly to General Libutti, and redeployed back to Fort Campbell, Kentucky with selected members of the staff, some of the augmentees, and personnel who needed to attend various schools and courses. The remaining ODAs merged with this downsizing; for instance, ODA544 and ODA542 combined their assets into one team. The unit now consisted of only fifty-eight to sixty personnel, approximately. The assigned SOCCENT staff officer on the JTF also redeployed after September, before the Thanksgiving holiday period.

An Increase of Incidents

There was another incident involving aircraft hit with bullets. When one occurred at Baidoa, General Libutti cancelled food flights to the airfield for one week. The message was loud and clear; there were no further incidents at Baidoa.

Increasing incidents like these and others were noted by General Libutti. To preserve force protection, he ordered the teams to remain on the airfields for the remainder of the mission and not to attempt visits into the surrounding areas. Whenever a close call occurred by threatening militia, the JTF commander restricted any further relief flights into these areas until the local warlords, clan elders, and NGOs could guarantee security for the aircraft.

This option became necessary on the 2nd of October when clan militia assaulted the ICRC warehouse in Baidoa, wounding four Somali workers, on the 15th of October when militia factions blocked the runway at Belet Weyne with mines, and on the 17th when faction fighting was reported between Colonel Jess and General Morgan near Bardera.[33]

Sergeant First Class Greene was on a mission with SFC Mark Bisceglia from ODA546 (medic) and the USAF Combat Controller; he recounts the nature of the bizarre sights and sounds at one of the airfields, before he experienced an incident (written in a paper later used in one of his college classes, titled "A Day in Paradise"):

> . . . Once we land, things change. On close inspection, the vehicles and guns are nothing more than rusting, junked leftovers. Even the jets have been stripped of all their useful parts and left behind like scary lawn ornaments. All of them have become monuments to an earlier time and the former communist government.
>
> We take up a position on a dirt berm along the western edge of the 10,000 foot north-south tarmac runway. Last night's sudden thunderstorm has turned the dusty Martian red soil into a thick slimy paste. It sticks to the soles of our boots and anything else that comes in contact with the ground.
>
> The air is hot and muggy, full of the smell of burnt jet fuel as the C-130 cargo plane taxies down the runway to its assigned off-loading station. The faint smell of cooking fires drifts from the town. The occasional sickeningly sweet smell of human bodies reminds us of the death that is around us. Sweat runs down our backs into the tops of our pants and the day is just starting.
>
> As I look across the runway to the east, I see the tail of an MIG-21 just above the rim of its revetment. Beyond that I gaze upon the beaten gravel road to the town located northeast of our position. From what I see, the place is a mixture of dilapidated concrete and brick buildings intermixed with shanties made of stolen or discarded junk and traditional African huts.
>
> I look to the south, heat shimmers off the runway, giving it a mirror-like sheen. Down across the runway, the C-130 sits like a fat, squatty green bird. Its four engines hum in a low drone. The

> workers form a human chain that runs in and out of the aircraft's tail. They hurriedly unload the precious cargo of badly needed food into their battered, old Russian-built MAZ trucks.
>
> The ground around me is littered with unexploded cannon shells of all sizes. The shells, for lack of the right caliber weapon to fire them from, have been beaten out of their casings and the powder charges carried off . . .
>
> Local tribesmen guard the airfield. They are a strange looking mob dressed in old army uniform and traditional attire, armed with automatic weapons of all description. In small groups they are scattered around the airfield parameter. They are alert and ready to defend the runway and the relief supplies that are as precious as gold to the starving masses flocked around the area. The whole place looks like it should be in a "Mad Max" movie.

A dystopian world indeed. While the two SF sergeants and the CCT relaxed on a dirt berm as the aircraft continued to unload, they heard the crump of what sounded like an artillery shell. A black cloud arose about 200 meters from their position. They thought it could have been a large anti-aircraft shell:

> We believe it came from a 57mm because the shell exploded and then we heard the gun go off. At the time we evacuated the airfield. That was the biggest evacuation, we just ceased operations. I determined it to be about a 57mm because it probably was a fuzed shell because it went right off at the end of the runway. The NGOs said that an RPG had been fired at the food warehouse and it probably was the RPG self-detonating. The only problem with that story is that the food warehouse was three kilometers from the airfield and an RPG detonates at 920 meters.[34]

They immediately alerted all aircraft to suspend operations and pick them up. Sergeant First Class Greene, Sergeant First Class Bisceglia, and the young Combat Controller gathered their bags and the radio and hastily boarded. The evacuation was safely conducted.

Meanwhile, many in the media and in the Congress were keenly interested in the mission; the JTF began hosting a parade of delegations and VIPs visiting Kenya. One of the visits they did not mind was when the CENTCOM commander, General Hoar, and his Command Sergeant Major, stopped in accompanied by Ambassador Theros. Brigadier General Libutti chose Oddur for him to visit as a typical mission. One SF team with their USAF Combat Controller supported the flight.

General Hoar observed a C-130 unloading supplies onto food trucks and then participated in the security convoy into town. He may have been amazed at the refugees who lined the route eager to see such a prominent visit. In town, the Somali police from Oddur maintained order while General Hoar and Ambassador Theros met with local officials, ICRC staff, and the UN delegation, which had just arrived via a twin engine Cessna at the airfield. During his visit, the CENTCOM Command Sergeant Major took the opportunity to visit with the populace of the village. The trip was successful and gave the CENTCOM commander a good feel for the scope of the operation.[35]

Daily summaries of the mission by the B-team illustrated September's activities. Between September 1st and September 30th, the SF teams participated in food flights and airfield assessments at Baidoa, Belet Weyne, Oddur, and a special trip to Mogadishu to support a Department of State liaison team with communications (conducted by CWO2 Bob Head, the battalion operations Technician, the last week of

September). On September 13th, Major Listoe sponsored a congressional delegation.

The first "real exciting" day occurred on September 18th; the mission C-130 flown that day with ODAs 545 and 546 operating near Belet Weyne was hit by several rounds of small arms fire. As a response, Brigadier General Libutti halted the mission for about a week until receiving assurances from the NGOs and locals it would not occur again.[36]

As the food flights began running smooth, Brigadier General Libutti desired an expansion in the number of airfields where food could be delivered. This decision would have put a strain on the already stretched 18 Delta SF medics to support, but the expansion never occurred. To handle the next rotation of an SF company for continuation of the mission, three of A Company's medics would extend to support them.

The JTF aircraft flew other missions, unilaterally (without food deliveries). At the extreme use of air assets, a C-130 was used one day to fly low level and conduct a "recon" of suspected Aideed militia while the SF team used binoculars and eyeballs to identify anyone on the ground.

The ground landing team mission also supported unilateral airfield assessments for the JTF when not accompanying the C-130 food flight aircraft. On one such mission, the commander of ODA524 joined the CCT airmen and others for a flight aboard a bush-piloted Cessna Twin headed to Bale Dogle for an airfield assessment. For the long flight, extra fuel containers were stowed aboard.[37]

The pilot landed at Bale Dogle and began his taxi onto the cross aprons when a jeep with a mounted machine gun suddenly appeared and wheeled around to block the aircraft. Gunmen on board jumped down to surround the aircraft. Soon, a large cargo truck with a Bofors heavy machine gun pulled in behind the Cessna, effectively blocking the aircraft in position. The SF team conferred among themselves and decided to step off the aircraft and find the local leader. Perhaps if the purpose of the mission was explained to him, things would go smoother.

The then Captain Steve Moniz explained the situation they faced:

> We remained calm as we departed the aircraft and greeted one of the militia who appeared to be the man in charge. No one from the militia could speak English. We then attempted to converse in Italian and in Arabic, using my team members who spoke these languages, but to no avail. The supposed leader only spoke Somali, and sent for someone who could speak English. As the spokesmen arrived, with the capability of broken English only, I explained the mission with heavy emphasis on getting the airfield looked at to facilitate the arrival of humanitarian supplies.
>
> The militia leader understood the importance of the help to people in the surrounding area, and allowed the CCT airmen to finish the assessment while his militia provided local security. I then spoke to him of potential work as the "security force" for Bale Dogle for future food flight landings. The team's work was ended in about an hour, and we departed on an amicable basis with the militia. No one wanted to imagine the alternative of Bofors rounds going through the Cessna and igniting the fuel containers aboard.[38]

On September 14th, the United States assisted with the move of 40 Pakistanis to Mogadishu, followed two weeks later with an airlift of 500 peacekeepers. To ensure the security of the airlift, U.S. Navy Phibron 1 with the 11th MEU (SOC) positioned themselves ten miles off the coast to intervene if needed. USAF air traffic control personnel were sent in to assist with the airlift.[39]

As more relief supplies filtered through, the new game in town became extortion. Each NGO food flight location paid exorbitant amounts of protection money to "lease" the location from local thug militias. Baidoa alone was paying $20,000 weekly to ensure the NGOs security. Then there were rental fees for gunmen, trucks, and technicals.

The head of the Office of Disaster Assistance, William Garvelink summed it up:

> If you flew into Baidoa in a C-130, which holds about twelve tons of food, you'd load up two or three trucks and one of them would always disappear. Plus they charged landing fees. Plus they charged a fee for each truck to get out the airport gate.[40]

October 1992

October began with the estimated amount of food getting through at only 40% of what was needed. Talk in the United States began to center around the idea of sending more troops to Somalia. Senator John Lewis and Senator Nancy Kassebaum were the most vocal. The leadership of CARE even recommended turning Somalia back into a Trusteeship to solve the problem.

During this deployment period the company was visited by the 5th SFG(A) commander and his Command Sergeant Major who rode along on one of the relief flights and praised the teams for their work. When Colonel Bowra met for discussions with General Libutti, the CJTF commander also commended the unit for the valuable information collected by the teams and the overall success of not losing any aircraft or crew during the flights.

It was apparent by mid-October that the mission would continue beyond its intended sixty- to ninety-day projection. The 5th SFG(A) would require deploying another company-sized rotation. Unfortunately, the 5th Group was to learn an additional rotation would not be reimbursed by JCS funds. Colonel Bowra requested higher command assistance in making the rotation possible and offered shifting 5th SFG(A) priorities around to find the necessary funds.

The request was approved on October 15th. The rotation plan would consist of downsizing the current deployment from five ODAs to three ODAs and was scheduled from November 10–17, 1992. Company C would replace Company A. ODB560 (Company C) was scheduled for deployment on the 10th of November, with two ODAs following on the 13th, and the final ODA arriving on the 17th of November.[41]

On October 23rd, the UN sponsored an initiative to develop a renewed 100-Day Action Plan for Accelerated Assistance to Somalia. Even the most optimistic members of the committee agreed nothing would be accomplished without addressing security concerns. In the end, the plan was worthless due to under-staffing, lack of resources, and no credible way for UNOSOM to address the security problem.[42]

With the election looming in November, President George H. W. Bush was slowly coming around to the need for troops, also. He was influenced by his own study of the problem, the inefficiency of the UN and the constant barrage from the media and Congress. Surprisingly, even though there were a few pessimistic hold outs (who basically thought getting involved in Somalia would be a morass while there were no valid American security interests), key members of the JCS, his staff, and the State Department thought an intervention would be the right policy. President George H. W. Bush certainly understood only the United States could provide a large body of professional troops, fast, while the capabilities of other member states were suspect.

Seeing the writing on the wall, in the ending days of October Aideed characteristically ordered all UN troops removed from Somalia; Somalis would fix Somalia's problems, not outsiders. In a knee-jerk response from the UN, the Special Representative Sahnoun was removed (he resigned) as one of the few, and effective, representatives sent to Somalia. He was replaced by Ismat Kittani, an Iraqi-Kurd.[43]

November and the New Special Forces Company Rotation

On November 10th Pakistani peacekeepers were able to take over the airport and establish a security perimeter. Anticipating a Presidential decision to intervene in Somalia, CENTCOM issued a warning order to I Marine Expeditionary Force on November 21st.

On November 11th, four CARE employees were killed when their truck convoy was ambushed. Thirty-three out of thirty-four trucks were lost. Operation Provide Relief continued, and in November, Brigadier General Libutti turned over the JTF command to BG Paul L. Fratarangelo (USMC). The New Port and the Mogadishu airfield came under mortar attack while relief ships were trying to deliver humanitarian aid, which were ultimately run off by the shelling.[44]

Company C, 2/5th SFG(A) was alerted in early November for deployment to Mombasa as a replacement for Co A. The unit deployed with four Operational Detachment Alphas (ODAs 562, 563, 564, and 565) on November 12th. Each team was required to have a minimum of a Team Leader, a medic, and a communications sergeant. Some cross-leveling occurred throughout the company from the two teams not deploying (plus, Major Carroll was gaining three medics from the B-team, which were deployed at the time).

Due to some unforeseen circumstance, no aircraft were available to fly into Fort Campbell; the unit was directed to move by bus and flatbed cargo trucks to Charleston, South Carolina. The contingent numbered 40 pax.[45]

After being released for three hours to spend time with their families, the unit gathered back at the battalion area. It was a glum, November day, chilly and with rain. Company C (ODB560) was commanded by MAJ Lelon Carroll, known as "Lee" to his colleagues. He had been able to go to Mombasa earlier and assist the staff, so he was pretty knowledgeable about the mission for his company's ninety-day rotation. Sergeant Major Harris served as the company Sergeant Major.

Upon their arrival to the JTF, Company C coordinated with the outgoing teams and signed for essential equipment remaining in Mombasa (primarily ammunition, explosives, and grenades). Since Major Carroll was well informed of the mission requirements, the teams spent their initial time gleaning information about the clan militias from the outgoing teams. Company A also got them bedded down in the Intercontinental Hotel.

Major Carroll assumed operational control of the mission on November 17th. The company easily slipped into the routine already established by A Company. Each team would work two days on and one day off. Such a routine soon became monotonous, but when there was time off, the teams could avail themselves of the tourist attractions Mombasa was famous for in the region–beaches, pools, shopping for local carvings and gems, or playing volleyball. They also soon availed themselves of the variety of bars and restaurants in the area.

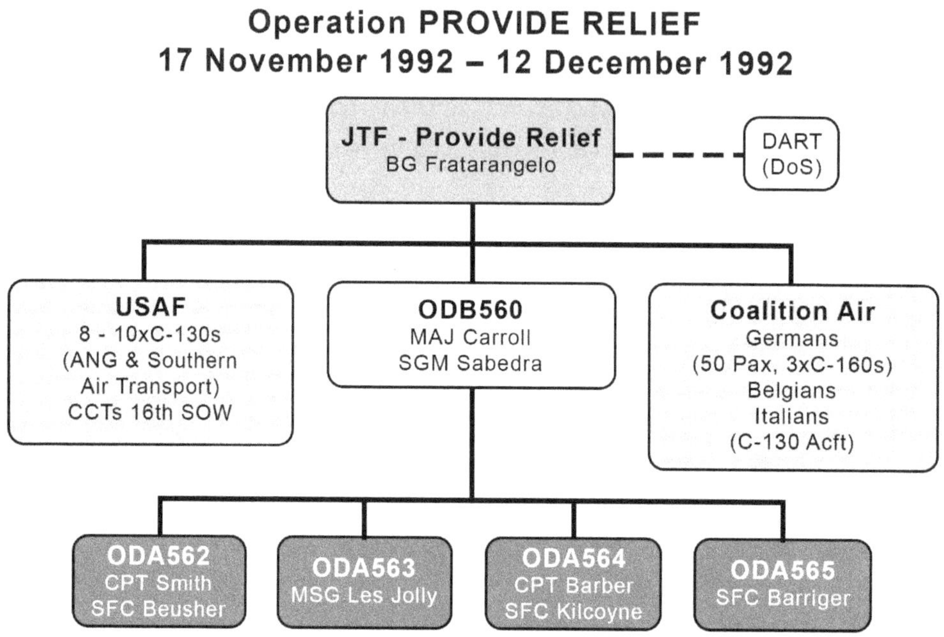

As had Co A, as the mission became routine, Major Carroll eventually combined ODA563 and ODA565 into a composite team, allowing him to send people home for school and other commitments.

As the end of the month neared, all were aware of the announcement and planning for a robust intervention into Somalia, Operation Restore Hope. The scope and thrust of Carroll's mission began to focus on airfield assessments and the handover to incoming forces at each of the airfield locations.

On November 25th, the Assistant Secretary of Defense-Special Operations and Low Intensity Conflict (ASD-SOLIC) sponsored a National Security Council forum and provided a briefing to review their policy recommendations to President George H.W. Bush as the situation in Somalia continued to deteriorate. NGOs operating in Somalia reported an increase in volatility of lawless acts. NGOs and UNOSOM troops were intimidated by the actions of clan militias. The increase in violence soon became intolerable. Lack of ability to increase the distribution of humanitarian aid ensured the world would continue to watch horrific images of the dead and dying in the media.

The NSC's advice to President George H. W. Bush was basically, "Let's go, go big, do it fast, it is the right thing to do." With General Powell's agreement and the CENTCOM commander on board, Bush signaled his approval to intervene in Somalia with U.S. forces. George H. W. Bush felt the operation could be completed prior to Clinton's inauguration in January and would be a show of his "New World Order" to bring peace and stability to an area of the world while he was still President. The Clinton transition team was duly informed of the President's decision.

On November 26, 1992, the ODB was tasked to provide security for the Commander of TRANSCOM, General Fogleman, when he visited Oddur, Somalia. The number of aircraft provided to the JTF had actually increased by then (or were being pre-staged for Restore Hope).

One week prior to the Marine landing in Mogadishu, the Special Forces company in Mombasa was informed it would serve as a basis for the Army SOF component assigned to Restore Hope. Lieutenant Colonel Faistenhammer was going to return with his staff as the component commander and bring additional teams with him to perform as Coalition Warfare Teams to various coalition forces assigned to the Unified Task Force, UNITAF.

On December 4, 1992, Company C contributed a four man team to provide security for Air Force engineers conducting an airfield survey at Bale Dogle Airfield. They were met by the local Somali airfield security force who greeted them by pointing their weapons at them. The security force departed the airfield and was unable to conduct the assessment at that time. Operation Restore Hope began on December 9, 1992.

Results of Provide Relief

USAF humanitarian food relief flights delivered approximately 23,000 tons of much needed aid. The operation eventually grew to using around 41 C-130s, flown by twenty-four organizations (from active USAF and Reserve NG Squadrons, Airlift Wings, and Composite Wings in U.S.). With around 370 to 400 personnel in the JTF, the operation lasted from mid-August to February 28, 1993 and exceeded the humanitarian relief achieved during the Berlin Airlift, with 2,000 flights made into Somalia. Given the turmoil, it is impossible to count the impact of lives saved, but certainly tens of thousands of Somali men, women, and children were saved from starvation.[46]

Bruce Watts, one of the senior SF Warrant Officers on the Provide Relief mission, summed up the contributions of the Army Special Forces, writing later in his paper, "Special Forces Support to JTF Provide Relief":

> The five ASATs flew an average of thirty missions each logging over 300 flight hours per team. In fact, the Special Forces troopers had more flight hours than any other U.S. personnel in the area of operations. A crewman aboard one of the C-130s exclaimed, "They ought to give those guys wings!" Although the teams were never called upon to respond to an emergency, several incidents became "close calls." Daily uncertainty at the Somali airfields made the security contingencies imperative . . . The overall success of Provide Relief can be measured by the significant reduction of human suffering in the region where over 23,000 tons of relief supplies were delivered by U.S. personnel over a three-month time period. The Special Forces troopers enjoyed a close working relationship with their sister service [USAF] in the JTF and paved the way for future humanitarian assistance operations in denied areas.[47]

Endnotes

1. Piasecki, Eugene. The History of U.S. Army Special Operations Forces in Somalia. Fort Bragg, NC: Draft

Research Paper prepared for the USASOC History Office, August 2007. Chapter 3, pp. 1–3.

2. Hirsch, John L. and Robert B. Oakley. Somalia and Operation Restore Hope: Reflections on Peacemaking and Peacekeeping. Washington, DC: Institute of Peace Press, 1995, pp. 17–18.
3. Hirsch and Oakley, pp. 18–19.
4. Ibid, p. 19.
5. Ibid, pp. 20–21.
6. Ibid, pp. 21–22.
7. UN Secretary General Report S/24343.
8. Rutherford, Kenneth R. Humanitarianism Under Fire: The U.S. and UN Intervention in Somalia. Sterling, VA: Kumarian Press, 2008, p. 30.
9. Livingston, Steven. "Clarifying the CNN Effect: An Examination of Media Effects According to Type of Military Intervention. Research Paper R-18, Boston, MA: The Hoan Shorenstein Center, Harvard University, John F. Kennedy School of Government, June 1977, pp. 1–2.
10. Rutherford, p. 41.
11. Ibid, p. 38.
12. Harned, Glenn. Stability Operations in Somalia 1992–1995: A Case Study. Carlisle Barracks, PA: Personal Monograph Series PKSOI (19950612 009) United States Army War College PKSOI: War College Press, July 2016, p. 15.
13. Poole, Walter S. The Effort to Save Somalia August 1992–March 1994. Joint History Office, Office of the Chairman of the Joint Chiefs of Staff, Superintendent of Documents, Washington DC: U.S. Government Printing Office, 2005, pp. 7–8.
14. Ibid, pp. 8–10.
15. Ohls, Gary J. "Somalia . . . From the Sea." Newport Papers #34. New Port, RI: U.S. Naval War College, Jul 2009, p. 54.
16. Interview with LTC Steve Moniz by COL (Ret.) Joseph D. Celeski, conducted on 29 Nov,2006 in Raipur, India.
17. Draft magazine article from CW3 Richard A. Detrick on his Provide Relief rotation, provided to the author.
18. Steve Moniz interview.
19. Information provided by Wendell Greene during interview with COL (Ret.) Celeski, 2 July 2005, Nashville, TN.
20. Celeski, Joseph D. "ARSOF in Somalia: 1992–1995." Research paper prepared for the USASOC History Office, Ft. Bragg, NC, 2006.

21. Steve Muniz interview.
22. Interview between COL(Ret.) Joseph D. Celeski and Wendell Greene, 2 Jul 2005, Nashville, TN.
23. Ibid.
24. Wendell Greene interview.
25. Detrick article.
26. Interview between MAJ Kent R. Listoe, CO A, 2/5th SFG(A) Commander and debriefer from 44th Military History Detachment, conducted on 13 May 1993 at Ft. Campbell, KY. Interview on file with USASOC Historian archive section.
27. Greene interview.
28. Interview with Steven P. Moniz, ODA542 Detachment Commander and MSGT Daniel J. Kaiser, ODA542 Team Sergeant on the roles and missions of ODA542 in Operation PROVIDE RELIEF. Interview conducted by the 44th Military History Detachment on 10 May 1993 in the 2/5th SFG(A) Classroom at Ft. Campbell, KY.
29. Detrick article.
30. Interview between SSG Glenn Wharton and 44th Military History Detachment in the 2/5th SFG(A) classroom on10 May 1993, at Ft. Campbell, KY.
31. Moniz interview, 10–11.
32. Greene interview, 5.
33. Celeski, p. 26.
34. Greene, Wendell M. "A Day in Paradise." Undated English composition paper provided to the author.
35. Information and pictures provided the author during the month of October 2006 by BG (Ret.) Mark Hamilton (USAF) who served on the CJTF staff during Operation PROVIDE RELIEF.
36. Summary of CJTF-Provide Relief SITREPS for the period 01–30 Sep, 1992.
37. Steve Moniz interview.
38. Ibid.
39. Ohls, p. 61.
40. Rutherford, p. 53. From Susan Rosegrant's "A Seamless Transition: United States and United Nations Operations in Somalia, 1992–1993." Kennedy School of Government Case Program (Cambridge, MA: President and Fellows of Harvard College, 1996), A, p. 9.
41. Celeski, p. 27.
42. Rutherford, p. 55.
43. Ibid, p. 57.
44. Ibid, p. 65.
45. The unit's alert, preparations, deployment and arrival in Kenya were provided in detail from Charles B. Smith's draft book, "Honor Bound: A Special Forces Detachment in Somalia," written in 1996. A copy was kindly provided to the author, with permission and photographs for use in this research.

46. Haulman, Daniel L. "Provide Relief." Wikipedia, accessed on July 30, 2020.

47. Watts, Bruce R. "Special Forces Support to JTF Provide Relief." Undated paper on file with USASOC Historian, Ft. Bragg, NC., pp. 5–6.

3

Operation Restore Hope: Phase I, December 1992

> I was not eager to get involved in Somalia, but we were apparently the only nation that could end the suffering.
>
> — General Colin Powell, CJCS

President George H. W. Bush's decision to deploy U.S. forces to Somalia in the waning days of his administration, a place of no vital security interest, was a humanitarian response to the deteriorating situation in Somalia and the UN's inability to solve the problem. All policy discussions eventually led to the fact that only the United States could solve the chaotic problem of Somalia; it was frankly the only country that could help the UN recover such a sinking ship. On November 25, 1992, the Assistant Secretary of Defense-Special Operations and Low Intensity Conflict (ASD-SOLIC) sponsored an NSC forum on the Somalia crisis and provided a briefing on their policy recommendations to the President. Meanwhile, the situation in Somalia continued to deteriorate. The Non-governmental Organizations (NGOs) with operations in Somalia were concerned over the increase in lawlessness. This increased violence could hamper future humanitarian aid and, consequently, have a detrimental effect on the Somali people.

President Bush agreed to intervene in Somalia based on advice from his National Security Council, and with concurrence from General Powell, as well as the CENTCOM Commander. President Bush's goal was to complete the mission prior to the inauguration of the newly elected President Clinton.

Now breathing a sigh of relief, on December 3, 1993, the UN Security Council passed Resolution 794, stating ". . . use all necessary means to establish a secure environment for humanitarian relief operations in Somalia." The resolution also stated an operation of this nature would be conducted under Chapter VII of the UN Charter–Peacemaking. This situation would be unique, and a first for the UN. Chapter VII required agreement from the sovereign nation where the UN intervention was to occur, but there was no "sovereign state" in Somalia. The resolution also set the table for the United States to lead the operation. The UN acquiesced in its own role of leadership by the phrase ". . . offers a member state to lead an international force."[1]

President George H. W. Bush chose a robust response, driven by Chairman Powell's admonition that if the United States was going to go, go big and get it over with quickly. This involved building a coalition force to work under an American-led task force. The Joint Chiefs of Staff alerted U.S. Central Command (CENTCOM) on the 1st of December; they in turn alerted the 1st Marine Expeditionary Force (I MEF) to lead the operation. On December 3rd, LTG Robert B. Johnston, the I MEF Commander, was designated as the commander of the Combined Joint Task Force–Somalia. Brigadier General Anthony C. Zinni was chosen as his operations officer.

On December 4th, the President addressed the nation announcing his decision to deploy troops to Somalia, led by USCENTCOM, and named Operation Restore Hope. He further outlined the size of the force and major military units designated to conduct the operation, followed with a list of the coalition member states who agreed to participate with the United States. In his statement, he explained to the American and Somali people the reason for intervening and his intent for how he wanted the operation conducted:

> First, we will create a secure environment in the hardest hit parts of Somalia, so that food can move from ships over land to the people in the countryside now devastated by starvation.
>
> Second, once we have created that secure environment, we will withdraw our troops, handing the security mission back to a regular U.N. peacekeeping force. Our mission has a limited objective: To open the supply routes, to get the food moving, and to prepare the way for a U.N. peacekeeping force to keep it moving. This operation is not open-ended. We will not stay one day longer than is absolutely necessary.

> Let me be very clear: Our mission is humanitarian, but we will not tolerate armed gangs ripping off their own people, condemning them to death by starvation. General Hoar and his troops have the authority to take whatever military action is necessary to safeguard the lives of our troops and the lives of Somalia's people. The outlaw elements in Somalia must understand this is serious business. We will accomplish our mission.
>
> We have no intent to remain in Somalia with fighting forces, but we are determined to do it right, to secure an environment that will allow food to get to the starving people of Somalia.[2]

To diplomatically balance the military operation, President Bush chose Ambassador Robert Oakley, the former U.S. Ambassador to Somalia (1983–1984), as his Special Representative to Somalia. Ambassador Oakley sent a dire warning now as the U.S. Special Envoy. He had read the reports from Ambassador Hempstone in Kenya about the "hell" of Somalia, and remarked:

> The true threat that Somalia will become a tar baby does not lie in a Vietnam or Beirut-type guerrilla war or terrorism, rather, it lies in an implied neo-colonial attitude, and in unrealistic/idealistic objectives and missions stemming from this attitude, which would require very considerable foreign involvement and major expense over a long period of time before there could be any hope of "success."[3]

CENTCOM's goal to achieve President Bush's Somalia policy objectives was to deploy American and coalition forces in order to provide a secure environment for humanitarian organizations to do their work and feed the suffering to alleviate starvation. There were a few caveats: the Combined and Joint Task Force (CJTF) would not be under UN command; all coalition nations desiring to participate would come under the operational control of the CJTF, not their national commands; there would be no nation-building (mission creep); the CJTF would not operate into northern Somalia; and the CJTF would not enforce disarmament of the clan factions. Finally, the CJTF and UNOSOM would conduct a handover and transition once Restore Hope's mission objectives were met. The operational guidance predicted mission completion by January 1993, in deference to President Clinton's inauguration on the 20th.

The media was keen to announce the deployment of U.S. forces to solve the humanitarian crisis in Somalia in support of the UN. The New York Times News Service released the following from the Virginian-Pilot newspaper:

U.N. favors U.S. command of military force in Somalia
The Security Council wants to have close ties to the operation.
New York Times News Service

> UNITED NATIONS – The Security Council reached a broad agreement Tuesday night that an American general should command the new multinational force it plans to send into Somalia to disarm the warring factions there and ensure that international aid reaches that country's starving people.
>
> But the council is continuing to discuss how much control the United Nations should exercise over such a military enforcement operation, with most members pressing for the Security Council to

be given greater powers of oversight than it had during the Persia Gulf war.

"The United States is likely to be commanding the operation," said Britain's delegate, Sir David Hannay. He spoke of a "broad convergence" within the council in favor of an operation under American command but with close links to the United Nations.

Hannay said these links would include a requirement that the commander report regularly to the Security Council, which would also decide when the operation should be terminated.

Several other council members also appeared ready to accept an American commander. "Who commands does not seem a major problem," said India's delegate, Chimaya Gharekhan.

Only China made clear that it is unsympathetic with any use of outside force in Somalia. But other council members believe Beijing will abstain instead of using its veto.

On Tuesday, the United States began discussing the text of a resolution authorizing an operation in Somalia with Britain, France, Russia and China, the other four permanent members.

What will the Somalia force do?

The likely tasks and U.N. involvement of the force expected to be approved to safeguard relief shipments in Somalia, as outlined by Secretary-General Boutros Boutros-Ghali in a letter Monday to the Security Council:

- Confiscate heavy weapons of the fighting factions and put them under international control.
- Disarm the fighters.
- Ensure the safety of international relief personnel and the 500 U.N. peacekeepers already in Somalia.
- Establish a cease-fire.
- Periodic Security Council review of the operation.
- Force would be replaced by U.N. peacekeepers when fighters are disarmed and heavy weapons are brought under international control.
- Security Council or the secretary-general attaches a small liaison staff to the field headquarters of the operation.
- Security Council would send some of its 15 members to visit the operation periodically.

The Virginian-Pilot
Wednesday, December 2, 1992

CENTCOM issued its deployment OPORD on the 6th of December.

The objectives for military force were sequential. First, open the ports and airfields for delivery of aid. Next, secure major nodes for the delivery of aid (designated as the Humanitarian Relief Sectors–HRS), and then open up the lines of communication between the airfields and ports out to nine designated humanitarian relief sectors, while simultaneously providing security to relief convoys, food stockpiling, and assist nongovernmental organizations (NGOs) trying to conduct their work. The center of gravity was Mogadishu, with its large port and airfield, followed by major airfields and other ports in southern Somalia.[4]

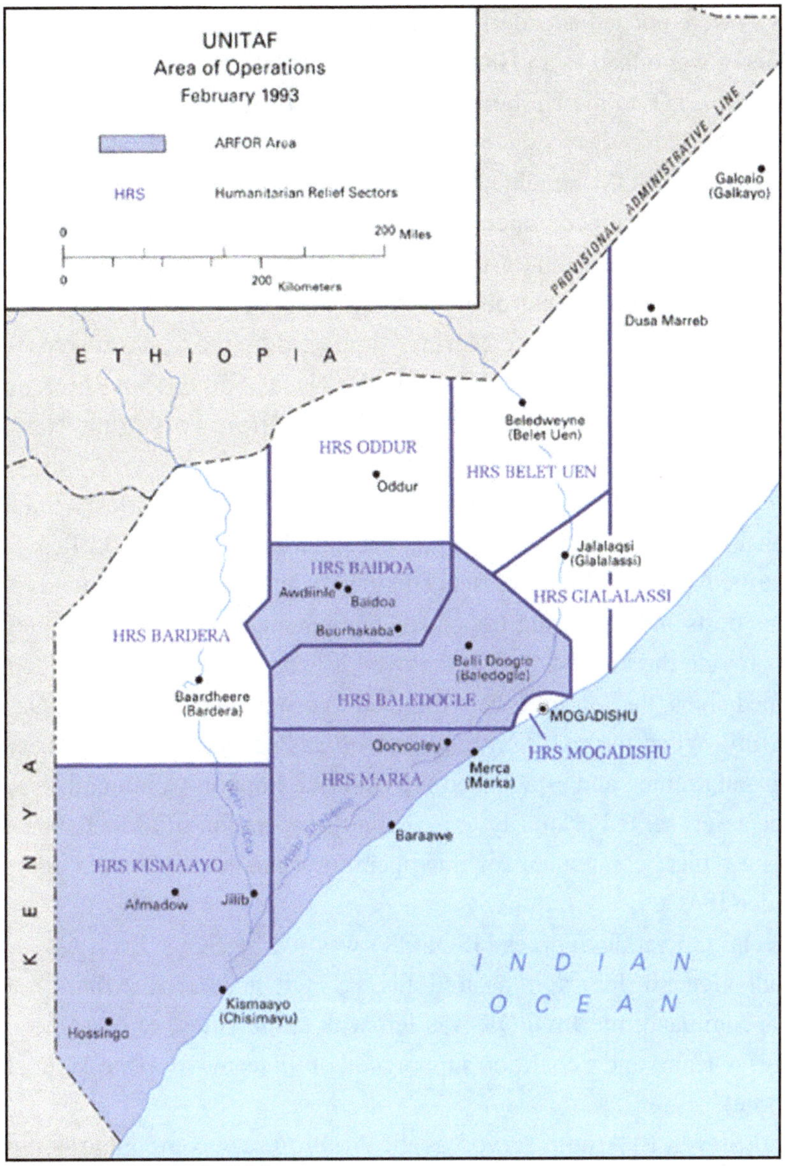

HRS map from *The United States in Somalia, 1993 –1994,* the U.S. CMH Pub 70-81-1, by Dr. Richard W. Stewart, page 12.

The OPLAN issued by the 1st MEF was a four-phased plan designed to achieve the following objectives:

- **Phase I**: Secure HRS Mogadishu, followed by HRS Baledogle and HRS Baidoa
- **Phase II**: Secure and expand out to the other HRSs–Oddur, Belet Weyne, Bardera, Gialalassi
- **Phase III**: Continue expansion and securing of remaining HRSs (Kismayo); improve and secure major Lines of Communication (LOCs) between Mogadishu and the HRSs; expand operations to assist NGOs throughout all HRS sectors
- **Phase IV**: Handover and transition to UNOSOM II

All of the HRSs were opened ahead of the anticipated schedule, but not necessarily in the order of the planned phases. Merka was added as an HRS on January 1, 1993. In mid-December 1992, the command changed its name from the CJTF to the Unified Task Force (UNITAF) as a better reflection of the coalition composition.

Forces still under UNOSOM remained under the control of the UN. This created a parallel chain of command but did not seem to impede operations. The UN contingents served primarily in Mogadishu, helping to secure the port and airfield, and establishing checkpoints and strongpoints throughout the city. UNOSOM peacekeepers also conducted patrolling to assist UNITAF.

The bulk of U.S. ground forces were Marines supported by the USS *Ranger* Carrier Battle Group (CV-61) and the USS *Tripoli* (PHIBRON 1 ARG) as the Naval Forces (NAVFOR) combined with the forces of a U.S. Army Division, the 10th Mountain Division (LI) from Fort Drum, New York, as the Army Forces (ARFOR).

U.S. Marine Forces (MARFOR), commanded by Major General Charles Wilhelm, consisted of the 15th Marine Expeditionary Unit (MEU), organized as a Special Marine Air Ground Task Force (SP-MAGTF) with its associated Marine Air Wing (MAW) assets and logistic and support commands. Given the nature of the operations—seizing ports and delivering troops and equipment—the I MEF was chosen to lead with the capability to quickly provide the bulk of forces deployed ashore ("From the Sea" doctrine). With support from four pre-positioned ships, the Marines were directed to provide the common logistic support for both themselves and the Army. When the coalition requirement was added, this system became a bit stretched until the Army was up and running and established the UNITAF Support Command.

U.S. Air Force assets served primarily to assist in the operation of airfields and aerial delivery and were not employed as a typical component for this operation (thus no Air Forces component or Joint Air Component Commander-JFACC).

The Joint Special Operations Forces (JSOFOR) was led by SOCCENT, a sub-unified, functional component. Lieutenant General Johnston desired his PSYOP and Civil Affairs capabilities assigned immediately under his command; the JSOFOR was left with employment of one Special Forces company with five Special Forces A-teams and a coalition support cell of SF teams from the 1st, 3rd, 5th, and 10th Special Forces Groups (Airborne).

Major General Steven L. Arnold served as the Army Forces component commander assisted by his Deputy, BG Lawson W. Magruder III. The 10th Mountain Division (LI) deployed with a brigade of light infantry and its aviation support and logistics elements. The Brigade was one battalion short in order to meet the cap on personnel deployments; 2-87th Infantry and the 3-14th Infantry deployed without their sister battalion.

U.S. Marines would deploy first using equipment from pre-positioned maritime vessels; the Army awaited their equipment aboard Fast Sealift ships, which were scheduled to deliver equipment in the latter part of December. The 10th Mountain assets began arriving by air on the 23rd of December.

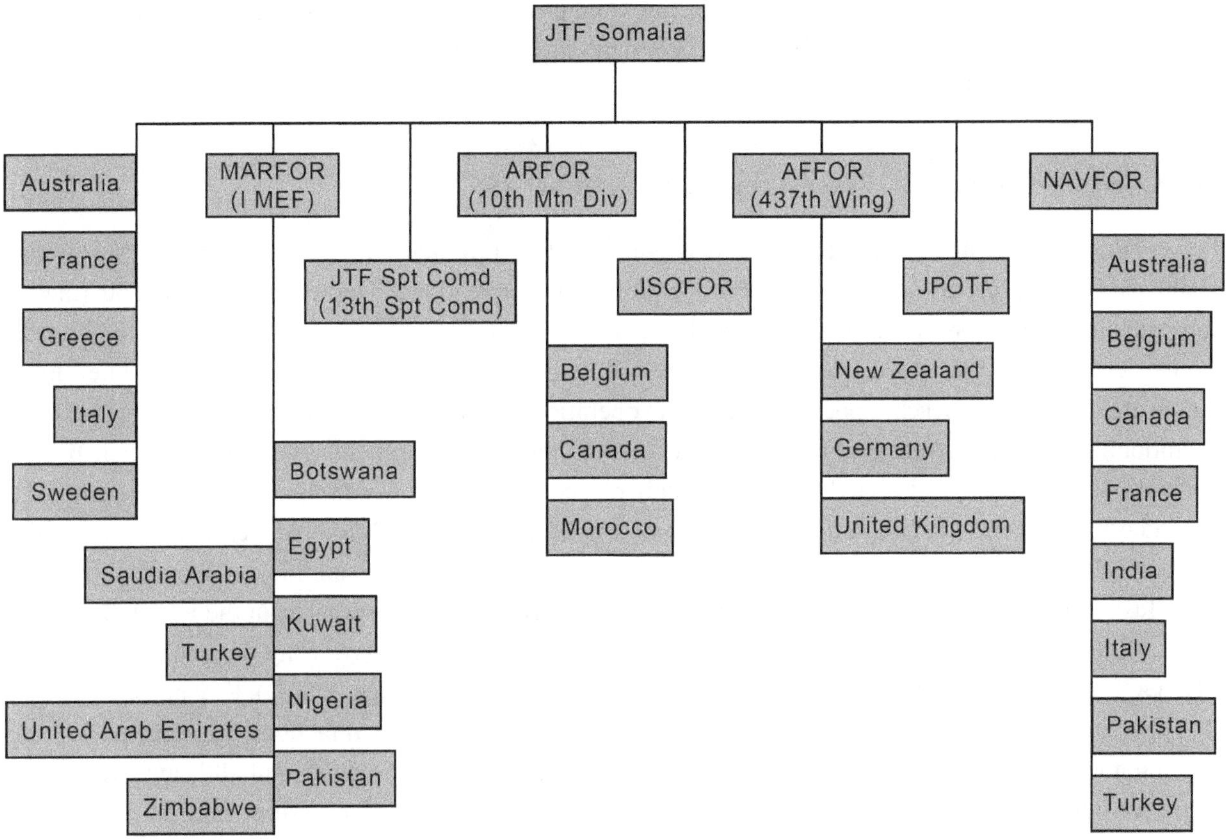

UNITAF Task Organization as of 1 January 1993. As depicted in Glenn Harned's *Stability Operations in Somalia 1992–1995: A Case Study*. Carlisle Barracks, PA: Personal Monograph Series PKSOI (19950612 009) United States Army War College PKSOI: War College Press, July 2016, p. 48.

Coalition forces arrived in all sizes. France, Italy, Canada, and Belgium were the largest contributors. Some countries only deployed small contingents entirely dependent on American support. Smaller units were generally assigned to security duties in the Mogadishu HRS; their contribution, however, was symbolic and important.

Other countries with a moderate level of force initially deployed to Mogadishu but waited for the opening of HRSs by Marines or Army soldiers and then deployed out to assist. New Zealand and the United Kingdom provided air support with C-130s only. About 23 countries provided troop strength. In all, the United States deployed around 28,000 troops and the contributing nations about 17,000 (from over 20 countries). The deployments began around mid-December for many nations. By the end of December, if anyone was going to serve in UNITAF, they were already in Somalia somewhere.[5]

Everyone provided engineer support if they could. The engineers of the Task Force were responsible for the improvement of communication lines between Mogadishu and the HRS bases. This effort involved improving roads (most of them dirt or hard-packed earth) and repairing bridges as needed. The lines of communication were also color-coded. For instance, the road from Mogadishu to Belet Weyne was coded Route "Orange." Often this task included the dangerous mission of mine clearance. Simultaneously, engineers constructed base camps and other facilities and provided much-needed, purified water. Improving and repairing airfields became a constant task. In two months, this monumental challenge was accomplished with over eight airfields up and running.

The Combined Joint Task Force Arrives, 9 December 1992

Prior to arrival of U.S. forces, Ambassador Oakley flew into Mogadishu on December 7th and established his office. Special Forces (SF) operators from Operational Detachment–Bravo 560 (the SF company, or B-team) provided his security and sniper support.[6] On the 9th of December, Navy SEALs from the carrier battle group landed ashore in the early hours of the morning amid the media and klieg lights pre-positioned on their beach, tipped off before the operation. The marines followed up with an amphibious operation. A company of the 2nd French Foreign Legion Parachute Regiment, flying in from their base in Djibouti, was the first to follow and was immediately placed on security operations at the K4 traffic circle. By the end of the day, the 13th MEU, under the command of COL Gregory S. Newbold, achieved their first day objectives: the airport, chancery, and seaport were in UNITAF hands. It was on this day the title Combined Joint Task Force was changed to Unified Task Force–UNITAF–after suggestions from the UN staff.

Oakley and Johnston met with Aideed and Ali Mahdi on the next day, using the good offices of the CONOCO compound (where Oakley also established the U.S. Liaison Office–USLO). Key in the discussion was the elimination of the Green Line (a DMZ of sorts between the two rival factions, splitting the city and running north to south), a cease-fire, no visible weapons in the hands of Somalis, and all heavy weapons to be placed into cantonments under UNITAF or UN control. The two clan leaders assured both gentlemen they would not interfere with operations.

Ambassador Oakley saw his role as leading the political line of effort for UNITAF on matters concerning disarmament, reconciliation, and restoring internal security institutions (police). Previous intelligence summaries indicated there were at least 15 armed factions operating in Somalia. Aideed and Ali Mahdi informed the General and Ambassador Oakley of a threat from the Somali National Islamic Front (NIF), not under observation by the coalition. Considered a terrorist threat by the West, the Front was supported by Iran and Sudan. Intelligence would be the short pole in a long tent plaguing both the UN and the UNITAF; HUMINT sources (human intelligence) were almost non-existent. Perplexingly, no one thought to ask Major Carroll and his SF teams what information they had gleaned while on Operation Provide Relief.

Six days later, Mogadishu was totally secured (three weeks ahead of schedule). Only scattered, minor incidents occurred. A few renegade militiamen with two Technicals, who apparently did not get the word from their bosses, fired on Marine AH-1 Cobras passing by and ruefully regretted it when their vehicles were destroyed by 20 millimeter cannon fire.

The Marines and coalition forces on hand spread out into the city to conduct a sweep for any remaining dangers and to enforce UNITAF's rules. On the 13th of December, the Marines of the 15th MEU occupied Baledogle and then pushed on to Baidoa by the 16th. In short order, the Kismayo, Oddur, Bardera, Gialalassi, and Belet Weyne HRSs were under UNITAF control, at least 30 days, if not more, ahead of schedule. It was so quick that by the 28th of December, UNITAF planners began to put together transition plans for the turnover to UNOSOM II. All the while, UNITAF continued to work alongside the 4,000 UNOSOM peacekeepers in Mogadishu.

As minor clashes occurred between the forces in Mogadishu and random Somali gunmen, the MARFOR Commander ordered Operation *Clean Sweep*, conducted between December 27th and the first week of January, to confiscate or destroy any weapons seen in the hands of Somalis. This process often included door-to-door searches.

Seven authorized weapons storage sites (AWSS) were designated to store Technicals and heavy weapons (tanks, artillery, rockets). Two were assigned to Ali Mahdi and five to Aideed's militia. The coalition, working under the operational control (OPCON) of MARFOR, was assigned a sector and patrol area within the environs of Mogadishu, often walking beside a Marine patrol. The Italians had their own sector in northern Mogadishu, an area controlled by Ali Mahdi.

Ambassador Oakley assumed the role of initially preparing the HRS for the incoming UNITAF contingent. It was a simple, three-step process. First, PSYOPs delivered leaflets and pamphlets from the dissemination teams, or loudspeakers, announcing the arrival of the Task Force to the area (at least two days prior). The State Department Special Representative then remained at the location to meet with village and town councils, speaking with the assembled elders, and letting them know what to expect and what UNITAF expected from them. Often, Special Forces A-teams went ahead of the Ambassador to conduct an assessment of the area prior to his arrival.

Next, UNITAF forces moved in and secured the area as announced. This process was then followed by delivering humanitarian relief supplies as immediately as could be mustered (on day three or four of Oakley's opening procedures). The Civil Military Operations Center (CMOC), with input from the Disaster Assistance Response Team (DART) and primary NGOs, made the decisions as to what aid went to what location and in what order.[7]

UNITAF adopted different approaches in their occupation plan. In Mogadishu and Kismayo they chose an amphibious assault with Marines and coalition contingents, followed up by air and land transported reinforcements. At Baledogle, Baidoa, BeletWeyne, and Merka, the coalition began their occupation using air assault insertions. Ground movements (overland with vehicles) were used at Oddur, Bardera and Gialalassi.[8]

The Joint Special Operations Forces (JSOFOR) for Restore Hope

Stripped of the Psychological Operations and Civil Affairs (the 8th and 9th PSYOP battalions, the Dissemination Battalion and three fourman Civil Affairs teams from the 96th Civil Affairs Brigade), and with no special operations fixed-wing or rotary-wing aviation apportioned for the Joint Special Operations Task Force (JSOTF), SOCCENT was left with nine Army SF teams to handle the coalition's integration into the anticipated nine HRSs. An additional four teams under ODB 560 were there to conduct special operations missions (Special Reconnaissance and area assessments). COL Thomas D. Smith, the SOCCENT Operations Officer, commanded the JSOFOR with a small staff, headquartered at the Embassy compound. Thus, it was really not a *joint* SOF Task Force; only the JSOTF staff was jointly manned. It should have been designated as the ARFOR. Regardless, other liaisons were requested for UNITAF, but with the JSOTF collocating with the UNITAF headquarters at the Embassy compound, there was no need.

The application of SOF for the operation could be characterized in four parts. First was the pre-deployment preparation once Operation Restore Hope began on the 9th. Two measures took place–first, the coalition teams traveled to the home country of their assigned counterparts and deployed into Mogadishu with them (and some went straight into Mogadishu). Their mission was to help integrate the coalition into UNITAF operations by providing liaison, communications links, and, if needed, calls for air support (which dictated using U.S. procedures). Some of the coalitions were just receiving American-supplied equipment and also needed familiarization and training prior to employment in their security sectors.

Two of the SF teams already in theater deployed directly into Mogadishu to conduct initial link-up. The Coalition Support Teams (originally named Coalition Warfare Teams from their role in Desert Shield/Storm) for Pakistan and Saudi Arabia, tasked to the 5th Special Forces Group (Airborne) to fill, came from teams already assigned to Major Carroll's B-team in Mombasa. The two CSTs detached from Operation Provide Relief in Mombasa and flew via C-130s into Mogadishu.

There would be nine CSTs making up the Coalition Support Team cell, based on the number of HRSs. The CST cell of the 10th SFG(A) comprised three provisional, six-man teams assigned to Belgium, France, and Italy. The 3rd SFG(A) provided two SF teams to assist Morocco and Botswana. The 1st SFG(A) was assigned support to the Canadians and Australians. In the CST role, all the teams were OPCON to the JSOTF. To support the 10th Mountain Division, the 5th SFG(A) was tasked to provide a Special Operations Command and Control Element (SOCCE) for liaison. ODB 520 was designated for this mission and deployed to Fort Drum in early December. When the Division was told to cut the numbers, the SOCCE requirement went away (along with one of their three Brigades) and Major Jesmer and his staff redeployed back to Fort Campbell, Kentucky.

The second use of the SF was reorienting the Airborne Security Teams on Operation Provide Relief to begin landing with their vehicles on the airfields instead of circling overhead and providing security in order to conduct HRS airfield assessments ahead of UNITAF occupation. Combat Control Teams (CCTs) and the USAF REDHORSE Engineers accompanied them on this mission. This measure was considered key to ensure the airfields were serviceable and any threats identified, neutralized, or removed. In this role, the four SF teams remained under the OPCON of Major Carroll in Mombasa. The B-team and its four ODAs would not transfer under the control of the JSOTF until later in December.

All of these activities by the U.S. Army SOF were completed in Phase I, the securing of Mogadishu. In Phase II, the assigned coalition teams deployed with their counterparts as they moved to operate in their HRSs, in accordance with the opening schedule of the Operations Plan (OPLAN). ODB 560 moved to Belet Weyne and conducted SF missions to support the Canadians (but not as CSTs; they would be tasked to conduct SF missions unilaterally in support of Canadian operations). The pace of securing the Humanitarian Relief Sectors was quicker than anticipated, as both Phase I and Phase II were completed ahead of schedule in early January of 1993 (a great present for President George H.W. Bush when he visited with the troops over the New Year's holiday). Other than the 5th SFG(A) Special Forces units already in theater, all other assigned CSTs were redeployed back to their groups at their home stations. The mission for them was only to assist their counterparts until UNITAF was established in the nine sectors.

In Phase III, the 5th SFG(A) SF teams operated to conduct special reconnaissance and area assessments. In some cases, they were the only independent, mobile force available to the HRS commander to determine threats, identify other NGOs operating in the HRS, and conduct meetings with village elders. During Phase III operations, SOCCENT redeployed home, leaving the command and control of the remaining SF Company with the 2nd Battalion, 5th Group commander, LTC William Faistenhammer, who now donned the hat as the SOF Army Forces component commander (ARSOF).

As tensions arose in February when General Morgan and Colonel Jess's clan militias clashed in Kismayo, Lieutenant Colonel Faistenhammer deployed a SOCCE and two ODAs, with the 10th Mountain Task Force reinforcing the Belgians. This mission lasted about a month, while the teams remaining at Belet Weyne conducted missions in the HRS. Some out of sector missions to support the Marines in Bardera and the Australians in Baidoa were executed by ODA526, becoming a sort of "roving" team for support to other sectors.

The final utilization of the SF teams was transition and redeployment in anticipation of turning over sectors to UNOSOM II forces. After redeployment of the JSOTF and the CSTs, the ARSOF was replaced by ODB 520 in April; they would conduct missions for the UNOSOM II U3 Directorate and provide support to the 10th Mountain Quick Reaction Force (QRF).

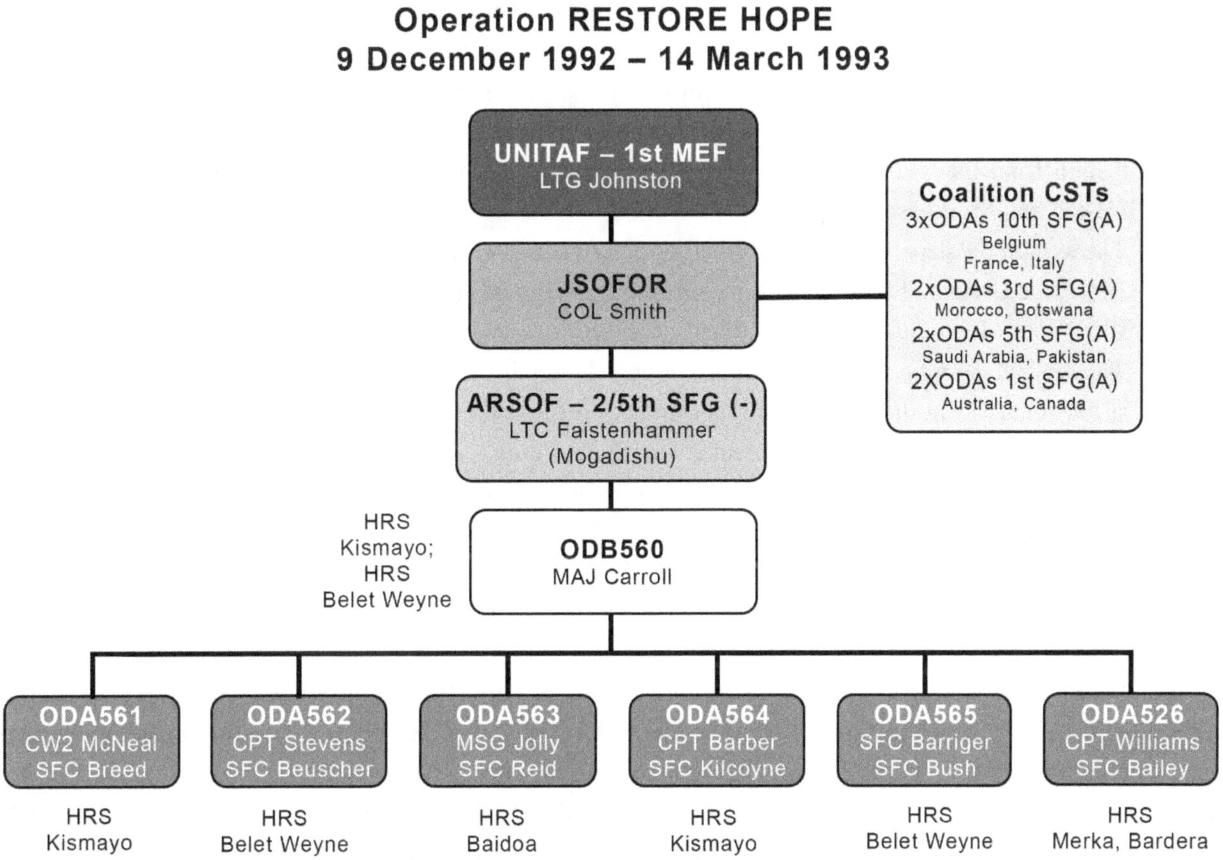

The 10th Mountain Division SOCCE and Establishment of the JSOTF

In early December, the 5th SFG(A) was alerted for participation in Operation Restore Hope. The Group Commander tasked 1st Battalion to fill any additional requirements, since Major Carroll and his company were already in theater. Major Jesmer, the Company B commander, was selected to deploy with additional forces and serve as a SOCCE to the 10th Mountain Division. He would bring an additional ODA to Somalia to serve as one of the CSTs to either the Pakistani or Saudi Arabian contingent. ODA526 was selected; they had Desert Storm experience as a Coalition Warfare Team to the Saudis. Major Jesmer selected CPT Kevin Murphy to serve with him on the SOCCE, or if required, serve on SOCCENT's JSOTF staff.

Major Jesmer relayed Colonel Bowra's guidance to Murphy, ". . . to find a decent location for the SFOB [the 5th Group headquarters] in order to expand it into the CJSOTF." Sergeant First Class Flick accompanied the SOCCE as the communications sergeant. Murphy and Flick departed Fort Campbell as the SOCCE advanced party and reported in to the G5, Major Stahl, at Fort Drum. They were informed the Division was desirous of a liaison relationship with the 5th SFG(A); they saw the intelligence products from the teams on Operation Provide Relief as better than any other source. The G2 hoped the SF could answer some of the Division Commander's RFIs (Request for Information) on the environment in Somalia, the existence of any military bases, helicopter landing zones and questions about the clan factions.

Brigadier General Magruder, the assistant division commander, planned to launch an assault Command Post (CP) to Somalia, and asked for a member of the SOCCE to be part of that advanced element. Once Major Jesmer arrived to Fort Drum with the rest of his SOCCE staff, he chose Captain Murphy to accompany the advanced CP. Jesmer soon had his operations set up in the 10th Mountain conference room. Sergeant Shell from the 112th Signal Battalion set up radios and established communications with Major Carroll's B-team in Mombasa.

The assault CP departed on a snowy winter day via C-141 aircraft. It was fully packed and included the assault CP, the Division Long Range Surveillance Detachment (LRSD), and two vehicles. The main body would follow with Major General Arnold.

The Assault Command Post arrived in Mogadishu on the 13th of December. They established the 10th Mountain headquarters in the old commissary annex on the Embassy compound. A temporary camp with tents was erected until EOD could inspect and clear the building.

The SOCCENT JSOTF was already at the compound. Captain Murphy met up with MAJ John Brush, Lieutenant Colonel Raines, and Sergeant First Class O'Brian, assembled as the SOCOORD to UNITAF (SOF liaison and coordinating element). Captain Murphy assisted members of the advance CP to get the headquarters established with communications; he bunked with the LRSD and performed duties as Major General Magruder's security detail.

Once, while talking to the SOCCE at Fort Drum, Captain Murphy and some members of the assault CP started taking sniper fire. Brigadier General Magruder came out to see the situation for himself and asked ". . . who and how was communications being made to Fort Drum?" Apparently, efforts on the part of the 10th Mountain signals unit were not successful. Captain Murphy offered the radios at the JSOTF; Brigadier General Magruder immediately ran over to borrow the SATCOM (satellite communications) radio line so he could talk with the division about the incident. The JSOTF staff assisted by getting the division's G5 on the line to talk to the general. As a result of this secure, instant comms, Magruder looked Captain Murphy in the face and told him ". . . you are going to go everywhere I'm going and bring this radio." Captain Murphy remembered years afterwards when, out at JRTC, Brigadier General Magruder talked to others about this story and the Special Forces teams reliability and dependability for operational commanders on the battlefield.[9]

Inspecting and Securing the Airfields

From December 10–12, 1992, ODA565 secured Baledogle airfield and provided security for Air Force Engineers conducting an essential airfield assessment for the 10th Mountain Division's follow-on forces.

On the 10th, supported by a pair of USMC AH-1 Cobras, the situation was far different from any earlier missions. It would be one of the first "engagements" between UNITAF forces and Somali gunmen. Major Carroll and other members of the B-team were assisting ODA565. Sergeant First Class Barriger summarized the opening of the operation: "ODA565's plan was to conduct a RAPIDS (rolling offload while aircraft engines are running) infiltration from a C-130 with two DMVs and eight personnel with an Air Force CCT on a motorcycle. ODB 560 would conduct C^2 and security around the aircraft on the ground until the airfield was secure. The team had air cover from two Marine Corps AH-1 Cobra gunships that were relaying information from the air to the team about potential hostile positions and strength."[10]

Per Major Carroll's request, the pilot first conducted a low-level pass over Baledogle to assess the ground conditions. Two Technicals with 106-millimeter recoilless rifles were spotted. Carroll formulated a quick reaction drill: they would hot land, roll off the aircraft, and move on line toward the Technicals, while the aircraft returned to the air. It was imperative to disarm the Somalis.[11]

Captain Smith interviewed Major Carroll upon his return from the mission:

> The insertion was unopposed, and the Hercules hurled itself back into the air without incident. Ken Barriger got his team into formation, and the troops moved down the airfield purposefully, weapons at the ready. The Somalis made no reaction, simply sitting in their gun jeeps at the end of the runway and watching the crazy Americans walk toward them.
>
> It was at this point that somebody in Barriger's detachment let off a burst of automatic fire. The tracers ripped through the air above the heads of the Somalis, and the militiamen scrambled for cover, abandoning their vehicles.
>
> Carroll halted the advance momentarily, trying to determine what the hell was going on. Not wanting to lose the advantage, however, he quickly pressed on. There should have been no firing, but luckily the Somalis hadn't fired back.[12]

As ODA565 secured the gun jeeps, Major Carroll attempted to speak with the Somali gunmen and let them know they had to give up their weapons. As a concession, he offered to let them keep their vehicles. At first, the leader of the militia group looked a bit perplexed and refused to hand over the weapons. However, understanding the Americans were armed and meant business, he soon agreed. The 106mm recoilless rifles could not be dismounted; SFC Barriger's men merely removed the firing pins, rendering them inoperable. Also present was one technical without a weapon and one Land Rover with an M2 .50-caliber machine gun. These weapons were gathered up and moved to the team's location.

The Somali pointed to a shed where he agreed to park the trucks out of the way. Once located, Major Carroll was surprised to see 106mm ammunition stored. ODA565 confiscated the five rounds. Carroll decided to eliminate the problem of the truck and ammunition. If the Somalis found some new firing pins, the Technicals would just threaten the next friendly force sent in to the airfield. Not one of the Somalis appeared to share his thinking but walked off into the bushes out of range of the upcoming destruction.

Without explosives, grenades, or anti-tank weapons, the vehicles could not be destroyed. There was no way to load them and send them back to the JTF at Mombasa. Major Carroll knew the Somalis would probably just get new firing pins and the vehicles would be back in action tomorrow. Major Carroll contacted the AH-1 flight to check if they were in his vicinity; soon, they passed overhead. Smith remarked, "Needless to say, the Somali militiamen were impressed, and probably very grateful that their leader had put up no resistance. A pair of Cobras, hanging over your head and bristling with cannon and rockets, can be a very sobering or a very cheering sight, depending on who you are."

After a situation brief by Major Carroll, the pilots agreed to destroy the ammunition stocks and the Technicals. After clearing the area of all Somalis, the Marines accomplished the mission, expending all their ammunition.[13]

After the airfield was secure, the ODB was ferried by vehicle to the airfield control tower to set up command and control (C2) and have a better vantage point to direct 565's roving vehicle security force. An Air Force Red Horse Element was brought in via C-130 to assess the asphalt surface of the runway to determine the type of aircraft it could take. By late afternoon on the 10th of December, all elements extracted back to Mombasa by C-130.

Also on December 10, 1992, Major Carroll ordered ODA562 to be one of the first Coalition Support Teams to deploy to Mogadishu for Operation Restore Hope. Captain Stevens discussed the mission with Major Carroll; it appeared the Marines and the Pakistanis were not working well together. In some good news, the Pakistani commander requested a Special Forces team as his liaison with the difficult Marines; he had remembered the Army Special Forces performance from the Gulf War. Stevens was told to be ready to fly into Mogadishu on the 14th of December and meet with the Pakistani commander. His team would follow shortly thereafter.

An extremely rare photo of one of the first clashes between Special Forces and clan militia at Baledogle airfield, December 10, 1992. Shown is a USMC Cobra providing fire support to destroy captured technicals (*photo courtesy of Gary Ramsey*).

With the loss of ODA562, Major Carroll carried on his Provide Relief mission with the two teams remaining. The amalgamated team was split back out, now giving him three in preparation for other mission taskings for Restore Hope.

The unit returned to Baledogle over the next three days to complete the assessment. ODA565, members of the B-team, and engineers from the REDHORSE unit did not encounter Somali gunmen on these days. Over the three days, 33 weapons were confiscated. With the area seemingly quiet, the REDHORSE engineers were able to bring in dozers to clear vegetation from the sides of the airstrip. While they worked, ODA565 used their winches on the DMVs to remove other obstacles. On the final day at the airfield, they got a taste of the media when CNN filmed their weapons confiscations. The CNN news crew mistook them for Marines when they broadcast the episode on the news.[14]

The Marines arrived on the 13th, transported by helicopters to "seize and secure" the airfield. The Green Berets welcomed them in. Soon, soldiers from 10th Mountain Division (LI) relieved the Marines.[15]

In what was the last mission of the Army Special Forces supporting Operation Provide Relief, ODA565 and ODA564 conducted an airfield assessment of Kismayo on the 17th of December. Part of their task was escorting Chief Warrant Officer 2 Bell's composite team of French speakers from the 10th Special Forces Group (Airborne). Bell's team was assigned as the Coalition Warfare Team to the Belgians, scheduled to occupy Kismayo with the Marines via a seaborne landing. Between the two teams, a total of 18 weapons, eight cases of small arms, and a small amount of 57-millimeter recoilless rifle rounds were seized.[16]

On December 19th, ODA565 flew a special night mission in support of a SEAL team in Kismayo. On the 20th of December, Marines and Belgian paratroopers secured the airfield and the port area of Kismayo, followed by more forces from the 10th Mountain Division (LI) on the 22nd.[17]

Back in Mogadishu with the JSOTF and Coalition Support Teams

Captain Murphy worked to affect the linkups between the arriving CSTs (Chief Warrant Officer 2 Bell from the 10th Group for the Belgians and ODA562 for the Pakistanis) and their country contingent, all the while keeping the 10th Mountain staff informed on the status of SOF in Mogadishu. Captain Murphy found some confusion from the Marines and the ANGLICO (Fire Support Team for Naval and Marine aerial or ship gunfire) when he informed them a Special Forces CST would be working with the French and moving into the field with them. The team would not await instructions from the Marines before moving off.

Planning began for the 10th Mountain Division and the Belgian operation to Kismayo (scheduled to occur around the 18th of December). The SOCCE and SOCA team (Special Operations Communications) were requested in support once they arrived in-country. Murphy remembered, "Scuttlebut around the staff had it that the Belgians wanted to conduct an airborne operation into Kismayo to top the recent, and well televised, airmobile and airborne jump the Canadians had into Belet Weyne, but the command did not want the USMC 'out-spectacular-ized' and cancelled the Belgian request." In the end, the operation became an amphibious operation with a commensurate air landing at Kismayo airfield.

Soon after, the headquarters element for the Canadians arrived. With them was CPT Danny Alvis with a six-man CST. Captain Alvis stood out from the other detachment leaders when he decided to wear a Canadian airborne uniform, but he kept a U.S. Army tape above the pocket. The remainder of the 1st SFG(A) CST arrived shortly thereafter and set up communications; they were soon operational with the Australians.

As UNITAF learned of the capabilities and performance of the SF CST teams, they passed a request to the JSOTF to ensure that all of the coalition maneuver units needed an SF team. The Moroccans, Kuwaitis, and Batswana were in-country and needed CSTs. This need had not been originally planned for, so Captain Murphy passed this requirement on to Major Brush, the SOCOORD from SOCCENT. By now, ODA526 had linked up with the Saudis.

Captain Murphy was passed information from the rear as to the status of the deploying SOCCE and other SOF elements. Major Jesmer was scheduled to depart Fort Drum via air and arrive in Mogadishu around the 26th of December; he was bringing a Humvee, large tent, and three other members of the SOCCE (Berry, Flick, and one more). The SOCA team was not scheduled to arrive until January 1, 1993; the vehicles were on a ship and were being escorted by Sergeant Delasky. The same day, Colonel Smith, the JSOTF commander, was scheduled to arrive with his vehicle plus two staff members from SOCCENT.

As the operation continued to grow, Captain Murphy received guidance for the SOF–there would not be any SF special reconnaissance (SR) requirements as of yet. All the CSTs were going to be C2'ed by the SOCOORD (until the JSOTF stood up and became operational), unless the coalitions were chopped to ARFOR.

Captain Murphy found himself involved in the details of finding real estate for the arriving SF. At this time he observed there was ". . . a mad shuffle for real estate." The land management was handled by the USMC Headquarter commandant. Units and disparate personnel were going around the Embassy compound spray painting "Xs" on buildings and marking out their turf. The LRSD moved into the building identified for the SF ODB and became squatters but were eventually moved out once the SF Company deployed in from Kenya. Captain Murphy kept in touch with Colonel Smith every day prior to his arrival in-country and also had the additional task of finding a suitable building, hopefully with electricity, for the SOCCENT contingent; one was found near the 10th Mountain divisional headquarters.

Captain Murphy received word that Colonel Smith plus three SOCCENT staff members were now due into Mogadishu on the 23rd of December; he received another notification they were coming on the 19th of the month, with the rest of SOCCENT due in on the 29th. Once Colonel Smith arrived, he was apprised of the situation concerning the status of the CSTs, reconnaissance mission requests to date, and the totality of 10th Mountain Division aviation assets. By this time, there were at least six CSTs operating: Belgium, Pakistan, France, Botswana (MSG Eddie Goodrich from 3rd SFG), Saudi Arabia, and Canada. Some initial tensions arose between the Belgians and the U.S. commander for the Kismayo operation, because the Belgian commander, Colonel Jacquemann, gave the warlord Omar Jess a map which contained all the coalition positions and operational information (apparently as a gesture of good will). Captain Murphy remembered picking up and escorting Colonel Jacquemann to the Embassy compound for a meeting with Lieutenant General Johnston. Soon thereafter, Colonel Jess lost his power position in Kismayo; he had been conducting joint patrolling with U.S. forces, but Brigadier General Magruder soon put an end to that.

Captain Murphy remembers the CNN factor on television had a great bearing on the planning and deployment dates of the coalition–pictures of starving children contributed to the pressure of pushing UNITAF assets to secure other sites as quickly as possible.

The next day after Colonel Smith's arrival, he moved to consolidate the SOCCE and the SOCOORD element into the JSOTF. He also immediately informed the members of his staff that they were prohibited from going directly to UNITAF from that moment forward. When Task Force Kismayo requested the SOCCE for the Kismayo operation, Colonel Smith disapproved the measure with the intent to keep the status quo on the amount and activities of SOF to a minimum–an increase in more work or more SOF was not desired. Soon, what appeared to be bad blood and relations broke out between the SOCCENT staff and the UNITAF and Army Task Force staffs. Friction was especially heavy between COL Paul Eaton, the 10th Mountain G3 and Colonel Smith on what the perceived role for SOF should be in Somalia. Colonel Smith and LTC Charley Raines (3rd SFG) flew to Oddur and Kismayo to visit with the CSTs to get a first-hand look at what SF was doing for the operation.

When ODB 560 deployed to Belet Weyne to support the Canadians, Captain Murphy flew out to the HRS to have initial discussions with Major Carroll and assist the SF teams with their communication crypto fill. Colonel Smith, no longer needing the SOCCE, selected Captain Murphy as the Operations Officer, J3 Air, and Movements officer. Colonel Smith now clearly assessed that the JSOTF was not capable or functional as a JSOTF – there was no real operational control over the SF (or PSYOP and CA) deployed in theater and, for the most part, personnel on staff soon found life boring. Colonel Smith soon began to see the only role for SF was in providing coalition forces with secure, reliable SATCOM communications. Kevin Murphy distinctly remembered Colonel Smith's comment," I do not want to provide high paid radio operators to move around with the coalition forces."

On December 21, 1992, the 2/5th SFG(A) (-) received a warning order to end Operation Provide Relief and prepare to move to Mogadishu in support of UNITAF operations–Restore Hope. On December 27, 1992, the ADVON for Company C departed from Mombasa to Mogadishu, followed by the remainder of the company, arriving in Belet Weyne on the 30th of December, the site of their new Advanced Operating Base (AOB). The Green Berets now switched from humanitarian operations to contingency operations alongside U.S. military and multinational partners, under the operational control of UNITAF.[18]

In early January, Colonel Smith began work to extract the SF CST teams from Somalia (the team requirement continued to grow; Colonel Smith refused to fill out anymore CSTs).

The SOCCENT commander, BG Bill Tangney, concurred with Colonel Smith's assessment and said so in follow-up message traffic to the ARFOR and UNITAF commanders. In short order, the 10th Mountain SOCCE at Ft. Drum was told not to come forward, stand down, and return back to home station, although ODA526 did get through as one of the needed CSTs, deploying along with the 10th Mountain's main body. As Phase I and Phase II of the operation were complete, the CSTs were disestablished and started to redeploy back to their home stations–most were gone by the 20th of the month.

The 5th SFG augmentees to SOCCENT, along with Captain Murphy and Sergeant First Class Flick, were told they were also ". . . not needed" and were sent home by Colonel Smith around the 15th of January. Colonel Smith, with the SOCCENT staff, closed down the JSOTF and turned over SOF operations to Lieutenant Colonel Faistenhammer, now the ARSOF commander. Lieutenant Colonel Faistenhammer deployed with his battalion headquarters to establish Forward Operating Base 52 (FOB 52). The JSOTF departed Somalia in late January 1993.[19]

The Coalition Support Team Cell – Opening the Humanitarian Relief Sectors

The nine Coalition Support Teams were organized with six personnel each, consisting of at least a team leader (Captain or Warrant Officer), a medic, and a communications and weapons Sergeant. Except for the 5th SFG(A) teams already in theater, the assigned CST linkedup with their coalition counterpart in the country of origin, then deployed with them to Somalia, or flew straight into Mogadishu to affect the linkup. Some of the teams were equipped with HMMWVs or the SF-modified Desert Mobility Vehicle (DMV). Those without transportation borrowed vehicles or rode into the HRSs with their counterparts. All of the teams were required to send daily situation reports back to the JSOTF. As an organizing method, the teams would be covered sequentially in the order in which each HRS was established and occupied. All the teams began their mission with their counterpart in Mogadishu.

Mogadishu Humanitarian Relief Sector

Coalition SF CSTs
15 Dec 1992 – Jan 1993*

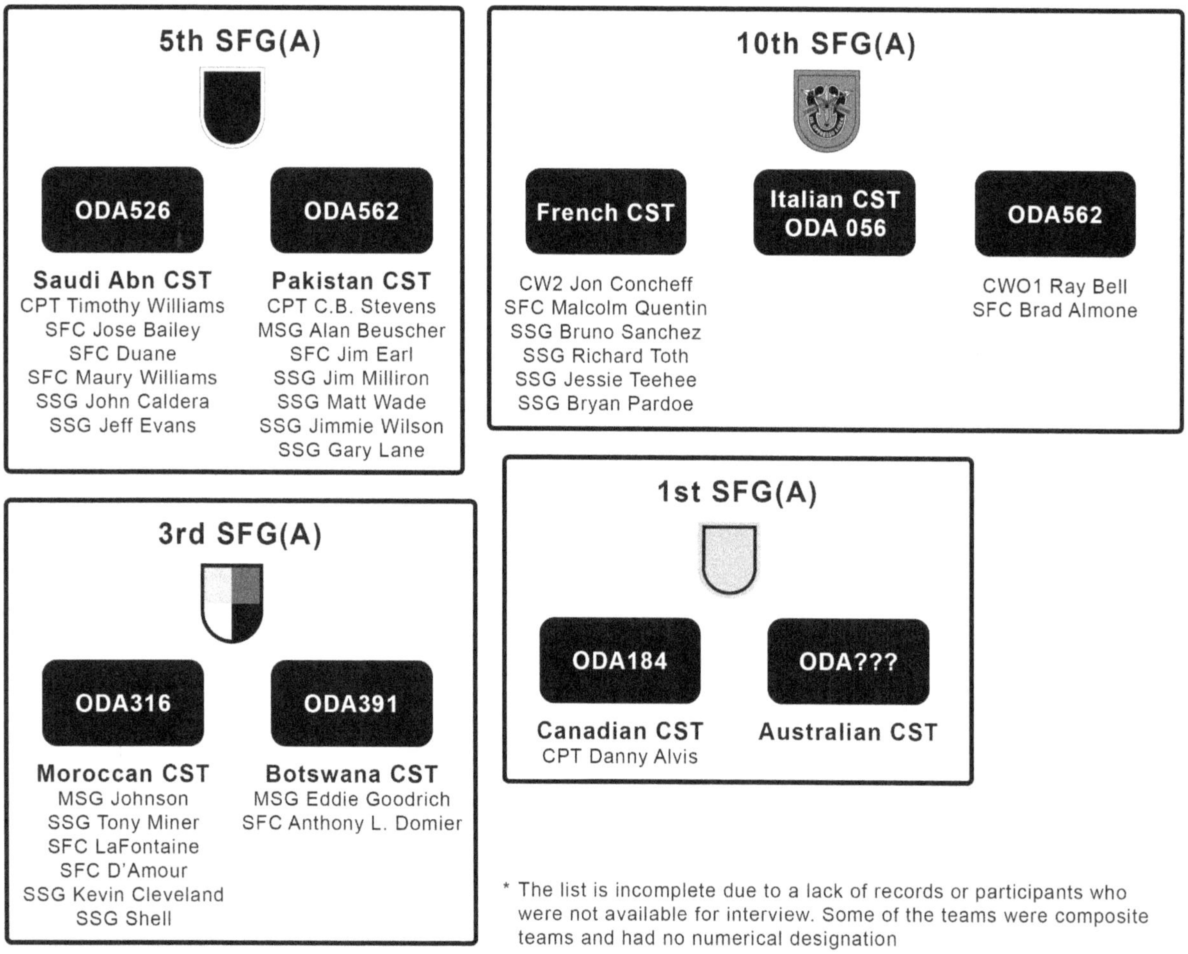

* The list is incomplete due to a lack of records or participants who were not available for interview. Some of the teams were composite teams and had no numerical designation

ODA184 – Canadian CST. The team of six men was notified for deployment on December 7, 1992, with the mission to support the Canadian Airborne Battle Group. They identified some additional implied tasks during their missing analysis–serving as part of the area security in the assigned humanitarian sector and providing satellite communications capabilities and medical aid amongst any UNITAF units in the area.

The team conducted pre-mission activities and traveled to Pettawawa Canadian Forces Base (the home of the Canadian Airborne). The Battle Group departed for Somalia on December 13, 1992, on Canadian military air transport.

ODA316 – Morocco. ODA316 was a four-man team designated to serve as CST to the Moroccan contingent based at the Mogadishu Airfield. The team was composed entirely of senior NCOs. The team departed Fort Bragg via C-141 to Dover, then Spain, Cairo West, and on into Mogadishu. On their arrival, they were picked up and reported in to SOCCENT at the Embassy compound. Some confusion existed as to their mission, and the team spent three days at the compound working out these particulars. Once their mission with the Moroccans was clarified, they returned to the airfield and linked up with their coalition contingent. However, after three or four days on the airfield, the team assessed the Moroccan mission as one of merely presence, and the Moroccans were not going to do external patrolling or convoy escort missions. Master Sergeant Johnson, the team leader, was able to procure a Humvee, and the team began light patrolling outside the perimeter to gather assessment data for SOCCENT.

A breakthrough occurred for the team when they combined their assets with those of ongoing civil affairs missions. The team deployed with the Italian marines to Quorleey, along with the assigned CST from the 10th Group, ODA056, to assist the Save the Children organization with food aid. They remained in the area and linked up later with ODA526 in the Merka area, continuing to assist with area assessments and facilitating food aid operations. After the New Year, the team was informed by SOCCENT their CST mission was complete. The team members redeployed back to the United States in early January of 1993.

ODA391 – Botswana. The detachment mission received a mission tasking to deploy to Somalia and provide liaison and coalition support to a battalion-sized element of the Botswana Defense Force, commanded by LTC Thulanganyo Masisi. The team task organized and departed Fort Bragg, NC on December 18, 1992, without vehicles. ODA391 flew on C-5As to Dover, then Torrejón, Spain, then cross-loaded onto a C-141 for a flight through Cairo West in Egypt. The team arrived on Mogadishu airport on December 21st. The team coordinated with SOCCENT and linked up with the Botswanan battalion, procuring a place to hootch and logistical supplies. In the early days of their deployment, the team relied on bumming rides from various UN and coalition vehicles traveling the airport roads.

The Botswana camp was named Camp Higgins, in honor of USMC COL Higgins who was murdered by terrorists in Beirut. The Botswana battalion consisted of around 300 personnel, organized into three infantry companies, a mortar platoon, and a small Special Forces platoon. The battalion commander and the battalion operations officer were well trained and professional, and had attended a variety of infantry courses in the United States, including the Special Forces Qualification Course. The soldiers of the battalion dressed in British battle kit and were equipped with Belgian FNs, AK-47s, and Israeli Galils in the Special Forces platoon. The mortar platoon doubly served as a heavy weapons platoon and was equipped with RPGs and Uzi sub-machine guns. For vehicles, the Batswana utilized the Cadillac Gage V-150 armored cars mounted with Belgian-made light machine guns. The specialty sections were mounted with Israeli RAM vehicles, and the support personnel rode in unarmed versions of the Israeli RAM.

Throughout the short deployment, the 3rd SFG(A) CST commented on the high standards of the unit (most members were airborne qualified) in their vehicle and weapons maintenance, patrolling procedures, reaction to fire, and in overall appearance, including their physical conditioning. The Botswana contingent had responsibility for about a fifth of the airfield perimeter security, as well as responsibility for escorting humanitarian relief convoys through Mogadishu and into some of the immediate surrounding environs. While at the airfield, the battalion worked diligently to enhance their perimeter through sandbagging and barbed wire, almost an impossible and monumental task. The ODA assisted with advice and input on several security measures for emplacement.

ODA391 accompanied the Batswana during their security escort missions by riding along in their vehicles. The operational area for the battalion was significantly large and included the Mogadishu city proper, surrounding environs, and the specific roads utilized to convoy supplies to various outlying relief centers. During several mounted missions, the team (SGT Anthony L. Domeir provided much of what occurred during his debrief upon return to the States) was impressed by the tactical discipline displayed by the battalion when they often came under sniper fire while crossing a bridge on the outskirts of Mogadishu. The Batswana operated with professionalism and restraint and were accorded the highest respect from the Somalis out of all the coalition contingents operating there.

These several forays also allowed the team to meet with local people, elders, and NGOs and observe local conditions, including the fact that there were robust corn fields and animal herds outside of town. This useful information was passed to SOCCENT in their daily situation reports (SITREPs). Although constrained by a lack of SATCOMs (satellite radios), desert uniforms and boots, and vehicle capability, ODA391 successfully accomplished their entire mission task.

In what would be the last combat action for the CSTs, ODA391 assisted the Batswana in their mission planning and capture of a key bridge held by enemy clan forces who were defending the bridge with armed technicals, thus opening a key line of communication for the coalition.

As the situation changed and they were no longer required as a CST, SOCCENT stood them down and released them from their mission in the first week of January. They departed the Botswana sector in military vehicles and boarded a C-5A on January 8, 1993, for a flight to Saudi Arabia, then Rhein Main. The team continued on to Dover Air Force Base in Delaware and rode a commercial bus to Fort Bragg, North Carolina.

ODA526 – Saudi Arabia. Special Forces Operational Detachment–Alpha 526 (ODA526) was notified on December 16, 1992, to prepare for a Coalition Support Team mission to the Saudi's 5th Airborne Battalion serving with UNITAF in Mogadishu. CPT Tim Williams was the Detachment Commandeer of the six-man team; the members of the team were Jose SFC Bailey the Team Sergeant, SFC Duane, SFC Maury William, SSG John Caldera, and SSG Jeff Evans.

By the 20th, the team moved to Charleston, South Carolina and linked up with the Coalition Support Team from the 10th SFG(A); they flew to Rheine-Main AFB in Germany, then through Cairo West. Since ammunition was prohibited on USAF cargo flights, theirs was confiscated; the first item of business was procuring small arms for their weapons while at the Cairo West Egyptian Air Base. On the 24th, they flew into Mombasa then transferred aboard a C-130 for the flight into Mogadishu.

The teams parted ways with the 10th SFG(A) CST who went off to look for their counterpart while Williams' team procured transportation and linked up with the Saudi contingent in assigned positions at the airfield's main gate. Captain Williams was pleasantly surprised to meet up with Prince Fahd, the unit J3, and a friend from their time together during Operation Desert Storm. After a generous welcome, the Saudis provided the team with tents, food, and water. Coincidentally, this same day, the SOCCENT staff drove in from the Embassy compound to pick up Colonel Smith, the Joint Special Operations Forces

(JSOFOR) commander who had also just arrived. The staff issued ODA526 some additional ammunition, aware of their arrival.

The team established their camp on the airfield perimeter and conducted coordination and liaison with the Marines. It was a time of confusion, and very little was known of who supported whom; ODA526 deftly scrounged as much material and maps as they could get their hands on.

Captain Williams spent his next few days at the Embassy compound with the JSOTF staff, conducting mission analysis. The CST task was pretty well-defined; the 5th SFG(A) had conducted this role with coalition counterparts a few years earlier during Desert Storm. ODA526 was tasked to provide liaison to other American units (mostly the Marines) and be a source of communications with aircraft used for any fire support missions. A daily requirement consisted of a situation report (and assessment) of the coalition's operations.

Captain Williams and the team improved their position while the remaining Saudi forces continued to pour into Mogadishu (eventually 700 troops with vehicles and equipment). They began foot patrols up to 100 meters out from the airfield perimeter, in accordance with the operating procedures the Marine command established for security. The Saudi 5th Airborne did not participate and were satisfied to merely serve in a defense position and act as a deterrent by their presence. They would, however, provide a reaction force if tasked.

A platoon of Kuwaiti soldiers was assigned to the Saudi defensive sector. They too anticipated a Special Forces CST to be assigned to them; Colonel Smith refused because the team was already stretched too thin to split in half.

During this early period of Restore Hope, incidents of gunfire were beginning to break out with more frequency between rival factions and between the Somalis and UNITAF. On the 27th, the team reported gunshots in a village north of the Saudi position. At another time, when the team was returning to the airfield perimeter from a meeting at the Embassy compound with the JSOTF, their two-vehicle convoy thought they were receiving gunfire; no one was hit and the patrol did not return fire. Later that day, a riot ensued in front of the Saudi position, composed of about 150 people with stones, knives and sticks. When the Saudis confiscated some knives, the crowd threw rocks at them.[20]

On the 29th of December, the Egyptian contingent assumed the duties from the 5th Airborne Battalion for security at the Mogadishu airfield main gate, freeing up the Saudis to begin some vehicular patrolling outside the wire. The A-team anticipated future operations for the Saudi 5th Airborne extending to Merka, but this did not occur. The team assisted the Saudis and Kuwaitis to the best of their ability and provided training in patrolling, security operations, and classes on the .50-caliber machine gun.

The potential existed for a U.S. Disaster Assistance Response Team (DART) mission of humanitarian food relief operations being assigned to the Saudi contingent (coordinated through the ODA). The request was for a security provision to truck convoys headed to Merka between the 6th and 14th of January 1993. Captain Williams coordinated the mission with Prince Fahd, the S3. ODA526 went to Merka on the same day the Saudi contingent scouted the region in anticipation of their upcoming humanitarian operation.

There was not much activity for the remainder of December; ODA526 conducted mini-reconnaissance and assessment missions for SOCCENT in and around Mogadishu–some vehicular and some by rotary-wing. They went out to small villages southwest of Mogadishu where the populace needed food, medical support, and engineering support. Early in January, SOCCENT desired to bolster ODA316 operating in the Merka area, and ODA526 assisted them in security operations during food supply handouts as well as conducting initial area assessments to assess the situation on behalf of the 10th Mountain Division in anticipation of the Division's future operations to open the Merka HRS. They went to the towns of Quorleey, Merka on the seacoast, and down to Brava during their reconnaissance.

ODA562 – Pakistan. In early December of 1992, Major Carroll called in Captain Stevens to inform him of a new mission for the team. His SF team was being selected as one of the 5th SFG(A) Coalition Support Teams to support the Pakistanis once Operation Restore Hope began. OD562 was chosen because they were one of the few teams Major Carroll had which was led by a Captain (most counterpart officers were either Lieutenant Colonels or Colonels). Major Carroll directed Stevens to catch a flight over to Mogadishu to meet with the Colonel Asif, the Pakistani coalition commander, spend a day getting requirements and details of the mission, then come back to Mombasa to prepare the ODA for deployment.

The Pakistanis deployed with four battalions, totaling 880 men. They were the 6th from the Punjab regiment, the 7th from the Frontier Forces, the 10th from the Baluch Regiment, and the 1st Battalion from the Sind Regiment. The Pakistani peacekeepers in UNOSOM were separate and were not counted among these.

The U.S. Marines were occupying the airport in force after their December 9th landing. Stevens was taken to the MARFOR headquarters to announce his arrival and mission. After speaking with a few officers at the headquarters, he was pleased to meet Lieutenant General Johnston in person when he stepped momentarily into the room. General Johnston was pleased to hear of the coalition support mission and asked Captain Stevens to return with his team the next day. The Marine operations officer directed him to drop his radios (he called them "high-speed" communications) and his radio operators off at the operations shop, then report to the Pakistanis. Stevens remained silent, choosing to ignore any such thing. He was told, however, he could find some "special ops guys" led by a Colonel at the Embassy compound, where he should be able to coordinate for lodging. Once again, Stevens silently ignored the advice; no advisor or liaison officer worth his salt would live separated from their counterpart.[21]

Stevens hitched a ride from a gentleman he guessed worked for the State Department, who was dressed in civilian clothes and armed with a pistol; the civilian dropped him off at the Embassy compound and pointed out the building where he thought SOF personnel were operating.

Stevens was surprised to find a Sergeant and a Colonel (Colonel Ford), both Civil Affairs Officers, presumably leading the JSOFOR until SOCCENT could arrive. Colonel Ford was responsible for the request for CSTs but was surprised to see Captain Stevens two weeks earlier than anticipated. Having satisfied the reporting in requirement, Stevens moved back to the airport to meet with the Pakistanis.

Captain Stevens found the Pakistani headquarters located in some ruined buildings of an aircraft servicing area, at the northern edges of the airfield. This was the former site of the Somali Air Force. He arrived at the Pakistani guard post outside their headquarters and was accompanied by Major Alvi, one of the operations officers. Stevens met the commander, got as many facts on the mission as he could, then spent the night before returning the next day to Mombasa.

ODA562 deployed using two C-130s to load their pallets and two DMVs (Desert Mobility Vehicles, SF-modified Humvees) on the 23rd of December. The team chose a small, fenced-in compound atop the sandy ridge on the south side of the airstrip adjoining the beach, to erect a tent and furnish it with cots, chairs, tables, and other items scrounged from Marines and other coalition contingents. The remainder of the company still at Mombasa with Major Carroll was not anticipated to arrive until sometime around Christmas.

The A-team performed mounted reconnaissance to ascertain the situation, traversing most of the major city streets. They noted heavy weapons, gun jeeps (Technicals), population activities, and the conditions of major intersections. Captain Stevens annotated his observations on a map after each recce and passed the information to the UNITAF intelligence section. They were able to view Aideed's compound during one of their trips. He was surprised the J3 and J2 never thought to task the team for route recons or assessments during those early days of the Unified Task Force, when very few units had the skills or means to get out beyond the wire of their compounds. Of note, the Somalis appeared more hostile and more openly armed the further they got away from Marine and coalition positions: "One day we even passed a tank rumbling up the street!"[22]

Smith wrote home to his wife about his observations of the city (the team filmed their recce around the streets of Mogadishu, providing one of the few, and rare, glimpses of the early period of Operation Restore Hope).

> Anything of any conceivable value has been looted. And everyone has a gun. They hid them all before the Marines landed, but they are quickly appearing again because the Marines have made no attempts to disarm these people. The other day I saw a multiple rocket launcher and an artillery piece being towed down the street, and the day before that a tank passed us going in the opposite direction. You can hear shooting throughout the city at night and occasionally watch tracers arc across the sky.[23]

The mission to conduct coalition support only lasted a few days. They were surprised to be pulled off the Pakistanis to work for a U.S. Navy Commander who was a liaison officer to the UNOSOM headquarters. Disappointedly, this tasking was only to provide the Naval Officer a personal taxicab to get around, and life became boring.

They were soon placed under the command and control of the JSOFOR (Colonel Smith from SOCCENT). Colonel Smith pulled them from the CST duty; it was clear the Pakistanis were going to conduct a static, passive presence mission at the airport and were settled in to the routine of the job, not much to the team's liking.

An officer on the JSOTF staff approached Captain Stevens one day with an idea for a mission. The Marines had told him that their reconnaissance platoon had discovered a cache of weapons just a few miles north of the Embassy, but lacking further orders and the resources to destroy it, they had left the cache intact. He wanted ODA562 to escort him to the location to take a look. At first, Captain Stevens refused; the UNITAF was clear about not taking weapons out of the hands of Somalis if they weren't being used in a threatening manner. However, Stevens was informed the Marine reconnaissance detachment saw no Somalis or guards at the site. Now it was a free look-see to pinpoint the location; no weapons or arms would be taken from the site. Stevens wrote:

> Beginning to feel rather guilty about how little we were really doing, I eventually agreed to take him there. He had gone over the map with the reconnaissance platoon, and knew the location of the cache to within about 300 meters. With that and a description of the area from the men who had discovered the cache it wouldn't be too difficult to find. However, when he showed me the spot on the map I discovered that the only way to get there was to pass right by the gate of Mohamed Farah Aideed's main compound.

They departed the Embassy in the morning, bypassing a large crowd and a Technical. The route took them by Aideed's neighborhood. They located the site, which contained several small, rubbled buildings. There were a few rounds of heavy ammunition and an RPG rocket or two scattered around in the dirt, and a rusty old single-barreled 40-millimeter anti-aircraft gun. A further search turned up more barrels and some scattered artillery rounds. While departing, Captain Stevens noticed a few Somalis hidden in a bush alongside the road, popping their heads up to observe. Not taking any chances, he fired a few warning shots from the .50 caliber in their general direction and scared them away.

The Marines went back some time later and did find a large weapons cache but in a different location. They ended up capturing or destroying a D-30 howitzer, some heavy mortars, and a Milan anti-tank missile system.[24]

One of the most unusual requests came to the JSOTF from the UNITAF staff for a mission to conduct a reconnaissance outside the city to ascertain where the Somali National Alliance had removed its Technicals. Colonel Smith considered it absurd and did not sign on to the mission–it involved a 200 square kilometer piece of the map and was heavily infested with mine fields.

On New Year's Eve, ODA562 was back with Company C. They were headed to support the Canadians at Belet Weyne.

Humanitarian Relief Sector Baledogle – December 13, 1992

With Mogadishu secure, the first HRS operation took place in Baledogle. UNITAF saw its occupation as a key steppingstone to all the other HRSs. Its most important feature was the ex-Somali Air Force Base northwest of the town of Wanlaweyn, with an improved concrete runway. Although suffering damage during the civil war, the base had some skeleton of structures for barracks, a control tower, and large hangers.

The Wanlaweyn district is located in the Lower Shebelle region. With trade and exposure to foreigners from the earlier colonial period, the region was considered cosmopolitan. Before the civil war, the economy was good with trade in agricultural and livestock products.

From Mogadishu, one gets to the airfield by first driving to Afgooye, where a bridge crossed the Shebelle River. The next major stop was the city of Wanlaweyn. Wanlaweyn is located about seventy kilometers from Mogadishu. The predominant clan is the Digil clan. In 1991, the USC/SNA forces of Aideed dominated the region, in partnership with Colonel Jess's Somali Patriotic Movement (SPM) and the militia forces of the SNA under Ali Tur (Issaq). In February 1991, this alliance fell apart as Aideed moved to dominate the political spectrum from his power base in Mogadishu. The fight occurred in Afgooye; Jess took his SPM towards Kismayo, and Ali Tur moved his forces further northwest and north towards Baledogle.

On the 13th of December, elements of 2/9 Marines from the 1st Marine Division and French Legionnaires conducted a helicopter assault on the airfield. Medium Helicopter Squadron 164 supported the movement and the base was secure within 48 hours. The Marines and the French were relieved on the 15th of December by elements from 2-87th Infantry of the 10th Mountain Division, who arrived in three C-141s. Condor Base was soon established.

Baledogle Air Base (DOD).

ODA184's first mission with the Canadian paratroopers was the move to occupy the Baledogle airfield (via C-130) before their launch to take HRS Belet Weyne. Upon their arrival, the team participated in helping to establish defensive positions and then established coordination links between SOCCENT in Mogadishu and with the 10th Mountain Task Force elements on the airfield. In the following days, the team participated in base camp security missions. It was a drastic change between the cold, snowy days of Washington State and the heat, dust, and humidity of Somalia.

In early January 1993, soldiers from the Royal Moroccan Army arrived to assume the security mission for the Humanitarian Relief Sector (primarily, the airfield). They were placed under the operational control of Army Forces (ARFOR) on the 12th. The Moroccans deployed with one of the larger contingents among the coalition, about 1,000 troops of the 3rd Motorized Infantry Regiment. The 3rd had sufficient wheeled vehicle and light armor assets to conduct outlying patrols in the HRS. The King of Morocco and the Moroccan people were very proud of their contribution. They wanted to make a significant contribution while not becoming a burden to UNITAF. The unit was self-supporting with supplies flown in from Morocco. Their best asset was establishing a large hospital to assist the Somali people.

ODA316 did not participate in this mission with the Moroccans. After only spending a short time with them at the Mogadishu airfield, the Detachment Commander borrowed a vehicle for the team and went off to find other humanitarian missions in the area (they ended up in Merka for a short time before their redeployment back to the States).

The Moroccans, under the leadership of Colonel-Major Omar Ess-Akalli, ran one of the best and quietest sectors during the UNITAF period. Due to their performance, they were placed under the control of UNITAF during the beginning of March, with responsibility as the sole contingent running the HRS.[25]

Humanitarian Relief Sector Baidoa – December 16th

Baidoa is located in Somalia's southwest Bay region, 240 kilometers west of Mogadishu. The Humanitarian Relief Sector, with the capital city of Baidoa, comprised over 17,000 square kilometers. It lay in the drainage basin of both the Juba and the Shebelle Rivers on the Huddar Plateau; the town itself is fed from a natural spring. It had a serviceable airfield, 10,000 feet long, outside of town to the southwest. Baidoa was one of the feeding stations serviced by many NGOs and was also a recipient of the food flights during Operation Provide Relief.

During the civil war, it earned the nickname as the "Triangle of Death" as it was purported that more Somalis died from the war and famine there than in any other area of Somalia–an estimated 500,000 in total.

Baidoa grew as a road junction along Somali trade routes. In ancient times, Baidoa was part of the Ajuran Empire and later ruled by the Geledi Sultanate. Locally, the people called it Baydhabo. In 1992, four urban villages grew around its suburbs. Much of the local architecture reflected the Italian colonial period before Somalia's independence.

When UNITAF arrived, there were a couple hundred thousand inhabitants (estimates were 180,000),[26] many of them internally displaced people (IDPs), refugees from the war, making it the densest city in southwest Somalia. It was predominantly populated by the Rahanweyn sub-clan of the Digil (associated with the Somali Democratic Movement–SDM). Even though it was a semi-arid area with open savannah terrain covered with scrub and cactus, the fertility of the soil provided enough for it to be a large agricultural center and market, if not for the near-genocide level destruction wreaked on it from the attack on pro-Barre forces and the occupation of the region by the USC/SNA (Aideed faction). The period of major drought magnified the horror.

With the available airstrip and its centrality as a market town, Baidoa was a magnet for numerous humanitarian relief organizations. The ICRC, UNICEF, World Food Program, Irish

Concern, GOAL, CARE Australia, and *Médecins Sans Frontières* (MSF), as well as many other NGOs, were all present to alleviate the humanitarian catastrophe occurring in the Baidoa region. Most of the NGOs were located in buildings with warehouses along a major road in the city; thus, it took the nickname "NGO road." Among the NGOs operating in Baidoa was the Islamic Relief Agency, operated by Islamic fundamentalists with suspected support from Saudi Arabia. The fundamentalists would pose a problem to Australian efforts during their tour in the region to form town councils and committees of governance from the existing elders. An Australian officer who served in Somalia wrote the following.

> These outside interests were Islamic fundamentalists who were receiving backing from elements in Saudi Arabia, Sudan, and Iran. They were not in a position to assert any authority at that stage, but elements had already been engaged in armed confrontation with the force, operating under the guise of the Islamic Relief Agency. Their agenda was to undermine any Western presence and cause casualties where they could. This threatening element was removed after a final confrontation on the night of 21 February 1993 when an Australian soldier was wounded during an incident in which 66mm rocket fire was required to suppress an attack on a night patrol through the town of Baidoa.[27]

The threat in the Baidoa region was from the iron rule imposed upon its civilians by the Somali National Alliance (SNA), titling themselves as the Somali Liberation Army. This force operated as lone gunmen and small groups of bandits. NGOs long complained of the $5,000 fee charged to any food relief flight landing at the airfield.

On December 16th, 530 Marines from the 15th MEU (SOC), the "Hammerheads," and 140 French Legionnaires departed Baledogle with 70 vehicles and air support from AH-1 Cobras and UH-60 Blackhawks for the 160 mile road trip to occupy Baidoa. There was little opposition as they arrived in force, and only a few Technicals fleeing away were seen. The following day, a massive convoy of trucks delivered relief supplies as Marines secured the route.

The Marines soon imposed a strong hand on the local leaders and focused on security, mostly thwarting looting and banditry. Anyone seen carrying a weapon soon had it seized. Many of the nongovernmental organizations, living on the "NGO Road," complained they still suffered violence from the local SNA, but the Marines did not see their protection as their primary mission. Once the line of communication was opened to Mogadishu and the airfield secured, the Marines settled in to local patrolling of the town in the absence of any other form of law and order. The French contingent pushed on to Bardera. On December 27th, Marines from 3/9th replaced the forces from the 15th MEU.

Food and humanitarian supplies began to reach the town, it was relatively quiet and, as far as the Marines were concerned, mission accomplished in HRS Baidoa. On January 16th, responsibility for the HRS was turned over to the 10th Mountain Division (LI). In mid-January, the Marines were ready to turn over the sector to coalition forces, designated as the Australians.

Operation Restore Hope, Phase I was now complete, almost 30 days ahead of schedule.

Endnotes

1. Ohls, Gary J. "Somalia. . . From the Sea." Newport Papers #34. New Port, RI: U.S. Naval War College, Jul 2009, p. 212.

2. Harned, Glenn. *Stability Operations in Somalia 1992–1995: A Case Study*. Carlisle Barracks, PA: Personal Monograph Series PKSOI (19950612 009) United States Army War College PKSOI: War College Press, July 2016, pp. 24–30.

3. Rutherford, Kenneth R. *Humanitarianism Under Fire: The U.S. and UN Intervention in Somalia.* Sterling, VA: Kumarian Press, 2008, p. 82, citing reflections from Ambassador Oakley in his paper, "Urban Area During Support Missions" as part of a conference on urban operations held by the RAND corporation in March of 2000.

4. McGrady, Katherine A. W. "The Joint Task Force in Operation Restore Hope." Alexandria, VA: Case Study CRM93-114, the Center for Naval Analysis, March1994CRM, pp. 93–114.

5. _________. "The Blue Helmets: A Review of United Nations Peace-Keeping." Third Edition. New York: United Nations Department of Public Information, 1996, p. 295.

6. Mroczkowski, Colonel Dennis P. (USMC (Retired)). *Restoring Hope: In Somalia with the Unified Task Force, 1992–1993: U. S. Marines in Humanitarian Operations.* Washington, D.C.: History Division, U.S. Marine Corps, 2005, p. 52.

7. CRM Task Force, p. 80.

8. Ibid, pp. 83–84.

9. Somalia Interview between LTC Kevin Murphy and Retired Colonel Joseph D. Celeski conducted on May 30, 2007, in the USASOC Historian's Conference Room, Ft. Bragg, NC.

10. Conversation and interview between the author and SFC Barriger in the spring of 2002 held at USASFC(A) Headquarters, Ft. Bragg, NC.

11. Stevens, Charles B. "Honor Bound: A Special Forces Detachment in Somalia." Houston, TX: Author's Draft Copy, 1996, pp. 34–35.

12. Ibid, pp. 31–32.

13. Ibid, pp. 31–32.

14. Ibid.

15. Ibid.

16. Ibid.

17. Celeski, Joseph D. "A History of SF Operations in Somalia: 1992–1993." Draft Research paper prepared for The USASOC Historian, Ft. Bragg, NC. Extract from chapter titled,"Operation Restore Hope (UNITAF)–SF Operations 10 December 1992 to 8 April 1993," pp. 37–38.

18. Executive Summary for C/2/5 for Operation PROVIDE RELIEF, on file at USASOC Historian, Ft. Bragg, NC.

19. Murphy interview.

20. CPT Tim Williams, Personal Daily Patrol Journal and Commo Journal for ODA526, SITREP 004 entry, 27 Dec 1992. Kindly provided for the author's use in the preparation of this manuscript.
21. The story ODA526 serving as the Coalition Warfare Team (later changed to Coalition Support Team in deference to being on a humanitarian mission) with the Pakistanis was derived from Charles B. Stevens' personal work, "Honor Bound: A Special Forces Detachment in Somalia," written as a Draft book in Houston, Texas in 1996. C.B kindly provided a copy of the work, along with pictures of ODA526 to use in this manuscript.
22. Ibid, p. 115.
23. Ibid, p. 107.
24. Ibid, pp. 115–119.
25. Mroczkowski, pp. 112–113.
26. Brocades Zaalberg, T. *Soldiers and Civil Power: Supporting or Substituting Civil Authority in Peace Operations During the 1990s*. Amsterdam: 2005, pp. 218–220.
27. Kelly, Michael J. (MAJ). *Peace Operations: Tackling the Military Legal and Policy Challenges*. Australian Government Publishing Service, Canberra: 1997, Chapter 8, p. 16.

4

Operation Restore Hope: Phase II, December 1992 - January 1993

Here, they were not engaged in a conventional war with clearly demarcated battle lines, allies and foes. This was neither war nor peace, and each day they had to feel their way, with constantly shifting directives from above . . . It was a taste of the messiness of war amid civilians, the classic guerrilla warfare for which they had been trained.

— Linda Robinson, *Masters of Chaos*

With the Cold War over, the New World Order of multilateral arrangements and new market economies began to merge. President George H.W. Bush signed the North American Free Trade Agreement (NAFTA) along with Mexico and Canada. McDonald's opened in China. Russia began to privatize businesses which were formerly under state rule sponsorship.

Media interest in Somalia began to increase as Operation Restore Hope began. Previous data indicated that from less than 300 stories about the crisis in Somalia in the month of July, by the 30th of December, the stories about Somalia rose to over 2,000.

With Phase I of Operation Restore Hope well ahead of schedule, UNITAF began Phase II operations to expand their activities beyond Mogadishu and occupy each of the other Humanitarian Relief Sectors. These activities were conducted according to the sequence outlined in the OPLAN, although in some cases there were minor changes in the order of occupation based on the availability of the coalition contingent's arrival in Somalia. In Phase II, the following Humanitarian Relief Sectors (HRSs) were secured between December 20, 1992, and January 3, 1993: Kismayo, Bardera, Oddur, Gialalassi, Belet Weyne, and Merka.

In Mombasa, ODB560 was detached from Operation Provide Relief and placed on orders to serve in Operation Restore Hope. With this action, all Army Special Forces units in operations were placed under the operational control of SOCCENT's Joint Special Operations Task Force (JSOTF). The Joint Psychological Operations Task Force (JPOTF) from the 4th PSYOPs Group led by LTC Charles Borchini served UNITAF directly through the J3, Brigadier General Zinni. The unit consisted of 125 personnel from the 8th and 9th PSYOP Battalions: the PSYOP Development Center (PDC–leaflets), the PSYOP Dissemination Battalion (PDB), and loudspeaker teams from the 9th PSYOP Battalion. Company C, 96th Civil Affairs Battalion also deployed to serve under UNITAF control through the Civil Military Operations Center (CMOC), with six Civil Affairs Direct Support Teams (CA-DST) of four personnel in each team. The PSYOP and CA units, although not part of the Joint Special Operations Forces (JSOFOR), served a crucial role in the UNITAF plan to assist coalition and U.S. forces during each phase of the operation.

Kismayo – December 20, 1992

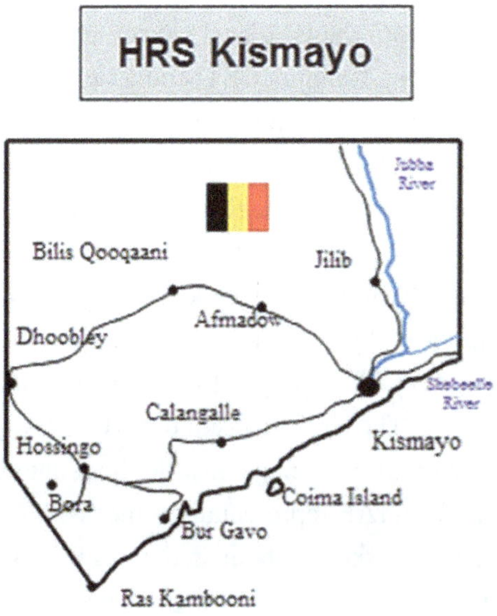

Kismayo is the capital of the autonomous Jubal region. It sits at the mouth of the Jubba River where it drains into the Indian Ocean. The surrounding area is geographically part of the lower Jubba Valley. Kismayo started as a small fishing town, but its location was excellent for many of the ancient trade routes crisscrossing the area; the town grew as a major port and trade center, exporting bananas and agricultural products. It was also well situated for internal Somali trade–three roads intersected at Kismayo: from Mogadishu, from Bardera, and from Afmadow.

Kismayo is the second largest port in southern Somalia with a capacity to berth four ships. The somewhat modern port was constructed by the United States in the 1960s. Kismayo's concrete airfield, located ten kilometers to the northwest of the city, was the former site for training the Somali Air Force. The Somali Navy was stationed at the berthing jetty.

The Ajuran Sultanate, taking advantage of the rainfall from the two monsoonal seasons and water from the Jubba River, established several crop plantations in the area. Those driving in the vicinity of Kismayo viewed lush fields of crops on both sides of the main roads leading into the city. The Italians called the Kismayo region the "Italian Somaliland" when they were ceded this portion of Somalia by the British in 1925 to thank Italy for its contribution as Allies in World War I.

Kismayo is a hot, arid place. It averages temperatures in the 80s during the spring. Although not a great spot for tourism, it did attract Somalis from most of the clans to live there. Sub-clans from the Darod included the Marehan, Ogadeni and Majeerteen. There were also populations of Hawiye and non-clans, like the Bantu and Hartis. The Ogadeni revolt in Kismayo against the former regime resulted in the first formation of the Somali Patriotic Movement (SPM).

As Aideed and the United Somali Congress drove off the tyrant Siad Barre's forces, Aideed's strategy was to continue ridding Somalia of Siad Barre, his former military forces, and the Darod clan influence. The Somali National Alliance attacked Colonel Jess's Somali Patriotic Movement (SPM) in Afgooye in February 1991 as a sort of back-stabbing on Jess, who previously supported Aideed. Aideed wanted control of Afgooye. The militia forces of General Morgan (Marehan clan) and Colonel Jess retreated to Kismayo rather than face the onslaught. Colonel Jess occupied Kismayo, pushing General Morgan and his subordinate COL Barre Adan Shire Hiraale out several kilometers north of the city, up the Jubba River.

Intelligence estimated Colonel Jess's strength as three to five-thousand fighters; General Morgan was believed to have 1,000 fighters.

The 10th Special Forces Group Belgian Coalition Support Team

The Belgian Coalition Support Team (CST) led by Chief Warrant Officer Doug Bell consisted of the Chief and five other U.S. Army Special Forces operators, all French speakers. They deployed from Fort Deavens, Massachusetts on December 10th aboard a flight to Brussels, Belgium. The detachment conducted a linkup with the Belgian 1st Parachute Battalion of the 1st Para Regiment and continued mission planning. The 10th Group Belgium CST would be responsible for ensuring the communication linkages between the Belgian battalion and U.S. forces, to include fire support assets from both jet fighters and rotary-wing attack helicopters.

COL Marc Jacqmin led the Battalion. He was a very experienced officer who had served in previous assignments with the Regiment in Africa. Colonel Jacqmin was bringing his three line companies of airborne infantry, a mortar platoon, an attached recon company (equipped with Scimitar light tanks) and additional platoons of heavy weapons, Engineers, and EOD (approximately 850 personnel). The task force was supported by an aviation detachment with four Alouette III helicopters. The battalion and the 10th SFG(A) Belgian CST deployed to Mogadishu via Belgian military aircraft later in December 1992.

The Unified Task Force (UNITAF) alerted the Belgians and the 10th Mountain Division of the operation to enter HRS Kismayo next in the sequence of Phase II activities. In preparation for the occupation of the port city, UNITAF representatives met with Colonel Jess and General Morgan to announce UNITAF's intentions. UNITAF asked for both faction leaders to agree to declare Kismayo an open city; militia forces would move off to a safer distance while their heavy weapons were to be placed into cantonments outside the city limits.

A disturbing incident occurred in Kismayo that made it more than important to get the HRS secured as soon as possible. Between December 8th and the 19th, Colonel Jess's forces purportedly massacred an estimated 100 prestigious leaders of the Kismayo community. The gunmen went door-to-door, pulling out civilians identified for ethnic cleansing, justified in Colonel Jess's speech at a rally on December 6th when he exhorted, ". . . the town needed to be 'cleared of people who would cause trouble.' The Darod were hated as the other." The myth of Darod dominance needed to be eliminated. NGOs discovered mass graves and mutilated bodies as evidence of the deed. In some strange logic, Jess thought eliminating these people would put him at a political advantage with UNITAF, since the dead were Morgan's people. Taking and occupying the HRS was now crucial to prevent further killings.

When Ambassador Oakley conducted his pre-visit to the HRS to speak with Colonel Jess, he commented about the atrocities when asked by the media. He basically restated the position held by the United States that America was not an occupying power and had no authority to arrest Somalis. The situation appeared to follow one of the lines in the Somali proverb about clans, "I and my clan against Somalia."

ODA565 and ODA564, still in Mombasa on Operation Provide Relief, conducted an airfield assessment of Kismayo on the 17th of December. Part of their task was escorting Chief Bell's composite team of French speakers from the 10th Special Forces Group (Airborne). Bell's team was assigned as the Coalition Support Team to the Belgians, scheduled to occupy Kismayo with the Marines via a seaborne landing. Between the two teams, a total of 18 weapons, eight cases of small arms, and a small amount of 57-millimeter recoilless rifle rounds were seized that day.

On December 19th, ODA565 flew a special night Airborne Battlefield Command Control and Communications (ABCCC) aircraft mission in support of a SEAL team operating in Kismayo. On the 20th of December, Marines and Belgian paratroopers secured the airfield and the port area of Kismayo, followed by more forces of the 10th Mountain Division (LI) on the 22nd.

Under the control of MARFOR, the Marines of 2/9th, 15th MEU (SOC), along with the Belgian 1st Parachute Battalion conducted an amphibious operation on the port of Kismayo. The Operation was supported by the USS *Tripoli* PHIBRON (Amphibious Squadron) and the French frigate, the FS *Dupleix*. The FS *Dupleix* served as the advanced force loaded with U.S. Navy Seals from Naval Special Warfare Task Unit–Alpha (NSWTU-A) and Force Recon Marines, arriving off the harbor of Kismayo on the 18th of December. Using combat rubber raiding craft (CRRCs, called Zodiacs), the SEALs provided the necessary hydrographic surveys and beach landing area reconnaissance. Although unopposed, the SEALs had unknown small arms fire directed at one of the locations ashore that they had just evacuated; the same situation occurred the next day. The SEALs did not return fire at any point.

The Marines aboard the USS *Juneau* (one of the ships in the USS *Tripoli* PHIBRON) moved ashore at 0700 hours along with two platoons of Belgian paratroopers. The port was quickly secured without incident and the landing force was joined by reinforcements of Task Force–Kismayo under the command of the Assistant Division Commander of the 10th Mountain, BG W. Lawson Magruder III, bringing a ground convoy from Mogadishu with U.S. Army vehicles and Belgian light armor.

The contingent deployed from Mogadishu on December 21st in a military, wheeled convoy and drove south along the coastal road to reach Kismayo. The Belgian aviation detachment flew into Kismayo airfield and established operations. The ground force headquarters collocated with the USMC and 10th Mountain elements at the seaport, which was still damaged at this time. The 10th Mountain contingent and the USMC contingent were already there to assist by providing security in the area, along with the Belgians. Initially, the area of responsibility included Kismayo, the port, the airfield, and the environs out to a 60-kilometer radius. This was later changed to a 200-kilometer radius, which proved difficult for the Belgian battalion to effectively gain control over.

The mission for the 10th SFG(A) CST was to provide combat liaison to the Belgian contingent while assisting and coordinating information gathering during the Belgians' combat patrols. They also provided ground truth and situational awareness on the threat from local clans to the Amy Forces component (ARFOR).

Task Force–Kismayo worked to set up local councils, improve security through street patrols, impose a "no weapons" policy, and assist the Somalis in establishing a security police force. Not long into Phase III of Operation Restore Hope, the Kismayo HRS would become one of the problematic sectors due to constant maneuvering and clashes between Colonel Jess and General Morgan. On the third week of January, General Morgan began his move south to take back Kismayo from Jess.

The threat in the area came from the Somali Liberation Alliance (the SPM and the SNA), led by COL Ahmed Omar Jess. Previously, the city had been held by General Siad Hersi Morgan's Forces who had pushed out Colonel Jess's army but, while pursuing him north to Mogadishu, overextended his lines. Colonel Jess, reinforced by Aideed, took this opportunity to counterattack and re-seize Kismayo. He eventually pushed General Morgan and his forces west to the region of the Kenyan border after vicious fighting. The SLA occupied a compound in Kismayo, guarded by several gunmen and defended with a ZSU-23-4 gun mounted on a trailer inside the compound. Other forces consisted of city security forces responsible for the various checkpoints on the edge of the city. There was a contingent of secret police dressed in civilian clothes, but nonetheless armed and mounted on a variety of pick-up trucks.

A short time after arrival, the battalion commander deployed forces into the city sector and secured the port, city, and airfield. The battalion conducted military operations on urban terrain (MOUT) refresher training and sniper training after analysis of the environmental conditions they were facing. The CST conducted the liaison and coordination with other U.S. forces and put in place the procedures for close air support (CAS).

As the mission progressed, four objectives were assigned to the Belgian contingent.

- Disarm random gunmen or bandits
- Store confiscated weapons and equipment in the designated holding compound
- Serve as a staging area for the arrival of follow-on 10th Mountain Division forces
- Assist follow-on units of Australian and Indian forces

The 10th SFG(A) Coalition Support Team conducted area and refugee camp demographic assessments throughout their deployment, talking and meeting with numerous NGOs in the Kismayo region. Their efforts were instrumental in providing information on the medical, humanitarian, and engineering needs in the area. Although the team did not get in any serious combat engagements, they experienced nightly harassment fire from random shooters and snipers, as did all forces assigned to this area. Their combat effectiveness was only limited by having been formed as an ad-hoc CST, which hampered them from receiving an ODAs worth of combat equipment, not to mention the lack of time during pre-deployment activities to conduct combat drills and develop contingency SOPs as a regular team would.

The team was notified in early January of the end of their CST mission. The Belgian CST traveled overland via military convoy to Mogadishu, then flew out on military aircraft through Saudi Arabia, Rhein-

Main, and then Dover. The team transferred to a commercial bus and closed out at Fort Devens, Massachusetts on January 10, 1993.

Humanitarian Relief Sector Bardera – December 24, 1992

The Bardera Humanitarian Relief Sector (HRS) is split north and south by the Jubba River valley. The Gedo region, although lush along the river valley, is a plateau with increasing low mountains as one travels north toward the Ethiopian border. To the west lay Kenya's border with Somalia, including the northwest province border line under dispute with Gedo. To the east, the road leads to Baidoa; to the south is the road to Kismayo. The capital of Gedo is Garbaharrey. The town of Bardera has a 1,300 foot dirt-compacted runway which was in high use by the NGOs and the food flights for Operation Provide Relief. There was no other safe way to enter the town due to the surrounding threat from mines. Temperatures ranged from the 90s to the low 100s; there was rain during the spring and again in the fall, which provided enough precipitation to sustain agriculture.

Due to famine and war, the population of Bardera was estimated at 6,000 inhabitants; NGOs recorded about 8,000 others in refugee camps outside of town, living destitute in an area nicknamed "The Italian Village." The river split clan affiliation. To the west of the river lived Darod clansmen; to the east, Rahanweyn. This area was heavily fought over during the civil war when Aideed's Somali National Alliance attacked General Sayeed Hersi Morgan's Somali Patriotic Movement (SPM) militia.

After seizing Mogadishu and driving out Siad Barre and his forces (who fled west to the Kenyan border into the 21st Military district), Aideed looked to consolidate his USC/SNA power and saw the opportunity to rid himself of Darod clan influence as the former Somali military fled west. He also desired to track down Barre and rid Somali of his influence and the Somali Patriotic Movement once and for all. Aideed attacked the Gedo region and drove off General Morgan's forces in what was titled the "War of the Militias"–not civil war, with the dictator gone, but internal, ethnic clan-cleansing. Somalis who helped or acquiesced and did nothing to challenge Morgan's forces while they were in the region were singled out for collective punishment or death as traitors by Aideed's clansmen: the Darods, Majeerteen, Marehans, and Ogadenis. Siad Barre fled to Kenya.

In August 1992, Aideed established his Somali National Alliance headquarters at Bardera. He was able to form a new alliance with Colonel Jess to battle against General Morgan. In October of 1992, General Morgan's men drove out Aideed's SNA, who fled back to Mogadishu. Morgan's henchmen took total control over Bardera to run a lucrative extortion business, charging NGOs thousands of dollars for each food flight delivered. When his thugs got out of control, probably high on khat, someone fired a missile at one of the food flights which resulted in all food flights being cancelled. The humanitarians were stuck in town without any way to alleviate starvation; most were afraid to leave, not wanting to take the risk on mined roads. One Somali described the plight of the inhabitants, who were dying at the rate of an estimated 300 a day, as, "This is the death people."

Bardera, meaning "tall palm trees," sat astride the Jubba River in the Gedo region. It was one of the media named "cities of death" for the almost daily Somali casualties due to starvation (DOD).

The plan to open the Bardera HRS moved rapidly as major contingents closed their forces in on Mogadishu. After the Marines and Belgians secured Kismayo, next on the Phase II schedule were Bardera and Oddur for opening on Christmas Eve.

COL Emil R. Bedard led the 1/7th Marines of the 1st Marine Division (MARDIV) on a road march from Baidoa with about 800 U.S. Marines (a reinforced battalion). Bardera was notorious for its large surrounding minefields left over from the Ethiopian War and those added since the civil war. This setup included the mining of major roads leading into and out of town. Engineers conducted minesweeping ahead of the lightly armored vehicles (LAVs) and trucks, along with an accompanying humanitarian relief convoy. Overhead, they were escorted by Marine helicopter gunships; U.S. Navy F-14 fighter jets from the Carrier Battle Group preceded them by two days to conduct reconnaissance.

There was only a small element of the United Somali Party (General Morgan's forces) in town which faded away as the Marines occupied the airfield and Bardera itself with no opposition.

Colonel Bedard ran a tight ship; soon, the HRS was quiet and starvation eliminated. Young Marines, working within the UNITAF Rules of Engagement, clamped down to seize weapons openly carried while on patrol to provide security to Nongovernmental Organization warehouses and feeding stations. The engineers had a huge task to clear the line of communication from primarily Russian-made mines.

It was here that the first death in UNITAF occurred on December 23rd when retired Special Forces Command Sergeant Major Lawrence N. Freedman was killed by a Russian landmine while working as a contractor for the U.S. government. Three U.S. State Department security personnel riding along with him in the vehicle were severely wounded, requiring medical evacuation to Europe.

The 1/7th Marines were relieved at the end of January by the 3rd Amphibious Assault Battalion. During their tenure, SF ODAs from ODB560 at Belet Weyne assisted them with zone reconnaissance and area assessments. It was in Luuq (in the northern portion of the region) where ODA526 would run into a disagreement and, almost, an altercation with the al-Itihad al-Islaymia (AIAI–The Islamic State), considered an organization affiliated with other Islamic terrorist groups.

Due to a suspected threat from Colonel Jess's militia maneuvering toward Bardera to challenge General Morgan's militia, the Marine force directed the Special Forces team to conduct a reconnaissance out of sector south to the village of Sacco Uen. The threat never materialized.

The 3rd Amphib was relieved by the Botswanan contingent in mid-April, who deployed in preparation for the turnover to UNOSOM II. The details of ODA526 in the Bardera HRS are covered in the next chapter, Phase III of the UNITAF plan.

Humanitarian Relief Sector Oddur–December 24, 1992

The city of Oddur was situated 260 kilometers northwest of Mogadishu near the border with Ethiopia. The predominant clan in the region was the Rahanweyn. Ogadeni clansmen also lived in the border region. The airfield at Oddur was serviceable for C-130 aircraft operations. The Oddur Humanitarian Relief Sector (HRS) would fall under the responsibility of the French forces contingent. The first of the French forces, a company of the 2nd French Foreign Legion Parachute Regiment from Djibouti, arrived in Mogadishu on the 10th of December via French transport aircraft and immediately moved to take up positions and secure the K4 traffic circle. This key piece of terrain served as the intersection of streets between the port area, airfield, and the Embassy and UNITAF compounds.

The entire French contingent deployed under the name of Operation "Oryx," commanded by Major General René de l'Home and consisting of approximately 2,300 troops. The forces were task organized from two motorized rifle battalions: 5th RIAOM (Combined Arms Overseas Regiment) and the 13th Foreign Legion DBLE (Demi Brigade), all supported by the 4th Armored Squadron with their light tanks and the 3rd Company from 2nd REP of the Legionnaires Parachutists, which provided a unique special forces capability to the commander. The infantry ground force utilized M113 Armored Personnel Carriers from the 2nd French Marine Regiment for their mobility operations. The entire force was supported by the aviation assets of the 4éme Division *Aeromobile* (Attack Helicopter Regiment) and the 5éme Division *Aeromobile*, equipped with SA 330 Puma helicopters. The task force included supporting elements of the command group, engineers, logistics unit, an intelligence unit, military police, and communications assets.

In the initial stages of Operation Restore Hope, the French task force came under the operational control of the Marine Forces component (MARFOR). Their first combined operation with U.S. forces was launched on the 15th of December to secure the airfield at Baidoa. Task Force "Hope" consisted of the 15th MEU along with the French 2d Foreign Legion Parachute Regiment and elements of the 13th Foreign Legion Demi-Brigade.

The Task Force deployed with ground and air assets from Mogadishu and secured the airfield at Baidoa by the 16th of January. Soon, humanitarian relief convoys began flowing into the region. With French help, the task force established patrolling and security operations in the city, bringing the first phase of Operation Restore Hope to a close.

On the 18th of December, UNITAF issued Task Order #8 to General de l'Home to secure HRS Oddur around Christmas. However, the remaining forces of the Task Force were still arriving in Mogadishu, so the order was amended by several days to provide General de l'Home more time to assemble his forces. Not deterred in the mission, Major General de l'Home began his movement by truck convoy to Oddur on Christmas Day, anticipating a quick closure of his remaining forces from Mogadishu (shown above, courtesy of Jon Concheff). The 5th Combined Arms Overseas regiment and the 13th Demi-Brigade (FFL), supported by an organic support unit and a company of U.S. Marines TACON (placed under tactical control) for the operation, closed into Oddur to spend Christmas night.

In short order, General de l'Home deployed his forces throughout the Oddur HRS establishing outposts in Wajid, Ceelgasass, and El Berde. On the 29th of December, UNITAF acceded to his request for a boundary extension eastward to include the town of Tiyegloo.

The mission for the French in Humanitarian Relief Sector Oddur was to provide humanitarian convoy security escort, security patrols, collect abandoned weapons, and further securethe surrounding towns to provide stability for relief agencies and their employees. The predominant threat in this region was from bandits and hooligans equipped with gun-jeep style vehicles and the "Technicals," often crew-served by four to five bandits. An abundance of abandoned light arms, munitions, and mines littered the area as well. The largest clan in this area was led by the former Commanding General of the Somalia Army, General Adib. The other predominant clan leader was General Mohammed Hersi Morgan, the son-in-law of the formerlydeposed dictator.

General de l'Home deployed his units north to the border (and also included the town of El Berde, predominantly populated by Ogaden clan), Tiyegloo to the east, and Wajid to the southwest. French battalions were assigned to each sector and further broken down into companies and platoons to cover smaller towns, eventually occupying about 20 sites. General de l'Home kept a small Quick Reaction Force (QRF) of infantry and attack helicopters located at the main French headquarters in Oddur. Each of the French units worked to provide security for humanitarian operations, establish local Somali governance, and monitor the activities of the various militia clans in the region. Additionally, the French conducted mine clearing and other engineering projects. Where it would work, the French assisted the Somali leaders to establish local police forces.

The border region would prove problematic. Refugee flow across the border exacerbated tension with the dominant Rahanweyne clan as non-clan refugees entered the area. The Ethiopian Army also operated in the region, potentially creating a condition whereby a clash could occur with the French UN-contingent. French and U.S. Special Forces were deployed to the border region to assist General de l'Home with his operation.

By the end of January, the HRS was secure and a sense of stability for Somalis in the region brought back many of the NGOs, most notably the French NGO *Medecines Sans Frontieres* (Doctors Without Borders). By February, General de l'Homme was able to downsize his contingent and send about half of the French forces home.

U.S. Special forces teams contributed to the success of the French efforts in Somalia and in HRS Oddur. The 10th SFG(A) initially supported the French task force with a coalition support team during their first weeks in Somalia. After the redeployment of the 10th SFG(A) CST, ODAs from ODB560 maintained combined operations with the French in the HRS during the remainder of Operation Restore Hope.

The 10th Special Forces Group (Airborne) French Coalition Support Team

The 10th SFG(A) was tasked in early December to form a Coalition Support Team (CST) to support the French task force. The USMC contingent operating with the French in Somalia lacked French language skills, so the Special Operations Command–Central (SOCCENT) assessed the need for SF to conduct the mission. The members of this team were alerted between the 6th and 9th of December and were task organized by all SF military occupational specialties, absent an SF engineer. All the members of the team were picked from throughout the 10th SFG(A) based on their French language skills. The team was led by Chief Warrant Officer 3 Jon Concheff and would later include two Marines attached as radiomen, Corporals Shaver and Seiber. The team was required to deploy on December 11th for linkup with the French military forces in Europe, so any opportunity for the newly formed team's pre-mission training slipped away. After crossing several hurdles at the airport in getting their equipment and weapons through customs, the team deployed to Paris, linkedup with their counterparts, and received their first update briefings.

The contingent deployed via a French Air Force DC-8 to Djibouti on December 13th, where the team had an opportunity to work out some coordination issues and plan more. The team leader also had the opportunity to meet with General de l'Home, who remarked he was happy to have the team on board and that he hoped to depart soon for Somalia once all arrangements for forces and logistics were in place. Chief Concheff also opened initial coordination with Colonel Smith, the JSOFOR commander in Mogadishu. The CST's remaining time at the French base in Djibouti was well spent by conducting communication checks, drawing ammunition from the French, and zeroing and firing their weapons. On December 18th, the team flew with French forces to Mogadishu on a C160 aircraft.

Upon their arrival, the team ran into the SF liaison to SOCCENT who informed them ammunition and supplies were awaiting them at the Embassy compound once they had an opportunity to report in to Colonel Smith. The team remained overnight (RON'ed) in a hangar at the airfield and then moved to the French Embassy to await further transit out to the field. Select members of the team reported to SOCCENT at the Embassy compound where they received their communication crypto and intelligence updates from the SOCCENT staff, then returned to the French Embassy to update the team.

Over time, a coalition convoy was slowly assembled in the courtyard and the 21-vehicle convoy departed for Baidoa on December 21st with the members of the 10th SFG(A) French CST spread throughout the column. General de l'Homme had intended to keep the CST with him in Mogadishu but later decided they would be more useful to the French operations planned for Oddur; this decision surprised Major General Wilhelm who was not aware of USSF attached to the French for the operation.

Upon arrival at Baidoa, the French commander conducted final coordination with the USMC at Baidoa for onward movement to Oddur (Xoddur). Ambassador Oakley dropped in to look around during this period. The team departed on December 24th with the French column along with a U.S. Marine company task force, escorted by USMC AH-1 Cobras, U.S. Navy F14s, and a French Navy ship, the FS *Atlantique*. No incidents occurred other than taking a wrong turn, costing the convoy time to turn around. A couple of mines were located during the road march and marked, and the column pressed on. At one point, the entire column stretched out over 28 kilometers. Upon arrival in the Oddur region, the CST linked up for operations with the 3rd Company of the 2nd Regiment *Etrangere des Parachutistes* (R.E.P.) of the French Foreign Legion and spent the night in the town of Waajid.

On Christmas day, the unit moved into Oddur without incident. C-130 and C-160 aircraft immediately started landing on the dirt airstrip even before it could be secured. The SF team later assisted an airfield survey team to properly record the conditions of the runway for further updates and effective use. The CST settled in to the task of assisting French forces with their security operations. The team was laagered in an abandoned military barracks on the northeast side of the runway. One member of the detachment took the opportunity to ride along with some French Foreign Legion troops in the conduct of security patrols east of the town while Chief Concheff met with U.S. Civil Affairs personnel, led by U.S. Army Major Nelson, to coordinate CMOC activities.

Throughout the remainder of December, the members of the team performed their liaison duties and also accompanied French patrols in the surrounding region. The area was littered with small arms, mines, and rocket-propelled grenades (RPGs) of Soviet manufacture; a continuing mission for UNITAF was to try and remove these weapons from the humanitarian sectors. The locals were very helpful in pointing out the location of many of these cached items.

On January 1, 1993, the team split into three-man sections to conduct a helicopter insert with the French into the village of Moragaby. Intelligence had indicated there were "Technicals" and arms stored there. The plan involved a French armored column on the ground conducting a link-up with the air-land element in a minor hammer and anvil operation. The operation succeeded; however, no Technicals or heavy weapons were discovered during the operation, although several grenades, mines, and small arms were confiscated.

On another mission conducted January 2, 1993, three members of the CST accompanied French forces on a reconnaissance to the Ethiopian border. The force conducted an insertion by Puma helicopters into the village of El Berde. There was no opposition and, after discussions with the local leaders and elders, it was decided to remain in place with forces and occupy an old abandoned colonial police fort. The fort was searched and a thorough cleaning of its buildings ensued while the team remained outside. On this same day, the team received word from SOCCENT that the CST mission was terminated and they were being recalled. They were ordered to return to Oddur for follow on movement to Mogadishu.

In the morning, the team took the opportunity to tour the village and visit with the Somali citizens throughout the market area. The team returned to the fort in the early afternoon and awaited the extraction helicopter. At 1700, they lifted off and headed for Oddur, leaving the two USMC communicators in place per orders from the USMC contingent at Baidoa. The following day, they packed and said their final goodbye to General de l'Home. The team caught a ride on two Pumas headed for Mogadishu that morning.

After out-briefing SOCCENT, the team was dismayed to hear their flight was delayed. Sergeant First Class Quentin, the team sergeant, took matters into his own hands and found a KC-10 on the airfield, which was soon departing. With approval from the SOCCENT J1 Personnel Officer, the team boarded and departed on January 4, 1993.

The flight took them to Saudi Arabia, then on to Seymour Johnson Air Force Base in North Carolina. The team took a bus to the Raleigh-Durham airport and flew to Logan Airport, Boston. They were met by members of the 10th Group and transported to Ft. Devens. Although conducting a difficult mission of only 23 days in duration, the outstanding success of the team in enhancing relations between French and U.S. forces was noted. During their time in Somalia, the French CST team conducted daily interface with the French Command staff and provided communication links to SOCCENT and the Unified Task Force. They assisted the French with daily coordination for logistics and also coordinated for two medical evacuations for French personnel. The team conducted combined operations with their counterparts, participating in three combat assaults. Their knowledge of the French language was also extremely useful in document interpretations between English and French. Last, their vast knowledge of munitions and mines was extremely useful to the French ordnance disposal teams' efforts.

Humanitarian Relief Sector Gialalassi–December 27, 1992

The town of Gialalassi is located 115 kilometers by road from Mogadishu and 116 kilometers from Baledogle, making it one of the contiguous and close HRSs to Mogadishu. Overall, it would be a quiet sector under the responsibility of the Italian Task Force. Six main Humanitarian Relief Organizations (HROs) were operating out of Gialalassi.

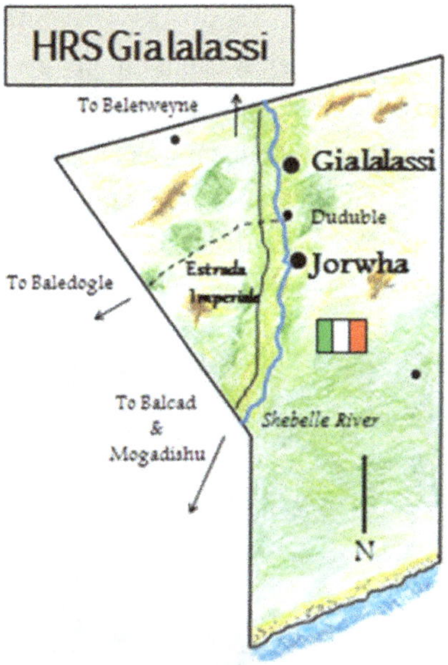

Gialalassi is in the Hiren Province and had a population under 20,000 (Hawiye clan). The town is built along the Shebelle River and is an agricultural area, green and lush with vegetation. Mysteriously, in the middle of starvation, UNITAF troops found fields of crops and sorghum as they entered the HRS.

The Gialalassi sector is generally a flat, plains-like plateau, with some scrub and small pockets of scattered forests. It had one large packed-dirt airfield improved enough to handle C-130 cargo aircraft; there was another unimproved airstrip primarily used by the small aircraft flown by the NGOs. The USC controlled the sector, but not in any appreciable strength. Like most of the other regions and towns, banditry was more prevalent than clan fighting.

Gialallasi had a deep history of Italian influence during the colonial period, reflected in the road signs and architecture of buildings.

On December 16th, UNITAF issued a Tasking Order (TASKORD) for the Italians to secure Gialalassi. The Italian contingent did not arrive in Mogadishu until the 26th; the start of the operation was reset to the 27th. The San Marco Marine Infantry moved from working with the U.S. Marines to back under Italian command.

The HRS operation opening was designed to open with a road movement north up the *Estrada Imperiale* built by the Italian colonizers in the early 1930s. The road followed the Shebelle River north. Led by General Gianpietro Rossi, the Italians deployed one of the two companies of the *Folgore* ("Lightening") Brigade with 160 paratroopers. They were supported by logistics, mortar, and armored vehicle elements (this included M-60 tanks).

The Italians soon found out the Imperial Road was in vast disrepair from neglect over the years and heavily damaged from the civil war. This situation would require a large amount of engineering work to improve the line of communication between Mogadishu and Gialalassi. Even with constant delays from bad road conditions, the Italian convoy escorted food relief trucks to deliver to Humanitarian Relief Organizations (HROs) in Gialalassi.

The Italians occupied the airstrip by early evening and set up camp for the night. The next morning, they established small patrols and checkpoints in town and began the process of handing out food relief from the convoy.

U.S. Army Military Police and Marine and Italian helicopters supported the movement, along with U.S. Navy fixed wing fighter jets. U.S. Army and USAF engineers were standing by to assist with airfield, road, and facility improvements.

The 10th Special Forces Group (Airborne) Italian Coalition Support Team

Chief Warrant Officer Two Graybill commanded the 10th SFG(A) six-man team (ODA056) chosen to conduct coalition operations with the Italians. He was notified of the mission on the 6th of December. Each member of the team was chosen for an appropriate specialty (a Team Sergeant, two communications NCOs, a medic, and another operator who spoke Italian well). Several operators in the unit wanted to go with the Chief to Mogadishu, but time was of the essence and Chief Graybill finished assembling his hand-picked detachment by the 13th.

They were informed the detachment would be working with the paratroopers of the *Folgore* Brigade. The team conducted pre-mission training, zeroed their weapons, and caught a flight out of Boston to meet with military transport at Charleston. They flew with some refuel stops along the way and eventually ended up at Mombasa, in Kenya, where they trans-loaded on the 23rd of December onto a C-130 for the trip into Mogadishu.

All this time into the mission, the detachment anticipated working with the Italian airborne unit; when the *Folgore* Brigade was delayed in their arrival, the CST was assigned to the San Marco battalion (the Italian naval infantry already in Mogadishu). The San Marco bivouacked in a compound at Old Port, east of the Green Line; the detachment procured a tent and supplies from the Italians and hoped to get some form of transportation from the SOCCENT staff, to no avail. The Italians had none to spare, either.

The San Marco battalion consisted of three Marine infantry companies along with organic support and logistics sections. Italian SEALs were attached to the unit to give them a special operations capability. An Italian support ship, the ITS *Vesuvio*, supplied the unit daily using landing craft to go between ship and shore.

The battalion was equipped with M113 APCs and lighter tactical wheeled vehicles similar to armed jeeps. The unit was armed with 106mm recoilless rifles and .50 caliber heavy machine guns. Light weapons were mostly of Beretta-made rifles and pistols. The unit was well equipped with night vision devices for most of their crew-served weapons.

The San Marco battalion's mission was to patrol the eastern portions of Mogadishu. Additionally, they were tasked to provide security and escort for Italian VIPS. As a highly trained unit, they were postured to react to any contingencies, if needed.

ODA056 assisted the battalion in its initial establishment of the camp wherever they could improve security. In this early period, rules of engagement for forces in the city were very restrictive and limited the opportunities, or desire, of units to conduct patrols for security reasons. As time went by, more and more patrols became routine. The team was very impressed with the urban warfare acumen and nighttime patrolling skills of the battalion. During this initial time with the battalion, no skirmishes occurred between the Italians and the Somalis, but everyone soon got used to random gunfire in that area of the city (some spilling over into the compound). Everyone observed one attempt by the Somalis to mortar their position, but the shells flew harmlessly into the bay.

Before the Italian main body arrived, Chief Graybill took the team on a trip with the San Marco Marines to escort a food relief convoy to Merka, trying to break the boredom of trivial duties at the compound. On another day, the Chief was surprised to see Mohammed Farah Aideed come into the Italian compound to use the satellite communications radio (SATCOM).

As the end of the month neared, Chief Graybill sought every opportunity to get out of Mogadishu on a mission. For some inexplicable reason, the team was not tasked to participate in the opening of the Gialalassi HRS on the 27th.

On December 30th, the team traveled to Baledogle to assist in coordination meetings between the French and the 2nd Brigade of the 10th Mountain Division on the matter of security and escort for an NGO food convoy of 33 trucks. The next day the team participated in providing security for President George H. W. Bush's visit to a feeding center near Afgooye when the Somalis were receiving food aid.

Like the other CSTs, when Operation Provide Relief Phases I and II were completed by early January, ODA056 was removed by SOCCENT from their mission with the Italians. They departed the Italian compound and spent the night near the Joint Special Operations Task Force headquarters (JSOTF); in the morning, they redeployed back to the United States arriving on the 10th of January 1993.

During their tenure the team collected and reported area intelligence, prepared security assessments, provided vital communications links between Ali Mahdi and UNITAF, and also enhanced mil-to-mil relationships with the Italians through language and interpreter support. The team was also able to facilitate and participate in combined operations with the Italian SEALs.

After the Italians' success in opening the HRS, General Johnston assigned them the northern portion of the Mogadishu sector, predominantly in the territory controlled by Ali Mahdi. They were one of the few coalition contingents in Mogadishu with their own sector. The Italian Embassy was located in this sector, also. In time, the Italians were given primary responsibility for an extended HRS which now included the northern sector of Mogadishu running north into the rest of HRS Gialalassi. To control the modified HRS boundary, General Rossi established his headquarters about 20 miles outside of Mogadishu in the town of Balcad. At first, he considered using the Italian Embassy in Mogadishu. General Johnston frowned on the initiative; it would look too much as if UNITAF was favoring Ali Mahdi. UNITAF had to appear neutral to all the factions and not mix military lines of operations with the political lines of operation. (Above, Italian vehicles on Mogadishu airport, courtesy of Moe Elmore.)

The *Folgore* Brigade expanded its security operations throughout the sector using packaged patrols by size—from small to large—dependent on the situation, establishing checkpoints and conducting sweeps to confiscate heavy weapons and any openly-carried weapons. The Italians put an emphasis on tackling the bandit problems in the HRS and went to great lengths to ensure food warehouses and NGO feeding sites were protected. There were no recorded clashes with any of the clan militia during the occupation.

The brigade combined their security operations with civil-military projects. These tasks included digging or repairing water wells, improving roads, opening schools, and even setting up a post office where both Italians and Somalis could send letters either to Mogadishu or back to Italy. The Field Hospital *Centaur* established itself as a central hospital to not only serve the Italian forces but also care for the Somali people. Medical clinics were set up in other areas to handle minor cases of sickness which did not require evacuation to the main hospital.

Italian paratroopers garrisoned the towns of Gialalassi, Balcad, Jawhar, and Mogadishu with forces titled A, B, and C. Each force ranged from 150 to over 400 men. From each camp, forces expanded outward to secure their sub-sectors. Over 1,000 paratroopers were positioned at Balcad to serve as a reserve contingency force (mostly light armored units with rapid mobility).

The San Marco battalion continued operations with the U.S. Marines in northern Mogadishu. This combined effort was organized as Task Force Columbus with the mission to keep arms out of the Karaan market. This assignment posed a problem between the two countries due to culture, language, and different operating styles but, for the most part, was effective in enforcing UNITAF's mandates. The HRS remained fairly quiet for the remainder of Operation Restore Hope Phase III activities.

Humanitarian Relief Sector Belet Weyne–December 28, 1992

The city of Belet Weyne (or Belet Uen) lies in the northern portion of the Hiran district, split into halves by the Shebelle River, and is 210 miles from Mogadishu. It is a flat plain with moderate hills. To the north is Ethiopia, accessible by a road leaving the city headed to the northwest border town of Fer Fer, an Ethiopian Army outpost. The Ethiopian Army had been on the border since the end of the Ogaden War with Somalia; the inhabitants of Fer Fer were Ogadeni Somalis. This area of the border was still under dispute by the Ethiopians with Somalia.

The Hiran has a hot, desert climate and is very arid. The mean temperature is around 95°. The areas along the Shebelle River were used for agricultural production.

The residents of Belet Weyne are predominantly Hawadle, a Hawiye sub-clan. USC forces (SNF and SSDF) captured the town during the civil war. A small faction of Aideed loyalists garrisoned the town. Like other HRSs, the problem of starvation in Belet Weyne was due to looting, extortion, and banditry for the most part. NGOs were also charged thousands of dollars to allow their food flights to land at the airport, located east of town (a compacted dirt strip, but C-130 capable). The paved road was a unique feature, one of the few in the area, between Belet Weyne and Mataban, and known as the "China Road." The main road to Mogadishu was named the Route "Orange" line of communication.

The threats in this region were minefields, heavy weapons, and the congruence of three armed factions: the Somali National Front, the Somali Salvation Democratic Party (SSDP- Issaq), and other militia forces aligned with the United Somali Congress and Somali National Alliance (USC/SNA). Combined with the tensions over the border dispute with Ethiopia, UNITAF needed a strong military and political presence to operate in this HRS.

Because the Indian contingent was very late in arriving, the opening of HRS Belet Weyne was passed to the Canadians. To open and secure the HRS on the 28th of December, the Canadian Airborne Battle Group commanded by COL J. Serge Labbé was assigned the mission. COL Labbé established his headquarters as the Canadian Joint Forces Somalia (CJFS) in Mogadishu and selected LTC Carol Mathieu to lead the Battle Group and establish the HRS. The Canadians named this mission "Operation Deliverance."

The Canadian Airborne Regiment Battle Group consisted of around 1,000 troops assigned to three Commando battalions and one Royal Canadian Dragoon Squadron (Company A). The Canadian government sent the logistic and supply ship, the HMCS *Preserver*; the ship also provided a rest and relaxation platform for the Canadian soldiers while in Somalia.

Belet Weyne–December 28, 1992

On the 28th of December, a small force from the Regimental Battle Group and soldiers from the U.S. 2-87th Infantry (10th Mountain Division) conducted an air assault using Canada's 427th Tactical Helicopter Squadron equipped with CH135 Twin Hueys and U.S. Army UH-60s, supported additionally by Marine CH-46s. The operation was led by pathfinders followed by the insertion of infantry in a rotary-wing assault.

Once the airfield was secure, the Canadian main body flew in by C-130s with the remainder of the troops (reaching about 1,300 soldiers at the peak of the operation). ODA184 from the 1st Special Forces Group (Airborne) conducted its movement with the main body.

After two days of securing the town and declaring the HRS open, the American infantry contingent returned to Baledogle. The HRS was divided into four sectors, one for each of the maneuver units: Zone 1 for the 1st Commando Battalion (the western portion of the HRS), Zone 2 for the 2nd Commando Battalion

(Belet Weyne and its environs), and the 3rd Commando Battalion securing the southwest sector. Squadron A, the Royal Canadian Dragoons, were assigned a smaller sector in the northeast. A total of seven outpost camps were established to enhance security patrolling throughout the HRS (for instance, the Dragoons established Camp Holland in their sector).

The Canadians were equipped with the MOWAG six- and eight-wheeled armored vehicles (Grizzly, Bison, and Cougar). These came in different variants: mounting a cannon, mounting a mortar, and one variant for infantry transportation mounted with light machine guns.

The challenge in the area would come from the USC/SNA, mostly. COL John Hussein led militia around Belet Weyne (Aideed supporter), and Colonel Omar Jaua commanded the 1st Division of the USC/SNA in the Galciao region. Colonel Labbé developed an overall security strategy for the sector based on its complexity. First, make contact with the local leaders and enforce placing heavy weapons into cantonments; this included contact with the Ethiopians at Fer Fer. To conduct long-range reconnaissance and assessments of the Ethiopian border, the Canadians requested U.S. Army Special Forces to assist them. Major Carroll (ODB560) was assigned to support the Canadian Regimental Battle Group and headquarters with his five SF teams, pushing the teams out depending on the commander's intelligence requirements. They conducted their first meeting with the Ethiopians at Fer Fer on January 5, 1993. A big task in this first phase was identifying landmines laid during the civil war. Incidences of banditry would also have to be addressed and stopped.

ODA184 was not equipped with vehicles, nor were the Canadian tactical vehicles sufficient for the long-range, route reconnaissance requirements which had been levied on other teams. Instead, the team chose to gather their assessment data through participation with ongoing Canadian security patrols when they ventured out to confiscate weapons or deliver food aid. The team proved valuable in their knowledge of various makes and types of mines found in the area and passed off this data to the Canadian forces. Through their discussions with locals during the patrols, the team was able to find out the Somali secret concerning marking minefields: three sticks laid out in a triangular shape around the mine. This information helped to potentially save many soldiers from stepping on mines.

The ODA was also limited in its operation with the Canadians due to a lack of satellite radios throughout the task force. There were no antennae kits or solar battery charging panels to boost the range and efficiency of the existing FM radio capability.

The next line of effort was civil military operations to ensure food delivery and protection of humanitarian infrastructure and employees of the various NGOs, followed by the rebuilding of schools and opening medical facilities.

Third and finally, Labbé directed his forces to work with village councils and elders to promote some form of self-reliant governance, pursue reconciliation and disarmament, and get a police force up and running.

In the first week of January 1993, SOCCENT notified the CST team of its end of mission. The team was disappointed, as they had just gotten started at Belet Weyne, and the relationship and coordination with the Canadians was superb. The team left Belet Weyne on January 10, 1993, using Canadian military air passing through Germany and then on to Canadian Forces Base Trenton in Ontario, arriving there on January 14th. ODA184 continued their travel that day and arrived at their home station in Ft. Lewis, Washington.

The Canadians turned the HRS into a successful operation. When the Indian contingent had still not arrived by transition time to UNOSOM II, the Nigerians and Italians relieved the Canadian forces at Belet

Weyne. Just prior to their arrival, Colonel Labbé was successful in getting the HRS boundary pushed out to Galciao, based on the area occupied by clan leaders affecting the HRS eastern border.

Humanitarian Relief Sector Merka–January 1, 1993

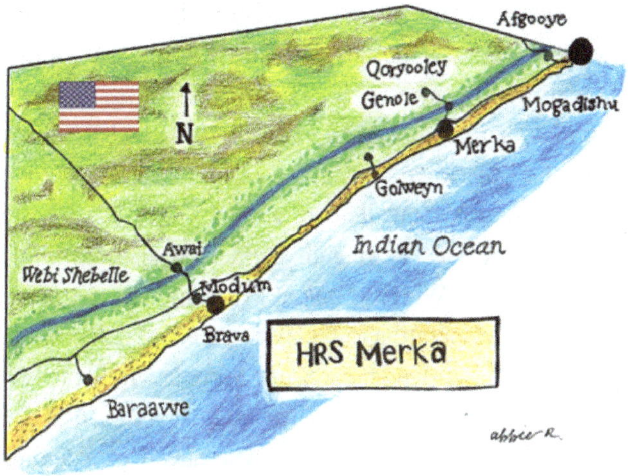

After the operation to open Belet Weyne, the Phase II objectives for opening the major HRSs with UNITAF was complete (December 28th). Occupying Merka had been delayed when the Saudi Arabian contingent changed their commitment to remain in a security role in HRS Mogadishu instead of taking the lead for the HRS.

The port of Merka and the towns along the lower Shebelle River were not part of the plan, but the CMOC and the Disaster & Assistance Response Team (DART) discovered the region had not received any humanitarian supplies for over six months. It was not for a lack of trying on the part of the NGOs; the area was rife with banditry and the mayor of Merka was considered one of the most corrupt politicians in the lower Shebelle region–looting and humanitarian aid theft became his new income as he, and other mayors, established their fiefdoms. Merka and Brava were the two largest towns in the sector and some of the few recipients of any food and goods because they sat along the main road between Mogadishu and Kismayo. At Merka, ships ceased to deliver humanitarian relief supplies.

The road was treacherous and unimproved; any convoy required rented security to navigate through a number of roadblocks set up by local bandits who exhorted tolls for each passage. Although Merka was only located 109 kilometers from Mogadishu, all of the aid organizations often experienced trips of up to five and six hours to get through. Several small villages off the beaten path were suffering tremendously.

Merka is a coastal town with a small port and was once a famous tourist resort (the nearby Sinibus beach), built on a slight escarpment off the water. It is humid and windy, which contributed to dust clouds along any road used by vehicles. The town was established by the Koofi and Bimaal sub-clans of the Dir clan, considered a war-like people. The Bimaal clan was famous for its rebellion against the Italian colonizers, implementing a guerrilla war for almost 30 years, between 1896 and 1926. This history with the Italians was the main reason UNITAF did not assign them the sector.

The water of the port is not deep (averages five feet) and is fraught with reefs; offloading a ship required lighterage using smaller, short-draft vessels. The port was well known as a place to export cattle and agricultural products, and most notably bananas and sorghum (Merka was called the "Port of Bananas" by local Somalis). Other goods traded included ivory, slaves, and animal hides. The lower Shebelle River valley provided fertile soil and moisture (seasonally) to allow for the establishment of large banana plantations, connected to Merka by a small railroad. The town was serviced by a small airfield; further to the north was the K50 airfield (named based on the count of kilometers fifty miles from Mogadishu). K50 was notorious for its nest of bandits due to it being an aerial delivery spot for the khat trade. In the early nineties, the estimated population was around 70,000 inhabitants; some of these were refugees from the civil war. Refugees were ignored by the locals and lived a very destitute life.

Merka is surrounded by the Hawiye clan, aligned with the United Somali Congress. General Warsame (the name means "bearer of good news") chaired the Southern Somalia National Movement (SSNM) and was aligned with the Somali National Front (SNF) and Ali Mahdi in Mogadishu. General Warsame's claim to fame was serving as the Director of the Somali Military Academy during Siad Barre's regime and, during the civil war, was the leader of the Somali Army's rampage in Somaliland. He was accused of genocide on the people of Hargesia (Somaliland) when he leveled the town with artillery. There was also a fundamental Islamist faction in the area, the al Itihad al-Islamiya (AIAI). Prior to the arrival of UNITAF, the SNA controlled the town of Merka but had been overpowered and pushed aside by AIAI fundamentalists. The USC/SNA also controlled Brava.

UNITAF Fragmentary Order #8 directed the Italian Forces Somalia to be prepared to secure Merka (Objective 10). With the Italian Force focused on moving in and consolidating around the Gialalassi HRS, it was clear the operation to secure the Merka HRS would not take place until after the New Year. Additionally, no good amphibious landing area existed near Merka; the best place for such a landing was several kilometers south of the town along the coast. The UNITAF staff found their solution once the Canadian and 10th Mountain Forces secured Baledogle and Belet Weyne ahead of schedule; the 2nd "Commando" Brigade of the 10th Mountain Division (LI), now pausing back at Bale Dogle, was ordered to seize Merka. UNITAF tasked Italian Forces Somalia to provide forces to the Canadian Commando Brigade no later than the 30th of December in support of the operation. The Commando Brigade began planning COAs on the 29th, tasking the 2/87th Infantry, the Aviation Brigade, and a company of soldiers from the San Marco battalion to conduct the operation: seize and secure Merka. The combined ground and air assault to Merka was launched on the 31st of December.

Brigadier General Arnold, the 10th Mountain Division (Light) commander, sent his forces on each and every mission with his "four nos" directive: no crew-served weapons, no technicals, no roadblocks, and no bandits.

Task Force 2-87th consisted of two companies of infantry (Company A and Company C). It had its organic 81mm mortars and an attached Brigade Long Range Reconnaissance Unit (LRSU). The battalion was commanded by LTC James E. Sikes Jr. Company B remained behind at Baledogle, Condor Base, to act as a reaction force.

The 2/87th conducted the air assault with its two companies (one company remained behind to secure the airfield at Bale Dogle) and quickly secured the port and airfield without incident. The battalion established their headquarters and base camp outside of town on the site of an old police barracks and awaited the ground convoys from the battalion trains and a food convoy destined for Qoryooley being escorted by the Italians, departing from Mogadishu. The provisional SF CST to the Italians from the 10th SFG(A) participated in this convoy escort to provide liaison and close air support to the Italian company. Once the convoy arrived safely, the Italian naval company departed and returned back to Mogadishu.

The contingent from Saudi Arabia was originally tasked to open the HRS on December 27th. For some unknown reason, the Saudi Arabian government demurred in the assignment, preferring to remain at the port area and airfield in Mogadishu, closer to logistics.

On the 2nd of January, the battalion hosted President George H.W. Bush. He met and talked with as many soldiers as he could and shook their hands before departing to Baidoa.

The 2/87th conducted operations throughout the HRS, patrolling extensively throughout the Shebelle River valley, eradicating roadblocks and deterring banditry.

The lower Shebelle River valley was about 300 kilometers in length, and very populated. It was considered the "breadbasket" of Somalia. Field after field of agricultural products ran the length of the valley, causing many soldiers to wonder why the Somalis were starving. Even cornfields stretched as far as the eye could see. The area was abundant in water wells, creeks, and irrigation canals built up over a couple hundred years. After leaving the coastal escarpment, vast forests stretched northward.

The significant terrain in this region was human terrain, made up of internally displaced populace and refugees. Several large camps of refugees in stick huts or shelters made from the ubiquitous blue tarps issued by the relief agencies dotted most of the small villages. Surprisingly, most of the local Somalis were indifferent to their plight (ignored by others in the "not from my clan" syndrome).

Most of the militia factions relied on logistics and arms from numerous caches in the town of Afgooye. Although Afgooye fell just over the boundary of the Merka HRS into the Baledogle HRS, the illicit activities at Afgooye supported numerous factions in the Merka HRS. The 2/87th conducted an air assault on Afgooye and seized several large arms caches at K50, most buried underground in huge shipping containers. The caches included 106-millimeter recoilless rifle rounds, mortar rounds, RPGs, and a vast amount of small arms. They immediately followed the Afgooye raid with numerous cordon and search operations in the villages of Kurtenwaarey, Baraawe, and Qoryooley.

Master Sergeant Johnson of ODA316 found little to do at Mogadishu after somewhat of a letdown in working with the Moroccans at the airfield. The Moroccans desired to conduct a static presence mission as part of UNITAF. With the team now idled, Johnson looked for other opportunities to contribute. When he discussed the matter with SOCCENT, he learned of the impending trip to Merka and volunteered to participate (he spoke with some of the Civil Affairs units at the Embassy compound and learned of the ongoing CMOC mission). After some wrangling, Master Sergeant Johnson was able to include his team with the Italian naval infantry company from the San Marco battalion assigned to assist the 2-87th with pushing relief aid into the sector.

The detachment first worked to assist the Save the Children NGO at Qoryooley to deliver food aid. They operated in this area for a short time then linked up with ODA526 near Merka to assist with area assessments and lend a hand to any relief organization needing help.

Around this time, ODA526 was also out of work; the Egyptian contingent assumed the duties of the Saudi 5th Airborne at the Mogadishu airport main gate. Captain Tim Williams tried a few limited patrols outside the perimeter but, when the Saudis were not that enthusiastic about it, also looked for more work for his team. He too heard of the expanding humanitarian mission to Merka on the 29th of December.

When it was first thought the Saudis would execute the occupation of HRS Merka, to be performed the 6th through the 14th of January, Captain Williams began anticipating the trip after it was confirmed by the Saudi S3, Major Fahd. Captain Williams' team worked with a small contingent of Saudis to conduct some limited scouting around Merka in preparation (the deployment of the Saudi contingent did not happen).

Once the mission was underway for the Merka HRS earlier than anticipated, SOCCENT tasked ODA526 to bolster ODA316's mission with the 2-87th Infantry and the Italian Marines. Prior to UNITAF's insertion into HRS Merka, the team departed overland to the area on the 4th of January, borrowing some vehicles from the 5th Saudi Airborne vehicles and M60 machine guns to use on the vehicles. Once they arrived in the region they sought out the 3rd SFG(A) team and conducted a linkup at the village of Qoryooley, spending the night in the Save the Children NGO compound. The night passed with no incidents but the team often heard random gunfire, fortunately not aimed at their compound.

In the morning, both ODAs drove to the local airfield to monitor an incoming flight delivering food aid. They watched as a UN-chartered C-130 offloaded bags and pallets of food to the NGO. All of the material was hand carried off the aircraft, with the ODAs providing security until it was all loaded on trucks for transport. They were assisted by a small contingent of 2-87th Infantry, supervised by Major Stanton, the battalion S3. As the infantry recovered back to Merka, the ODAs moved off and began their area assessments of the surrounding smaller villages after assisting with handing out food in Coryooley.

They conducted a reconnaissance to Kurtunwaarey and Arbowoherow. Kurtenwarrey was reached by driving down a potholed dirt road and crossing the Shebelle River. The town was constructed of cinder block buildings. There was a warehouse for aid relief supplies storage, run by the International Committee for the Red Cross. Due to mud and standing water on the roads, the team could not complete their assessment of Atafuuroq or Buulo Warbow. They returned to their RON site at the Save the Children compound while witnessing a riot from Somalis as the International Red Cross was attempting to pass out food. Again that night, random gunfire punctuated the silence.

The next day, ODA316 conducted a recon mission outside of Qoryooley; they received random gunfire twice during their drive. ODA526 conducted a reconnaissance and assessment of the villages of Abdi Ali and Furuqieey, followed by inspecting the armory in Merka. At the end of the day, Captain Williams stopped off at the 2-87th Infantry headquarters to pass on the information the ODA had gleaned to date. He was informed the battalion would be leaving shortly as they anticipated a relief in place by the Saudi Arabians (the original mission to open the HRS). The remaining mission for 2-87th was checking out the town of Baraawe.

Captain Williams volunteered ODA526 to conduct the reconnaissance to Baraawe and departed to the town on the 7th of January. Upon their arrival they noticed the absence of any militia activity and no Technicals roaming about. Meeting with the mayor of the town, they assessed the population to be pro-U.S. and aligned with Ali Mahdi. The mayor did confirm they were in vast need of humanitarian relief supplies. The detachment spent the remainder of the day surveying local NGOs, who expressed their concern as they had heard reports of fighting somewhere between U.S. forces and Somalis, leaving U.S. soldiers dead and several Somalis captured and killed. Captain Williams assured them he had heard no such thing from his intelligence sources to calm their fears.

ODA526 spent a peaceful night in the town with no incidents (once again, only hearing random gunshots nearby, the new normal for Somalia) and reported their observations to the 2-87th staff S2 intelligence section the following day. By this time, ODA316 was informed by SOCCENT their mission was over; they returned to Mogadishu and redeployed home to Fort Bragg.

On the 8th of January, ODA526 conducted reconnaissance of the Shebelle region. They conducted an assessment of the village of Mardhabaan, followed by a move to link up with the long-range reconnaissance unit attached to the 2-87th, in Baraawe. They passed the results of their day's efforts to the LRSU. They then stopped by the infantry headquarters and passed off the results of their patrol to the battalion intelligence officer, the S2 Captain Mike Klein.

On the 10th of January, ODA526 returned to Mogadishu. When they stopped by to debrief SOCCENT, they were tasked with a new mission: move to the Bardera HRS in support of the U.S. Marines.

By February, HRS Merka was quiet and humanitarian relief activities were in full swing. On the 1st of March, UNITAF adjusted the Merka HRS boundary to include the town of Afgooye. Banditry was down, heavy weapons and Technicals were in cantonments, and relief supplies were getting through. The battalion did seize one bizarre "Technical," an old Italian *Oto Melara* 6614 armored car belonging to a cranky Somali. The battalion refused to return it to the elderly gentleman.

On the 9th of April, the 1st Brigade ("Warrior"), 10th Mountain Division (LI) assumed control from the 2nd Brigade for the HRS. On the 28th of April, the Pakistani 6th Punjab Regiment took over the Merka HRS in preparation for their mission with UNOSOM II. The Brigade would see the Special Forces team later onduring February operations in HRS Kismayo.

Operation Restore Hope had successfully accomplished the first two phases of the plan ahead of schedule. Phase III objectives began–expanding the security within each sector. President George H. W. Bush conducted a trip to visit the troops over the New Year holiday.

5

Operation Restore Hope: Phase IIIa, January 1993

In the guerrilla areas, the governing authorities should commence what we shall call a territorial offence. As in the cases of territorial defense and consolidation, territorial offence will require assignment of small military detachments to a large number of specific zones. Although these detachments should establish local operational bases, they should not be garrisoned in posts. Rather, they should continuously "nomad," using "whirlwind" (tourbillon) type tactics–as the French describe them.

— John J. McCuen, *The Art of Counter-Revolutionary War*

By the beginning of 1993, the starvation in Somalia had been addressed and the situation throughout southern Somalia was stable and improving. President George H.W. Bush thanked the troops during his four-day visit over the New Year. The nation turned its attention to the upcoming January 20th inauguration to swear in President-elect William Jefferson Clinton into office. When Iraq once again refused to allow UN weapons inspectors entry into Iraq, U.S. forces fired over forty cruise missiles at suspected weapons factories in Iraq; Iraq conceded and allowed the weapons inspectors back into the country. In the continuing breakup of the old Warsaw pact, Czechoslovakia broke apart into two newly independent, democratic countries—Slovakia and the Czech Republic. In another sign of lessening old East-West tensions, the Chemical Weapons Convention sponsored by The Hague in Geneva was signed mid-January. The world media continued to support and focus on the successes of the Unified Task Force in Somalia. The Unified Task Force (UNITAF) began phase III operations, essentially mopping up in the Humanitarian Relief Sectors (HRSs) and putting the house back in order to turn operations back to the UN.

The Special Forces Company in Mombasa anticipated moving its operations into Somalia to support Operation Restore Hope. ODA562 was already in Mogadishu, having departed earlier when tasked to serve as a coalition support team for the Pakistani contingent in Mogadishu. The ODB560 advanced party (ADVON) flew into Mogadishu on the 27th of December to meet with Colonel Smith at the Joint Special Operations Task Force (JSOTF). Major Carroll, his Company Sergeant Major, the company operations sergeant, and the team leaders and their team sergeants learned of their new mission to support the Canadians in HRS Belet Weyne.

Major Carroll operated in Mombasa with four of his Special Forces teams, but as downsizing to match the mission requirements of the Airborne Security Augmentation Teams (ASATs) continued, along with attrition from his men when some were sent back to the United States to attend schools or fulfill other commitments, the teams averaged about six or seven Special Forces operators each. Major Carroll consolidated the unit and spread out the remaining operators among the teams, giving him an effective maneuver force of three teams on the ground.

He received his orders to deploy his company to Belet Weyne to work alongside the Canadian Airborne Regiment. At first, the Canadians anticipated deployment to Bossaso to support the UN; they painted all their vehicles white and marked them "UN" for visual recognition. When the Unified Task Force began its mission, there were no projected operations intended outside the nine Humanitarian Relief Sectors in southern Somalia. The Canadian commander chose to deploy the Battle Group to the Belet Weyne HRS.

With the Coalition Support Team (CST) mission in Mogadishu winding down, SOCCENT also chose the Belet Weyne sector as the most important place to deploy any SF capability. The number one concern on the part of UNITAF was surveillance of the Ethiopian activities along the border; the Ethiopian outpost at Fer Fer was only thirty kilometers from Belet Weyne. UNITAF had heard some troubling reports that the Ethiopians were conducting cross-border operations into Somalia to support various clan factions opposed to Aideed and who were willing to challenge any Islamic fundamentalist threat against Ethiopia. This remained a festering problem, since the tensions of the Ethiopian-Somali war over the Ogaden Region in 1977 still remained a concern. The region was a large conduit for Somali refugees living in Ethiopia and transiting into Somalia, purportedly providing support and arms to separatist factions opposing Ethiopia. It was notoriously famous as a corridor of illicit trade activities, such as smuggling. Fer Fer, on the border near Belet Weyne, was populated by Ogadeni-clan Somalis (the clan population around Belet Weyne was Hawadle).

There was also a vicious stew of United Somali Congress and Somali National Front forces. In addition, the USC in the area stood in opposition to Aideed's Somali National Alliance (SNA). Even some Aideed factions were opposing him to pursue their own goals in the region. In short, it was an area still wrapped in clan fighting. Unfortunately, the eastern boundary of the Belet Weyne HRS ended short of where the fighting was occurring. UN Operations in Somalia (UNOSOM) were highly interested in the Galciao region and still wished to push their operations into northern Somalia. Anything they could ascertain early about the area was of the highest interest, even though the Canadians and Unified Task Force (UNITAF) did not want to exceed their mission parameters.

Little else was known about the situation in the remainder of the HRS outside Belet Weyne. The SF teams had proven their ability to conduct long-range reconnaissance missions and provide area assessments to the UNITAF intelligence database, where no other sources were able to penetrate, perfect for the employment of SF in this sector. Importantly, the Canadian contingent was also well-sourced to provide a base of operations for the SF Company and provide logistics.

The relationship between American Special Forces and the Canadian Parachute Regiment was strong. It began with the combined 1st Special Service Force formed in World War II as a Canadian-U.S. military unit, gaining its fame through conducting special operations and commando-like raids, and earning the nickname given to them by German forces in Italy as the "Devil's Brigade." Each year the organizations took turns hosting one another in their country to celebrate this association and to honor the remaining Veterans ("Menton Days"). SF teams fit hand-in-glove with the Canadian paratroopers deployed throughout the Humanitarian Relief Sectors.

The C Company ADVON returned to Mombasa, prepared the unit for onward deployment, and flew into Mogadishu for movement on to Belet Weyne. Major Carroll reorganized the unit back into four SF teams, anticipating more operators, equipment, and vehicles arriving from the States. ODA562 remained with the CST mission to the Pakistanis and ODA526 continued to support the Marines in Bardera. Major Carroll flew into Belet Weyne with the company staff on the 29th of December. The teams followed, deploying within two days via C-130 into Mogadishu then on to Belet Weyne.

On January 3rd 1993, CWO2 Ronald E. McNeal, Detachment Commander for ODA561, received a 2/5th SFG(A) tasking directing him to prepare his mounted ODA for overseas deployment in support of Operation Restore Hope. The order also alerted additional team members to fill out the roster in Somalia and replace Major Carroll's shortages in personnel. ODA561 had remained behind at Fort Campbell during the Operation Restore Hope mission and would now rejoin the company.

ODB560 in the Belet Weyne Humanitarian Relief Sector

Major Carroll now had his four teams from Mombasa with him to maneuver with. He arrived with a mixed fleet of Humvees, some of them Desert Mobility Vehicles (5th Group modified Humvees) and some cargo Humvees. For the DMVs, half of them mounted the .50 caliber and half mounted the MK-19 40-millimeter automatic grenade launcher. The DMV variant was referred to as a "hatchback." There was also the plain cargo-version Humvees which had been scrounged or belonged to the support sections in the battalion. If utilizing a "soft-top" cargo Humvee, Squad Automatic Weapon light machine guns (SAW) were used for firepower. Normally two SF operators rode in each vehicle along with one translator, comprising a mounted patrol. Before more DMVs arrived from the states, each vehicle often carried four to six operators, depending on its configuration.

Initially, conditions at the airfield were harsh and logistics were severely limited. Drinking and bathing water was rationed. After a week, as more support flowed into the area, the Canadian commander designated a new camp location located on rising ground beyond the east side of Belet Weyne and across the Shebelle River, about two miles from the airfield. The new location was near a small suburb called "Village Kilometer 115" and sat near the main road running south to Mogadishu. It was a dry, dusty piece of sand, surrounded by scrub and low ridges, but the Canadians were able to occupy an old nongovernmental organization (NGO) building to house their headquarters. There were a few other buildings, mostly in ruins. Major Carroll had his company utilize the concrete pads (foundations) of buildings which had been razed to set up their tents.

Carroll knew his first priority for the unit was surveillance along the Ethiopian border and to make the attempt to contact the Ethiopian military manning the border post at Fer Fer (estimated at five hundred soldiers). The goal for UNITAF was to confirm or deny reports, mainly from the NGOs in the area, of Ethiopians operating cross-border in civilian clothes, or conversely, ascertain the activities of the three factions in the area raiding north across the border into Ethiopia—the USC, SNA, and SNF.

The area was known to be heavily mined. Thousands of mines and multiple minefields were used during the Ogaden War in the 1970s. The factions were still fighting one another and laying more mines. They were also using Technicals with heavy weapons, along with tanks and artillery.

The AOB (Advanced Operating Base—the SF Company designation for field operations) immediately set up camp near the airfield and became operational with the first missions conducted by ODAs. The ODAs began combined security patrols inside of town with the Canadians, both vehicular and dismounted, in the immediate area of the airfield and into the town of Belet Weyne.

Major Carroll initially prioritized his unit's efforts on the tactical missions assigned to the teams and the meetings with Ethiopian military on the border. Carroll divided his mounted patrols into search areas approximately twenty to forty kilometers square. Two to three patrols were out operating at any given time. A mounted patrol was designed to last for up to six to eight days, given the route and the tasks in the mission order. Sufficient fuel and water were loaded to support the length of these operations; if the team needed to conduct extended operations, the B-team staff or an idle SF team ran resupply missions by driving to a rendezvous sight picked out in the patrol sector.

ODA564 members recalled that teams were designated for reconnaissance patrols through a warning order developed by the B-team. Major Carroll received guidance from both the Joint Special Operations Forces commander and through discussions with the Canadian commander as to his objectives. In short order, mission analysis was conducted, the team briefed back Major Carroll on their course of action, and the team continued preparations for the trip. Sufficient ammunition, water, fuel, and food, plus some fudge factor for a day or two more, was loaded onto the vehicles. The teams were not supported from the Canadian helicopters for their trips unless a delivery was being made to the field where they were collocated with a Canadian unit. Then they could radio in their wish list to the B-team to have it delivered to the helicopter before take-off. Ideally, the team departed with one DMV mounting the M2 .50-caliber machine gun and one DMV with a Mark 19, 40mm grenade launcher. The light anti-tank rocket (LAW) was available for issue to the teams if they saw the need. All personnel carried their personal weapons: the M16A2 rifle (5.56

millimeter) and an M9 pistol (9 millimeter). Hand grenades and smoke grenades were also loaded; hand grenades were useful for destroying mines in place.

The medic on the team packed more than the team needed, anticipating treatment and first aid for Somalis throughout the journey. With the vast amount of mines and unexploded ordnance in the area, the engineer brought extra explosive materials (TNT and C-4) to destroy whatever the team found (within reason). As part of the kit, teams were issued mine-detectors. Unfortunately, on ODA564's first trips out into the sector, they were not issued one and had to use hand-probing whenever it was necessary to locate a mine (at best, they just blew them in place). All mines found intended for removal were first checked by the engineer for anti-handling devices before lifting or moving them.

The PRB-3 Belgian antitank blast mine, the TM-46, the TM-57, the PPMISR-2 Czechoslovakian-made mine, the antipersonnel style called Italian V, the American-made M19 and M16, and even an unidentified plastic case blast mine were found. There were no nationalistic patterns used for laying the mines, although some teams found the minefield patterns to be laid according to American military doctrine for obstacle emplacement. In one incident, ODA564 found 200 PVR-3s stacked neatly and another time found 3,000 multiple types of mines (35,000 tons of explosive force). The Somalis laid some mines with no detonators or fuses, merely as an attempt to scare off the opposing forces.[1]

The first mission in HRS Belet Weyne was executed by ODA563, led by MSG Les Jolley. On the 3rd of January, his patrol was ordered to attempt a linkup or contact with the Ethiopians at Fer Fer, about twenty miles away to the north. The patrol drove to within four kilometers of the border outpost, awaited permission from Colonel Smith at SOCCENT to make contact, then sent a note written by their translator to a local Somali who walked it to the Ethiopians for delivery; he returned with news that the Ethiopians had agreed to meet the next day. They only asked for U.S. military forces to approach the position with weapons turned to the rear, then dismount and approach the outpost on foot. Master Sergeant Jolley returned to camp and reported this information to Major Carroll.

Major Carroll rode with ODA562 in the morning and reached the outpost around 0900 hours. They met the Ethiopian S3 Operations Officer of the border unit. They were able to exchange information on activities in the region over hot glasses of tea. The Ethiopians described their mission as halting cross-border infiltration into their country and described how the SNF conducted raids to steal camels in Ethiopia. The S3 also informed Major Carroll that the USC and SNF often clashed with one another, even exchanging fire in artillery duels. Most of this activity was occurring in Balenbale, to the east about sixty kilometers away. They also informed Major Carroll about General Ghani, operating in central Somali near the town of Abud Wak, north of Dhusamareeb on the Ethiopian border. General Ghani's SNF force was made up of ex-Somali soldiers from the National Army.

The two delegations ended their meeting over lunch and exchanged pleasantries. Major Carroll and the patrol returned to camp to brief the Canadians. He surmised his next objective was to confirm the intelligence and meet with Colonel Hussein, the leader of the USC/SNA in the area.

ODA563 provided escort and security to additional meetings, now with both Major Carroll and Lieutenant Colonel Mathieu. During their drive to and from Fer Fer, they also spotted mines and minefields and recorded the various trails and roads seen during their reconnaissance. The Chinese Highway was the only road leading northeast; the locals remembered the Chinese who built the road. It appeared to the team as impassable due to a partially blown bridge and heavy mining of the bypass. The Chinese Highway would have to serve as one of the best routes selected as it was the only paved road of substance in the region and the only road leading to Matabaan and Balenbale, where ODAs would have to operate in the future.[2]

ODA564's initial missions included escorting Major Carroll for the meetings in Fer Fer with the Ethiopians (Major Carroll's third trip). Sergeant First Class Kilcoyne, the Detachment

Commander, would ultimately make three of these escort and security trips with the team. MSG Kilcoyne explained the procedure once they arrived at the border post.

> We would talk with the guard and tell him why the unit was there. He would tell us to wait in a hut while another soldier located their commanders. We drank tea with the commanders and sometimes had lunch, exchanging information. They shared information concerning the Somali fighting up north. We were the third element to link up there after ODA563 made initial contact, followed by ODA562.
>
> The same lieutenant colonel was there as when the ODA563 team was there and he would always be dressed in civilian clothing wearing an Eastern semi-automatic bloc pistol on his hip. Speaking through translators they would go from English to Somali and then translate from Somali to Ethiopian dialect, sometimes needing three translators. The people in the sector were both Somalis and Ethiopians. We never saw their actual forces except for guards and the people that escorted the unit and the commander. They were disciplined. They had a lot of AK's and AKM's but no uniforms, looking like a militia instead of an army.
>
> The Ethiopians were extremely interested in getting the border cleared of mines in the sector of Fer Fer. There had been a couple of wars there. Both sides laid mines but there was no record of where. They also wanted to know if equipment and vehicles were in cantonment areas. They took me and Captain Barber northwest of Belet Weyne to the small village of Fallabab to show us where mines were. They were still visible in the ground and rusting. Because the people are nomadic there was no way to tell how long the mines had been there. Their history is oral. They only reported the location of the mines, not defusing any. It was linear, a triple belt.[3]

ODA565 led by SFC Ken Barriger, a six-man team, also conducted the escort mission to Fer Fer and wider area reconnaissance and assessments in the Belet Weyne HRS.

The U.S. Army Special Operations Forces Mission

The SOCCENT Joint Special Operations Task Force began preparations to end its mission in Somalia and deploy home. On the 8th of January, the two-man SOF liaison team to UNITAF departed. SOCCENT began its transition from coalition support operations to employing the Army Special Forces in the following manner during their remaining time, per UNITAF's guidance:

> The Joint Special Operations Forces (JSOFOR) plans and conducts Special Operations in Somalia to support Unified (UNITAF) Task Force Humanitarian Relief efforts.

Also included in the mission were the following operational objectives:

1. Make initial contact with indigenous factors and leaders.
2. Provide information to UNITAF on potentially hostile forces to aid in force protection operations.
3. Provide area assessments to assist with planning for future relief and security.
4. Conduct border surveillance along the Somali-Ethiopian border region to determine the nature of Ethiopian military activities (ODB560's primary mission).[4]

Mary Nemeth, a reporter for the Maclean's Weekly Newspaper, wrote about the large amount of mines and ordnance recovered by the Task Force in her article on January 18, 1993 ("Coming Back to Life"). The U.S. Army Special Forces teams assisted in this endeavor.

"Coming Back to Life"

In Belet Uen, 330 km north of Mogadishu, Canadian soldiers last week located what may be one of the largest single weapons caches discovered by coalition troops: an armory that included more than 3,000 hand grenades, 300 rocket-propelled grenades and 27 long-rang multiple-rocket launchers. With no place to store the cache, which belonged to a local faction of one of Somalia's largest warring parties, the United Somali Congress, the Canadians simply padlocked the building. But Lieutenant-Colonel Carol Mathieu, commander of the 85 Canadian and 55 American troops stationed in Belet Uen, pointed out that rival forces loyal to deposed president Siad Barre, in areas not pacified by coalition forces, are not being disarmed. "If we disarm these people, they will just be more vulnerable to attack."[5]

ODA561 Arrives

Sergeant First Class Kim J. Breed, the Detachment NCO; Chief Warrant Officer 2 McNeal; and the other members of ODA561 left for Somalia on January 12, 1993, arriving in Mogadishu on the 13th of January. Both had been to Somalia earlier in their careers during the conduct of a Foreign Internal Defense Military Training Team (MTT) to run a Somali airborne course. Colonel Smith kept them in Mogadishu for three days, needing some extra hands to help guard some Somali VIPs. Thoroughly discouraged with that task, they hopped a C-130 for the ride into Belet Weyne on the 16th.

Extra vehicles (DMVs) and additional SF operators from the 2/5th were also on the deployment from the United States to fill out Major Carroll's shortages in Company C. To help with the operations, CWO3 Doug Jackson also arrived (C Company Operations Technician). Major Carroll was ably assisted by the "Chief," a very wellexperienced B-Detachment Tech. Chief Jackson took one of his separate tasks as traveling with the Deputy Commander of the Canadian regiment to assist in conducting meetings with NGOs and elders. Chief Warrant Officer 3 Jackson also oversaw the support and running of the Company's operations in the field. ODB560 was now firmly established in HRS Belet Weyne with six operational detachments.

The team collected the equipment and supplies needed to begin missions and became operational by the 18th of January. The first area assessment mission took place south in Bulabarde along the Shebelle River. The team only sought out general information on this first mission, such as the weather, population makeup, and health and welfare of the people. The mission lasted eight days before their supplies began to run out. During their assessment, they found the people short on food due to the current dry season but not starving as they had heard. The NGOs were providing subsistence levels, but as was the case with undernourished people, the hungry constantly asked for more food.

The only threats to ODA561 during their recon patrols were land mines. While out on forays, they could see old mine holes where warring factions would hastily plant mines upon retreating from opposing militias; after the skirmish was over, the mines were recovered. The team noticed all the mines they found were old and not recently laid, which in its own way provided the comfort of knowing they were not riding into an ambush.

During all their assessment missions, ODA561 found the Somali people polite and friendly; no one was hostile to them. The team experienced no engagements with gunmen, bandits, or militias.[6]

The SOCCENT Transition with 2/5th SFG(A) – From JSOFOR to Army SOF

As the SOCCENT JSOTF Commander and staff completed their remaining time before transition to the Army Special Operations Forces contingent, the 2/5th Commander, Lieutenant Colonel Faistenhammer, deployed on January 12th to Mogadishu with his battalion staff to serve as the ARSOF Commander (SOCCENT was scheduled to completely depart by the 31st of January), less the PSYOP and CA contingents. Most of the CST missions were complete, and the few remaining CST detachments from the other SF Groups departed by the end of January. FOB 52 (-) was now the Army SF headquarters, making Lieutenant Colonel Faistenhammer the senior SOF advisor in theater to General Johnston. It would now be the ARSOF responsible for all missions with regards to Major Carroll and his SF Company.

Bill Faistenhammer Jr. explained his role and how decisions on the various missions were formulated.

> The decision to launch me and the FOB (-) (Forward Operating Base—an SF Battalion configuration in the field when deployed) was made as I saw the need to create a buffer between the UNITAF staff and Major Carroll and the B-team, now that the JSOFOR was gone. My senior leaders agreed with that assessment. We occupied the JSOFOR building on the Embassy compound when they left, next door to UNITAF.
>
> I developed missions for the teams and the company in a unique way. Each day BG Zinni worked out with his staff in an area for weightlifting, generally around 1600 in the afternoon. Since I'm a big weightlifter, I went with the staff to do the same thing with me as Zinni and his folks were leaving. This gave me the opportunity to develop a relationship and have several small discussions concerning the role of the SF and where they could be best utilized. He'd state his concerns, I'd provide my input. I found him to be good in allowing me free reign to decide how to accomplish an objective for UNITAF.
>
> One thing was clear; UNITAF was very interested in information on activities along the Ethiopian border. I think that was the number one reason the B-team was sent to the Canadian HRS.

There were other times I ran into the Generals of the Marines or the Army. They would ask if I could help out in an HRS with an A-team; their forces could use them. The teams were well known for their assessments in HRSs earlier on. This is how the mission with the Marines in Bardera developed, and also the sending of Major Carroll as a small SOCCE to Kismayo, with two teams. When the situation developed in Kismayo, General Magruder asked for teams. Again, I wanted a buffer between the 10th Mountain Task Force staff and the teams, so I sent Carroll as the SOCCE (-) with a small staff to handle that while the teams conducted their missions.

Plus, the six teams at Belet Weyne were all standing on top of each other and I wanted to get some of the teams out of the base doing something.

We all at Fort Bragg would have liked to have more forces in this Phase, but my other company was on an exercise in Qatar, along with my S-3.[7]

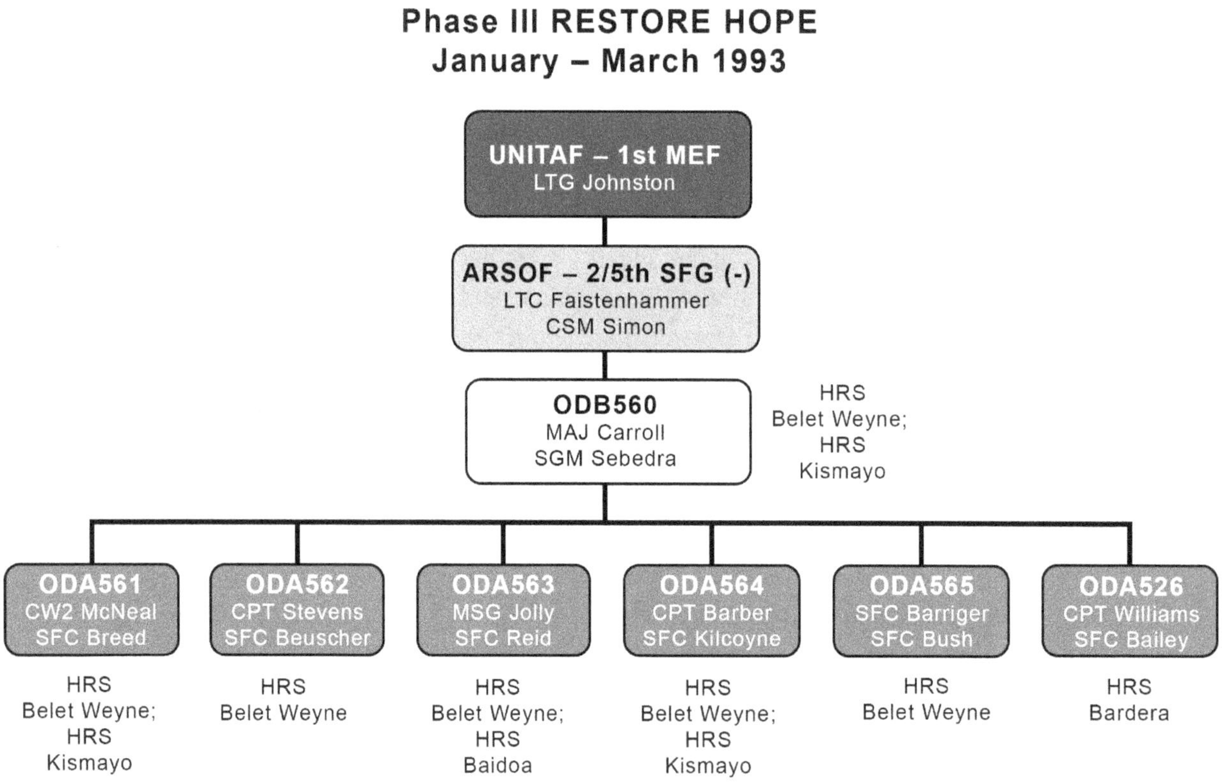

More importantly, SF would be employed as intended and trained—to conduct Special Reconnaissance (SR), Direct Action (DA), and Unconventional Warfare (UW). Foreign Internal Defense (FID) is an additional doctrinal mission in special operations, but for the remainder of Restore Hope the requirement did not surface, as the Special Forces teams were deployed with Marines or coalition partners who were very capable and did not require military training from the teams.

Special Forces Activities in Humanitarian Relief Sector Belet Weyne

ODA562 relieved ODA563 for the Ethiopian contact mission in Fer Fer in mid-January. The information on land mines and routes was passed on to them, and ODA562 moved to linkup with the Ethiopians. All the meetings were cordial.

On another reconnaissance, ODA564 and ODA565 found a demolished bridge in the hills along the Chinese Highway. Apparently, this route was used by Siad Barre's retreating forces who attempted to destroy the bridge during their flight. Over time, local Somalis established a bypass route near the bridge, but due to factional fighting, someone mined the road. One of the local village elders complained to the detachment on the lack of humanitarian relief supply trucks reaching his village. When the detachment explained the situation, the Somali elder went with the team to the site and pointed out where the mines were buried. ODA565 destroyed three of the mines closest to the road, opening access and gaining a high degree of rapport with the Somalis in that area.

The Addis Ababa Agreement

Between the 4th and 15th of January, political effort doubled as the end of Restore Hope was in sight. The UN sponsored fourteen factions in Addis Ababa for a conference to work out reconciliation, disarmament, and ceasefires. All the leaders of the factions signed the agreement. It was about this time, mid-January, when U.S. forces reached their peak strength—25,426 military personnel.[8]

ODA526 – The "Ranging" Special Forces Team

ODA526 would become the "ranging Operational Detachment–Alpha" of Operation Restore Hope. The modern term for Rangers originated in old England. Lords, barons, and kings employed constable-like men to "range" their holdings to prevent theft and hunting on their properties. The men they employed "ranged" throughout landholders' properties, in movements akin to modern day military sweeps, to apprehend lawbreakers. These bodies of men, as they became more organized, wore uniforms, and became ever more crafty and adept at their trade (like good game wardens), were called Rangers. The early colonists in America employed units of Rangers to conduct the same security activities around their isolated towns and villages, but in their case against a new threat—the indigenous natives.

The early form of frontier warfare relied on self-reliance by settlers to conduct their own defense against native marauders, establishing forts for protection whereby citizens could rally as they formed militia volunteers for defense. To forestall an attack from the enemy, "ranging" companies patrolled between forts. COL Benjamin Church was considered America's first Ranger, chartered by the Plymouth Colony Governor to employ an independent, full-time company of experienced frontiersmen to roam into enemy territory during King Philip's War (1675–1676). In the French and Indian War, COL Robert Rogers employed his Ranger unit, "Roger's Rangers," in accordance with the rules he promulgated for their employment, titled *Rules for Ranging*.

Much of what the SF ODAs conducted during Phase III of Operation Restore Hope in reconnaissance and area assessments was very similar to "ranging." Today, this utilization of Green Berets is part of the SOF doctrinal concepts used in Special Reconnaissance (SR). ODA526, with an understrength team constantly scrounging for vehicles, set the record for ranging across southern and western Somalia HRSs.

After completing its mission as the Coalition Support Team to the Pakistanis, followed by supporting operations in HRS Marka, SOCCENT tasked ODA526 to support the Marines in the Bardera HRS. Captain Tim Williams gathered as much information as he could from the Marines concerning the HRS then sat down with the team to formulate a Course of Action (COA) satisfying the Marine request for an SF team—to conduct an area assessment and be prepared to conduct a linkup with General Morgan and his forces south of Bardera. SOCCENT approved the course of action.

The next step required reporting in to the Marine Regimental headquarters in Baidoa and providing the commander with a briefing on the operation's concept. COL Emil R. Bedard commanded the 1/7th Marines in Baidoa. ODA526 flew by helicopter out to Baidoa, briefed Colonel Bedard, and sought his guidance. Colonel Bedard apparently thought the team was going to be a CST for an incoming coalition contingent scheduled to replace the Marines in the sector; in essence, they were there to conduct liaison between the Marines and a coalition contingent.

Captain Williams explained that this task was not the mission given to him by SOCCENT, which was to work with the 1/7th to conduct reconnaissance and assessments by participating in their patrols and talking to Somalis. Colonel Bedard understood and he directed them to work for the battalion on the ground, but they had to have a Marine security detail with them at all times. There was some other bad news. The Marines could not provide any vehicles (Williams gave his borrowed Saudi vehicles back to the 5th Saudi Airborne Regiment when his CST duties were completed). The team would have to request a ride with the Marines whenever they were going out on a mounted patrol.

ODA562 was initially hampered in effectively conducting area assessments and meetings with local leaders in the Bardera HRS due to a lack of their own vehicles. Pictured here, a member of the ODA rides with two attached Marine Corps radio operators, Corporal Scheiber and CPL Shaver, in a borrowed Humvee from the Marine regiment in Bardera (*Courtesy of Jose Bailey*).

Colonel Bedard agreed he wanted the team to conduct an area assessment of the HRS. He also desired the team to ascertain the potential for the locals to receive coalition training in order to build self-defense forces, or a security force, with the goal to challenge lawbreakers and take on armed militias. Williams anticipated this task would require working with the Regimental S-2 to find those areas of the population who would be amenable to developing a security organization and also speak with villagers and elders to get a picture of the threat in their area. He knew from experience with Operation Provide Relief that the NGOs were often the best source of this information, so he planned to include meeting with them and getting updates. Williams knew it would take him three or four days of gathering this information to make his first recommendation for where to begin the training of a security force.

There were logistical requirements to consider when raising a force. How many? How was it to be organized? What types of uniforms and weapons would be issued, and where would they come from? How would members of the force be paid? The team was trained in the unconventional warfare task of organizing and training an irregular or guerrilla force, so they needed an appreciation of what they had to work with before they could make an estimate of how long the training would need to be conducted.

With Colonel Bedard's guidance in hand, the team took a Marine helicopter to Bardera on the 14th and reported into Lieutenant Colonel Getz, the 1/7th Marine Battalion Commander, when they arrived. Captain Williams pitched his mission and intent for how to support the Marines. Apparently, Lieutenant Colonel Getz was not a big fan of special operations or was not used to a unit conducting independent or semi-independent operations in his area of responsibility. He told Williams there was no talking to local leaders, there were no vehicles available, and they could only travel when a Marine convoy or security patrol was leaving the camp. Lieutenant Colonel Getz also did not understand that the team worked under the operational control of SOCCENT and only under the Tactical Control (TACON) of the marines; in other words, SOCCENT ran the mission while the Marines only had local control and direction, meaning they could not change the mission.

In time, everyone came to an understanding. It was not anyone's fault, just a clash of two different military cultures. The ODA soon received a tent to live in, rations, and a vehicle. The Marine J2 intelligence and counter-intelligence sections were on board with Williams' assessment concept. Remarkably, SOCCENT felt the mission could be accomplished in three days, to be followed with a flow of more SF and Civil Affairs activities into the HRS. Colonel Smith at the JSOTF anticipated the team's return shortly along with a debriefing to develop the next step in introducing special operations in the HRS.[9]

Captain Williams was a well-respected and competent Detachment Commander with a calm demeanor and great interpersonal skills. He could get along with anyone and was not deterred from accomplishing his mission. As he and the team worked to fit in, they were accepted more and more by the Marines; the ODA was now able to ride with the commander or members of his staff to local meetings with the NGOs and village leaders. The number one topic in these meetings was the threat of banditry.

One incident occurred the day the team arrived in Bardera which did not help to calm things down with the populace. The Marines seized a shipment of khat delivered to the airfield and burned the load. The locals were not happy. That night, three bandits stole $50,000 from an NGO. Everyone suspected it was COL Barre's henchmen, since he controlled the crime and banditry in the city. However, the town population viewed him as a hero since he stopped Aideed's forces when the town was attacked. There was no way the Marines would get any participation from Somalis to pin the crime on him.

One time, Lieutenant Colonel Gertz asked the team to provide night security in the CARE compound, and four team members volunteered. The next day, Captain Williams accompanied Lieutenant Colonel Gertz for meetings with the local NGOs, clan leaders, and the regional warlord, Colonel Barre. However,

Captain Williams was still only allowed to observe the meeting and could not be an active participant in any discussions. Unperturbed, he continued to work on that dilemma. That night, the ODA members continued to conduct their security and protection mission for the CARE compound, and although several rounds were fired throughout the night, only a few hit the compound walls, causing no further damage.

As the situation improved between Captain Williams and Lieutenant Colonel Getz, the detachment commander was given permission to start conducting personal interviews to meet the MICON (Mission Concept) for the area assessment. Captain Williams and his team spent the next couple of days in meetings with the police chief and local elders, allowing the SF team to gather useful information for future humanitarian and Civil Affairs operations.

On the 15th and 16th, ODA526 met with several NGOs from a variety of different organizations. Lieutenant Colonel Getz personally orchestrated the meetings, but on one occasion the team was able to talk one-on-one with the Police Chief, so progress was made towards assessing the creation of a police auxiliary force. COL Barre was always present for Lieutenant Colonel Gertz's meetings (many on the staff assessed Barre as playing the Marines and NGOs against each other).[10]

Williams described one of the meetings when Barre was present.

> I met with Major Abdi Barre (chief of police), Lt. Abdullahi Abokar (the Deputy Commander), and Cpt. Osman Hussein (police coordinator of the district), asking them to consider what size force was needed, what were the biggest problems in the town, what equipment was present and needed, and also the experience level that was desired.
>
> The Police Chief and the Mayor indicated they would need basic training for police recruits. They provided me a letter of their concerns, number one being the provision of weapons for the policemen and the authority to carry and use them.
>
> . . . They told me of their belief the Somali police force was effective before the Marines arrived and that most of the difficulties out in the villages were because of the disarray of the police force before. As to equipment, their list included three vehicles, four radios, uniforms, nine hundred thousand shillings a month, per man, with the requirement for twenty men on duty and forty men on call (the Police Chief guaranteed me he already had sixty-one men ready to go). I reported all this to LTC Gertz and awaited his authorization to begin the program.

During these contacts, ODA526 began to learn about Islamic fundamentalists attempting to infiltrate Bardera. There was high interest in command about this threat, but as of yet, no hard intelligence was available to provide SOCCENT with proof. Colonel Smith asked Captain Williams to fly to Mogadishu with any information he had available the next time he was free. Before this could happen, SOCCENT turned over its command to Lieutenant Colonel Faistenhammer's FOB 52 (-). Since Williams now worked for the FOB 52 (-), the team was placed under the command of Major Carroll and ODB560. Lieutenant Colonel Faistenhammer was also interested in the fundamentalist threat and tasked them with stopping by Mogadishu to brief him before they transferred to Belet Weyne.

Meanwhile the team continued to conduct its mission with the Marines. The detachment mainly helped provide security to the ICRC feeding centers and guarded the CARE compound at night. One firefight occurred between the Marines and Somali gunmen on the 19th of January. The gunmen soon learned a valuable lesson about charging at Marines. The Marines fired back–one Somali was wounded and one captured. As banditry increased, the NGOs were awaiting the proposed establishment of the security force, considering it better than nothing.

On the 20th, the team was able to catch a ride on a C-130 headed into Mombasa, where they caught a C-12 into Mogadishu. After briefing Faistenhammer, the team remained in Mogadishu awaiting new mission orders. Go back to Bardera or go to Belet Weyne with the SF Company? Faistenhammer ordered them back to Bardera after discussing the matter with the Marines.

ODA562

On the 14th of January, some members of ODA562 at Belet Weyne were sent back to Mogadishu to serve as liaison officers for SOCCENT to UNITAF; SOCCENT pulled its assigned liaison officers to send back home. When the task was completed, they returned to Belet Weyne and went back to performing the scheduled security escort for Major Carroll when he visited the Ethiopians; if Major Carroll was not present, Captain Stevens conducted the face-to-face meetings (liaison) himself. ODA563 handled the tasking while ODA562 team members were in Mogadishu.

For the meeting on the 15th, Major Carroll and Captain Stevens were hosted by the Ethiopian commander, Lieutenant Colonel Menassi. One goal Carroll wanted to accomplish was to get an agreement for the Canadian commander to start attending the meetings, if the Ethiopians agreed. By this time, Captain Stevens was assigned a Somali interpreter whose name was Abdenar Darman. Mr. Darman was an American Somali who volunteered to work with the U.S. Army during Restore Hope. His knowledge of Somalia and his Somali dialect capabilities were extremely helpful to all the teams during his assignment with ODB560. The meeting was productive. The political officer attending the meeting assured Major Carroll they looked forward to Lieutenant Colonel Mathieu's attendance; they were aware he was the commander of Canadian forces in Belet Weyne.

As the entourage boarded the vehicles to return back to base, they were approached by a distinguished looking older Somali, who was one of the elders of the predominant clan in the area. He informed Major Carroll of Somali National Front (SNF) activity just up the road in Matabaan. He cautioned Major Carroll that he would have to bring back all his Technicals to the area if the threat from the SNF in Matabaan could not be stopped. He had to protect his people. Carroll returned to Belet Weyne to inform Mathieu. Meanwhile, ODA562 drove to the intersection on the Chinese Road to place an observation post for monitoring bandit activity (coincidentally, as told to Stevens by his interpreter, they were in the exact spot where the Somali Civil War started when United Somali Congress forces attacked Siad Barre's military forces).

Major Carroll received permission from Lieutenant Colonel Mathieu to move to Matabaan and try to establish contact with the SNF. Stevens explained, "Matabaan was forty seven kilometers northeast of Belet Weyne, as the crow flies, on the Chinese Highway. The people of Matabaan were loyal to the United Somali Congress, the faction led by General Mohamed Farah Aideed. Matabaan fell within the Canadian HRS, but until that day none of the coalition forces had been that far north."[11]

ODA563 met Captain Stevens and his team, combined the two forces, and moved to Matabaan on the morning of the 16th. They used the bypass at the blown bridge on the Chinese Highway and made good time with their four Humvees, averaging forty to forty five miles per hour. Stevens noted the flat terrain, cut with low ridges and wadis, dotted with scrub so typical of the entire sector, once one got away from the Shebelle River Valley. As they reached the outskirts of Matabaan, they were met by a delegation of elders who spoke with them about the SNF attacks on the village of Guri Jiri, using armored cars and Technicals. Soon, a meeting was arranged with COL "John" Hussein, the faction leader of the United Somali Congress (anti-Aideed).

The meeting was to be of significant importance to UNITAF. The two ODAs were about to pull off the first personal meeting with a USC clan leader in the Belet Weyne HRS. Colonel Hussein was considered a hero, very Patton-like, as one of the commanders who stood up to Siad Barre's forces during the civil war. Stevens noted his command presence when he arrived at their location.

> We did not have to wait long for the colonel to arrive. The front must not have been very far away. He showed up in a battered old Toyota Land Cruiser, dressed in blue jeans and an American camouflage BDU shirt. A black and white keffiya was wrapped tightly over the top of his head in the style U.S. soldiers called a "do rag." When he stepped out of his truck I could see that he was tall, nearly as tall as I, and seemed to be trim and fit. His eyes were covered by a pair of brown-lensed sunglasses. A bushy moustache thrust out under his wide, flat nose. His skin was the color of black coffee, a deep, dark black that lightened up to brown around the edges. There was an aura of command about him that made me take notice. He stood erect, head thrown back, tapping a swagger stick against his left thigh.[12]

Colonel Hussein invited the SF into a building for the meeting. Colonel "John" described the situation along his military front, ranging from Dhusamareeb to Guri Ceel down to his flank at Galcaio. Apparently there was significant heavy fighting still ongoing in the region. He informed them the Somali National Front was in Balenbale. The SNF commander in the region was General Ghani, headquartered in Abud Wak, who was a former Siad Barre loyalist. Colonel Hussein ended the meeting asking what the coalition intended to do to protect his men and the people in the region. The two ODA leaders made no commitments and promised to go back to Belet Weyne to discuss the matter with their commander, assuring Colonel Hussein the Canadians would "probably" come as they became more settled into occupying the HRS. When Captain Stevens returned to the base camp, Major Carroll understood the importance of such a high-level contact and agreed with Stevens that the ODA should return the next day to Matabaan. It was a decision which would cement ODA562's and ODA565's role in HRS Belet Weyne for the rest of their deployment on Operation Provide Relief.

The Special Forces patrol returned to Matabaan the next day and spent the night. In the morning, Major Carroll and Lieutenant Colonel Mathieu arrived at the head of the 2nd Troop of Company A, Royal Canadian Dragoons. Major Kampmann, the Squadron Commander, ordered Lieutenant Purvis, the troop commander, to garrison the town until a larger force could arrive. The Canadian contingent would grow to include almost all of A Squadron Dragoons, establishing Camp Holland in Matabaan.

ODA562 – The Long Patrol

Upon the arrival of reinforcements from Fort Campbell in mid-January, ODA562 was finally equipped with all four of its organic DMVs and Humvees and all ten men on the ODA.

The Medic on the team, SSG Jeff Davis, had been deployed to Kuwait during Saddam Hussein's threatening moves toward the Kuwaiti border and was replaced with an A Company Medic, SFC Robert Deeks.

Two other significant meetings with local militia leaders were on the schedule—General Hubaro, Colonel John's subordinate who self-styled himself as a General, operating near the airfield and the Shebelle River, and General Ghani. General (Colonel) Hubaro ran Colonel Hussein's rear area near Belet Weyne like a thug and was highly suspected of controlling the local bandits and illicit activities plaguing Belet Weyne. The coalition forces met with him on the 19th of December and warned him about the complaints from the locals and announced their intent to clamp down on criminal activity from the bandits.

ODA562 returned to Camp Pegasus, now ordered by Major Carroll to attach to the Dragoons at Matabaan. The team packed for the move and returned to Matabaan. Some armored "Grizzlies" with Lieutenant Colonel Mathieu put an end to the damaged bridge on the Chinese Highway when they were crossing it as they returned to Belet Weyne. The bridge collapsed from their weight. Now anyone using the road would have to bypass the bridge and navigate unknown minefields near the bypass road.

The next formative meeting with Colonel Hussein was held on January 17th to outline the details for a ceasefire on the front and conduct a prisoner exchange between Colonel Hussein and General Ghani. During the meeting, Sergeant First Class Deeks checked the prisoners out for any medical problems. Later, after lunch in a local restaurant, Colonel Hussein invited the SF team for a tour of Cali Xasan to see the site of an SNF massacre. This was also the location of the prisoner swap. They were one kilometer from the front.

On the 19th, the Dragoons closed their force into Matabaan; ODA562 hosted two medical officers who had flown up from Mogadishu to assess hospitals and clinics in the area, especially those caring for wounded Somali soldiers under Hussein's command, located near the front lines in Dhusamareeb, the site of an old tank repair depot.

ODA562 visited Dhusamareeb with the two medical officers (Doctor Mount and Captain Holder) on the 21st of January. Before leaving town, the A-team assisted the Canadians with running to ground some local bandits (later identified by Colonel Hussein as some of his men who were absent). Major Carroll was unaware of the out of sector trip because Stevens could not get radio contact with him that day, but thought the recon was of importance to the area assessment.

Colonel Hussein escorted the team with his security detail, first stopping in Guri Ceel to meet with and introduce Stevens to the Governor, Doctor Ismail Halane, the governor of the Galgaduud region. The governor was going along with the medical site visit to Dhusamareeb.

During the drive they passed numerous military vehicle wrecks and abandoned tanks lying littered about the plains, evidence of the heavy fighting during the civil war years. As they rolled into town, they first passed Barre's former armor base. Two American tanks and one Soviet-made 100-millimeter anti-aircraft cannon adorned the entrance, now as rusting derelicts. The Americans were surprised by the greeting from the townspeople. A huge banner was spread across the street with "Welcome American Forces to Galgaduud." In the meeting place, an old police post, hung a picture of the now former President George H. W. Bush. They even met some American NGO pilots flying for the UN relief agencies.[13]

Camp Holland, the field location of Company A, Royal Canadian Dragoons established at Matabaan in Zone 4 of the Canadian Parachute Battle Command's Belet Weyne HRS. ODA562 moved to the camp to support the Dragoons with area reconnaissance, key leader meetings of the USC and SNF, and mine clearing. See American flag and tents on left of photo (*Courtesy of Canadian Joint Forces Command*).

The team then escorted the two Canadian doctors to the hospital. Stevens described the scene as follows.

> Inside, the wards were surprisingly spacious and clean. Most of the patients were gone because there were no beds and little food. Most were outpatients, living in the homes of relatives or friends. Those who had to remain in the ward lay on thin mattresses on the floor, in the cool dimness. New window shutters had been made from tank ammunition crates. The smell of rot and disease that I had come to associate with medical care in Somalia was distinctly absent.
>
> We saw six patients: two battle casualties from Cali Xasan and the rest victims of vehicle accidents. There was a baby that had been run over by a truck at Cali Xasan. The tread marks of the tire were still visible on his back. The hospital administrator, who was giving us the tour, had performed surgery on the child. He was not a doctor, he told us, but had been practicing medicine for eighteen years out of necessity. He had once been an assistant to a surgeon. The baby appeared to be in excellent health.[14]

They all returned to the police post. What happened next pleasantly surprised the Americans and Canadians. Colonel Hussein had just established a radio call with General Ghani in order to arrange a ceasefire meeting and extended the invitation for Colonel Mathieu and Major Carroll to attend. This moment was a tremendous breakthrough, not anticipated during a "medical assessment tour." When Stevens informed Major Carroll, Carroll was surprised to learn the team was out of sector without permission and recalled Stevens back to Belet Weyne. Stevens thought what they accomplished would outweigh any violation of crossing the HRS boundary into USC territory. "I could seek forgiveness later," he wrote in his journal.

With Lieutenant Colonel Faistenhammer's approval, urged on by UNITAF who saw the importance of the event, Stevens planned to return with Colonel Hussein. Without much explanation, General Johnston intervened and called off the meeting. He apparently saw it as a UN and State Department function, not in his line of military operation. He would urge State Department officers to affect the negotiations. The decision was then changed for a meet with Ghani. General Ghani was contacted by radio to set up the meeting day, to be held at Abud Wak, the headquarters of the Somali National Front.

As things began to look better in Somalia, the Clinton administration began to take its eyes off the ball. Certainly there were several other foreign policy challenges to deal with–Bosnia, Russia, and North Korea, to name a few. The Somalia policy began to drift. The Clinton administration was not known for having experts with knowledge of Somalia, and these were the people surrounding Clinton. Les Aspin, the newlyappointed Secretary of Defense, was considered not as focused; he began his role by guiding the Defense Department through a drawdown in light of a peace dividend. The troops in Somalia began to have the same attitude. The Marines of the 1/9th had already shipped home. Other troop units were packing up and ready to leave. The Joint Chiefs of Staff soon put a halt to any other departures. There was still the issue of when the UN would take over the operation (UNOSOM II), which did not appear to be happening any time soon (the UN wanted a date in June, the Americans much earlier; May 4, 1993, was the date finally settled on). UNOSOM military planners and the UN Security Council began dragging their feet on setting a date for the transition.

ODA564 and the "Bridge Trolls"

As a result of the information they were gleaning from the Ethiopians, ODA564 under Captain Barber's leadership conducted a reconnaissance to El Gaul on the 26th of January to confirm suspected minefields. Barber had his detachment locate on some high ground outside of town, which afforded a good view. After a halfhour in this position, they could not detect any threat. The team mounted up in the two Desert Mobility Vehicles (DMVs) and drove into town. They immediately saw what they had not detected from their observation position—an old Italian heavy truck with a ZPU-4-23, 12.7-millimeter heavy machine gun mounted in its bed. The Technical was manned by three Somalis. The ZPU-4 began to swing their way.

The team rushed the truck and captured the driver and his two cohorts. No return fire was discharged by the team, but upon inspection of the Technical they found all four guns (3 individual weapons in addition to the mounted machine gun) with ammunition locked in the chambers and ready to fire. They removed two of the Somalis at gunpoint and sent them on the road; the driver they kept, forcing him to drive the Technical back to Belet Weyne. After the team took their trophy picture of the capture, followed by others gathering around and climbing aboard, they dismantled the weapons. The Canadians parked it in their growing cantonment junkyard of Somali vehicles.

The Detachment continued its assessment at El Gaul. That night, north of El Gaul, ODA564 came across a truck that was stopped on the road at night, so they stopped to talk with its passengers. The people in the truck said they had just been held up in El Gaul. It was a common tactic of bandits in the area—block a road and extort a toll under the threat of being beaten or shot.

Barber directed the team to cut their lights and wear PVS-7 night vision goggles. He also directed loading the grenade launcher with illumination rounds. The frightened Somalis found earlier by the team warned them there were three bandits, armed with AK-47s. The Detachment Commander and Team Sergeant explained later in an interview what happened afterward.

> As the suspected bandits heard the SF vehicles approaching, they took off running. When they heard our ODA's vehicle they began to run up, but when they realized we were Americans they took off running. Sergeant Kilcoyne fired a few warning shots and yelled for them to stop but they kept running. They ran over the bank and into the wadi. Because we had illuminated the area it was like daylight. Sergeant Kilcoyne fired two more shots of parachute flares between them as they ran to try and get them to stop. They kept going. Our team never saw any weapons so they of course did not shoot anyone. We did a thorough sweep underneath the bridge there where we thought we would find some weapons. We also went along the edge of the village, but never caught them.

The Unified Task Force rules of engagement were as follows—seize any Technical with a weapon on it but let small arms go if not threatened. The policy could change daily, so sometimes the team would drive up to a village and people would run into the brush with AKs thinking they were going to be seized.

The next day the Canadians searched the village and found weapons and mortar rounds. Having been at this same town twice since seizing the ZPU and then another time finding American made anti-armor mines, the elders who were being questioned said they knew nothing and the incident must have been kids out messing around. They thought the three they chased were probably teenagers.

A Somali male with an AK-47 had been captured before at this same bridge. The modus operandi was to man the roadblock with about seven personnel on the bridge. Three people were close up on the approach side of the bridge as the unsuspecting victim(s) came in; the bandits illuminated the far side of the bridge. The Canadians assessed that the bandits' intent was to stop the vehicle approaching, if it would not stop on the near side, they would have personnel with guns on the far side of the bridge stop them. They would take old vehicle wheels, without the tires on them, and lay them across the road to make it look like they had mines across the road.

ODA562 at Camp Holland

Between the 25th and the 29th of January, ODA562 conducted assessments and recon patrols to Guri Jiri where they assisted the Canadians with destructing some mines and hunting bandits on the American Highway. They were asked by Major Kampmann to conduct an Ethiopian border reconnaissance at Balenbale.

After a meeting in Matabaan with Governor Halane, Major Carroll and the team were approached by a distinguished looking gentleman. It was COL Hogolof, the Division Commander of the Galcaio sector and Aideed's cousin. He urged them to hold a meeting with General Ghani; they would host the delegation in Abud Wak. Major Kampmann arranged a radio call with General Ghani. The General seemed surprised to hear Kampmann was currently conducting an operation in Balenbale and warned him it would be dangerous. Kampmann informed him the operation was already under way to clear the SNF from Balenbale. Fortunately, this disagreement did not deter General Ghani from a future meeting.

ODA562 would not participate in the meeting. Major Carroll saw the meeting as the Canadians' responsibility, now that the initial contact had been secured by the SF. He ordered the ODA back to Belet

Weyne, with no further involvement in anything to do with General Ghani. ODA562 followed a small convoy of Dragoons headed to Cali Meere on the 28th of January, with the team departing the convoy and turning off at the road junction towards Guri Jiri. They spoke with some locals at that location as the Somalis described a recent attack from the SNF, using a 37-millimeter gun during their assault on the village. ODA562 then departed to link back up with the Canadian patrol. A Bison armored vehicle the Sergeant Major of the Regiment was riding in hit a mine, damaging the hull and suspension on the right side. Neither Sergeant Major Sloan nor the crew was injured. ODA562 spent the remainder of the day assisting the Canadians in sweeping the area for mines.

They did some light reconnaissance around Fer Fer and conducted a night over-watch position near the border to catch some of General (Colonel) Hubaro's bandits. After returning to Matabaan, the team set up an observation post at Qhat Qat, after the Canadians heard of SNF and bandit activity in that area.

> We set out after it had gotten good and dark. Barely four miles south of Qhat Qat, less than ten miles from the Canadian camp, we came across what looked like a possible roadblock. I hadn't expected to find anything for another twenty miles or so. As there were no people, hostile or otherwise, in evidence, I dismounted to check it out. The blocked way ahead of us turned out to be nothing more than a large bush which had been dragged into the road. Gary Ramsey had dismounted from the trail vehicle, and he pointed out a light about 50 meters away in the brush. I picked it up in my goggles, too, and made out a pair of heads and shoulders above the bushes peering in our direction.
>
> No weapons were visible, so I didn't fire on them. Instead, I motioned for Ramsey and Edwards to take up station on line to my right, and we started moving toward the two Somalis. I could see the two people standing still, looking back in my direction. I raised my rifle to my shoulder to be ready to return accurate fire immediately should they open up on us.[15]

The bandits, alerted by some noise they heard, took off to steal another day. After a final mission to escort Colonel Labbé, the overall Canadian Battle Group Commander, to Fer Fer to meet with the Ethiopians and Governor Halane, Captain Stevens' A-team was replaced with Captain Barber's ODA564. ODA562 returned to Belet Weyne for some well-needed rest and refit. Their "long patrol" of January 1993 was over, for now.

ODA526 Returns to Bardera

On January 26th the team flew back to Bardera but this time was equipped with vehicles. With the capability to roam and operate independently, it was now possible to conduct wider assessments of the HRS. Captain Williams developed a robust schedule to cover all four points of the compass—north to Garbaheerey, east to Dinsoor, south to Sacco Uen, then west to the Kenya-Somalia border. In accordance with his guidance from higher up, the ODA would attempt to linkup and meet with General Warsame, pinpointing his location and the location of any of his forces.

Their northern thrust would extend to Luuq, to ascertain the extent of the Islamic fundamentalist control. They planned the Dinsoor trip when asked by NGOs operating there to accompany them on a relief mission, with a follow-on trip to Dinsoor. The only worry in this area was reports of the roads being heavily mined. Information from the locals indicated the roads were clear, so the trip was on to Dinsoor. Dinsoor claimed to be a neutral town, but the USC still clashed with the SNF in the area.

ODA526 led the small relief convoy on the 29th, operated by the French humanitarian organization ACIF—the agency to relieve famine. There were two ACIF employees in charge of their relief aid. The convoy left Bardera, passed through Uffrow, and drove almost four hours to the southeast to reach Dinsoor. Upon arriving on the outskirts of town, they were met by two gunmen who wisely considered not challenging the ODA and fled. The convoy continued to the airfield outside Dinsoor, noticed no other threat, and proceeded into Dinsoor.

Dinsoor was located in the Dooy District, which comprised 41 villages populated by the Dabarro sub-clan. Estimates of the district's population were 20,000 inhabitants, along with 2,000–5,000 refugees. Dinsoor's clans and cultural influence ran to the coast near Merka. One of the factions operating in Dinsoor was the Somali Democratic Movement (it had its headquarters in Merka).

Williams asked the gathered crowd for a meeting of elders; when assembled, they chose Abi Mussa, a local chief, as their head spokesman. He confirmed he was the local leader of the Somali Democratic Movement (SDM). Some in the crowd contested his claim of primacy and mentioned Mohammed Nur Aliyon, an SNF supporter of Aideed, as the symbol of local power. Mussa went to great lengths to deny any faction was in control—Dinsoor was neutral.

Of importance to the locals, as the talks began, was repairing the road to Baraawe on the coast rather than the road to Bardera. This section would be more important for trade and markets in Dinsoor. They claimed they were victims of propaganda making the area seem dangerous because of their taking sides during the civil war and in the following clan wars; due to this perception, they could not get humanitarian aid, and no NGOs would work in the area. They welcomed the presence of U.S. troops.

Captain Williams learned about more troubles around Dinsoor as he listened to the concerns of the apparent chief of the elders and NGO spokespeople.

> The chief made a plea for food, schools, and transportation for the farmers' crops during harvest. The hospital needed a new roof. Looters stole the one from before. Other complaints consisted of lack of food drops to the area (in the fall an airdrop was made but only nineteen bags of grain survived the drop) and trouble with bandits. The town had an armed police force of fifty men and security for the town itself was considered good by the inhabitants. The chief commented that the people of Dinsoor supported a national government and a national police force.
>
> NGOs suspected rival clan faction involvement in the banditry–either the SNF in Bardera or the USC in Baidoa. The dividing line between Warsame with COL Barre and General Morgan and Gabbeo (a political leader) ran north-south through Dinsoor. The Somalia Patriotic Movement (SPM) and the SNF/SNA with Jess were all part of the Democratic Alliance of Somalia allied against the USC (the bandits operating in the area always claimed allegiance to one or the other faction).

The team wrapped up its assessment of Dinsoor and prepared for the move to Sacco Uen. The HRS higher command now had a pretty good appreciation of the situation based on the information the team gathered.

The Gathering Storm in Kismayo

With UNITAF's increasing success, Aideed must have been getting worried about the loss of his influence and power in Kismayo. Although Colonel Jess's forces occupied the town, Aideed had not rid the area of the Marehan and Darod clan influence (Jess' ethnic cleansing in December was halted upon UNITAF's occupation of the sector). General Morgan was relatively close by, near the Kenyan border, and itching to retake the city. Till now, Belgian and U.S. Army forces had kept a lid on the pot.

Major Martin Stanton was the S3 Operations Officer for the 2/87th Infantry at Baledogle. He remembers early negotiations with the two warring factions, assisted by the attached SF ODA.

> Of all the HRSs in those early days, Kismayu was one of the quietest. . . . A truce of sorts had been worked out between the two warring factions of Mohammed Hersi Morgan and Omar Jess. This was thanks to the efforts of the special operations forces (SOF) detachment and the UNITAF negotiator, Col. Mark Hamilton, a brilliant officer who had a tremendous reputation from conducting previous negotiations in El Salvador.[16]

Maybe sensing UNITAF was focusing on plans for a transition with the UN (the 1/9th Marines had already redeployed home, while other American units were lining up to go), it would be a propitious time to regain his lost territory before UNOSOM II settled in to replace the forces in southern Somalia.

Morgan still retained about a thousand of his fighters, mostly ex-regime Army personnel. He also had armor and heavy weapons, backed by artillery, even though he was constrained to get his hands on all of it since he was forced, along with Jess, to put his assets into cantonment areas. Colonel Jess mustered 3,000 to 5,000 fighters, according to intelligence estimates, split out among the areas surrounding Kismayo and others settling into camps to watch over the cantonments of his heavy weapons.

The 1st Belgian Parachute Regiment and the 3-14th Infantry Battalion of the 10th Mountain Division continued in control of HRS Kismayo. Both units served Brigadier General Magruder as Task Force Kismayo. The Task Force brooked no violations of the ceasefire and held both Colonel Jess and General Morgan to their signed agreement to cease fighting during the Addis Ababa conference.

For whatever unknown reason motivating General Morgan, on the 24th of January he began his move south against Colonel Jess, arriving first to an isolated outpost roughly thirty kilometers outside of town at the village of Birhaane. His forces attacked the unit overseeing the TF Kismayo-supervised cantonment and ran them off. Generals Johnston and Magruder immediately saw his actions as a violation of the Addis Agreement and felt they could not go unpunished. Six U.S. Army attack helicopters, infantry, and Belgian paratroopers supported with their lightlyarmored Scimitars (equipped with the 30-millimeter RARDEN cannon), which made fast work of turning Morgan's stored heavy equipment into scrap metal. Six Technicals and four artillery pieces were destroyed; a rocket launcher and armored personnel carrier were damaged. Morgan withdrew his forces and retreated, but not without complaint from Morgan that UNITAF supported Colonel Jess.[17]

In late January, Lieutenant Colonel Faistenhammer tasked the B-team for one ODA to support the Australians in Baidoa. Major Carroll chose ODA561, and they deployed on the 29th.

The situation for UNITAF was looking good overall. A great amount of work in all the HRSs paid off. The populations were beginning to feel secure and participate in the normal activities of a civil society. There was one exception –Kismayo.

Based on the progress being made east of the Belt Weyne HRS, Colonel Labbé submitted a proposal to extend his boundary to Galciao, and if not, remove Matabaan from his sector. General Johnston agreed with the principle of the Colonel's request but, for now, restricted the requested extension area to only friendly reconnaissance activities.

Also on the 29th, ODAs 561 and 564 were back in the base camp at Pegasus in preparation for the move with Carroll's SOCCE (-) to Kismayo. The two teams, along with the Special Operations Command and Control Element (SOCCE), were going to begin conducting unconventional warfare (UW). They conducted linkups and gleaned intelligence on the two warring factions in Kismayo. As a result of the rapport they established with the two leaders, Colonel Jess and General Morgan, the teams moved into their camps to live with them for a short while and serve as their liaison to TF-Kismayo.

ODA565 teammates later commented on their utilization as Special Forces just doing what they were designed to do.

> Special Forces pride themselves on being able to adapt to any situation and live under any condition, and task organizes no matter the operation at hand. SF works well with other indigenous forces. They are trained to communicate with local people and also on how to treat the people. That is why they were chosen for Somalia. The mission was not out of their realm because the mission was a demobilization operation. If a country regains its own power and strength and needs help the SF is the very hand in hand they need to help them. This is what they are trained to do.[18]

Endnotes

1. Material for ODA564 gathered from the team's after-action reports and interviews with Captain Barber and SFC Kilcoyne conducted by the author at the 5th SFG(A) compound, Ft. Campbell, KY in 2006.
2. Description of the activities of ODB560 and its teams summarized from personal interviews of the operators in the company and Major Carroll's and 2/5th's after-action reports. This includes materials from CW3 Richard Dietrich and from a draft book of the travels and missions of ODA562 written by CPT Charles B. Smith, provided to the author.
3. Kilcoyne and Barber interview.
4. LTC William L. Faistenhammer, "After Action Report: Operation RESTORE HOPE," Somalia, 23 April 1993.
5. _______________. *In the Line of Duty: Canadian Joint Forces Somalia 1992–1993.* Quebec, Canada: Canadian Land Forces Command, National Defence, 1994, p. 85.
6. Sergeant First Class Kim J. Breed, ODA561Team Sergeant and CW2 Ronald E. McNeal, Detachment Commander interviews by the author conducted at Ft. Campbell, KY in 2006.
7. Telephone interview and e-mail exchange with Bill Faistenhammer on September 7, 2020.
8. Rutherford, Kenneth R. *Humanitarianism Under Fire*: *The U.S. and UN Intervention in Somalia.* Sterling, VA: Kumarian Press, 2008, pp. 105–106.
9. From Tim Williams' ODA526 Patrol notes and Commo Log provided to the author to use in the preparation of this manuscript.
10. Ibid.
11. Stevens, Charles B. "Honor Bound: A Special Forces Detachment in Somalia." Houston, TX: author's draft copy, 1996, p. 187.
12. Stevens, p. 194.
13. Ibid, p. 240.
14. Ibid, p. 242.
15. Ibid, pp. 287–288.
16. Stanton, Martin. *Somalia on $5 a Day: A Soldier's Story*. New York: Ballantine Books, 2001, p. 206.
17. Tucker, David and Ambassador Robert B. Oakley, *Two Perspectives on Interventions and Humanitarian Operations*, Carlisle, Pennsylvania: Strategic Studies Institute, July 1, 1997, pp. 76–77.
18. ODA565 consisted of Sergeant First Class Kenneth W. Barriger, the team leader/team sergeant, Staff Sergeant Richard W. Knapp (SF Engineer), Staff Sergeant Terry W. Sokolowski, detachment Medic, Staff Sergeant Thomas M. Leithead, detachment Engineer, and Sergeant First Class Albert W. Bush (Medic).

6

Operation Restore Hope: Phase IIIb, February - March 1993

(SOCCE Kismayo and Belet Weyne Operations)

Here, they were not engaged in a conventional war with clearly demarcated battle lines, allies and foes. This was neither war nor peace, and each day they had to feel their way, with constantly shifting directives from above. . . . It was a taste of the messiness of war amid civilians, the classic guerrilla warfare for which they had been trained.

— Linda Robinson, *Masters of Chaos*

In February 1993, Army Special Forces supported three major operations for UNITAF. The first was continuing to support the Canadians in Belet Weyne with half of the company headquarters and two ODAs, 562 and 565. CWO3 Doug Jackson, the Company Warrant Officer, ran the Company headquarters and assisted the two SF Detachments on their missions. The second emphasis was to support HRSs Baidoa and Bardera, using one ODA in each sector (ODAs 563 and 526). The third was employing a Special Operations Command and Control Element (SOCCE) led by Major Carroll and half of his B-team staff employing two ODAs to conduct area assessments, reconnaissance, and linkups for meetings with both Colonel Jess and General Morgan (ODAs 561 and 564).

Lieutenant Colonel Faistenhammer, the Army Special Operations Forces (ARSOF) commander, was asked to take a look at providing SF teams to Task Force Kismayo. The Belgians and the 3-14th Infantry lacked long-range capability to conduct deep reconnaissance and area assessments. Neither of the units had special operations forces. Most of the Belgian and U.S. Army soldiers were operating along major roads, tied down at times between outposts. Inside the city, they ran checkpoints and foot patrols. More important, Brigadier General Magruder, TF Kismayo commander, was tasked with meeting both of the protagonists in the area, General Mohammed Hersi Morgan and Col Omar Jess (Somali Patriotic Movement). Without the two agreeing to a ceasefire and some kind of reconciliation, violence continued in the countryside. Jess occupied Kismayo after Morgan was kicked out during the "Clan War" phase of the Somali Civil War, and Morgan wanted Kismayo back. General Magruder needed some skilled, discrete, and professional armed teams to make contact with both of them in order to begin negotiations.

Lieutenant Colonel Faistenhammer flew down to Kismayo on January 30th to discuss the deployment of Special Forces assets. After hearing the requirements from leaders on the Task Force, he returned to Mogadishu and issued orders to Major Carroll to deploy as a SOCCE with two ODAs to support operations in HRS Kismayo. Major Carroll remembered Faistenhammer's guidance when formulating his own course of action.

> 10th Mountain requested help from UNITAF for SOF support. One faction of Morgan's was about 15 km north of the city, and was threatening to run Jess out (which he actually did) and the 10th Mountain were in Kismayo. Then with our help, that faction was eventually disarmed and sent out back into the countryside (we flew them out by helo).[1]

Major Carroll chose four members from his B-team to form the SOCCE (-), leaving CW3 Jackson in charge at Belet Weyne and deployed with CWO3 Dietrich as his senior Warrant Officer for the SOCCE (CWO3 Dietrich was a recent attachment from the battalion staff). The SOCCE, along with ODA564 and ODA561, deployed to the port of Kismayo via C-130. They drove first to Mogadishu from Belet Weyne and loaded onto C-130s at the airport.

On January 30th, Major Carroll deployed to Kismayo and was placed under the tactical control of BG General W. Lawson Magruder III. Major Carroll established a base of operations on the quay near the maritime berthing docks, collocated with Task Force Kismayo. Major Carroll had a small building to work out of—everyone else lived in ARFAB tents erected next to the building. Advantageously, the unit was located near the Belgian logistical unit, and support would never be a problem.

They were ordered by Lieutenant Colonel Faistenhammer to support Task Force Kismayo for the conduct of area assessments, route reconnaissance, and to initiate contact with General Morgan's and

Colonel Jess' forces. If contact could be made, Brigadier General Magruder, the TF Kismayo commander, desired further coordination with the two rivals to negotiate a ceasefire and place their heavy weapons into cantonments. TF Kismayo did not have units with unconventional warfare capabilities and long range reconnaissance mobility platforms, so he asked Lieutenant Colonel Faistenhammer for assistance. The 10th Mountain Division infantry and the Belgians primarily focused their efforts in and around Kismayo, or in surrounding towns and villages. Traveling mainly along major roadways, the TF had not yet penetrated the countryside or spread to the Kenyan border where the rival forces of the SPM and SNA operated. The mobility inherent within the SF teams would be crucial to accomplish Brigadier General Magruder's objectives.

When Carroll arrived with the SOCCE, the 10th Mountain Division Military Police platoon controlled the hardball roads running north of Kismayo, while the Belgians controlled the town. The 10th Mountain had been tasked to open up the roads running north to Bardera and on to Luuq to facilitate humanitarian relief food convoys. This task necessitated providing security escorts for the convoys. With faction fighting in the areas alongside the main route, the 10th Mountain could not proceed with the planned mission. Intelligence was sketchy on the situation along the route; the SF teams were needed to conduct reconnaissance and area assessments to determine the feasibility of running food convoys through these areas. Additionally, the SF was asked to locate and neutralize banditry along the route. The Task Force commander needed to know the exact locations of the factions and their military status to monitor their activities.[2]

Little information on the situation west of the Jubba River was known and even less on the locations of Jess and Morgan's forces. They did know that Jess' forces occupied Kismayo and had a large base camp out to the northwest near Bir Xaani (this area was also the front lines for the Ogadeni clan based Somali Patriotic Movement). It was estimated Jess could muster 5,000 fighters and had a fleet of Technicals numbering an estimated 250 trucks.

General Morgan was near the Kenyan border, near Dhoobley, with an estimated 1,000 fighters and an unknown amount of heavy weapons and Technicals. Morgan had an associate, subordinate commander in the region, COL Shukri Weyrah. The teams knew they would not have a long time to settle in; missions would come fast and furious to fill in the intelligence gaps. Carroll's mission was clear: conduct area assessments and route recons, find Jess and Morgan, and facilitate negotiations between the two clan factions with the Task Force and UNITAF.

Colonel Gaddis served as the Task Force G3 and was responsible for selecting the area assessments and reconnaissance objectives. Major Carroll's teams would also work with COL Mark Hamilton, the chief negotiator for UNITAF, whenever opportunities arose to meet with Morgan or Jess. Major Carroll served COL Hamilton in the field during the negotiations.

On February 1st, Madeline Albright became the Ambassador to the UN. She was aware that the U.S. military could be a key element of national power to help further foreign policy initiatives and also help solve some of the intractable problems worldwide. The military did not see it that way; they were eager to depart from Somalia. Most of the sectors were quiet, people were fed, and the soldiers were becoming tired and bored. However, she was clear in her instructions about Somalia policy after administration pressure.

> Instructions were to negotiate the rapid handover of principal responsibility from the U.S. to the UN. The Pentagon was eager to call its mission a success and bring our soldiers home. Boutros-Ghali resisted, arguing that the world body was neither staffed nor equipped to take on another major new

operation . . . I told the Secretary-General he had no choice; U.S. troops would leave whether the UN was prepared to take their place or not.[3]

SF A-teams 561 and 564 went north that same day along the Jubba Valley to the town of Jalib, detecting some banditry in the area (under Colonel Jess' control). The teams discovered from locals that villages in the region were being raided for their food supplies. This information was passed to the Task Force, which subsequently used the information to deploy Belgian soldiers and U.S. Military Police to search the area and suppress bandits or find the location of armed factions who were looting villages.

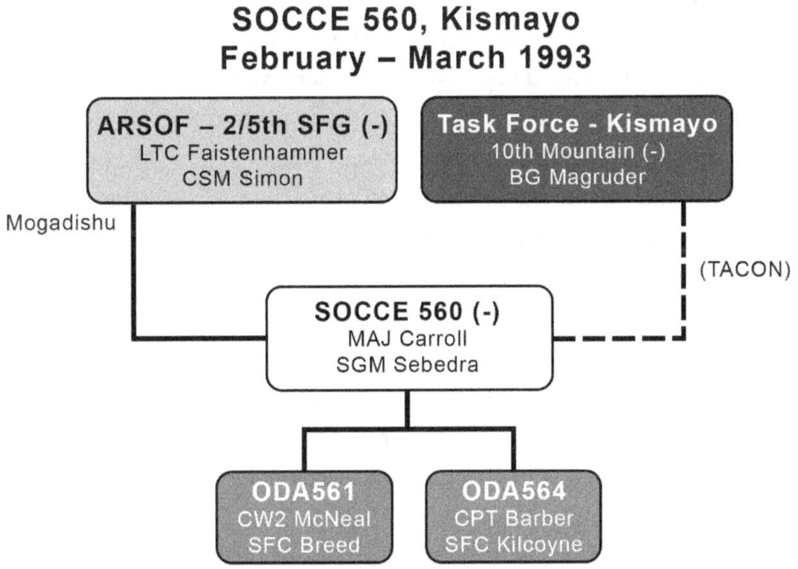

ODA564

ODA564 deployed to Kismayo via C-130 aircraft from Mogadishu and reported in to the SOCCE. The first mission for the detachment was the conduct of the February 1st reconnaissance north along the western area of the Jubba River for an area assessment to find the location of any bandits and locate Colonel Jess and his forces. Activities in this area were relatively unknown to the Task Force. The Belgians and the 10th Mountain infantry (3-14th) only operated about ten kilometers into the area; virtually nothing was known about the area along the Kenyan border.

Between February 1st and 4th, ODA561 also conducted an area assessment as their first mission, on the west side of the Jubba River. Although only a short reconnaissance, like ODA561, they also found many villages were being raided by the clan militias foraging for supplies.

The earlier clashes between Jess and Morgan on January 31st near Bir Xhaani forced Brigadier General Magruder to once again punish Morgan and force him back west to the Kenyan border. Magruder also told him to put his heavy weapons into a cantonment. Colonel Jess already stored his heavy weapons in the cantonment at Bir Xhaani. When Morgan did not comply, TF Kismayo launched an aerial attack using helicopters from the 3/17th Cavalry, flying on February 3rd and 4th with three AH-1 Cobras, three

OH-58s, and three UH-60s. Morgan was finally convinced General Magruder meant business. He moved his forces ten kilometers northwest of Bir Xhaani, placed his vehicles further back in a cantonment at Hoosingow, then established a base camp. TF Kismayo sent troops to garrison Bir Xhaani. This current disposition of clan forces gave a window of opportunity for UNITAF to reach out and begin disarmament talks with both factions (SPM and SNA). A ceasefire between the two factions was negotiated and agreed upon by the two commanders.

On February 5th, ODA564 conducted an aerial reconnaissance to the southwest of Kismayo in preparation for a ground area assessment conducted between the 6th and 7th. They used the results of their aerial recon to confirm the condition of roads and pinpoint villages in the area on their maps.

On the 8th, they were tasked with conducting reconnaissance north along the Jubba River valley. They established contact with COL Shukri Weyrah (Morgan's forces) near Bir Xhaani. One of their objectives was to determine the level of bandit activity. They had not yet achieved any measurable success in anti-bandit operations while at Belet Weyne, even though their procedure was to query all NGOs and villagers they met during their patrols. The capture of bandits proved elusive.

The situation in the area remained akin to irregular warfare. Villages and towns were support areas for whichever faction their loyalties lay with. Most Somalis were more than generous with information on their enemy but remained quiet when asked about the faction they supported. It would be challenging to get any information on the movement of the forces villagers provided with food, gas, intelligence, and medicine.

ODA564 returned to Kismayo. Major Carroll gave them their next mission—linkup with Morgan's forces along the front lines opposing Colonel Jess near Bir Xhaani and report back on the situation in the area. The priority intelligence requirement for TF Kismayo was knowing when and where General Morgan would launch an attack against Colonel Jess' forces. The Detachment successfully found SNF forces near Bir Xhaani and was able to establish contact with Colonel Weyrah.

SFC Kim J. Breed, the Detachment team Sergeant, described their early days in the Kismayo HRS.

> It would come down from 10th Mountain Division through . . . Major Carroll, our company commander, that they would want a certain section or an area checked out, an area assessment done to it, which is road trafficability, mines, if there was any bandits in the area, if anyone had seen any military. They were basically trying to find military that were moving through the area. And it would come down to us.
>
> We would already be loaded up; our vehicles were always ready to go within six hours after we got back in, five to six hours they were already repacked. And we'd take off, go out and do the mission. Report back at least twice a day. Sometimes it was more than that depending on what we found and the urgency of the message. And once we completed that, just drive back into Kismayo, stop. We'd run into a debriefing with Colonel Gaddis. He would have his full staff, his S-2, they would re-verify everything for all the message traffic, any little things that we might have left out, forgot or remembered later, and they would plot everything up on their board, kind of "OK, that area's clean–check." And then that would be it for the debriefing. We'd go back, get hot food, showers, and then usually by about, I'd say 7 or 8 o'clock that night, they would be ready for another mission. And we just kept rolling. Maybe twelve hours off at the most.[4]

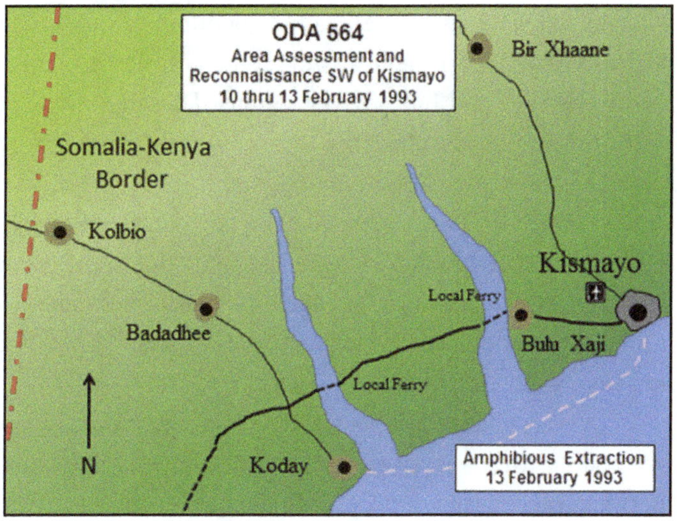

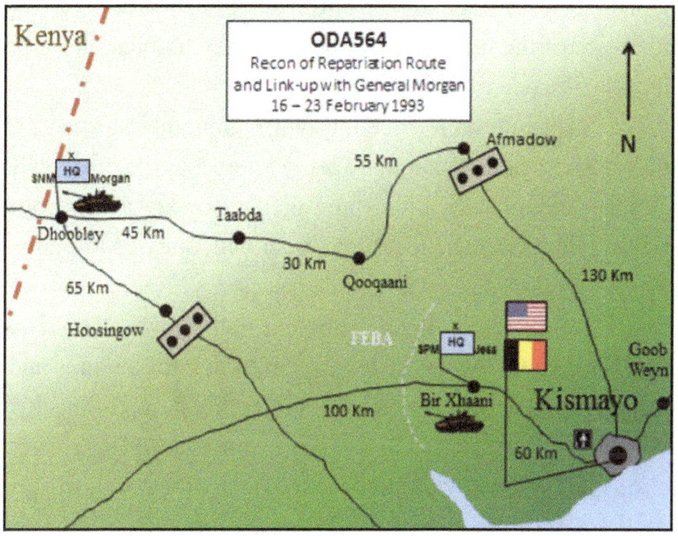

On the 10th of February, ODA564 launched an area reconnaissance and assessment of the area between Kismayo and the Kenyan border, in a circular pattern culminating down to the coastline. They met with Colonel Weyrah during the trip and spent some time in discussion with the commander, finding out from Weyrah he was having a hard time controlling his men from independently attacking Colonel Jess' forces. They established a successful rapport with him, and the team was able to camp near his forces until the 13th of February.

Each day, ODA564 rolled out of their patrol base and drove to Colonel Weyrah's camp, located approximately five kilometers northwest of Bir Xhaani. SFC Edward Mays, the team senior radio operator, explained the procedure, "We'd go down, we'd make a linkup, and people would tell us what they had been doing the day prior, and what they wanted. We would take this information and we'd go back to Bir Xhaani, and we'd send this information back via SATCOM [satellite communications radio]."[5] Major Carroll also flew out and spoke with Colonel Weyrah.

When the assessment was complete, the ODA met landing craft from the U.S. Navy Boat Squadron at the town of Koday, loaded their vehicles aboard off the beach and exfiltrated back to Kismayo on the 13th of February. Lieutenant Colonel Faistenhammer and Major Carroll flew down to the beach using an OH-58 to meet with Captain Barber's team and get a quick update on what they found. They rode aboard the landing craft with the team as it sailed on its four hour trip to Mogadishu.

The ODA took their updated information back to the Task Force in Kismayo. Major Carroll was impressed with what they had accomplished. It was just like a chapter out of the Unconventional Warfare Field Manual–a linkup with the "G" chief (guerrilla chief) and establishing rapport and presence with the guerrillas. Carroll wanted to build on this success and ordered Captain Barber to take his team back out and live at Bir Xhaani to provide continued information updates and serve as a liaison between Morgan and the commander of TF Kismayo.

During this portion of the mission, ODA564 was also asked to attempt to establish a linkup with General Morgan at Dhoobley. The Task Force also wanted to know the condition of the route from Bir

Xhaani to Dhoobley as well as the condition of other roads and villages in the area. Of importance to the UN and State Department was the route between Somali refugee camps through Dhoobley and on into the wider Kismayo and Jubba River regions. This area would be the route for the upcoming refugee repatriation. With this expanded checklist to accomplish, ODA564 would ultimately conduct a 500-kilometer route reconnaissance over a seven day period.

Captain Barber planned his "long patrol" route by traveling first to Afmadow, then turning east and performing recon on the villages and route out to the border town of Dhoobley. After contact with General Morgan, the Detachment intended to then turn southeast to Hoosingow, then continue on and catch the route back through Bir Xhaani east to Kismayo. ODA564 deployed on the 16th of February to conduct their mission, which lasted until February 23rd.

At Afmadow, Captain Barber was met by a helicopter sent by Colonel Hamilton, the chief UNITAF negotiator who was awaiting them at General Morgan's camp in Dhoobley. A meeting was scheduled between the General and Colonel Hamilton. Hamilton wanted some operators from the SF team to be with him to explain their presence and role in the area. Major Carroll was already with Colonel Hamilton in Hoosingow. The remainder of the team pushed on to Qooqaani, led by Sergeant First Class Mays, and were unaware of the minefield on the road they had just driven over outside of town (it was discovered on a later reconnaissance trip).

During their meeting with General Morgan, Hamilton and Carroll noted General Morgan's nervousness and hesitation to reply when Hamilton asked Morgan to assure UNITAF that he would no longer attack Colonel Jess' forces and pledge to stay away from Kismayo. At some point in the meeting, Major Carroll flew in the helicopter to meet with one of Morgan's sub-lieutenants, located about 20 kilometers from Hoosingow, to discuss the same issues.

Meanwhile, the detachment met with the residents of Qooqaani. The Qooqaani locals had no love lost for General Morgan and his forces due to the raids and pillaging on their town to supply his forces. The villagers spoke of their friends' and families' executions by Morgan's men and pointed out freshly dug graves where the bodies were buried. The elders readily agreed to supply the ODA with any and all information about General Morgan, then led the detachment to a weapons cache placed by Morgan's men (this was a risk on the part of the villagers who knew Morgan would probably retaliate against them).

Sergeant First Class Mays counted nearly 600 81-millimeter mortar rounds, mostly smoke rounds of U.S. manufacture, and 17 Soviet made 100-millimeter tank rounds. The tank rounds were still packed in containers and were in good shape. The boxes containing the mortar rounds were stamped with "Made in Kenya" and "modified in Kenya." The detachment took the ammunition outside of town, rigged charges on it, and destroyed it all.[6]

Sergeant First Class Mays continued on to Taabda with the reconnaissance. The team discovered few people present, even though there was a large lake nearby which could have provided water for drinking and crop irrigation. They chose to remain overnight (RON) near the lake. Dhoobley was their next destination (they were unaware the meeting was already over and Colonel Hamilton and Major Carroll had returned to Kismayo).

Now back with Captain Barber and Sergeant First Class Kilcoyne, the team established contact with Morgan's second in command, Colonel Achmed, since General Morgan was absent. Colonel Achmed led the garrison with about 100 troops.

The ODA toured the town conducting their assessment, apparently with the blessing of Colonel Achmed. They were surprised to find a well-equipped and maintained ICRC field hospital, staffed by doctors and nurses out of Kenya. It appeared the hospital operated without interference from Morgan since it was also treating his sick and wounded men. After visiting with the NGO running the site, the team moved three kilometers outside of town to meet a delegation from the Islamic Fundamentalists who were also running a relief camp. The team found them to be friendly and discovered they were more effective in getting relief assistance to the populace than Morgan's faction, which was reported as taking 40% of all supplies before the remainder trickled down to the needy.

The team noted the large amount of supplies and logistics stored in the town to support Morgan's army. Morgan chose Dhoobley as a strategic hub in his operations primarily because of its location along key routes in and out of Kenya, with feeder routes east into the wider Kismayo region. Morgan was not the only one who saw advantage in Dhoobley's location; along with being a key route for humanitarian organizations, it also attracted smugglers and drug traffickers.

The team noted the international flavor of the trade in Dhoobley. They found IV sets from Italy, food from Saudi Arabia, ammunition from various countries throughout the Middle East, and arms and ammunition from the Soviet Union. Most of the supplies and logistics Morgan used seemed to come out of Kenya. With their assessment of Dhoobley complete, the ODA departed for Hoosingow.

General Morgan returned. After a few days with him, Captain Barber and his operators assessed Morgan's forces as welltrained. The leaders were all ex-Somali Army officers, and most of them had attended foreign military courses. Ironically, those they interviewed spoke of their training in the 1980s during the USprovided security assistance for Somalia during the Barre regime (ironically, Army Special Forces provided several Military Training Teams during this period). ODA found Morgan's men sound tactically and employed guerrilla warfare tactics. Major Carroll rejoined them at this location to conduct a meeting with Colonel Weyrah on the 8th.

Most members of the Special Operations community are not around any media, nor do they speak with the press. This is for good reason—the need to maintain a high level of operational security when conducting a mission (OPSEC) is paramount. It is a form of force protection when one only has 12 men on a Special Forces A-team operating behind the lines or in non-permissive or semi-permissive areas against stronger enemy forces. No need telling adversaries the location of the team or their names, which could be researched to find where they live and the location of their families (this is why SOF adopts the procedure to not wear nametapes or U.S. Army tapes and unit patches as part of their "tradecraft"–that alone is a form of intelligence to an enemy or foreign agent).

However, there comes a time when it is beneficial to let the American public learn about the operators' deeds. Generally, it was policy from higher command that when conducting an approved meeting with the press or other media, leaders could use their names, but everyone else used a pseudonym to preserve their identity. The following press release appeared in the Washington Post during this mission. Sergeant First Class Kilcoyne, the ODA564 Detachment Sergeant, gave his pseudonym as "First Sergeant Killoran." The SF officers used their actual names and ranks.

NSA consortial agreement in Cyber education

Deep in the Desert

With a Somali Militia

U.S. Special Forces Team Tries to "Win Hearts and Minds" of Warlord's Inner Circle

By Molly Moore

Washington Post Foreign Service

BEER XAANI, Somalia – First Sgt. Bill Killoran sat in the glow of the evening campfire, sipping tea and listening to a Phil Collins tape with the senior colonels of one of Somalia's most feared militias, men accused of pillaging villages and slaughtering the families of rival clans.

For the past week, Killoran and seven other members of an elite U.S. Army Special Forces team have been living on the edge of a foul, snake-infested pond deep in the red Somali desert where they have tried to get close to the inner circle of warlord Mohamed Said Hersi Morgan's top field commanders.

"We try to befriend them," Killoran, a 33-year-old engineer from McLean, Va., said of the ragtag Morgan militia. "They could turn on us any time they want to–but if they did, they would have hell to pay."

Most U.S. Marines and soldiers have roared into Somali towns in noisy shows of force or rolled into the countryside as heavily armed escorts for relief convoys. But six Army Special Forces teams have slipped quietly into the desert on "defanging missions," collecting intelligence on feuding warlords and waging intense, unpublicized efforts to prevent bloody clashes between warring clans.

Living on packaged military meals supplemented by wild birds and dik-diks—agile deerlike animals that populate the desert—the U.S. soldiers have established acacia tree lookout posts to spy on rival clans and monitor raiding parties. They have scoured villages, pumping elders for details of troop numbers and locations, offering sacks of grain in return for cooperation and information. They have set up meetings with warlords, using nomads as note-carrying couriers.

"You don't just blow into a village, shake everybody down and lay them out in the street," said Maj. Lelon Carroll, commander of the Special Forces group assigned to Somalia. "We're trying to win hearts and minds and get information."

The Special Forces teams, whose members have received training in the Somali language and culture, also have found themselves acting as unofficial mediators and negotiators between warring factions. And no situation has been more prickly than the standoff between Morgan's forces and those of his archenemy, Col. Omar Jess, over the strategic port town of Kismaayo in southern Somalia. Each warlord claims the sprawling commercial center as his own, and they have been involved in a brutal tug-of-war over it.

While most other rival warlords have acquiesced to U.S. demands to cease their attacks, Morgan and Jess have clashed repeatedly despite coalition efforts to block their raids. Yesterday, in the latest fighting, several Somalis were killed and 21 wounded when Morgan's and Jess' forces battled for five hours in Kismaayo, according to news agency reports.

After U.S. Army commanders dispatched attack helicopters to break up a Morgan-Jess battle outside Kismaayo three weeks ago, Special Forces teams were sent into the desert near this abandoned cross-roads village 35 miles northwest of Kismaayo to monitor the movements and collect intelligence on the two clan militias.

For two weeks, Special Forces teams spied on Morgan's forces from crude observation posts several miles from a main encampment, monitoring raiding parties who stole into Kismaayo under cover of darkness to loot. Then, last Wednesday, the men of Charlie Company received an invitation they could not refuse. One of Morgan's senior commanders, Col. Mohamed Mahmud Weyrah, 57, who boasted to the American soldiers that he had killed his first policeman when he was 16, invited the Special Forces team to move into his austere camp.

"They're very cordial," Chief Warrant Officer Ron McNeal, who heads the eight-member Special Forces team based at Ft. Campbell, Ky., said the day his men joined the Somali encampment. "They're using us to try to have voice."

Morgan, Weyrah's boss, is perhaps one of the most despised Somali warlords in the eyes of American officials here. U.S. special envoy Robert Oakley has publicly derided Morgan as a cold-blooded murderer and has refused to meet with him. At the same time, Morgan and rival Jess also have rejected attempts by coalition forces to forge a ceasefire.

Morgan and his men have accused the Americans of siding with Jess in the battle for Kismaayo. When U.S.-led coalition forces arrived there, Jess happened to be in control of the city. The foreign forces stepped into the fray and ordered both sides not to move. Consequently, Jess' loyalists are living in the city and Morgan's forces are consigned to isolated camps outside the town.

"It's really tough to find a good guy," said Carroll, the Special Forces commander. "Each one has his own type of atrocities. It's ugly."

During the past two weeks, however, Carroll has come to know, and even empathize with, Weyrah: "He's sitting at a putrid, snake-infested water hole and trying to keep his men from deserting and going into Kismaayo," Carroll said. "As a person, he's a villain of the story; as a soldier, I can see some of the problems he's going through with no food, no medicine."

The U.S. Special Forces commanders—who seldom discuss operations with reporters—allowed a small group of American journalists to accompany the team on its recent desert mediation session with the forces loyal to Morgan.

Standing before Weyrah and his officers under the shade of an ancient tree at Weyrah's campsite, Carroll, a native Floridian, acted the diplomat, a stoic expression revealing none of his personal biases. Weyrah sat on a makeshift bench and wore sunglasses, a green bush hat and a faded olive-green T-shirt reading "U.S. Marines–Cobra Chopper–Hell's Fire." The irony behind the T-shirt was not lost on the U.S. soldiers: Three weeks ago, American Cobra helicopters destroyed several of Weyrah's armed vehicles during a battle with Jess' forces.

"We've been talking for days," Weyrah said, idly raking a stick across the sand. "If they [Jess' forces] want peace, they have to get out from our land. The only solution is to take it back by force. They have our houses and we have nothing. We're just staying in the trees."

Carroll nodded sympathetically, then replied through an interpreter. "I'm a soldier, you're a soldier. We can understand each other. My people here with you are not choosing sides. It's important for your forces and Jess' forces to remain apart. I know you're at this water hole for days on end, but we need more time."

Carroll paused, then added, "I believe the elders will get together. We have to give them the chance."

But out of Weyrah's earshot, Carroll expressed doubts to reporters. "We could use some help. We need something to give within the next three weeks. I'm not sure how much longer we can hold them."

In an effort to buy time, the Special Forces teams are providing supplies to Weyrah's troops even though relief agencies working in Somalia refuse to give food or medical aid to known militiamen.

"If we don't give it to them, they will get it from somewhere else," said Carroll. "Our function is to keep them there while negotiations are going on. If we keep them from moving, we save lives."

So far, however, negotiations have failed. Oakley was supposed to mediate a session between the elders of Morgan's and Jess' clans last week, but Jess' delegation did not show.

In a recent interview with several American journalists at his heavily guarded compound in Kismaayo, Jess said, "They [Morgan elders] do not have the right to come here. To bring them here would give them legality here. We would not like to talk with representatives of Siad Barre again." Morgan is a son-in-law of former ruler Mohamed Siad Barre, whose overthrow two years ago led to civil war and anarchy in Somalia.

Morgan, in his own interview with reporters, said his people too have rights to Kismaayo. "I'm not going to meet Col. Jess. He is the man who has blood on his hands . . . They have labeled me as being a murderer. I categorically deny all these accusations. I am simply out to defend my people."

The Washington Post
Tuesday, February 23, 1993

General Morgan seemed to enjoy ODA564's company and friendliness towards him. Their presence made him feel important to the Unified Task Force by being provided with his own Special Forces team. The team was treated extremely well.

General Morgan liked to talk and had no problem telling Captain Barber and the team about his exploits and ambitions during mealtimes. Of more interest to Barber, the General explained his tactics and some of his strategy when fighting Jess' forces.

General Morgan went on in great length to describe how he trained his soldiers and his prowess as an expert in training militia; General Morgan claimed he molded locals into very effective fighting forces. He told the team there was a small nucleus of about 300 veteran soldiers in his regular army; the rest were guerrillas spread out and living among the local villages. When he was ready for battle, he sent word out to local villages and they would come marching together, hit their objective, and then return to their villages. General Morgan bragged he could locally fight with 10,000 to 20,000 men. It was classic, Maoist insurgent doctrine.[7]

Earlier, one of the Humvees broke down, and the SOCCE flew parts out to the camp with a helicopter. The team repaired the vehicle and prepared to settle down for their liaison duties in Hoosingow. That very night (the 21st of February), General Morgan's forces attacked Jess in Kismayo. SFC Kilcoyne recalled the tense moment:

> After Morgan's forces took over the town, it was very important for us to get in contact with higher. We made commo; they told us that Morgan's forces had taken over the town. We asked Morgan about it. He said, yeah, his people, one of his commanders had moved in there and taken over the town. The general of the 10th Mountain Division, immediately, through our communications network, our SATCOM, talked to General Morgan. The actual real time communication there, which actually helped to defuse a very critical situation there . . . and they came through with the message. They wanted us to basically deliver an ultimatum to General Morgan saying, hey, you know, you must pull out of this town or we're basically going to attack you. And we talked to our command and gave them our feelings that we're 20 kilometers from anybody, you know, it would be probably four hours, what? About two to four hours before they could get gunship support to us and it didn't seem real wise to us for us to have to deliver this ultimatum to the General at this time. They agreed with us, and they said well, we'll call him. I believe it was round 1600 hrs (the time for the ultimatum), we'll call him on HF [high frequency radio] and deliver the ultimatum to him. And we said fine, we need to withdraw out of this town. And at a quarter till, we pulled out of town.[8]

Barber and Kilcoyne knew they had to get back to Kismayo; they could not appear to favor General Morgan during his attack. Under the pretense of wanting to inspect a minefield nearby, the team departed. At the minefield, they told the soldiers they needed mine-detection equipment and had to go back to Kismayo to procure some. They were able to extract from the area prior to the time set for Morgan in the ultimatum. Duped by this excuse, Morgan's men pointed out a bypass road around the minefield. They were met near Bir Xhaani by a helicopter ferrying Major Carroll to linkup with them and provide an armed escort back to Kismayo.

The Attack on Kismayo –February 21st–22nd

Colonel Weyrah and his men snuck into Kismayo on the night of February 21st to attack Jess' faction. Moving like urban guerrillas, going from house to house to use their auxiliary and underground supporters, they were successful in driving most of Jess' forces out of the city. Coalition forces were totally caught unawares. The battle raged for four hours.

In the morning, a local Somali female arrived at the main gate of the headquarters compound to describe the attack and sounds of gunfire throughout the night. She announced ". . . there was a new Sheriff in town."

Major Carroll and Colonel Hamilton jumped in a vehicle and drove to the center of the city to find Colonel Weyrah. Colonel Weyrah was pleased to see the two, hugging them like old friends. Hamilton reminded Weyrah of the rules of the 10th Mountain Task Force prohibiting clashes and gunfights in Kismayo. Soon, Brigadier General Magruder arrived and repeated the admonition. Weyrah became worried and sought assurance for the safety of his men if he moved.

Weyrah knew and trusted the SF; Major Carroll used his operators for the evacuation. The Belgians provided trucks for Major Carroll and his two ODAs who escorted Weyrah's force outside of town into the countryside beyond the airfield (about twelve kilometers away). Everyone camped for the night under the cover of some thick acacia trees and starlight. 10th Mountain Task Force's UH-60 helicopters arrived in the morning to transport Colonel Weyrah and his men back all the way to Dhoobley. Kismayo returned to its unique form of "normalcy." The Special Operations Command and Control Element (SOCCE) and ODAs returned to their primary mission.

Belet Weyne Humanitarian Relief Sector – February 1993

During the month of February, the Canadians continued to expand and consolidate their operations in the Humanitarian Relief Sector. Of significant interest to the Canadian National Commander, Colonel Labbé, was Zone 4 garrisoned by A Company, Royal Canadian Dragoons. Zone 4 was situated in the northeast of the HRS and was the scene of the most fighting still on-going between the United Somali Congress and the Somali National Front. The fighting occurred just outside the HRS boundary limit. When not fighting each other, both factions abetted lawlessness and banditry within areas under their control, making attempts by humanitarian relief organizations to alleviate suffering ineffective.

The USC (anti-Aideed faction, like Ali Mahdi's faction in Mogadishu) was concerned about pending largescale attacks from the SNF. They were eager to cut a deal with UNITAF and seek a ceasefire with General Ghani before this occurred. The SNA appeared to be jockeying for more territory administered under their control (Aideed's strategy) before the turnover from UNITAF to UNOSOM II. Aideed was keenly aware that UNOSOM II desired to expand into central and northern Somalia. The efforts of the Somali National Front in these areas were designed to forestall or prevent that objective and hold on to power.

By the beginning of February, the UNITAF controlled HRSs were operating effectively in accordance with Phase III objectives with the exception of three trouble spots, characterized by constant clashes between rival warlords: in Mogadishu between Aideed and Ali Mahdi, in Kismayo between Colonel Jess and General Morgan, and in Balenbale (just outside HRS Belet Weyne) between Colonel Hussein and General Ghani. Although Colonel Yusuf of the SSDF caused problems near Galciao, his activities against the SNF were not affecting any UNITAF administered HRSs.

To further stabilize the Belet Weyne, HRS required seeking permission to expand military operations, area assessments, and reconnaissance outside the boundaries of the HRS in order to determine the nature of the front lines and identify any threats to UNITAF which could spill over into HRS Belet Weyne. The contested area was also of great interest to UNOSOM; it was the UN desire to expand into northern Somalia to enforce reconciliation, ceasefires, and disarmament. If it was not to be done on UNITAF's watch, then certainly the UN would be left with an intractable problem during the UNOSOM II takeover.

Although outside the parameters for UNITAF operations, the Canadians and their attached Special Forces in HRS Belet Weyne were given authorization to move into the contested area and seek out the leaders of the two factions. In a sense, they were performing the UN's political line of operation with military lines of effort; the UN was incapable of penetrating into the area.

It would be a feather in the cap of the Canadian command if a peace conference could be arranged between COL John Hussein in Dhusamareeb and General Ghani in Abu Wak, with the intent to hold the meeting near Galciao. Special Forces ODA562 assisted Major Kampmann, the commander of the Dragoons in Zone 4, and Colonel Labbé, to make this event happen. Kampmann moved his force northeast and established his base of operations in Matabaan, just inside the HRS boundary. The Canadians named this new outpost Camp Holland. ODA562 moved to Camp Holland and remained attached to the Canadian Dragoons until their tour was complete with the Dragoons (mid-March).

In the beginning of February, Chief Warrant Officer 3 Jackson assumed the responsibility for AOB560 operations when Major Carroll's SOCCE departed to Kismayo. ODA562 and ODA565 remained with him to conduct assigned missions. ODA565 conducted reconnaissance missions with the Canadians to identify and remove mine caches and other dangerous ordnance, the largest one being over 2,000 anti-tank mines and 1,000 anti-personnel mines, most revealed to Canadian forces by the Somalis themselves. They continued to remain in contact with the Ethiopians at Fer Fer, attending scheduled meetings. They also assisted Canadian patrols to conduct routine humanitarian and medical assistance trips. Both teams assisted the Canadians to identify militia heavy weapons and get them moved into cantonments. On the 23rd of February, ODA565 departed to serve with the Marines in Baidoa, replacing ODA563 who then moved to Kismayo to reinforce Major Carroll.

At Camp Holland, ODA562 supported Major Kampmann and his Dragoons with an out-of-sector mission to establish contact with factions of the SNF and USC. ODA562 was the only SF team in the Belet Weyne sector authorized by Lieutenant General Johnston to perform this mission. It was during these operations the ODA suffered the loss of SFC Bob Deeks (SF Medic) due to a landmine.

There were a variety of activities performed by both teams. ODA detachment members participated in conducting medical assistance visits to assist NGOs. Both teams helped distribute PSYOP products, mainly the Rajo newspaper and a variety of leaflets and posters warning the Somalis about the dangers of mines and unexploded ordnance. When in camp, daily tasks included vehicle and weapons maintenance, going to the local firing range to maintain weapons proficiency, hosting members of the media and delegations from various relief agencies with transport to local villages, providing briefings to visiting VIPs, and when there was downtime, writing letters, reading books, having barbeques, watching movies, and the ever scarce chance to get on a satellite phone to call home to loved ones. Any outdoorsmen on the teams went hunting for local birds and game animals. All members of AOB560 agreed the Canadians took care of them extremely well with amenities and logistics.

One key factor on the American side was a desire to disengage any American forces from further coordination and meetings with the clan factions outside the boundaries of UNITAF responsibilities, once the SF initiated contact for the Canadians. American leadership was keen to not have any mission creep before departing Somalia. The trigger for pulling back on American activity would be when Colonel Labbé was able to meet with both the Somali National Front and the United Somali Congress leadership; the ball was then in his hands. At that point, UNITAF saw it as the UN's responsibility, along with the State Department, to take charge and broker a ceasefire and subsequent negotiations between the two factions.

The SNF in Balenbale seemed to be the problem child. They would soon receive attention from Canadian forces. The second goal was to reduce or stop the fighting along the front lines outside of Guri Ceel, near Cali Xasan.

ODA562 ranged far in supporting the Canadians. The team's initial reconnaissance began when they traveled with the Canadians to Matabaan, and then on to Dhuusamarheeb. In the pursuit of their mission, ODA562 also conducted area assessments in Giliansoor, Guri Ceel, and Adabo.

The Battle of Balenbale

The Canadians intended to contact the Somali National Front (SNF) in Abud Wak. A message of their visit was sent announcing their trip and their desire to meet with SNF leadership at the site of their faction's headquarters. Major Kampmann assembled a troop of Dragoons and a detachment of engineers riding in armored Cougars, totaling nine vehicles as a show of force. ODA562 accompanied the force with its two Desert Mobility Vehicles, one with a .50-caliber machine gun and the other with a MK-19 grenade launcher. The convoy assembled on February 10th at Cali Xasan and set out at 1000 hours toward Balenbale and then on to Abud Wak.

Before the visit, Major Kampmann was informed he could only meet with the rear detachment of SNF at Balenbale because the main body of SNF forces left for the front lines at Giliansoor.

Captain Stevens moved ahead of the column. He approached the outskirts of town, stopping momentarily near a decrepit building. He observed a large clearing to his front, then more abandoned buildings. He ordered his men to move quickly across the clearing and into the cover of some trees near the buildings. As they approached the position, they glimpsed a Technical and what appeared to be a body of SNF soldiers maneuvering around the south side of their position. The Somalis were observed carrying AKs and light machine guns. Captain Stevens pulled his team off the road into firing positions.

The Canadian column was behind him, to his left. Then everyone heard the crump of mortars, aimed at the Canadians. A Cougar fired one round from its 76-millimeter cannon in return fire. As Stevens was getting ready to engage, Major Kampmann came over the command net and ordered everyone to ceasefire. Apparently, he did not want to begin this first contact with the SNF by forcing a gun battle.

AK-47 rounds were now being fired at everyone. Meanwhile, the A-team saw a Technical whose crew was loading a recoilless rifle. Captain Stevens and his men took immediate action to prepare for return fire.

> That was the point when I finally thought that we might really have to fight. My men were all taking cover behind the vehicles and searching for targets. They were sorting out their fields of fire and ensuring that they were not going to waste rounds shooting at the same people. Matt had his sniper rifle out and was scanning the trees through the scope. "RPG," he called out, "one o'clock! . . ."
>
> . . . Al was coordinating the defensive arrangements while I took in the whole scene and tried to figure out what to do next. "Anybody see any crew-served? Spot 'em and mark 'em. Dave, Jim, you got 'em. Priority targets."
>
> Dave Davis was manning the MK-19 40 mm automatic grenade launcher, ready to lay down an explosive curtain across the treeline. He had a pair of binoculars to his eyes, scanning the trees from the vantage point of the DMV turret. "Nothing yet," he called back to Al. "Looks like about a platoon deploying on line. I count eighteen so far. They're spreading out both left and right. Range is three hundred meters. More coming up. I make over twenty five hostiles. Only one RPG so far."
>
> Jim Milliron was manning the .50-cal machine gun and supervising the deployment of all the crew served weapons, ensuring we deployed our maximum firepower and that in the heat of the moment nobody forgot about our two other machine guns and plinked away with an M16 instead. Tim Edwards grabbed the SAW and moved into the rubble on our left. Jimmie Wilson pulled the

> M60 out of Al's truck on the right and set it up behind the rear tire. Greaves traded his video camera for his rifle, then for Dave's rifle, which had an M203 grenade launcher attached to the barrel. Deeks loaded a grenade into his own M203. The two of them squatted by the rear quarter panel of my DMV. Aside from our vehicles and the small rubbled building there was absolutely no other cover. Matt was lying prone out in the open on the sand, his sniper rifle resting on its bipod.
>
> "Got somethin'," Dave called out. "Looks like an RPD, eleven o'clock." He put down the binoculars and swiveled the massive grenade launcher to bring it to bear on the target he had picked out.
>
> "I see him. Machine gun," Matt confirmed. "Looks like he's gettin' in firin' position."[9]

Captain Stevens dismounted, attempting to walk toward the militia with his arms raised to indicate they were not there to fight. The interpreter with the team was loudly shouting at the Somalis to cease firing, trying to explain the Canadians were just arriving for the scheduled meeting. AK-47 rounds continued to crack all around the Canadians.

As CPT Stevens walked slowly toward the SNF positions, a small body of militia gunmen were seen attempting to maneuver around the flank of the ODA. The detachment could see the gunmen carrying RPGs and RPK light machine guns as they were deploying. Captain Stevens returned to his vehicle. More mortar rounds landed among the Canadians and the AK fire increased. About 300 meters away, the gunmen broke to the left in what appeared to be a flanking maneuver on the Canadians. The interpreters with Major Kampmann, along with Captain Stevens' interpreter, kept yelling out that they were friendlies.

At this moment in the firefight, some town elders appeared and approached the interpreter. Once the elders understood who Major Kampmann was and why he was there, they ordered all the gunmen to cease firing. Calm was restored.

The elders explained it was a case of mistaken identity. They had not been notified of the Canadians arrival or the planned meeting. Contact with the SNF at Balenbale was now officially established. (It was later learned that the Canadian Cougar had a misfire and accidental discharge; it was not shooting back. The SNF not only employed mortars at the Canadians, but an M47 tank was later identified as firing its cannon during the skirmish.) A meeting was negotiated for the morning. All friendly forces returned to Matabaan.

Captain Stevens transported a CNN crew with him the next day to record the first meeting ever with coalition military forces and the SNF. The meeting was held about 800 meters outside the town under some trees. Major Kampmann represented the Unified Task Force; Captain Stevens attended the meeting. During the talks, Major Kampmann spelled out the rules of the new sheriff in town: stop fighting, put vehicles and heavy weapons into a cantonment supervised by the Canadians, Canadian forces would soon begin patrolling the area, and a system of communications would be established between the Canadians and the SNF.

During the meeting, ODA562 and the Canadian Dragoon troop provided security. Medics from the troop and the ODA spent their time in town conducting first aid for Somali civilians.

After the meeting ended, the delegations visited an SNF weapons storage park. It was there they discovered the operable M47 Patton tank which had fired on the Canadians the day before. Several vintage tanks were in the park (M41s and T-55s), along with heavy and medium Technicals mounting a variety of weapons. They found the 120-millimeter mortar which had been firing at them during the battle along with four D-48 85-millimeter Chinese anti-tank guns. Almost 75 mines of assorted types were stacked about.

An astonishing photo taken during the first meeting between UNITAF military forces in Belet Weyne and the Somali National Front (SNF) on February 11, 1993. MAJ Mike Kampmann, commander of the Royal Canadian Dragoons (in beret on right) and CPT Charles B. Stevens, ODA562 commander (in desert camouflage uniform), sit with General Ghani's men along with village elders of Balenbale for the historic *shir*. Major Kampmann sternly passed on the operating rules for the militia while the Canadians operated in the area (*Courtesy of C.B. Smith*).

ODA562 left the Canadians to continue with the CNN news crew on to Dhusamareeb, where CNN wanted to meet and interview Governor Halane and Colonel Hussein (Colonel Hussein was not present; his subordinate commander took the meeting). When the detachment was back in camp gathered around the video player to review the CNN tapes, Captain Stevens took quite a bit of ribbing when he was captured on film approaching the gunmen with his hands raised to indicate he was not carrying a weapon. The video clip soon became "the surrender clip."

Through the 27th of the month, ODA562 assisted with coordinating meetings for the Canadians with key players in all the towns, went once again to Fer Fer providing security for a meeting with the Ethiopians, and assisted the Canadians with organizing militia heavy weapons cantonments. They also participated in demolishing captured mines and ordnance. Meanwhile, ODA565 continued reconnaissance patrols within the Belet Weyne HRS. During this period, two Canadian armored vehicles hit mines near Qhat Qat, damaging their undercarriages. Thankfully, no one was injured in the explosions.

On the 25th of February, ODA526 escorted a representative of the State Department's DART for the conduct of an assessment trip northwest of Belet Weyne to the small village of Ferlibah, just west of the Shebelle River along the Ethiopian border. On the 26th, the team provided security to the Canadians when they met with Lieutenant Colonel Dahir in Dhusamareeb. Lieutenant Colonel Dahir was the commander of the Galciao sector, where the SSDF was in loose alliance with the SNF against the USC.[10]

SSG Gary Ramsey, the team medical sergeant, summarized ODA562's activities while with the Canadians.

> During the months of January and February 1993 the primary mission of ODA562 was to assist and advise the Royal Canadian Dragoons in demining operations, humanitarian aid, and locating and confiscating heavy weapons (technicals, tanks and artillery pieces). Locating and reporting on the Somali National Front (SNF) and the United Somali Congress (USC) operating in the area; the elimination of the Somali bandit threat in the area became a secondary mission for the Detachment. The Detachment and the RCD successfully demined the main highway from Belet Weyne to Dhusamareeb and the dirt road from Matabaan to Balenbale. The Detachment and RCD also successfully located and confiscated numerous "Technical" vehicles, tanks, and various artillery pieces, which were placed in the Canadian guarded compounds throughout the region. The Canadian and Special Forces medics operated daily medical missions in surrounding villages near Matabaan treating the villagers for illnesses, burns, TB, gunshot and shrapnel wounds.[11]

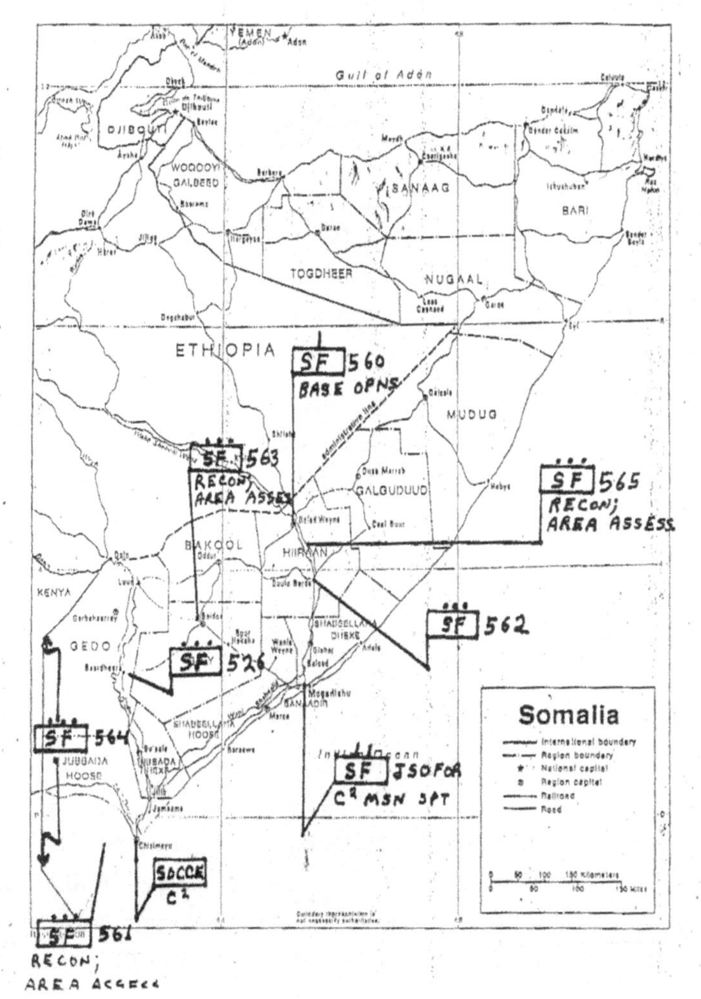

Hand-written annotations on a map of Somalia depicting 2/5th SFG(A) unit locations in February and March of 1993 during Operation Restore Hope, provided by LTC William Faistenhammer, the ARSOF commander, to insert in his 2/5th SFG(A) Annual History Report.

Humanitarian Relief Sectors Baidoa and Bardera – February 1993

ODAs 563 and 565 conducted operations in HRS Baidoa from late January to mid-March 1993. The Special Forces teams in HRS Baidoa provided valuable assessments to the Australian forces. However, at no time during their involvement in HRS Baidoa did the two ODAs ever have significant contact with the warring faction leaders.

In February, when increased acts of provocation against humanitarians occurred, the Australian commander reoriented his forces to protect NGOs operating in or near Baidoa. He planned to deter incidents and theft from bandits by positioning small groups of Australian soldiers inside NGO compounds and conducting 24-hour street patrols. This effort was successful in forcing most of the bandits out of Baidoa. Having now alleviated the suffering from starvation and protecting humanitarian relief infrastructure, it was time to cast the net wider in the HRS to secure the lines of communication from acts of banditry on convoys delivering food relief and other essential supplies. He initially began those operations in late January, focusing on Buurhakaba to the southeast in Area of Operation CATS (the main line of communication between Baidoa and Mogadishu) and the area to the southwest in his sector, the route to Oddur (AO TIGERS). The Australians divided the HRS up into eight Areas of Operation and assigned each one a code name; for instance, Baidoa was AO HAWKS.

ODA563

As ODA563 arrived to the Baidoa airfield by road march from Belet Weyne on January 30th, LTC David Hurley (Australia), the HRS Commander, tasked them to conduct an area assessment mission in the southwestern portion of his sector. They were attached under the tactical control (TACON) of the 1st Royal Australian Regiment (RAR) for operations. They were pleased with their accommodations, thinking they were much better than what they had experienced in Belet Weyne; this was not true for the "Diggers" of the 1st RAR who felt that living in cramped tents at the airfield, on cots and dirt floors, extreme heat, and with no amenities was any way to treat Australian troops on their first major contingency deployment since Vietnam.[12]

The threat in the area emanated from three major bandit gangs, followed by unexploded ordnance and mines. No significant militia forces were quartered in this area. ODA563 received their warning order to deploy to the Baidoa sector on January 24th. ODA563, led by MSG Les Jolley, spent three and a half weeks on this assignment before being replaced by ODA565.

The first area assessment mission was southwest of Baidoa near Qonsaxdheere, with a follow-on drive to Dinsoor (the western border of the HRS adjacent to the Bardera HRS). SFC Raymond E. Reid, the Detachment Medic, was on the ODA's night road march to Dinsoor in AO TIGERS.

> Then when, south of Consa Daheer (Qonsaxdheere), we slowed down and waited for nightfall, and then started traveling slowly down the road. And a vehicle came by, an indigenous vehicle. And we made an arrangement with them since they were going to Dinsoor. So we followed behind them without any lights on. And, if they were going to get stopped by bandits or anything, then we'd come to their aid.

> Well, it was a typical native vehicle and kept stopping and breaking down and a lot of problems with that. We followed them all the way to Dinsoor, or just short of Dinsoor that night. They got in there; there was no activity at all. And they told us that there was a lot of bandit activity on that road, particularly after the NGOS, the non-government organizations, had dropped off food anywhere in the area. And this was the next night right after that, so we expected something to be going on. But we never saw anything. We camped out that night and then the next morning went into Dinsoor and did an area assessment.[13]

Master Sergeant Jolly and Sergeant Reid took their interpreter to town with them and arranged a meeting with the elders. The other four men of the team tried their best to gather information from the locals and children playing around them to form their initial assessment. The team remained in Dinsoor until February 5th, reporting updates to the Baidoa headquarters each evening.

During their stay, a village elder surreptitiously met with Master Sergeant Jolley and the team outside of town to inform them of a village located about 50 kilometers to the west where a supposed nest of 200 bandits were camped (the town of Ganana). ODA563 took him up on his offer, but upon reaching the village and scouting the surroundings, they found everything abandoned.

When it was time to return back to Baidoa, the team departed Dinsoor by following an off-the-path route in order to assess three small villages. Again, Master Sergeant Jolley received information from the locals of theft and banditry rampant among the three villages, but nothing of significance was found.

After reporting to Lieutenant Colonel Hurley on the conditions they found in Dinsoor, Hurley sent them back three days later to refine their reconnaissance, as he intended to establish a patrol base in Dinsoor sometime in March. During their short trip, the team located an abandoned military camp at Manaas and spent one night with a U.S. Engineer battalion near Quansaxdheer (the site of an abandoned airstrip), gleaning any information they could from their fellow Americans on what they had seen in the area.

One of the overall impressions the team had from these trips was the lack of starving people; in fact, some towns looked very prosperous. It was later found out that towns of this nature were the locations of major gangs who enriched themselves and their clans with stolen and extorted goods. The one town the Detachment found in bad shape was reported to an NGO for a future visit.

In the following weeks, the team returned to Dinsoor and Ganana and met with as many village elders as possible. Although they discovered remnants of recent banditry, the team once again found no bandits in the area. They returned to Baidoa for rest and refit.

Between the 13th through the 22nd of February, ODA563 then conducted a reconnaissance trip and area assessment southeast of Baidoa in AO CATS and the area around the town of Buurhakaba.

On the 20th of February, intelligence sources warned the Australians of an impending attack by forces belonging to Colonel Jess (SPM), which were headed to Baidoa with 90 vehicles and 2,000 fighters. His staging area was predicted to be in Awdinle, directly west of Baidoa. The intent of the attack was to drive General Warsame and the USC anti-Aideed factions out of the area and recapture the town. Lieutenant Colonel Hurley alerted and deployed the 1st RAR into positions outside of Baidoa. ODA563 deployed with the Digger battalion. Ultimately, the intelligence sources proved false.[14]

ODA565 replaced ODA563 on February 23rd. The two teams conducted a quick orientation and transition, and ODA563 returned to Belet Weyne to prepare for a new mission. The team received a mission task (MITASK) to reinforce the SOCCE (-) in Kismayo. They deployed to Kismayo on the 2nd of March.[15]

ODA565 in Baidoa

ODA565, led by SFC Kenneth W. Barriger (Team Sergeant and acting Team Leader), were getting tired of the routine now established at Belet Weyne and hoped to conduct another mission. When not on patrol, any ODA sitting around in the base camp was subject to being used for administrative tasks or sat on standby as a rapid reaction force. Between February 1st and 22nd, ODA565 performed this unlucky task. To their relief, Chief Jackson gave them a new tactical mission—support the Australians in Baidoa.

After receiving the mission to Baidoa to replace ODA563, the team packed and on the 23rd of February conducted a road march on the 150 kilometer stretch of road to Baidoa, linking up with ODA563. They spent a short time with ODA563 to get a situational update and reported in to the 1st Royal Australian Regiment headquarters soon after. They also received their interpreter. Lieutenant Colonel Hurley's guidance to the team was to go conduct an area reconnaissance and assessment in Dinsoor. Hurley had some priority information requirements: What was the status of the population and where were they clustered? What was the refugee status? What was the attitude of the people towards coalition forces? Were they USC or SNA-affiliated? When conducting meetings with the elders, were there problems in their towns for receiving support from humanitarian organizations? Had any NGOs been in their area?

To support Lieutenant Colonel Hurley's objective to disarm any Somali not needing a weapon (with the exception of NGO security guards or local police forces who needed them), Sergeant First Class Barriger intended to check out the local arms market in Dinsoor for smuggled weapons and illegal sales. The team would not find any threat from clan militias during this mission. As with ODA563 and the Australian troops in other sectors, banditry and illicit activities were more prevalent than anything else. When the detachment arrived in Dinsoor, the first objective on their minds was to detect and remove any unnecessary weapons in the town. The team moved into the market area, a place notorious for trading, buying, and selling guns.

Somalia was a dumping ground for old weapons. Some of the weapons found by the team since their term in Somalia began included old British .303 Lee Enfield bolt-action rifles—once they found a 1918 vintage World War I rifle. Four weapons were confiscated on their way into Dinsoor and at the market area: an AK-47, one G380 rifle, one SKS, and one PPsH-41 Soviet submachine gun. A variety of mixed-type ammunition was also confiscated.

As they approached the central market, a shop owner appeared to be repairing weapons and reselling them. Members of the detachment began going stall to stall confiscating weapons. The mood turned ugly. During the team's interview, they described the scene at the market.

> . . . as we looked through the market, some of the team members came across weapons and it was a no weapons area of operation. And so these individuals confiscated the weapons. At that point, the townspeople really got upset and the team leader, SF Barriger, was in a town meeting with the commo sergeant (Sergeant Filio), and they radioed back to us saying there were some people that were really mad at what we had done and to call the team members in and go to their location, which

was only several hundred yards away. We did this, and we had a vehicle that drove by and fired several shots . . . Not at us, but pretty close to us, relatively, within 75 meters.[16]

No one was injured while the shooting took place, and afterward the team returned to Baidoa.

Meanwhile . . . back in Bardera with ODA526

With the humanitarian relief aid workers successful in their trip to Dinsoor, Captain Williams pushed on to the World Food Program's next objective, Sacco Uen. One representative still remained with the team to ride along. An aerial-delivered food drop was scheduled for the town, and the team was prepared to provide security at the drop location. Williams did not anticipate any troubles at the village located on the east side of the Jubba River, as Morgan's men were known to operate on the west side. The para-drops were successful; when the crowd got a bit testy and unruly, the WFP representative cancelled the third drop (no one looked like they were starving).

The team and WFP representative met with the village elders. It was an opportunity for the team to gather some information for the Marines. During the discussion, Captain Williams urged the elders to establish a security force to protect their own supplies—the coalition could not be everywhere, all the time. The elders felt banditry was their biggest problem, and promised to conduct a survey of their requirements and pass them on to the Australians.

Williams found that this area of the Juba river was inhabited by people of the Ogadeni clan, but there were also Rahanweyn and Hawiya mixed in. The population in the area was estimated at around 7,000, with an additional 1,000 refugees on the west side of the river. The region was under Somali Patriotic Movement (SPM) control, and the SPM district commander lived in Sacco Uen. The elders told him they knew of no SPM camps in the area, and that General Morgan was to the northeast near the Kenyan border, in the town of Liboi. Williams reported all of this intel to the U.S. Marine intelligence staff in Baidoa when the team returned.

He was informed by the intelligence staff that General Warsame was in Garbahaarrey preparing to conduct a drive against the Islamic fundamentalists in Luuq. Garbahaarrey was totally SNF controlled; it was the seat of government for the Gedo region. Only one Dutch NGO operated in the area, and word was getting out that the elders needed more humanitarian assistance; there were an estimated 7,500 people living around the town.

The team was directed to make contact with General Warsame and ascertain the humanitarian requirements in the area. They departed on the 5th of February, intending to drive straight through to Luuq, but when they reached Garbaaharrey and learned General Warsame was present, Williams stayed and sought a meeting with the General.

General Warsame extended his support for anything the team needed to make the mission successful but warned the road to Luuq was mined. He suggested a safe bypass at Beledxaawo (180 kilometers distance). Some elders informed Williams that the fundamentalists had erected a roadblock outside of the town, about four kilometers to the southeast, and about 15 or so men manned the roadblock. The men were backed up with Technicals mounted with machine guns and parked inside the town. That evening, General Warsame hosted the ODA for dinner. Williams recorded the proceedings in his patrol log.

> General Warsame ordered us a dinner and will talk with us after dinner. Meeting with General went very well, we talked at length. He is willing to meet with higher officials. Wants us to be his go between. He is worried about fundamentalists in Beledxaawo, Dolo and Luuq. Wants coalition help to disarm them. He does not keep his troops in camps—more like National Guard. They have approximately 150 police in Garbahaarrey—keep weapons in central location. No problem with SPM-friends—no problems to west. He came to Garbahaarrey to stop fundamentalists. SNF not interested in expanding, etc. Just want to keep area stable and willing to cooperate in every way with coalition forces.[17]

Beledxaawo was controlled by the al-Itihad al-Islamiya fundamentalist faction (AIAI) under the command of Musa Abdi Farrah, a former Somali Army Colonel. Farrah was assisted by an Imam, Sheik Mohammed Hagi Yusef. The AIAI were Marehan and claimed to have control over the entire region and all the fundamentalists in Somalia. Beledxaawo was a central headquarters for the AIAI and purported to be a fundamentalist training center. Unfortunately, no NGO could operate in the region without being under the control and close supervision of the AIAI, so few came to service the needs of the populace. The SNF controlled the region, but the al-Itihad al-Islamiya controlled the towns.

Not to be cowed, the intrepid operators headed toward Beledxaawo on the 6th of February, not knowing what troubles may arise. No coalition unit had directly met or experienced any major clash with Islamic fundamentalists during Restore Hope, and Williams hoped there would be no trouble during his attempt at first contact.

The ODA was indeed stopped at the roadblock. Everyone was tense. Williams sent a note ahead stating his desire to come into town and speak with the leaders. Shortly, a guard informed him that three men could drive in. Williams had the remainder of the team stand by on the radio in case he needed them. MSG Jose Bailey, the Team Sergeant, and Staff Sergeant Evans accompanied Williams. When they drove in, the gunmen on the Technicals did not get the word and pulled out to stop the Humvee. Once that situation was cleared up, Williams and Bailey were directed to an enclosed compound to meet with Imam Yusef. They pulled in and parked. About 150 armed men, pointing weapons at them, rushed out like ". . . ants on an anthill," said Williams. Jose Bailey told the story of what occurred next years later in his interview.

> My guys were magnificent and CPT Williams was downright fearless, especially our last confrontation with Islamic Fundamentalists (IFs) in the NW corner of our AOR (Tri-border area). Our original instructions from the JTF commander were to recce and if possible parlay with the IFs in Luuq. We stopped in Garbahaarrey where we had successfully befriended general Warsame (kept him out of the power struggle) between other forces. He re-routed us from the direct route from the south into Luuq due to the route being heavily mined. At first he tried to talk us out of going, then tried to give us an armed escort due to the hostile and soon to be found out unpredictable nature of the IFs; we of course declined on all counts.
>
> We headed NW to the tri-border area and Beledxaawo. We stopped about 200–250 meters from a checkpoint just outside of this village within pinpoint effectiveness of our M203 and M249 on loan from the USMC. . . . The CPT, myself, and SSG Evans drove towards the village with one of the IFs in our vehicle only to be intercepted by a technical mounting the biggest DShk pointed at us with about twenty other IFs all around. . . .

> We pulled into a large courtyard in front of this long, one story building with three or four technicals parked adjacent to it. They had M2s, DShks, 106 mm RR and PKMs mounted aboard; in the next instance from out of this long building a swarm of men armed to the teeth, bandit looking knuckleheads come running out like an ant nest that had been disturbed. They managed to form what looked like a firing squad around us with everything from AKs to RPGs and any other of a dozen foreign weapons types. They looked to be anywhere from 13–20 years of age and slightly agitated at best.[18]

While seated next to the Iman Sheik Yusef, Williams heard him say the team could travel no further. Providing food aid and negotiating with leaders in the region on any procedure for provisions was a political task, not a military one; UNITAF should propose a meeting with their diplomats to talk with him. All the while, the armed men kept their weapons pointed at the Humvee and the other two members of the team inside. Sensing he had an intractable problem, Williams radioed his men in the other Humvee providing over-watch at the roadblock and directed them to call the Marines for assistance if things got ugly. He told them he would fight his way out back to their position and for them not to attempt entering the town.

The meeting continued. The Sheik advocated for political recognition of the AIAI as a major player in Somalia (something Williams could not grant on the spot). The meeting abruptly ended, and Williams and crew returned back to the over-watch position with the remainder of the team. When offered an armed escort out of the area, Williams declined, yet the gunmen shadowed him down the road with a Technical.

When the ODA reached a sort of wadi, out of sight of the trailing Technical, they immediately pulled in and deployed into fighting positions. As the Technical turned the corner into view, they were stopped by the team and told to turn around. Thankfully, no armed confrontation occurred and the gunmen considered their "escort" duties complete, so they left.

For the remainder of February, ODA526 conducted assessments along the Jubba River near El Wak. They met with village elders and NGOs along the way to gather information for Lieutenant Colonel Hurley. They were suspicious upon seeing a large group of young Somalis in El Wak and suspected them to be either Morgan's or Warsame's men; the observation was reported higher.

The thrust of this recon patrol was to find bridges over the Jubba River. In this type of war in Somalia, a bridge was considered key terrain and needed to be guarded. Williams surmised if he could find bridges, he could find militia camps of Warsame or Morgan's forces.

The only bridge over the Jubba River between Bardera and Luuq was located at Bordubho and, therefore, was of significant tactical value to both the SNF and USC; they had fought over this bridge previously. The detritus of the battle lay around the village in the form of unexploded ordinance. To deny the bridge's use to the USC, the SNF militia removed the steel plates off the eastern half of the bridge, then mined the road to Baidoa to strengthen their defenses.

When the team located the bridge, the steel planks were back in place. The bridge was guarded by a 120-millimeter mortar and at least ten TM-46 and TM-57 mines along a six kilometer stretch of road. Locals in the area showed the team at least six of the mine locations, actually uncovering one of the TM-57s. A number of SEALAND metal containers were located in the vicinity as well. The area had all the markings of a potential military camp.

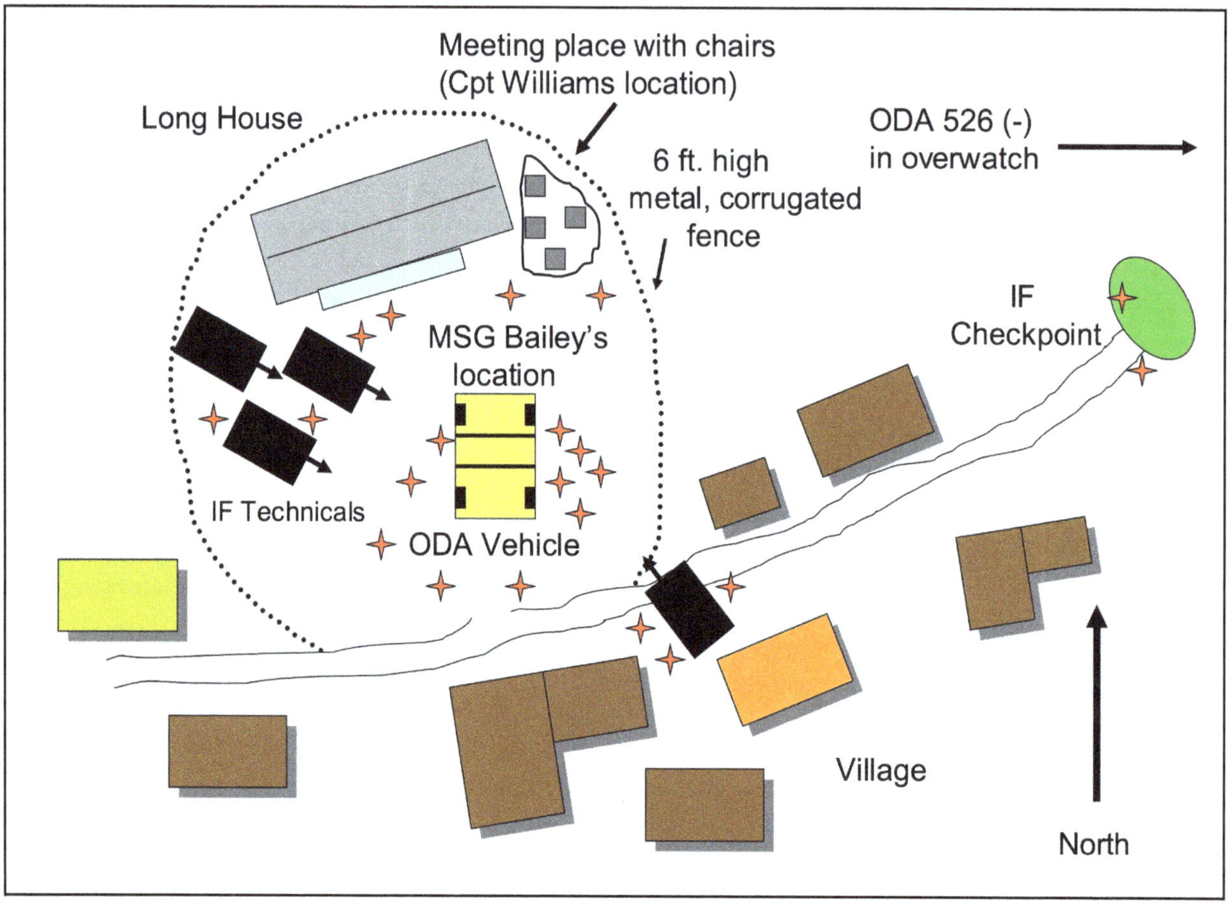

Diagram of the meeting between CPT Williams, Detachment Commander of ODA526, and Sheik Yusef, religious leader of the Islamic Fundamentalists near Luuq on February 6, 1993. Diagram developed from hand-written sketch provided by Jose Bailey (crosses indicate positions of gunmen).

General Warsame's forces were concentrated in the region, fearful of a possible USC/fundamentalist alliance attacking them from Luuq. The team noted everything suspicious in the area and sent the information higher. Afterwards, they drew off to find a secure piece of ground to establish a laager.

After a remain overnight (RON), the ODA continued south to cross the Jubba River at Buale, then drove up the west side of the river. The team remained in the vicinity of Buale, reporting on Morgan's activities. All indications pointed to Morgan not staging for an attack, at least not into the northern portion of the HRS. ODA526 assessed from their numerous discussions with the local populace that Warsame was most likely giving moral support to Morgan, even providing weapons from his cache at El Wak, but he was not providing Morgan any troops because he had his own hands full at the time with his ongoing operations in the north.

The team returned to Bardera for another mission break on the 25th, then finished their activities in February with another recon to Sacco Uen on the 28th.[19]

March 1993

The USS Wasp ARG arrived with its MEU (SOC). The flotilla parked off Kismayo, since the port and surrounding area was the consistent hot spot for UNITAF. On March 3rd, Ambassador Oakley finished his work as the Special Representative to Somalia. In the UN, conferences were held to draft a resolution for the turnover of UNITAF to UNOSOM II. Diplomats were arranging for a follow-on peace conference in Addis Ababa among the clan factions to finalize reconciliation, ceasefire, and disarmament details, pending binding signatures from the attending delegations.

ODA561

ODA561 conducted the liaison mission with General Morgan in Dhoobley from March 1st through the 5th. ODA561 worked well with General Morgan. Sometimes they sat and ate with him up to four times a day and spent long hours in discussions to glean information on his tactics and the status of his forces. The ODA assisted the coalition forces by providing a communications link to the General and his subordinate commanders. ODA561's work with General Morgan proved critical to UNITAF operations in that area.

ODA561 prepared for redeployment and conducted minor operations around the Kismayo area until their return to Belet Weyne on the 15th of March.

ODA563

ODA563 drove out to conduct the linkup mission with General Morgan (replacing ODA561), scheduled for February 5th. They stopped for the night at Hoosingow, arriving at Morgan's headquarters in Dhoobley the next morning. They were welcomed for the length of their stay to monitor his activities and conduct liaison between Morgan and the Task Force. Part of the monitoring duty was to oversee disarmament efforts by the General. General Morgan invited them to seize whatever weapons they thought were not appropriate. The team rounded up over 87 small arms, 208 mortar rounds, ten anti-tank missiles, and 150 pounds of assorted small arms ammunition. Their entire haul of seized ammunition was turned over to the 10th Mountain Division.

Major Carroll visited and discussed the need to begin moving heavy weapons into a cantonment, per the intent of the commander TF Kismayo, with General Morgan. That same day, Major General Arnold, the 10th Mountain Infantry Division commander, visited.

An entourage of visitors began a parade to Dhoobley. On the 7th, ODA563 escorted representatives of the UNHCR to four small villages around Dhoobley; on the 8th, a platoon of Belgian soldiers arrived to replace the team. The Platoon Leader was very interested in any of the locations where the team had found mines during the handover. The ODA returned to Kismayo and conducted debriefings with the TF staff. They were saddened to learn that just hours after their departure a Belgian vehicle hit a mine, killing five personnel.

ODA563 remained on the mission with the Belgians until their return to Belet Weyne.

ODA564

On the 1st of March, ODA564 participated in an aerial recon to ascertain faction movement around the towns of Afmadow, Qooqaani, and Bur Alle. Soon, they were back to liaison duties with Colonel Jess. The team deployed out to Jess' camp daily for about a week's period. It was on this mission they had their only altercation with the faction, on the 7th of March. The day prior, ODA563 had confiscated weapons and ammunition from Jess' men.

From March 3rd to the 7th, ODA564 was tasked to conduct a linkup with Colonel Jess and his forces outside of Kismayo. The team did not feel safe living in Jess' compound with him; he was still pretty much an unknown entity. Captain Barber moved his ODA into an over-watch position where he could observe Jess' camp (there was also a contingent of Belgian troops nearby if Barber needed reinforcements in an emergency).

While in position, around 5,000 people were observed walking down the road when shots rang out over the heads of the SF team. The team quickly maneuvered out of the line-of-sight from the gunfire and placed the Humvees into firing positions along the military crest of the hill. At that moment, the Belgian soldiers started to receive fire. Their commander requested armor support and soon reinforcements arrived riding in the Scimitars,which also took up over-watch positions near the team.

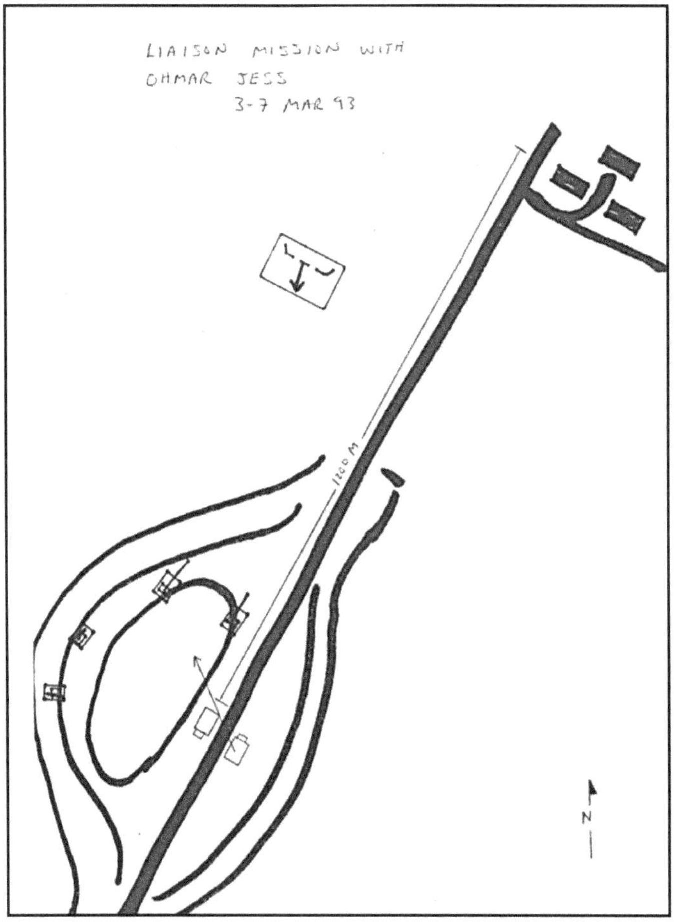

A copy of the hand-drawn map prepared by ODA564 depicting the incident in early March 1993 with Jess' militia when they were attempting to conduct a linkup with Colonel Jess at his compound (*courtesy of ODA564 AAR*).

Sergeant First Class Alan suspected the fire was coming from a truck on the road with perhaps a concealed machine gun—similar to a Technical. He maneuvered quickly to get a shot, but the rounds impacting the ground near him drove him back to the cover of the vehicles. The team did not return fire, fearful of hitting refugees. Captain Barber explained:

> We couldn't positively identify a target and what we could see, the refugees were down the road between us and them. And we didn't want to shoot over the refugees, the people that were evacuating the town. We didn't want to shoot over their heads. . . . The compound from which the fire came from had women and children in that compound. OK, I mean, there'd be a machinegun position and there'd be a woman and her kids right next to it. And we were approximately 1,000 to 1,200 meters away. You can be effective with an area target at that range, but there's no way that we could have returned fire and not hit civilians. So the moral codes they were (Jess' men) functioning by did not apply to themselves. They had no problems firing. There were thousands of people on the road between us and them. This hardball was in between our position and the compound. They fired over the civilian's heads. We just had different rules.[20]

The firing ceased and no one was injured. The team later learned the fighting involved elements of General Morgan's and Colonel Jess' factions shooting at one another. The local Somalis were angry over the disruption to their lives by being driven from their homes and took it out on the coalition. ODA564 completed its mission with the SOCCE and returned to Belet Weyne on the 15th of March. Captain Barber summed up their "irregular warfare" experiences in the HRS.

> I'd say that one lesson we learned, especially when we got shot at by Jess' forces, is that the situation there was very volatile and it changed, literally, within seconds. . . . We went in every day on a daily basis and nothing had really, you know, everything was positive and then, within a one hour time span, it went from us sitting down with these people and eating with them and drinking with them to them shooting at us. So the lesson we learned from that was that, regardless of how well things are going down, you can never let your guard down. We always had a plan . . ."[21]

The Belgians took independent control of HRS Kismayo on 5 March. TF *Kismayo* disbanded,but General Magruder left one infantry company and one Military Police platoon, along with theSOCCE, to assist the Belgians. The SOCCE was ordered to remain in support until the 15th of March.

The situation seemed relatively calm until General Morgan sensed an opportunity (with the 10th Mountain now gone) to once again attack Jess' forces. On the 16th of March, Morgan's forces infiltrated Kismayo to confront them.

With this violation of the ceasefire, UNITAF deployed the Quick Reaction Force to put down Morgan's incursion. Representatives at the peace and reconciliation conference ongoing in Addis Ababa visited both Jess and Morgan to reinforce the goal of the conference and returned to Addis with their findings. Jess and Morgan were banned from any operations in or near Kismayo. The arriving MEU (SOC) aboard the USS Wasp deployed forces to enforce the démarche and established security positions from Kismayo out to Dhoobley. General Morgan did not challenge UNITAF for the remainder of the operation.

With the media ever present, some of the exploits and experiences of CPT Barber and his team were captured by a reporter from the Army Times.

Coalition walks tightrope between violent warlords
By Kathrine McIntire

Army Times

KISMAAYO, Somalia – When Special Forces A Team leader Capt. William Barber [ODA564] walked into the town of Beer Xani, about 27 miles northwest of here, he smelled blood–literally.

"There were blood stains all over the place. The first two or three nights, the place just reeked of the stench of blood," he said. Barber and his team from C Company, 5th Special Forces Group (Airborne), on a mission to establish contact with Somali warlord Gen. Mohamed Said Hersi Morgan and his forces, had found their man.

Morgan has a reputation for ruthlessness. Fighting between Morgan's forces and those of Col. Omar Jess here has wreaked havoc on the entire region and terrorized the residents of this devastated city and the Jubba Valley.

Coalition forces entered the area in January and froze the factions in place. For Morgan, that meant that his troops were in the bush and forbidden to go to Kismaayo, which they maintained they had been wrongfully ousted from in the first place. It also appeared as if the coalition forces were legitimizing Jess, who had control of Kismaayo. Fearing an attack on the city from Morgan's forces, coalition forces set out to establish contact with Morgan.

Dialogue with Morgan and the other warlords who have carved up this desperate country during the past two years has become fundamental to restoring any kind of sustainable peace here, said Col. Mark Hamilton. Hamilton's job is to try to negotiate that peace.

'The guys with the guns'

"I have absolutely no problem in dealing with [Morgan] at all," Hamilton said. "I'm just trying to put together the circumstances for peace. It seems to me you'd better go to the guys with the guns first and tell them what you're about and tell them what you're going to do."

The job of establishing a dialogue with the warlords fell to Special Forces soldiers, uniquely trained to work with foreign military forces. "This [mission]has been a great opportunity for us to do what we're supposed to do. Doctrinally it's our job to link up with indigenous troops and establish liaisons with them," Barber said.

Alpha Teams were sent into each of the humanitarian-relief sectors to make contact with the controlling faction, said Maj. Lelon Carroll, C Company commander. Each team included a medic, engineer, weapons expert, communications expert, team leader and operations sergeant.

The teams gathered information pertinent to force protection and assessed the area. They also secured airfields, confiscated weapons, and detonated mines and ammunition.

In Beer Xani in early February, Barber's team established contact with Col. Weyrah, one of Morgan's chief subordinates, and a potential threat to Kismaayo. The team made daily contact with Weyrah and informed him of the intentions of the coalition forces. Several days later, another team [ODA561], led by CW2 Ron McNeal, replaced Barber's team, which moved farther west to Hoosinga, where Morgan had stationed himself. In a move that seemed to outsiders like sleeping with the enemy, McNeal's team moved into Weyrah's camp outside Beer Xani.

The idea was to create a trusting relationship with Weyrah and his forces. Days prior, U.S. Cobra attack helicopters had destroyed some of Weyrah's technicals that were approaching Kismaayo, and it was important that coalition forces be able to communicate with Weyrah. McNeal's job was to establish a rapport with Weyrah and his officers and keep both sides informed.

Keeping perspective

There's a danger that Special Forces soldiers can lose perspective when working so closely with characters like Morgan and Weyrah, who can be likable, if not charming, in person, and forget the atrocities the men have committed, Carroll said.

It's not something Barber is too concerned about. "When you go through town after town that's empty and you finally get to a town that's supposed to have 2,000 people and there's 200, and they talk about how one [person] stood here while they were executing people in town, and another one stood right over there, and they describe exactly who they were . . . we don't usually forget," Barber said.

While Hamilton tried to initiate talks between elders affiliated with Morgan and elders affiliated with Jess, it was important to keep Morgan from marching on Kismaayo, where many of his troops have homes and families.

Like Hamilton, McNeal thought the time might be right for negotiating a peace between Morgan and Jess. "A year ago, I don't think this would have been acceptable to them," McNeal said while camped with Weyrah's forces. "These people are hungry for what this means," he said, indicating the table and chairs set up at the camp. "Talking is so easy compared to shooting. They're sick and tired of war."

Weyrah gave his word to McNeal he would stay put for the time being, but he needed water and food for his troops. McNeal and the team members were arranging for the supplies. Morgan said he was willing to talk.

Things looked promising for negotiations until Feb. 22, when a Morgan supporter, Col. Gram, previously unknown to coalition forces, attacked Jess' compound in Kismaayo, ousting Jess from the city. However, the attack shifted the balance of power, perhaps for the better, Hamilton said.

"The circumstances are lamentable, but if there's a silver lining, there's a silver lining," he said. "It certainly was not a neutral city with Jess in it, that's for darn sure."

Violence flares

Violence in Somalia has halted relief work in the capital of Mogadishu and postponed the planned transition of the peacekeeping mission in Kismaayo from U.S. to Belgian forces. The violence erupted after warlord Mohamed Farah Aideed claimed the U.S.-led peacemaking force allowed followers of Gen. Mohamed Said Hersi Morgan to attack an Aideed ally in Kismaayo. A synopsis of recent events follows:

Feb. 22

Fighting erupts at 3:40 a.m. in the southern port city of Kismaayo as troops loyal to Morgan attack the Somalia Patriotic Movement compound of Col. Omar Jess. At dawn, U.S. and Belgian aircraft began monitoring the situation. On the ground, three Belgians receive minor injuries and one Somalian [sic] trying to enter the U.S. compound at the port is wounded by a Belgian soldier.

The withdrawal of U.S. troops is postponed and another 600 Americans are moved to the area. U.S. officials say the fighting also endangers peace talks set for March 15 aimed at reconciling rival clans.

Feb. 23

U.S. officials issue an ultimatum for Morgan to get his forces out of Kismaayo by midnight Feb. 25 or face military action. Morgan agrees to the removal . . .

Army Times
March 8, 1993

Belet Weyne – March 1993

Chief Jackson personally continued the Advanced Operation Base-level work to assist the Canadians in establishing rapport with the various faction leaders. Jackson accompanied the senior Canadian Joint Forces Somalia Commander (JFCS), Colonel Labbé, to the Galcaio region. Although this area was outside the boundaries of the Canadian HRS, the activities of the warring factions in this region highly affected the outcome of establishing peace and security around Belet Weyne. Colonel Labbé intended to assess the area pursuant to a possible request to extend the Belet Weyne HRS boundary further east.

The political command of the United Somali Congress was located at Dhusamareeb (this was not a military base for the USC, but rather served as a logistical base). The forward command headquarters for the USC was located in Gelsinor, located along the Chinese Highway between Dhusamareeb and Galcaio.

The headquarters of the Southern Somali Democratic Front was located in Galcaio, with another base in Habdubab, aligned with the Somali National Front forces in that location. The Canadians intended to visit each and every one of these bases and headquarters, requiring the passage through faction war zones and battle lines in several different locations.

Chief Jackson helped during the conduct of the forward reconnaissance. The goal was to coordinate the passage of lines into Galcaio and then into Abud Wak, the capital of the SNF forces.

The threat was high in this area, and the battlefield was mid-intensity. Both warring factions were mobile and had armored vehicles: a variety of tanks, U.S. made M47s and M41s and Soviet T-55s abounded in the area. Jackson and the SF members in his forward recon party even noticed converted rocket pods from MiG jet fighters mounted on jeeps.

The trip was without incident, even though the team actually crossed between two feuding elements on the Chinese Highway as they were headed to Galcaio. Warring faction tank companies could be seen in both lines, and it was clear the area in between looked like a no man's land. Mines could be seen everywhere during the journey.

For the remainder of the time, the split B-team under Chief Jackson remained in Belet Weyne awaiting the return of Major Carroll. They continued to support the efforts of the Canadian Battle group, provide service and support to the operational ODAs, and assist with other low level security operations when possible. Jackson reflected on his experience as a split B-team commander.

> You have to adapt quickly to the situation at hand. You've got to make some quick assessments of what needs to be done, do a mission analysis, and be able to react accordingly. For instance, with the conventional commander, in many cases, he sees what you have at hand. You've got to be able to tell him, "Here's what I can do for you." You've got to be proactive and keep in mind the safety of your individuals . . . of your soldiers. You've got to be able to advise that commander how he can best utilize you.[22]

ODA562 hits a Mine

Ms. Kim Maynard, a representative from the State Department's Disaster Assistance Response Team (DART), visited Belet Weyne as Ambassador Oakley's personal representative to conduct assessments on border villages and to speak with NGOs in the area. Ms. Maynard was accompanied by the Spanish humanitarian representative from the Civil Military Operations Command in Mogadishu.

She found opportunities to travel with ODA562 or the Canadians with Major Kampmann as they conducted trips to Matabaan and Guri Ceel and to the smaller villages off the main Chinese Highway.

On the 2nd of March, she coordinated with the Canadians for a trip to Matabaan, leaving in the morning. Ms. Maynard spent some time in the village assessing potential requirements for additional aid from the United States. Perhaps as a result of the attempt to co-opt the village, extremists had entered the town in late February and were still inciting anti-American sentiments.

As intelligence grew on this threat, ODA562 received a mission to conduct a recce to Balenbale with the Canadians to interview the locals and ascertain the validity of the threat.

The 3rd of March was similar to any other day in the Belet Weyne HRS—hot and dusty with clear blue skies. ODA562 and the Canadian security escort for Major Kampmann departed Camp Holland in the early morning hours. Stevens reflected on that day, "The 3rd of March was to be one of the most fateful days of my life, but like most fateful days it started out just like any other."[23]

ODA562 departed earlier than the Canadians, with the team loaded in two vehicles: one "clam shell" DMV and one "soft top" cargo Humvee. Captain Stevens, the DART representative, and other members of the team rode in the DMV. Master Sergeant Beucher rode shotgun in the soft-top with Sergeant First Class Deeks driving; SSG Jimmie Milliron rode in the back along with SSG Gary Ramsey; SSG Dave Davis manned the M60 machine gun mounted in the bed of the truck.

Around five kilometers from Balenbale, at about 10:00 a.m., Captain Stevens heard a distinctive crump from an explosion. A team member with Stevens yelled out, "They hit a mine!" He was talking about the second Humvee enveloped in black smoke and sitting 17 meters off on the left-hand side of the road in the brush, where it had been launched from the explosion.

Captain Stevens saw one body, which appeared blackened and badly burnt, lying in the middle of the road beside the destroyed Humvee. It was SSG Gary Ramsey, thankfully alive and moving his head and arms.

The medic, Bob Deeks, was lying on the running board of the truck on the side where the vehicle bore the brunt of the explosion, and suffered severe injuries from the blast. Detachment members grabbed the M5 medical kit bag and immediately began to treat Staff Sergeant Deeks' wounds and injuries. About that time, the Canadian column approached. Captain Stevens waved them off, worried there were other mines in the road. Major Kampmann was on the radio immediately requesting a Canadian MEDEVAC. Other Canadian soldiers in the column moved off the road into the brush to clear a helicopter landing zone.[24]

When the MEDEVAC helicopter arrived, the Canadian medics strapped Sergeant First Class Deeks on a backboard and put a set of military anti-shock trousers on his legs. Both Staff Sergeant Milliron and Staff Sergeant Ramsey had wounds from the mine explosion and boarded the helicopter, headed to the hospital at the main Canadian camp in Belet Weyne. Chief Jackson met them as they arrived at the camp and escorted them to the hospital.

The detachment overlooked a scene of carnage. Captain Stevens recalled:

> Small pieces of the vehicle were strewn all over, some as far away as two hundred meters, just behind my truck. The crater left by the mine was about a foot and a half deep and close to three feet across. The mine must have been a Belgian PRB Mk 3, as anything larger probably would have at least flipped the truck if not completely obliterated it. According to Matt and Al, who studied the tracks, it appeared that the wheels must have slid down from the verge and right onto the mine. A one in a million chance. My vehicle had straddled the mine; if I had been driving instead of Matt I would have been dead.
>
> . . . The engineers came to the conclusion that the mine had been emplaced the night before, but I was not so sure. They based that conclusion on two things. The first was that they believed the soil was too sandy and loose for the heavy Canadian vehicles to have been able to bridge over the mine for so long. The second was that they found indications that a mine had been removed recently from a spot just a few meters behind my truck. I cannot think of a reason, however, why the SNF would mine the road again. What purpose would be served by blowing up coalition vehicles and injuring our soldiers?[25]

The three team members were flown to the main hospital in Mogadishu. SFC Robert Deeks died of his wounds He was the first active-duty Army Special Forces operator to die in Somalia.

Staff Sergeant Ramsey and Staff Sergeant Milliron returned to duty after healing in the States. Meanwhile, the team remained in the base camp to conduct maintenance and refit, awaiting a new mission.

SFC Deeks' tragic death was reported in a release from the Associated Press:

> **Mine kills U.S. Soldier in Somalia**
> **2 Marines face hearings on whether shootings involved excessive force.**
>
> Associated Press
> MOGADISHU, Somalia – A U.S. soldier was killed by a land mine Wednesday, the second American to die in as many days . . .

. . . Six American soldiers have been killed in Operation Restore Hope, the U.S.-led effort to end clan strife and guard food intended for Somalia's starving.

The soldier killed Wednesday was a member of the Army's Special Forces, who died after the Humvee he was riding in struck a land mine en route to a meeting of Somali elders in a town 90 miles northeast of Belet Huen, Marine Col. Fred Peck said.

The soldier's name was being withheld until his family could be notified . . .

The Virginian-Pilot and The Ledger-Star
Thursday, March 4, 1993

ODA565 returned from their duties with the SOCCE in Kismayo. They provided a spare truck to Captain Stevens and provided their team medic. Now able to begin operations again, ODA562 returned to Matabaan on the 10th. The AOB had heard UNITAF needed to keep the border reconnaissance mission going to set the stage to venture further into north Somalia, at least up to the 20th of March; the next rumor from the battalion commander was April 1st.

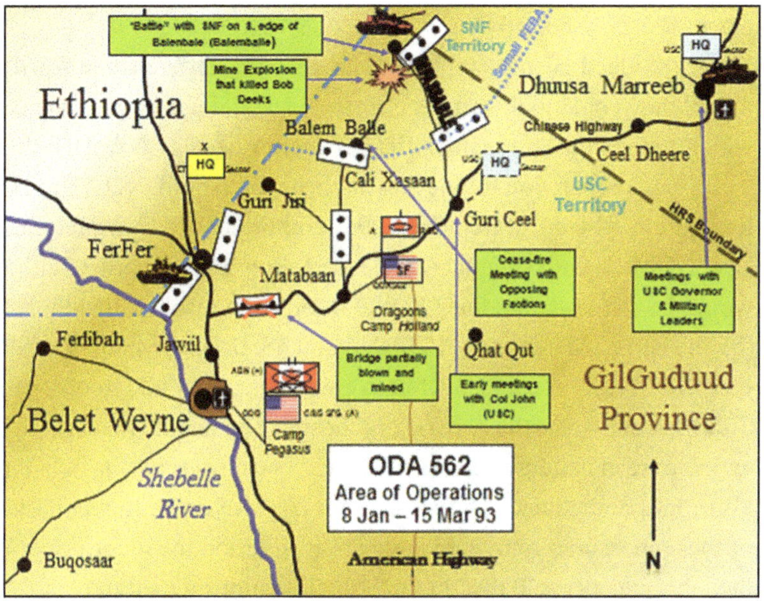

While the team was absent, the Canadians had discovered two large caches of mines north of Guri Ceel. The elders of Balenbale, Guri Ceel, and Matabaan were scheduling a meeting in Cali Xasan to resolve issues and lower tensions in that area. Canadian intelligence had also learned that COL "John" Hussein had been transferred to Kismayo due to all the trouble there. However, for the teams, life became routine with very little to do as everyone awaited the date to transfer the HRS to another coalition contingent.

On the 12th, Major Kampmann received the authority to extend his boundary and set up a substantial camp in Adado. Governor Halane hosted the elders meeting at Adado, and Major Kampmann arrived to inform the Somalis of his new move to the town. ODA562 accompanied Major Kampmann's delegation to Adado. Afterwards, the Canadians occupied their new camp. ODA562 departed, driving back through Dhusamareeb, intending to meet with Colonel Hussein before he departed to Kismayo, but he was gone; Colonel Roblay replaced him as the new representative for the USC.

On the 18th of March, Captain Stevens was ordered back to Belet Weyne. Colonel Bowra, the 5th group Commander, was visiting. All the team leaders present briefed the Group Commander on their operations, then took him on a trip to Fer Fer and a visit to a local village to observe a meeting between the elders and the Canadians.

The remaining time in HRS Belet Weyne was spent preparing for redeployment home, scheduled for the 28th of March.

The Addis Ababa Reconciliation Conference March 14–27, 1993

Between the 14th and 27th of March, Somali clan leaders and elders met in Addis Ababa for the "Addis National Reconciliation Conference" with the intent to design a nascent political system for governance. In the conference, they established a Transitional Council (TNC) as a ground-up approach towards an eventual national government, starting by building district and regional councils. Of course, Aideed did not participate in this decision, always adamant about there not being interference in Somali affairs from foreign nations (or from the UN, especially).

Aideed later attempted to run his own conference in April in Galciao; the UN refused to fund it and did all they could to thwart his efforts. He would not forget. One could even point to this snub as the catalyst which ignited the coming conflict between Aideed's SNA, UNOSOM II, and the United States.

In preparation for the inevitable handover of UNITAF to UNOSOM, the UN Security Council passed Resolution 814 on March 26, 1993, authorizing the transfer from UNITAF control to newly designated UNOSOM II, with Chapter VII Peacemaking as its charter. With the removal of U.S. forces, UNOSOM II would be left with less than 20,000 troops; the resolution authorized a build-up of five brigades, maximum (by the end of March, only 13,000 or so U.S. troops remained in Somalia). The estimate of forces needed for UNOSOM II for them to be effective was at least 30,000 troops. The leadership of the UN was still fixated on disarmament, since the Americans chose not to handle this task in the UNITAF mission and chose to ignore an increase in nation-building measures. The most robust task in the Resolution was the expansion of the UN effort into Somaliland. The security for the delivery of humanitarian aid also remained in the Resolution. All of these measures combined gave the Resolution the notoriety of being the first of its kind in the history of the UN (the efforts to disarm the Somali factions would inevitably lead to the mission's failure).

ODA526 March 1993

Captain Williams returned to Mogadishu in early March for mission guidance from Lieutenant Colonel Faistenhammer. Lieutenant Colonel Faistenhammer knew the battalion deployment would be over soon and the best utilization of ODA526 was to continue their support to the Marines in the Bardera HRS until the time for redeployment to CONUS.

The team continued the mission around Bardera and Garbahaarrey. After discussions with General Wilhelm and armed with new mission guidance, the team departed for Garbahaarrey on March 7th for meetings with General Warsame; on the 9th of March the team arrived to the town of Bordubho, where they found a large cache of munitions and a BM-21 rocket launcher. Detachment engineers rigged it all for destruction.

On the 11th, two children led the team to a cache of weapons 200 meters northeast of the water hole and the village. The cache consisted of 3,578 rounds of 23-millimeter (ZSU 23-4), five TM46 AT mines, three DM-11 AT mines, and 116 composite B explosive devices. The ODA sent one box of 23 mm, one sample of each type of mine, and a sample of the explosive back to Bardera. The detachment engineer rigged what remained for destruction, producing a spectacular explosion.

The ODA returned to Bardera for another rest and refit. On the 21st, Captain Williams reported to Mogadishu to seek the details of the upcoming redeployment schedule. The team would be going home soon. Williams returned to Bardera and began the preparations with his team to redeploy. Although packed, the ODA was able to support the Marines during an air assault mission into Salacce. This task would be the last mission in support of the Marines in HRS Bardera.

On the 30th of March, the team returned to Mogadishu to await their redeployment aircraft. The Marine regimental and battalion staffs, along with the commanders, praised the ODA for its work in HRS Bardera.

Departure from HRS Kismayo

Colonel Hamilton and the teams from the SOCCE continued to meet with Colonel Jess and General Morgan to continually make clear what UNITAF desired from their behavior. When the 10th Mountain prepared to depart Somalia, the HRS was left in the sole hands of the Belgian contingent. The Belgians were replaced by UNOSOM II forces in April.

The time for ARSOF (2/5th SFG) redeployment grew near. On the 15th of March, the SOCCE (-) and the three ODAs redeployed back to Belet Weyne. Major Carroll commented on the professionalism of the teams during this complex mission: "With rules of engagement, the guys could have killed numerous Somalis—SF maturity and training precluded that from happening. We all know that the center of gravity of the missions was humanitarian. Use of interpreters saved countless lives, especially our own. The Kismayo mission was one that required the patience of a Saint."[26]

By mid-March, ODA565 returned from Baidoa.

The End of RESTORE HOPE

UNITAF would soon transition to UNOSOM II, and many of the coalition nations were standing down their operations or conducting reliefs-in-place with newly arriving UNOSOM II forces. Once ODB560 was assembled at Belet Weyne, the unit held a memorial ceremony for Sergeant First Class Deeks. The unit conducted stand down activities the remainder of the month and prepared for redeployment. The Company B, 1/5th SFG(A) advanced party arrived and began coordination and planning for the arrival of their main party.

ODB520, 1/5th SFG(A), deployed to Somalia on March 27, 1993, with three ODAs: 521, 523, and 525. It was downsized in force from Major Carroll's contingent since it would be stationed in Mogadishu under UNOSOM II.

ODA526 remained behind at Belet Weyne to link up with their incoming parent company when they arrived. At the beginning of April 1993, Co B, 1/5th SFG(A), commanded by MAJ Dave Jesmer, relieved Co C, 2/5th SFG(A). C Company began their redeployment back to Fort Campbell, along with the 2/5th SFG(A) battalion staff and headquarters, and they all closed home station around the 8th of April.

Throughout the period of Operation Restore Hope, 2/5th accomplished the mission and met all of the operational objectives set for them. Specifically, region by region, the battalion achieved remarkable results throughout four Humanitarian Relief Sectors.

In HRS Belet Weyne:

- Established first contact with Ethiopian forces in Fer Fer and enhanced liaison between them and UNITAF; using this relationship gleaned valuable intelligence on Somali factions operating in the region
- Established contact with SNF forces in Balenbale, USC forces in Matabaan and Dhuusamarheeb (Dhuusamarheeb was outside the UNITAF sector), and with SSDF forces in Galciao (also outside the UNITAF sector)

In HRS Bardera:

- SF team established contact with General Warsame, the regional SNF commander, near Gharbaharrey
- Conducted reconnaissance to establish limits of AIAI influence near tri-border region and established contact with them in Luuq
- Identified several minefields and reported to higher, preventing future casualties
- Monitored Kismayo main MSR for any threats emanating from Gen Morgan and his forces

In HRS Baidoa:

- ODAs conducted successful liaison with Australian forces and enhanced mil-to-mil relationships
- Provided valuable expertise in the confiscation operations of arms, munitions, and mines

In HRS Kismayo (the most contested area):

- Conducted assessments along the Jubba River Valley
- Conducted reconnaissance and assessments west to Kenyan border
- Located General Morgan and his forces; collocated team members with both to provide situational awareness and liaison to CJTF commander
- Provided surveillance on Colonel Jess' cantonment

2/5th SFG(A) and its deployed forces drove over 26,000 miles while in country, captured over 277 weapons, and assisted in operations to locate and destroy over 45,300 pounds of ordnance. Throughout the operation, many lessons would prove valuable for later modifications of the SF-modified Humvees (named the Ground Mobility Vehicle System–GMV), as well as implementing ideas for improvement on various weapons, communication gear, and personal equipment.[27]

Throughout this period, the FOB (-), serving as the Army SOF component, developed and implemented a cohesive, non-linear battlefield collection plan and provided quality intelligence to UNITAF, CENTCOM, and several national level agencies. During Operation Restore Hope, the Joint Special Operations Forces command, and later its replacement, the ARSOF, met all of its operational objectives: contact was established with major regional leaders; potential threats to coalition forces were identified; and data on all accessed areas was provided. So successful were the Special Forces teams that, when Lieutenant Colonel Faistenhammer initiated redeployment activities on April 1, 1993, the Army Quick Reaction Force insisted on augmentation with SF teams. The UNOSOM II commander, Lieutenant General Bir (Turkey), foresaw a larger Special Forces role in the future of UNOSOM II and considered U.S. Special Forces a "must have." [28]

On March 31st, USN SEALs from the USS Tripoli ARG conducted a reconnaissance near Kismayo to look for an estimated 300 men from Colonel Jess' force, purportedly marching on Kismayo. The SEALs did not observe any troops and there was no subsequent fighting reported.

Ironically, General Said Hersi Morgan finally retook Kismayo in December of 1993 as U.S. Forces assembled on the Mogadishu airfield for withdrawal in early 1994. He continued to exert his influence in the area until 1999. Omar Jess became active in Somali politics and still is as of the date of this book. In 2018, he published his personal memoir about his experiences during the civil strife in Somalia.

As a slap in the face to U.S. soldiers, sailors, Marines, and airmen, the chain-of-command denied hostile fire pay to all U.S. military members, a policy backed up by higher commands and the administration so participation in a humanitarian operation with the UN would not be considered war-like. A Somali campaign ribbon was never struck, either. Those who participated were awarded the Armed Forces Expeditionary Medal, not even the Southwest Asia Medal.

When asked about his experiences in Somalia upon his return to the United States, Sergeant First Class Barriger reflected on the irregular warfare and unconventional warfare nature of the U.S. Army Special Forces' involvement in Somalia.

> But the reason that Special Forces is brought into play is because we're used to working with indigenous forces, we're more or less trained to go out amongst the local people and talk to these people, and we know how to treat them. I mean, your standard everyday Marine or Army person would not understand different cultures and things like that as well as we will, because we've gone to so many different countries
>
> . . . We have lived in those conditions. And that's why we were chosen for the mission. And that mission was not out of our realm because that mission was very much close to what a demobilization operation would be, if a country were to regain its own power, and strength, and it needed help to demobilize certain aspects of the country. And it's very hand-in-hand with what we're trained to do.[29]

Endnotes

1. E-mail between Lelon W. Carroll and the author on January 8, 2007.
2. Celeski, Chapter 2, Operation RESTORE HOPE (UNITAF): SF Operations 10 December 1992–8 April 1993, pp. 71–72.
3. Rutherford, Kenneth R. *Humanitarianism Under Fire: The U.S. and UN Intervention in Somalia.* Sterling, VA: Kumarian Press, 2008, p. 107.
4. Oral interview notes from ODA561, Company C, 2/5th SFG(A) conducted by the 44th Military History Detachment in the 2/5th SFG(A) classroom at Ft. Campbell, KY, 13 May 1993. Oral interview notes of all the detachments from 2/5th SFG(A) participating in Operation RESTORE HOPE, p. 16.
5. Oral interview notes from ODA564, Company C, 2/5th SFG(A) conducted by the 44th Military History Detachment in the 2/5th SFG(A) classroom at Ft. Campbell, KY, 13 May 1993. Oral interview notes of all the detachments from 2/5th SFG(A) participating in Operation RESTORE HOPE.
6. Celeski, pp. 35–36.
7. Ibid. pp. 28–29.
8. Ibid. pp. 39.
9. Stevens, Charles B. "Honor Bound: A Special Forces Detachment in Somalia." Houston, TX: Author's Draft Copy, 1996, pp. 331–332.
10. Stevens, pp. 333–340.
11. Gary Ramsey's personal note to COL (Ret.) Joseph D. Celeski describing his activities in Somalia operations in 1992 and 1993.
12. Breen, Bob. *A Little Bit of Hope: Australian Force Somalia 1993*. Victoria, Australia Battalier Books Pty Ltd. published under Echo Press, 2018, p. 129.
13. Oral Interview of ODA563 Co C, 2nd Bn, 5th SFG by the 44th Military History Detachment on 13 May 1993 in the2/5th SFG classroom at Ft. Campbell, Kentucky, p. 21.
14. Breen, pp. 160–161.
15. Mission highlights briefing files from the five ODAs of Co C, 2/5th who participated in Operation Restore Hope. These files contain annotated calendars and sketch maps of key operations performed beach ODA.
16. Oral Interview of ODA565 Co C, 2nd Bn, 5th SFG by the 44th Military History Detachment on 12 May 1993 in the 2/5th SFG classroom at Ft. Campbell, Kentucky, p. 17.
17. Dailey patrol log recorded by CPT Tim Williams, detachment commander for ODA526, on operations of the team December 1992 through March 1993. A copy of the patrol log was personally provided to the author by LTC (P) Williams at Hurlburt Field, FL during the JSOU SOF pre-command course in 2005.
18. E-mail from CSM Jose Bailey to retired COL Joseph D. Celeski, 23 February, 2002. COL (Ret.) Celeski followed up this e-mail with a personal interview at CSM Bailey's house at Ft. Bragg, NC.

19. From Tim Williams' ODA526 Patrol notes and Commo Log provided to the author to use in the preparation of this manuscript.
20. Oral interview notes from ODA564, Company C, 2/5th SFG(A) conducted by the 44th Military History Detachment in the 2/5th SFG(A) classroom at Ft. Campbell, KY, 13 May 1993. Oral interview notes of all the detachments from 2/5th SFG(A) participating in Operation RESTORE HOPE, p. 32.
21. Celeski, pp. 31–32.
22. Oral history interview with ODB560, Co C, 2/5th SFG(A), conducted by the 44th Military History Detachment in the 2/5th SFG(A) classroom, 12 May 1993, p. 31.
23. Stevens, p. 417.
24. Ibid, pp. 418–419.
25. Ibid, pp. 422–424.
26. Personal note to author on 17 January, 2007.
27. 7-2/5th Memorandum for Record, dtd 23 Apr 93, SUBJECT: After Action Report–Operation Restore Hope, Somalia. Executive Summary, pp. 1–3.
28. Memorandum: Executive Summary–Operation RESTORE HOPE, Headquarters 2d Battalion, 5thSpecial Forces Group (Airborne), Ft. Campbell, KY, dtd 23 April 1993, signed by the battalion commander, LTC William L. Faistenhammer.
29. Oral Interview of ODA565 Co C, 2nd Bn, 5th SFG by the 44th Military History Detachment on 12 May 1993 in the 2/5th SFG classroom at Ft. Campbell, Kentucky, pp. 33–34.

7

Operation Continue Hope: March - June 1993 UNOSOM II

> The world may have bitten off more than it could chew in terms of trying to bring the Somalis to a government. . . . People who look at the Somali situation now with even a small amount of optimism operate from the premise that it may be that now the foreigners are actually leaving Somalia, the Somalis themselves may be able to get together, close the door inside the family and say, 'Okay, now it's time to wrap it up and make a government.'
>
> — Dan Simpson, U.S. Special Envoy to Somalia

A transition date between UNITAF and UNOSOM II was finally settled upon—May 4, 1993. The UN decision and appointments were seen as favorable to the United States, which now saw it as its duty to canvass other nations to support the effort in Somalia.[1]

President Clinton was adamant about turning over the Somalia mission to the UN. America was not challenged by any major national security threat, except for the North Koreans withdrawing from the Nuclear Non-Proliferation Treaty and denying inspectors access to their sites (March 1993). The challenges were more domestic—the great Mississippi and Missouri River floods in April put pressure on the administration to show its capabilities with FEMA. On another item, Clinton appointed the first female Attorney General, Janet Reno, who would ultimately be caught up in the Waco compound siege to tragic effect, resulting in the administration's first scandal.

Meanwhile, in Bosnia, the UN declared Screbrenica a "safe area" to prevent ongoing acts of genocide against Bosniaks. This action would prove ineffective and resulted in the massacre of hundreds of men, who were buried in mass graves. Thankfully, there was no threat from China or Russia during this time. The American media focus was on the transition of U.S. forces over to the UN's UNOSOM II in April.

The United States designated Admiral (Retired) Jonathan T. Howe (USN) to replace Kittani as the UN Special Representative to Somalia on March 19th. Robert Gosende (an experienced State Department Foreign Service Officer who had previous experience in Somalia from 1968–1970) was appointed as the U.S. Special Envoy. From the Department of State, Ambassador April Glaspie was seconded to UNOSOM II.

LTG Çevik Bir (Turkey) was chosen on March 15th to lead the UNOSOM II military force, with MG Thomas M. Montgomery (USA) as his Deputy, appointed by CENTCOM on March 9th and arriving in April. The Joint Chiefs of Staff agreed to leave a residual force to ensure the success of the UN mission. An Aviation Brigade from the 10th Mountain Division, with its helicopters and one infantry battalion, would serve as the UNOSOM II Quick Reaction Force. A Marine Expeditionary Unit (MEU-ARG) afloat would also remain in the regional waters for backup. The 10th Mountain requested some SOF capability—one SF Company with two mounted teams would fill the bill. Other than that, the United States was adamant about not getting involved in nation building or any of the politics associated with Somalia.

Why agree to a QRF for UNOSOM II? The French, even serving in one of the quietest and safest Humanitarian Relief Sectors, threatened to leave the mission if their operations could not be backed up by a credible QRF with a decent response time (in other words, having reliable rotary-wing assets). They also canvassed the United States for a Marine Expeditionary Unit-type organization ashore (later solved with the infantry battalion of the 10th Mountain aviation brigade). No American armor was considered for the Quick Reaction Force, as Major General Montgomery felt the situation was not that dangerous. Armor would look like a hostile escalation, so it was deemed not needed.[2]

The Quick Reaction Force mission per UNOSOM II's OPLAN 1 stated the following tasks:

- Deal with attacks or threats to UNOSOM II forces;
- Be prepared to support expansion of the mission to central and northern Somalia; and
- Provide a reaction force to support contingency operations.[3]

With some U.S. forces remaining (anticipated to depart by summer, then the MEU-ARG would be used until UNOSOM II's charter ended on October 31, 1993), CENTCOM was not desirous of U.S. forces under the UN's OPCON (Operational Control). To solve this dilemma, Major General Montgomery was designated as the U.S. Forces Somalia Commander (USFORSOM). The 10th remained under OPCON of CENTCOM for any use or major operation; Major General Montgomery would have TACON (tactical control–local control).

The U.S. Logistics Support Command also remained. There really was no other country capable of performing this role, credibly. It would be essential to give the UN a tool for any hope of success for their mission. The LOGSUPCOM was retitled as the UN Logistical Support Command (UNLSC), but in this instance was placed OPCON to Lieutenant General Bir, remaining in support of UNOSOM II until December 1993.

Thus, the overall U.S. commitment to UNSOM II was approximately 4,000 troops with 1,400 of the total to be ashore. In time, the plan envisioned redeploying the troops ashore back home and just leaving the off-shore MEU for contingency responses.

From mid-March through early April, the laydown of forces for UNOSOM II was as follows:

- Bale Dogle – Moroccans
- Baidoa – Australia
- Oddur – French
- Gialalassi – Italians
- Belet Weyne – Canadians
- Kismayo – the Belgians (who assumed responsibility on March 5th)
- Mogadishu and Merka – Pakistan on March 26th and April 28th, respectively
- Bardera – Botswana on April 18th[4]

The Canadians and Australians were set to leave in May with the transition to UNOSOM II and withdrawal of the United States; they felt, like America, that their mission to end starvation was complete. Pakistan asked the United States for 72 M113 armored personnel carriers and eight M-48 tanks as a precondition to providing more force in Somalia, and the United States approved with a lend-lease type deal. The Indians were another large contingent for replacing UNITAF, however, they would not arrive until September to man the Bardera HRS.

Company B, 1/5th Special Forces Group (Airborne)

Company B, 1/5th SFG(A), deployed to Somalia on March 27, 1993 with three ODAs: 521, 523, and 525. MAJ David G. Jesmer commanded the company with SGM Patrick Ballog as his Company Sergeant Major. ODA521, an understrength team, was split up to augment the B-team and support the other teams. Jesmer's fourth team, ODA526, was already in-country having served in the UNITAF rotation; after duties supporting the Marines in Baidoa, they moved to Mogadishu and linked up with their parent unit (headquartered within the Embassy compound). They would not remain long—after turning over any critical gear and their vehicles to the other teams, they conducted a re-deployment on April 7, 1993.

The company was assigned the role as a Special Operations Command and Control Element (SOCCE) providing support to the 10th Mountain Division (LI) Brigade; the Brigade was tasked to remain in-country as the UNOSOM II QRF. It was reminiscent of December 1992 when Jesmer's SF company was designated as a SOCCE to the 10th Mountain Division for the conduct of Operation Restore Hope. The unit moved to Fort Drum but, due to unforeseen circumstances, was cut from the total number of personnel deploying into Somalia. Major Jesmer was able to send at least the split-team from ODA526 as well as the company's vehicles. Hoping they would eventually deploy, they soon learned the SOCCE mission had been cancelled on New Year's. ODB560 would form the core of the SF contingent (Major Carroll's Company C of the 2/5th SFG, located at Belet Weyne HRS).

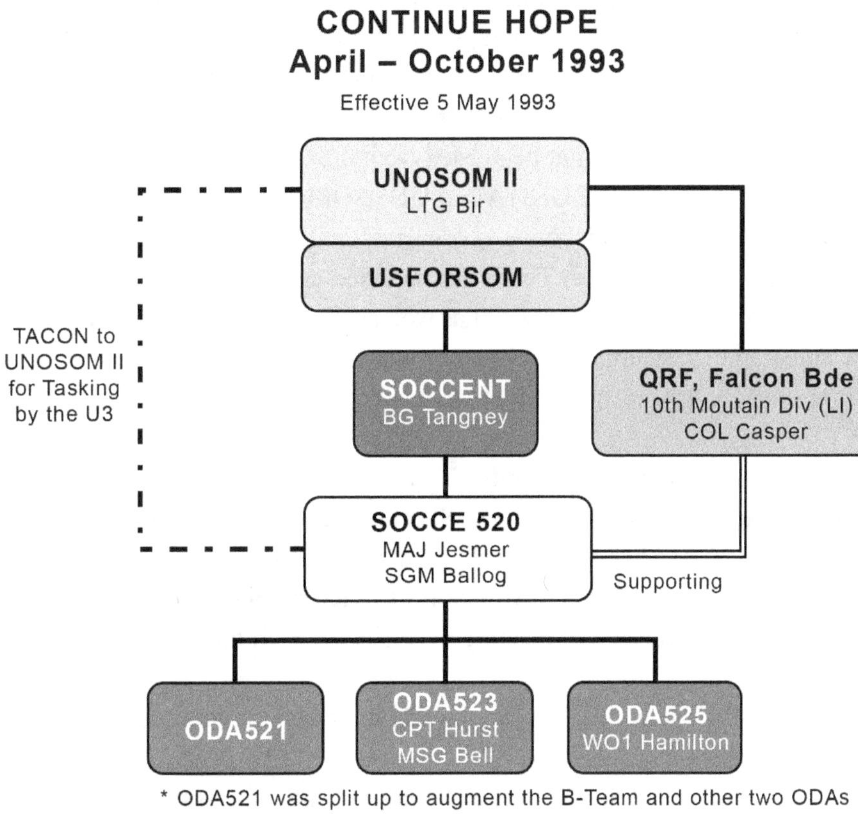

ODB520 replaced FOB 52 (-) as a downsized SF contingent as a result of a request from the 10th Mountain to have some Special Forces capability in their QRF role, beginning on May 4th when UNOSOM II took over military operations in Somalia from UNITAF. ODB520 served in this role as a SOCCE. A Special Operations Command and Control Element is defined in U.S. Army doctrine as follows.

> The SOCCE is the focal point for the synchronization of SOF activities with conventional force operations. The SOCCE is predominantly an ARSOF C2 element formed for land-centric warfare. It performs C2 or liaison functions according to mission requirements and as directed by the establishing SOF commander (JFSOCC or CDRJSOTF as appropriate). Its level of authority and responsibility may vary widely. The SOCCE normally is employed when SOF conduct operations in support of a conventional force. It collocates with the command post of the supported force to coordinate and synchronizes SO with the operations of the supported force and to ensure communications interoperability with that force.

Although a SOCCE is traditionally placed under OPCON of the U.S. Army organization it supports, the arrangement was a bit unique in Somalia. Company B, 1/5th SFG(-) was further designated as coming under the tactical control of the USFORSOM Commander and as taking guidance from the UNOSOM II U3 for missions. In fact, at no time did the SF Company conduct any mission in support of the 10th Mountain; they merely received logistics and administrative support from the Brigade QRF. The U3 requested that a four-man liaison element also remain in Somalia to support his staff, but that request was denied.

The first order of business was ascertaining just what role the company required. Major Jesmer spent some time with Lieutenant Colonel Faistenhammer (FOB52) and with the SOCCENT J3, Colonel Smith, to hear their ideas on effective use of the two SF teams. It was a bit strange to have such a small contingent for the vast amount of requirements which might be needed from the U3.

In preparation for the transition of UNITAF to UNOSOM II on the 4th of May, the UNOSOM staff drafted the operation order guiding the incoming contingents on their UN mission with descriptions of the phases required to achieve their objectives. The order included an SOF annex for the various contingents with these type forces deploying under UNOSOM II as well as the mission and intent of the UN commander. The UNOSOM II Campaign Plan with its SOF Annex would serve as a guide for the initial employment of ODB520's assets after the May 4th transition, as shown below:

> **Mission.** On order SOF conducts operations in Somalia in support of UNOSOM II objectives. These operations include: civil-military, SR [Special Reconnaissance], UN sponsored humanitarian activities, and psychological operations designed to assist in stabilizing the distribution of relief supplies and services; facilitate the distribution of relief supplies and services; and to develop an indigenous infrastructure capable of meeting the basic needs of the host population.[5]

Lieutenant General Bir's intent in the order clearly indicated the proposed campaign envisioned by the UN to not only consolidate areas under control during the UNITAF period but also continue to expand UN operations to central Somalia and northern Somaliland in an attempt to stabilize the entire country and bring it under some form of governance.

The campaign plan contained six phases, with SOF objectives for each phase. Phase I began on May 4th and consisted of all the activities UNOSOM II required to re-tailor the SOF in theater. ODB520 completed the task to transition from UNITAF operations into operations under UNOSOM II and began planning to develop courses of action to conduct required reconnaissance and assessment operations. Additional SOF tasks were to establish a Special Operations Coordination (SOCOORD) element within the UNOSOM II Headquarters and the establishment of both a Civil Military Operations Center and a UNOSOM II Psychological Operations Task Force (these proposals did not happen and were disapproved by the JCS).

Phase II consolidated the activities in the current Humanitarian Relief Sectors; at the same time, it employed SOF into central and northeast sectors of Somalia under the desire to expand UN operations throughout Somalia. The key population areas of Bossaso and Galcaio were to be the focus of this expansion. The campaign plan afforded three months to achieve these objectives, followed by SOF transitioning these areas to conventional UNOSOM II forces in Phase IV. Phase IV was planned for completion by September 1993.

Phase V repeated the SOF activities in Phase III–IV but with focus on the northwestern areas of Somalia. Once complete, SOF was to turn over all reconnaissance and assessment tasks in their sectors to UNOSOM II conventional forces and then assist them in implementing Phase VI, the withdrawal and redeployment of forces during the transition of all military and civic programs in each area to indigenous control.[6]

Major Jesmer at least wanted to add value with his small contribution.

The deployment to Somalia was worthwhile and provided numerous lessons learned for the Special Forces community. I want to note, however, that the unit deployed without having received a formal mission statement. Rather, a task organization (to include a SOCCE and two detachments) was fixed with a general mission to replace the Special Forces units already in country. As I understand it, the tasking was in response to a request by the CG, 10th Mtn Div (LI) to retain a Special Forces' capability with the Quick Reaction Force (QRF) under UNOSOM II in Somalia. Only after discussions with LTC Faistenhammer in Mogadishu did I understand his logic for determining that task organization. His concept was to deploy SF detachments ahead of UNOSOM II forces in the central and northern regions of Somalia to conduct special reconnaissance/area assessments. He also foresaw a likelihood of expanding the Special Forces force structure whenever UNOSOM II began preparations to deploy into Somaliland in the east. This concept was planned before a UNOSOM II concept plan was fully developed, however (indeed, the concept was briefed to the UNOSOMII leadership after my arrival for consideration—it had been briefed to the UNITAF leadership earlier) and had the situation not deteriorated so drastically in June, I believe we would have been terribly underemployed due to the lack of coalition forces to effect the UN expansion north.[7]

April

Major Jesmer saw his first order of business, particularly while still under the command and control of the JSOTF and FOB52 (-), as familiarizing his unit with the overall situation in Somalia through the conduct of reconnaissance and area assessments. This was going to be a month where the ODB worked for the JSOTF but was already receiving some guidance from the soon-to-be UNOSOM II staff. For a short period, Major Jesmer's unit was intermixed on UN missions while other SF teams and B-teams were still completing their missions under the UNITAF. For instance, ODA526, a team in Major Jesmer's company, was attached to Company C of 2/5th SFG(A) during Operation Restore Hope, and ODA526 would often work alongside Major Jesmer's teams, both with different mission purposes.

ODB520 focused their efforts on southwest Somalia—even though the future UNOSOM II desired to expand into central Somalia, they were still not in control of the region, and UNITAF stuck to its assigned HRS sectors. The JSOTF anticipated a quick drawdown as UNITAF departed and was not seeking an expansion of any new missions. Even so, each trip conducted could at least update the UNOSOM, CENTCOM and Defense Intelligence Agency databases, given Special Forces were one of the few units who could roam far and wide. In most cases, the information the teams provided higher command was the only intelligence gathered by any asset.

Several route recons were conducted to visit other peacekeepers remaining in their HRSs and to update area maps with the current condition of roads and terrain. These trips were conducted in Merka, Kismayo, Baidoa, Bardera, Belet Weyne, and Galcaio.

On one such trip to Baidoa in early April, ODA523 met with the local NGOs as part of their information gathering. Here, the team first learned of the presence of the Islamic fundamentalist organization, the al-Itihad al-Islamiya (AIAI), near Luuq. Meanwhile, ODA525 conducted their area assessment in Merka then moved back to Mogadishu, where they conducted a Medical Capabilities mission (MEDCAP) with the Samaritan's Purse NGO.

Major Jesmer and both the SF mounted teams deployed to Kismayo in mid-April to conduct a meeting with Aideed's financier, Osman Atto. Due to this performance by the Special Forces, Major General Montgomery changed his mind and decided to use them under his OPCON control, given their capability to extend their personal contacts down to the local warlord level. Although he did not succeed in switching them from OPCON under SOCCENT over to UNOSOM, a TACON arrangement was worked out so the B-team could get its mission taskers from the UNOSOM U3 (this command and control relationship would continue after the transition of UNITAF with UNOSOM II).

The trip to Kismayo to meet with the Belgians did not end on a good note, however. While there, the SF Company and its teams witnessed the harsh treatment of Somalis by the Belgian peacekeepers and reported their conduct higher. This resulted in a rebuke from the UN to Belgium, and there would remain ill will between the Belgians and Americans thereafter.

In April, the B-team assisted the efforts of UNOSOM to establish the Somali Security Forces (Police). They were skirting the fine line of legality in that U.S. military forces were not allowed to train the police of a foreign country (a funding illegality)—one SF weapons NCO did assist to alter some on-hand PPSH-41s from automatic weapon firing to semi-automatic weapon firing mode. The other concern by some on the UNOSOM staff was that the weapons could easily be reconfigured to automatic firing mode by someone with fundamental gunsmith knowledge.[8]

In the third week of April, ODA525, operating as a split team, was operating in the Baidoa Humanitarian Relief Sector and met with the nongovernmental organization Christian Relief Services (one of a few in the area, including the World Food Program relief workers). The team was informed by the relief workers there might be a possibility to arrange for a meeting with the elders of Luuq. The NGO had been told by some elders they were seeking some of the humanitarian relief like that delivered in other areas. They also stated they were willing to clear the road between Luuq and Berdaale to provide easier access and speed up any deliveries (which the fundamentalists did complete). The details of the "meet" with the NGO were incorporated into the daily INTSUM for April 14th and sent higher by Major Jesmer.

This report, and the earlier report by ODA523, piqued UNOSOM's interest. What was the situation of refugees and any NGO attempts to serve them in that area? Major Jesmer met with UN representatives and the UNOSOM staff and acknowledged he could run a reconnaissance and area assessment out to Luuq. While there, his team would attempt to meet with fundamentalist leaders. The al-Itihad al-Islamiya radical fundamentalist faction (the AIAI) was considered a terrorist organization in the region with links to al-Qaeda, so there was great interest from the intelligence community to glean any information on their activities (the AIAI also operated in the town of Merka). With approval from higher up, Major Jesmer tasked ODA525 with the mission.

It would not be the first time the SF had contact with the fundamentalists. Earlier, ODA526 experienced an altercation with them resulting in U3 placing the area in the vicinity of Luuq off limits to coalition forces (Luuq is located in the extreme northwest corner of HRS Bardera), setting a line of latitude on the map as the boundary—no patrols beyond three degrees, thirty minutes.

Botswana was scheduled to assume the duties of HRS Bardera once UNOSOM II became effective. They deployed into the region in April to establish their base and conduct transition activities with the outgoing UNITAF contingent. ODA525 went to Bardera on April 18th to hold preliminary coordination meetings with the Botswanan command and update them on the new missions for the incoming B-team rotation. While in Bardera, part of the mission of ODA525 was to seek any possibilities of reestablishing contact with the fundamentalists in Luuq. They were going to use the World Food Program NGO's good offices to arrange a meeting, but it had to be held south of the restricted maneuver line emplaced by the UNOSOM U3. The WFP representative, in consultation with the team, felt like the 24th of April might be the next possible window for the meeting; he would have to speak with the elders from Luuq, however, to confirm this meeting date.

Armed with this bit of information from ODA525, Major Jesmer met on the same day (April 18th) with the incoming UNOSOM II staff, Lieutenant General Bir, Colonel Ward, Ambassador Glaspie and Lieutenant General Bir's UN political affairs advisor to seek guidance on any intentions to reengage with the fundamentalists in this region. Could the SF confirm or deny the potential for UNOSOM II forces to deploy into this area after the transition with UNITAF? (Lieutenant General Bir and his staff deferred the decision to UNITAF, as they were not in command yet.)

Major Jesmer then met with Colonel Egan and Colonel Gregson from the UNITAF staff and explained the initiative to seek access to this area once more. He was cleared to conduct the meeting after Major General Zinni granted his approval on April 19th and further assured Major Jesmer he could disregard the maneuver restriction line. UNITAF had demonstrated during Restore Hope they would not be dictated to by any clan faction as to where and when they could go within an HRS; also, UNITAF certainly did not take marching orders from UNOSOM as to U.S. forces activities.[9]

On the 20th, Major Jesmer attended a seminar sponsored by the UNHCR concerning refugee repatriation and spoke with the representative from Bardera, Killian Kleinschmidt, on the possibility of co-opting the fundamentalists to work for UNOSOM. Major Jesmer was surprised to learn of an upcoming meeting with the fundamentalists and the UNHCR on July 24th and proposed his SF team participate, promoting their maturity and cultural awareness as an asset during the meeting. After receiving agreement on the proposal, Major Jesmer met with Lieutenant General Bir's UN political affairs advisor to seek guidance on what approaches the team should use if such a meeting occurred with the fundamentalists. From this meeting, he was able to provide guidance to the detachment commander with respect to any communications with Islamic fundamentalists.[10]

Major Jesmer's memorandum to ODA525 included this guidance and outlined the purpose of the meeting as ascertaining ". . . the intentions of the Islamic fundamentalists towards possible UN deployment of military forces into regions under their control." He wanted the team to emphasize to the fundamentalists they were under no obligation to participate in the implementation of the Addis Ababa accords, nor did the UN intend to impose a system of government in Somalia.[11]

He added further guidance to the team in his memo, as shown below.

> I wanted it reinforced that the UN was not there to impose any other religious or cultural norm on the people of Somalia, but merely wanted to establish a level of security needed to ensure humanitarian relief operations. It was to be made clear that UN forces would show up one day and seek to disarm the militias in the area and no militias would be allowed to impose their control into other parts of Somalia.

> I also directed the team to reinforce with the fundamentalists that SF and 5th SFG(A) soldiers had great respect for the Islamic religion and had worked in a variety of countries throughout the Middle East to assist Muslims. The team was there to help and to serve as a communications bridge between the IAIA and UNOSOM. Last, I wanted the team to show its firepower and be firm and to use their radios to call and talk with other forces and supporting air, so the IAIA knew there was military force backup for the team if anything went wrong.[12]

At Baidoa, ODA525 split-team was back to full strength when the other half of the team met them. The now present team leader was then briefed on the pending mission and met with Jeff Louis, a representative from the WFP NGO in Baidoa. Mr. Louis further apprised the Detachment Commander of the situation in Luuq and agreed to attempt coordinating a meeting at a village between Garabahaary and Luuq.

The meeting was successfully arranged by the elders and the UNHCR in the town of Luuq. A UNOSOM humanitarian officer present in Baidoa expressed his concern with showing up to the meeting with soldiers and armed vehicles when it was to be a neutral meeting about humanitarian relief supplies. Before the Detachment Commander became embroiled in his first attempt at diplomacy with the UN, he informed Major Jesmer on the issue of his team being present at the meeting. Major Jesmer's guidance was to defer to the UNHCR representative.

At this same time, the UNHCR was in negotiation with fundamentalists in Belhow, located near the Kenyan border in the town of Mandera. A meeting was scheduled, but it was not time sensitive. Meanwhile, the mission to Luuq still needed to be conducted.[13]

On the 23rd of April, ODA525 deployed to Luuq with a split-team (led by SFC Otto Hicks), taking along two interpreters to assist with the meeting. As they arrived around mid-morning, the demeanor and actions of the fundamentalists appeared extremely hostile to the team. The team met with Ali Hasan, the Chief of Police in Luuq. In the brief exchange between the team and the gathered elders, they were accused of entering the area without permission from the AIAI, and then told to put their weapons out of action. It was time to depart—the unwelcome team was pelted with stones by several villagers as they exited.

Based on the mistrust from the AIAI as to the team's intentions, a Technical vehicle with several gunmen, and mounting a .50-caliber heavy machine gun, trailed the SF team as an escort; the ODA repeatedly stopped to inform the Somalis they did not require a security escort. This charade went on for about 30 kilometers.

SFC Otto Hicks motioned for the team to pull into a draw right after crossing a stream bed to assess the situation. The team prepared itself for an engagement, not knowing the intentions of the gunmen. The security escort was still approaching the team. Hicks placed the vehicles in defilade with weapons oriented on the approaching threat from the Technical. While they were being followed, the driver of the Technical consistently sought to flank the ODA, with the gunmen in the rear pointing their .50-caliber weapon at the team. The Somalis, seeing the team displace in the draw, halted. Around seven to eight gunmen hopped to the ground and began maneuvering towards ODA525. While this occurred, the .50 caliber opened fire.[14]

The company Sergeant Major, SGM Ballog, related his knowledge of the incident after the event.

> On one occasion [earlier during UNITAF operations] SFC Otto Hicks led a team to a village using their DMVs—typical two-vehicle operation—and were shadowed by techs in Toyotas quite a

distance. He visually recon-ed the village and decided it was too hostile. The Somalis would signal their friendlies with gunfire as they shadowed Hicks (other units often thought they were "under fire" by these signals).
As ODA525 exfil-ed the area, the Toyota (w .50) took them under fire at a distance. Hicks sped down the road to a draw and pulled up into a hasty ambush. The Somalis came barreling down the dusty road unaware into the kill zone. They [Hick's SF team] killed the gunner, destroyed the vehicle and disabled the .50 caliber. Some Somalis escaped into the bush.[15]

Major Jesmer recounted:

Referring to ODA525's return to Bela How, SFC Hicks was not the Team Sergeant, but happened to be in the area of Bardera with a split team from ODA525 when we got the mission to attempt to co-opt the AIAI into working with the UN. The negotiations failed, and only then was the split team followed out of the town and back towards Bardera, and by only one Technical . . . when the two DMVs got close to friendly lines, near Gaarbaharey, SFC Hicks decided he needed to stop the Technical so that they wouldn't end up being caught between two opposing militias.

So, then he stopped his two vehicles in a wadi—one on the far side in a hull down fighting position, and the other to the flank in the wadi. Then he and SSG Clifford, as I recall, went back to the center of the wadi with the intent to explain to the Somali militiamen that they needed to stop following them. When the Technical pulled up on the near bank of the wadi, they deployed, and then opened fire. It was not meant to be an ambush, but rather to simply convince the Somalis, by show of force, to turn around. When the Somalis opened fire, ODA525 personnel returned fire with the MK-19, the .50 caliber, and the rest of their small arms.[16]

ODA525 had no casualties. The AIAI gunmen suffered one KIA and possibly three more wounded in this clash with the SF. This incident would later be recognized as one of the first clashes U.S. forces may have had with a radical Islamic organization with close ties to Osama Bin Laden.

All of ODB520's efforts in April to prepare for UNOSOM II operations led to a less than optimal assessment on the part of the B-team leadership—UNOSOM II was not ready for the transition and lacked the capability to expand to the north, and the roles and missions for SF were not very well defined. Enough forces were not on hand to consider attempting the mission of disarming the militias. The area had great potential for the conduct of Unconventional Warfare (teaming with clan militias who were challenging the SNA), but the Clinton Administration and the JCS were insistent on not taking or choosing sides in this arena. Aideed, correctly assessing the weaknesses inherent in the UN reassuming control, increased his anti-UN rhetoric on Radio Mogadishu. The B-team was tasked to develop a plan for a Direct Action raid on the radio station.[17]

May

By May 4th, the Australians and Canadians departed, leaving a huge gap in UNOSOM II capability, which was down to about 15,000 effective peacekeepers. UNITAF turned over its operations to UNOSOM II. Some of the nation states supporting the U.S.-led UNITAF were satisfied with their efforts to alleviate starvation in Somalia, particularly by April. This was the political mandate set by their politicians. When America turned over a successful mission accomplishment to the UN, Canadians and Australians left, not wanting to be under the command and control of a UN entity or rely on them for quick reaction forces or logistics.

On the way out of town to return to the United States, Major General Johnston and Brigadier General Zinni were riding through the streets in a HMMWV when General Johnston had the vehicles stop to give out pens and pencils to Somali children. Brigadier General Zinni wrote of his exchange with General Johnston after the act: "After his little act of charity, he slowly swung his gaze around. Something was obviously weighing on his mind. 'What are you thinking about?' I asked. He looked up at the bright sunny sky. 'I give this place thirty days,' he said, 'and then it's all going to hell.'"[18]

General Bir asked the departing UNITAF to give him the "worst case" assessments they were working on in the months before the transition date.

And so it began to "go to hell." The transition was going rough. There was not much enthusiasm from the remaining countries to aggressively conduct Chapter VII activities, nor was UNOSOM II effectively resourced or staffed to take on this monumental challenge. Even with pleading from the UN for UNITAF to extend into June, the Americans were adamant they had successfully accomplished the mission they were sent to Somalia to do and so refused to stay behind after the set transition date.[19]

In the United States, it appeared the Clinton administration moved on to other matters. Somalia lacked the media interest now that the famine and starvation was addressed, with its drastic portrayals in the newspapers and on TV now absent. For the American forces left in-country, the intervention began to take on aspects of a "forgotten war."

Aideed, now rid of UNITAF at least, spent the month of May moving arms and his heavy equipment back into the city. He was back to his plan of seizing western Somalia and the port of Kismayo to put it under Habr Gedir control. On May 6th, Colonel Jess attacked Kismayo in violation of the ceasefire. Jess's efforts failed after a counter-attack from Belgian forces, backed up with the QRF sent by Major General Montgomery, drove him out of town.

10th Mountain Division QRF Operations around Kismayo

As one of the first tests by the Somalis regarding the resolve of UNSOSOM II's capability to continue UNITAF's successful operation, SNA militia factions loyal to Omar Jess in the vicinity of Kismayo made an attempt to recapture Kismayo and drive the pro-Morgan Somali Patriotic Movement (SPM) faction out of the city. The Somali National Alliance (SNA) militia sensed the weakened military capability in the area with the withdrawal of the 10th Mountain Division (LI) forces, leaving only the Belgian contingent to secure the area. In the early morning on May 7, 150–200 armed SNA clansmen attacked Kismayo in a three-pronged assault to capture vital centers in the city and to purge neighborhoods of SPM leadership and followers.[20]

Belgian officers related their account of the incident in talks with 10th Mountain personnel. The Kismayo HRS was secured by the 2d Belgian Para-commando battalion consisting of five companies: three infantry companies, a support company, and an attached armored reconnaissance troop from the Para-Brigade's reconnaissance squadron. The armored troop was equipped with the CVR-T Scimitar, armed with a 30-millimeter chain gun. The armored troop was held in reserve at the port area while the three dismounted foot companies normally rotated into the city for their patrols.

The United Somali Congress (USC) command announced their intention to retake the town by publishing a decree asking the UN to evacuate Kismayo and not interfere with the militia's operation. The plan was developed by Osman Otto: announce a deadline for the UN to be gone, then enter the city at night to minimize civilian casualties.

Aware of the decree, the Belgian Para-commandos were not about to allow the action to take place. In the late hours of the 5th of May, the unit deployed into town with the 1st Company and set up blocking positions and ambush positions at key points in the city center. 2nd Company remained at the Port area to protect and secure the installation along with the 3rd Company and the Scimitars of the armed recce troop who formed the reserve. Shortly after midnight, the armed clansmen, led by Ali Hussein, entered the town on a single axis. The clansmen would break into several columns upon reaching the city park area, with one column to seize key buildings and another column to sweep neighborhoods to capture known Abgal clan leadership.

Once the clansmen were detected on the streets, the Belgians initiated their ambush, killing and wounding scores of the armed militia. When the Belgian 1st Company commander was wounded in the fight, the battalion commander committed the reserve, adding the 30 mm chain-gun fires to the engagement. As the fight raged, the Belgian contingent reported the event to UNOSOM II headquarters in Mogadishu. The 10th Mountain's 1st Brigade (the UNOSOM II QRF) was alerted for movement to Kismayo to assist the Belgians.[21]

The QRF formed with 1-22 Infantry, Task Force 3-25th Aviation, and the 10th Fire Support Brigade. SOCCE520 was alerted to provide assets from the B-team and its two mounted detachments. The mission initially was to drive the pro-Morgan SPM from the town, but by the morning of the 6th, the Belgians had completed their operation and driven off the militia. However, elements of the militia were still in the surrounding area and needed to be mopped up. The mission rapidly changed to conduct attack and sweeping operations in the environs around Kismayo to eliminate the remnants of the SPM attacking force. The SF ODAs infiltrated Kismayo ahead of the task force to begin their preliminary reconnaissance.

The task force deployed an initial force by rotary-wing lift on the 6th, followed by the QRF deploying over the 9th–10th of May. The QRF infantry companies and the battalion Tactical Operations Center deployed on Blackhawks to the Kismayo airfield, where the battalion TOC was established for the overall operation. Vehicles for the task force were flown in by C-130. Additionally, a ground convoy from the QRF launched from Mogadishu and drove to Kismayo. The SF B-team commander deployed to linkup with his two ODAs on the 11th of May; for some reason, he was stopped and detained by the Belgians near the Jilib checkpoint but, after verifying his role, was allowed to proceed.

The QRF and attached SOF contingent conducted operations in consort with the Belgian Para-Commandos, focusing on the area to the northwest of Kismayo. Mopping up operations lasted one week, and scores of Somali gunmen, hiding in the bush, were captured along with their weapons. The prisoners were sent back for processing to Mogadishu using C-130 transport.

The Belgian contingent requested to conduct their own independent sweep operation towards the town of Goob Uen. UNSOSOM II approved, and the QRF assumed responsibility for the security of Kismayo on May 13th, holding it until the return of the Belgians on May 16th. The QRF redeployed to Mogadishu during May 17th–20th.[22]

The mounted ODAs returned overland earlier to Mogadishu with their vehicles. A small altercation with the Belgians took place with ODA523 when it attempted to leave the port area to return to Mogadishu on May 11th. Belgian soldiers at the checkpoint had orders to not allow the ODAs departure and to stop any U.S. Special Forces vehicles, but after registering a complaint to the Belgian contingent by the ODA commander, the team was released for movement. The B-team and ODA525 departed on the 14th, arriving in Mogadishu on the 15th of May.[23]

The remainder of May 1993 was spent planning for future operations, particularly for the expansion north into the central and northeastern regions of Somalia. On May 18th, Major Jesmer received guidance from UNOSOM II headquarters on his role in the expansion operations into the north. SOCCENT deployed one planning officer to assist in the mission analysis; both he and Major Jesmer attended an NGO planning and coordination meeting to fine-tune the particulars of this mission, and on the 20th of May, the company was ordered to begin operations in the vicinity of Bossaso and Galcaio.

Short duration site surveys began in Hobyo, Galcaio, and Bossaso. In Bossaso, detachment members worked with local militia groups to set up reconciliation and election processes as well as to assist them in the development of training a police force. In Bardera, the unit conducted an assessment on the influx of refugees to that region.

SOCCE520 also made a special effort to establish supportive relationships with the various nongovernmental organizations (NGO) operating in Somalia. SOCCE520 personnel attended the daily NGO coordination meetings and provided briefings and handouts explaining how Special Forces could assist the NGOs in their work. SOCCE520 began participating on a regular basis in medical clinics run by the NGO Samaritan's Purse in and around Mogadishu. The NGO's primary clinic was set up in the corner of the former cinema along the green line. The team medics worked in other clinics throughout Mogadishu and its environs. In mid-May, four personnel from the detachments were heading back from a clinic north of Mogadishu and ran across a Somali who was being robbed after his assailants had shot him twice. The gunmen fled on the approach of the SF team members. They applied first aid and fashioned a splint for his leg from some cardboard they had on hand and drove him to the Swedish field hospital. The SF medics and B-team received a favorable press story for their actions during this event and for their work with Samaritan's Purse aid agency in an AP news article written by Paul Alexander, who spent the day with them as they worked on injured and sick Somalis at the cinema clinic. Major Jesmer described the utility of SF to the reporter.

> Our mission is to be the eyes and ears of the UNOSOM II force commander. . . . They [referring to his men] try and take advantage of the natural hospitality of Somalis. We spend a lot of time drinking tea, eating, and socializing. Sometimes we get accused of "sleeping with the enemy," but while you can gather technical information in other ways, you can't divine intention.[24]

As relationships strengthened, the detachments were offered meals, forward storage sites, and secure places to stay when traveling throughout Somalia and were given valuable insights about the situation in various regions by the NGO personnel living there.

An appreciation for Special Forces capabilities also resulted in requests for SOCCE520 assistance by the United Nations Humanitarian Officer for the conduct of area assessments and for providing medical assistance in the remote areas of Somalia. Different NGOs and UN agencies began offering medical supplies for SOCCE520 to use and distribute during MEDCAPs.[25]

The UN's efforts to consolidate and expand governance appeared to be going extremely well, and SOCCE's contributions were important. As warlord Mohammad Farah Aideed's influence diminished due to these activities, anti-UNOSOM rhetoric increased. One of the primary propaganda mechanisms for Aideed was his broadcast radio station in Mogadishu.

This was one area of Aideed's strategy where he enjoyed superiority—his propaganda efforts against the UN and UNOSOM II. Throughout the month of May, he exhorted Somalis to fight against the new form of colonialism from the UN and its interference in internal Somali affairs. His rhetoric was inflammatory. He accused the UN of attempting to establish a "trusteeship." He appealed to the idea of a pan-Somali region, evoking feelings of patriotism and nationalism for the population and urged Somalis to resist foreign domination (more xenophobic than not). His most aggressive rhetoric was to call for the killing of UN Peacekeepers and Americans (in May, peacekeepers and NGOs suffered an increase in attacks from bandits—*shiftas* in the vernacular—and aggressive militiamen).

The UN mission to Somalia was hard-pressed to counter the propaganda blitz, owning only one radio station with limited range, a drastically downsized PSYOP capability, and a newspaper and handbill press section plagued with a lack of skilled Somali journalists, constant machine breakdowns, and paper shortages. But something had to be done when the propaganda reached the point of calling for the killing of peacekeepers and Americans. The U.S. Policy Advisor to Somalia, Roger Gosende, urged Major General Montgomery to do something about it. The strategy would be as simple as "if you cannot counter the propaganda, then eliminate it." Radio Mogadishu (also called Radio Aideed) needed to go off the air.[26]

On May 19th, President Clinton issued PDD/NSC-6, clarifying U.S. policy for Somalia going forward. He reiterated that no U.S. forces would be under the UN's control. He agreed with the continued collection of heavy weapons into cantonments but did not go as far as promising America's support to further disarmament as spelled out in the UN Resolution. He agreed to support the establishment of some type of Somali Police Force. And, he would also provide Special Operations, if required.[27]

SOCCE Assessment on UNOSOM II Transition

About a month into his deployment, Major Jesmer was asked by the UNOSOM II U3, Colonel Ward, to provide an assessment of the overall contribution of his unit and canvassed him for any opinions he might have as to where the two teams and B-team would have the best effect. The assessment was due on May 5, 1994, the day of the handover between UNITAF and UNOSOM II. With the slow pace of new contingents for UNOSOM II arriving, Colonel Ward saw there would be some information and intelligence gaps which needed attention. One of the objectives for UNOSOM II was to move into northern Somalia (Somaliland) to expand the UN writ for reconciliation, governance, and security where heretofore they had not been able to expand in the region. Whatever Major Jesmer recommended for the northern initiative, it would have to remain on hold as the teams were still deployed in the Kismayo HRS with the Belgians and in Bardera HRS (with the Botswanans). The CENTCOM J2 was keenly interested in clan militia activity in the Kismayo region and any information about the area around Luuq, known to be populated with the al-Itihad al-Islamiya faction. Both Jesmer and the CENTCOM J2 (Brigadier General Hughes) were in agreement that the current team missions were more important than switching to operations in Somaliland.

There were lingering tensions between the Belgians in Kismayo and the SF team. Major Jesmer, however, felt the mission in that location was important enough to continue. The only other large factor shaping his operations was logistics and support from quick reaction forces of UNOSOM II; as UNITAF forces redeployed home, a large gap in capabilities developed. The largest deficiency would be in helicopter support; the Belgians had a limited number, and the Botswanans had none.

In his assessment, Major Jesmer deemed Galciao risk-free and ready for increased NGO involvement. The detachments conducted a few recons there and deemed the area safe and permissive.

The B-team's final activity would continue—conducting MEDCAPs to support the coalition whenever enough personnel were present in the Mogadishu B-team.

Jesmer did ascertain the need for Special Operations Teams—Alpha (SOT-As) from his overview, including the specialized communication teams assigned to Special Forces Groups. He had pressure from the intelligence cell at UNOSOM II to place their signals intelligence assets with the SF. Colonel Ward denied his request for the additional asset from the 5th SFG(A), and in return, Major Jesmer declined to use the UNOSOM II SIGINT officer's teams.

With the departure of UNITAF, Aideed saw an opening while UNOSOM II was at its lowest strength and once again began the propaganda campaign against the UN. His most effective tool, other than rallies at the stadium in Mogadishu, continued to be Radio Mogadishu. As this thorn in the side of UNOSOM II grew, planning commenced to take down the radio station and put it out of service. Major Jesmer's SOCCE was notified they would be part of the plan against the radio station, so the B-team began initial concept development and conducted coordination with the Malaysian QRF.[28]

The Defense Advanced Research Projects Agency Initiative

The company was the recipient of new equipment initiatives which would enhance their operations in Somalia. Prior to their arrival, the 5th SFG(A) Deputy Commander and force development NCO worked extensively to improve the sniper capabilities throughout the group through one of their weapons programs. New scopes, improved Barrett .50-caliber sniper rifles, and the SR-25 sniper rifle were issued to the teams in time for their deployment.

As a result of the mine incident outside of Belet Weyne, which killed SFC Robert Deeks, the Department of Defense sent funds to USSOCOM for force protection upgrades throughout the 5th Group, to include up-armoring the unit's vehicles for mine and small arms fire protection. The EOD Low Intensity Conflict program to improve counter-mine capability first assisted the Group with mine mini-flails and up-armored vehicles in Kuwait in 1992; they were asked once again to develop vehicle protection packages for installation on the DMVs and HMMWVs deployed in the Somalia Theater of Operations. In May 1993, EODLIC deployed a team to SOCCE520 to upgrade SF team vehicles with armored windshields, blast blankets, and armored tiles. This initiative would later save the lives of B-team members during a vehicle ambush on July 24th.[29]

June

As UNOSOM II got its feet under itself, peacekeepers began increased presence, patrolling and weapons storage site inspections. There were 18,000 troops from over nineteen countries active at the time.

On the 3rd of June, Major General Montgomery approved an action by the Pakistani contingent to conduct a purported inspection of a heavy equipment and Technical vehicle park located adjacent to the

Aideed radio station, located in their sector. While there, the Pakistanis could simultaneously drop in on the radio station and make an assessment for any further action needing to be taken against the facility. There was an agreement among the elders in Mogadishu that the radio station should be turned into a national radio station, not just a platform for Aideed.[30]

There were seven Authorized Weapons Storage Sites (AWSS) established throughout Mogadishu; five for Aideed's SNA and two for Ali Mahdi's militia (both of which had been shut down by the Italians). Aideed's weapons storage sites were labeled #1 through #5. The truck park near Radio Mogadishu was designated as AWSS #5. On short notice, Aideed was notified on June 4th of the pending inspection (a Friday, the Muslim Holy Day). The operation planners did not want to give him and his militia time to remove any of the equipment stored there. Aideed's advisor, Mohamed Hassan Awale Qaibid, received the letter delivered to Aideed's house and warned UNOSOM II that if the SNA was not given more time to prepare for the inspection, this action meant war.[31]

The Pakistanis were given the mission. Inspections throughout other sites would take place simultaneously. After inspecting the truck cantonment, the Pakistanis also inspected Radio Mogadishu, purportedly smashing radio equipment. After leaving the site mid-morning, the Pakistani escort column was ambushed by SNA gunmen along October 21 Road near AWSS #3, and Pakistani forces at feeding station number 20 were attacked by gunmen. The Somali National Alliance militia used crowds of women and children as their shield.

Reinforcements were hindered by several obstacles hastily thrown in their path. Italian helicopter gunships were called in to help with the rescue and accidently fired on the Pakistanis, adding to the confusion. The Pakistanis at the feeding station on National Street also experienced being shot at by gunmen shielded by angry mobs of civilians. Three Pakistani soldiers were killed in this incident. The platoon was rescued by Italian tanks when an earlier attempt by Pakistani armored personnel carriers failed to penetrate the mob, which threw obstacles in their way.

In the fighting that day, 24 Pakistani peacekeepers were killed and more than 50 wounded. Four journalists rushing to the scene were killed. Afterwards, the Somali mob mutilated some of the Pakistani soldiers' bodies.[32]

That same day, SOCCE520 personnel accompanied the UNOSOM II Ceasefire and Disarmament committee officials during an announced inspection of Authorized Weapons Storage Sites #1 and #4. Special Forces personnel assisted in categorizing the various weapons and developing target folders for future use if necessary. Despite the outbreak of fighting that day (three SF soldiers were pinned down by fire for seven hours), company personnel were able to collect valuable targeting information to develop detailed target folders for the follow-on operations. Using that same information, SOCCE520 briefed AH-1 pilots, AC-130 crews, and QRF commanders and staff on all targets prior to the seizure and destruction operations carried out in later operations beginning on June 12th.

On June 6th, the UN Security Council issued Resolution 837. The Resolution condemned the actions of the SNA/USC on June 5th and reaffirmed ". . . that the Secretary General is authorized under resolution 814 (1993) to *take all necessary measures* against all those responsible for the armed attacks . . ."

The Resolution also requested member states speed up the deployment of their promised forces to support UNOSOM II, more helicopters, and more tanks. Walter Clarke had been the Deputy Chief of Mission to the U.S. Embassy during the UNITAF era and wrote the following.

U.S. and UN officials faced the practical consideration that, around the world, thousands of peacekeepers were in vulnerable situations. Failure to act against a direct attack in Somalia, the Clinton administration felt, would put those forces in jeopardy. Finally, the U.S. and UN decision makers recognized that, given Somali culture, a forceful response was needed to stave off additional attacks.[33]

On June 10th, ODA525 deployed with the 10th Mountain QRF companies to recon a bunker complex outside Mogadishu. SF snipers with night capability provided protective fires during the destruction of equipment in AWSS #1 and AWSS #4 when Somalis fired at the QRF. Five armed Somalis were killed.[34]

Ambassador Howe wanted to hunt Aideed down and use American special mission units to conduct those operations. He was rebuffed by both the JCS and the CENTCOM Commander. However, AC-130H gunships were on the way.

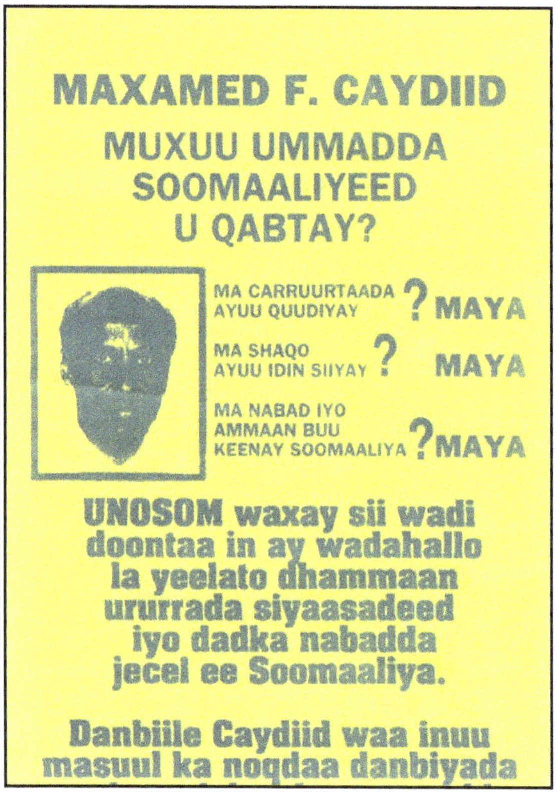

U.S. PSYOP-generated Aideed wanted poster (*DOD*).

President Clinton secretly approved the use of Special Mission Units in the SOF community but did not send them initially. The 75th Rangers and U.S. Army Delta Force, along with assets from the 160th SOAR, went into mission planning and rehearsals the same day the UN Resolution was released. The hunt for Aideed, dubbed "Elvis" by staff planners, was now on.

Endnotes

1. Rutherford, Kenneth R. Humanitarianism Under Fire: The U.S. and UN Intervention in Somalia. Sterling, VA: Kumarian Press, 2008, pp. 111–112.
2. Poole, Walter S. The Effort to Save Somalia August 1992–March 1994. Joint History Office, Office of the Chairman of the Joint Chiefs of Staff, Superintendent of Documents, Washington DC: U.S. Government Printing Office, 2005, p. 38.
3. Baumann, Robert F. and Lawrence A. Yates. My Clan Against the World: U.S. and Coalition Forces in Somalia,1992–1994. Fort Leavenworth, KS: Combat Studies Institute, 2004, pp. 105–106.
4. Harned, Glenn. Stability Operations in Somalia 1992–1995: A Case Study. Carlisle Barracks, PA: Personal Monograph Series PKSOI (19950612 009) United States Army War College PKSOI: War College Press, July 2016, p. 74.
5. Special Operations Annex to UNOSOM II Campaign Plan for Somalia (DRAFT), UNOSOM II Hqs, Mogadishu, Somalia, 041200C June 1993, pp. 4–9.
6. Ibid, pp. 6–9.
7. Memorandum dtd 18 September 1993, Subject: AAR-SF ODB520. Deployment to Somalia, 24 Mar–19 Sep 1993, signed by MAJ David G. Jesmer, p. 1.
8. Celeski, Joseph D. "Operation Continue Hope, April–August 1993." Draft Research Paper, 9 November 2006, p. 3.
9. Ibid, pp. 5–6.
10. Memorandum to J3 UNITAF, 28 April 1993, SUBJECT: Special Forces Detachment Meeting with Islamic Fundamentalists in Belahow, 23 April 1993. Memorandum provided to COL (Ret.) Joseph D. Celeski on 25 October2006 by COL Dave G. Jesmer from his personal collection of files on events during his Somalia tour. The memorandum was written by then MAJ Jesmer as a response to the incident which occurred between ODA525 and the Islamic fundamentalists of the IAIA on 23 April 1993.
11. Celeski, p. 6.
12. Memorandum for ODA525, 20 April 1993, SUBJECT: Message to Convey to Islamic Fundamentalists. Memorandum provided to COL (Ret.) Joseph D. Celeski on 25 October 2006 by COL Dave G. Jesmer from his personal collection of files on events during his Somalia tour.
13. Jesmer memorandum, April 28, 1993.
14. E-mail on the incident from COL Jacobelly to DCSOPS at USASOC after getting initial details from the B-team in Mogadishu, April 23, 1993. NC.
15. E-mail between COL Joseph D. Celeski and CSM Patrick Ballog, Monday Feb 5, 2002, Subject: Somalia.
16. E-mail between COL Joseph D. Celeski and David Jesmer on February 27, 2002.
17. April 30, 1993, commander's update memorandum from MAJ Jesmer to COL Bowra, 5th SFG(A) commander.

18. Clancy, Tom with General Tony Zinni (Ret.) and Tony Koltz. Battle Ready. New York: G.P. Putnam's Sons, 2004, p. 270.

19. Rutherford, p. 122.

20. United States Forces, Somalia AAR and Historical Overview: The United States Army in Somalia, 1992–1994, Center of Military History, U.S. Army, Wash. D.C., p. 125.

21. O'Mallaran, Jackson P. Outside the Wire: An Infantryman's Story. (Self-published, no date), pp. 115–121.

22. United States Forces, Somalia AAR and Historical Overview: The United States Army in Somalia, 1992–1994, Center of Military History, U.S. Army, Wash. D.C., p. 125.

23. Memo to Commander, Co B, 1/5th SFG(A), dtd 13 May 93 from CPT Michael R. Hurst, ODA523 Commander, Subject: Harassment Incident by Belgian Forces 11 May 1993. Copy provided to COL (Ret.) Joseph D. Celeski by COL Dave G. Jesmer, 25 October 2006.

24. Alexander, Paul. "Green Beret Soldiers Work in Clinic During Break from Field", AP news release, with photo, Associated Press Writers, Mogadishu. (No date but event occurred the 18th or 19th of May 1993–faxed copy of the news release provided to COL (Ret.) Joseph D. Celeski on 25 October 2006 by COL Dave G. Jesmer from his personal file collection on his experiences in Somalia.

25. Memorandum dtd 18 September 1993, Subject: AAR-SF ODB520. Deployment to Somalia, 24 March–19 September1993, signed by MAJ David G. Jesmer, p. 1.

26. Rutherford, p. 129.

27. Poole, pp. 38–39.

28. Memorandum for COL Ward from MAJ Jesmer, SUBJECT: SF Employment, dtd May 5, 1993.

29. Celeski, pp. 8–9.

30. Rutherford, p. 130.

31. Ibid, p. 131.

32. Baumann, Robert F., pp. 108–109.

33. Clarke, Walter, and Jeffrey Herbst. "Somalia and the Future of Humanitarian Intervention." Foreign Affairs,vol.75, no.2, March–April 1996, p. 80.

34. Commander's update memorandum to 5th SFG(A) commander, June 13, 1993.

8

Operation Continue Hope: June - August 1993 The Hunt for Aideed

> Somalis, as the Italians and British discovered to their discomfiture, are natural-born guerrillas. They will mine the roads. They will lay ambushes. They will launch hit-and-run attacks. . . . If you liked Beirut, you'll love Mogadishu. . . . We ought to have learned by now that these situations are easier to get into than to get out of, that no good deed goes unpunished.
>
> — Smith Hempstone, U.S. Ambassador to Kenya

The UN Security Council considered the ambush of the Pakistani peacekeepers an affront to its newfound authorities under Title VII. They loudly clamored for action to resolve the matter. Once again, America stepped into the mire as the only nation capable of accomplishing the task. The New World Order's crusade to bring peace to the region relied on the UN, backed by the United States. Somali irregulars could not slap the face of both the UN and the United States without damaging America's and the UN's reputations to go out and do good across the world and prove it was time to solve North-South global issues now that the East-West confrontation of the Cold War was over.

President Clinton was seeking an opportunity to move along his domestic agenda, even after weathering the Waco siege scandal. In June, there was a small crisis with the UN monitoring teams in Iraq (UNSCOM) when Saddam Hussein refused their entry into the country. This was President Bush's signature piece of the New World Order and Iraqi defiance could not stand. When it was discovered that Iraqi intelligence plotted to assassinate former President George H.W. Bush during his April trip to visit the troops in Kuwait, President Clinton ordered a cruise missile attack on the Iraq Intelligence Headquarters located in Baghdad.

Tensions between Europe and Russia continued to lessen as Russian troops finally departed Lithuania. By July 5th, INSCOM inspection teams were once again headed to Iraq. A peace accord was signed between the Palestinians and Israelis in mid-September, a crowning achievement by the Clinton administration.

As a response to Aideed and the Somali National Alliance's (SNA) attack on the Pakistani peacekeepers, UNOSOM II planned an aggressive set of moves to disarm Aideed's militia in Mogadishu. They also wanted to send a strong message to Aideed that activities of this nature would not be tolerated. The Special Operations Command–Central Command (SOCCENT) was tasked to deploy four AC-130H gunships (to be stationed in Djibouti); a small contingent of SOCCENT staff deployed to Mogadishu on June 9th and set up their command and control headquarters on the Embassy compound. Assisting with the upcoming strikes were SF Ground Liaison Officers (GLOs) assigned to the Air Force Special Operations Command at Hurlburt Field, Florida. Operational Detachment Bravo 520 was attached under the operational control of the small Joint Special Operations Task Force for the duration of SOCCENT's deployment.

The USAF Special Operations Detachment (AFSOD) from the 16th Special Operations Wing included air staff planners and targeteers to augment the JSOTF staff. All immediately began work to coordinate airspace de-confliction with the UNOSOM II U3 (operations officer) and build intelligence for the aircrews' target folders. UNOSOM II J2 intelligence and SOCCENT J2 intelligence personnel began reconnaissance flights over the proposed targets and took multiple photographs to brief the aircrews with. A picto-map was overlaid with grids and code words to speed up aircraft coordination with the GLOs and other forces on the ground during the strikes. All targets chosen had to be from the CENTCOM approved targeting list. After analysis by the targeteers on the most viable targets, the list was sent to Major General Montgomery for approval.

Some targets were not on the CENTCOM-approved list; in these cases, the staff worked to forward requests for their addition to CENTCOM. Any time the AC-130 gunships were to be used in Somalia, specific Rules of Engagement (ROE) were developed based on their weapons characteristics. The targets were all in a deep, condensed urban environment. Consideration was made for the two 20-millimeter Vulcan cannons and one 40 mm Bofors cannon on the gunship; with its high rate of fire, rounds were sprayed on the objective and could ricochet wildly among other buildings. Of course, the right weapon (the cannons or the 105-millimeter howitzer) had to be chosen to achieve whatever the desired target effect UNOSOM II had in mind (and then the type of round needed to be considered for use). Last, the priority of targets engaged would be a factor. Ultimately, the use of the 20 mm Vulcan was ruled out along with any use of a phosphorous, incendiary-type round.

Additionally, plans were prepared in the event of an aircraft loss or the need for recovery of aircrew. The airstrip at Baidoa was selected for use as an emergency airfield or for refueling, if needed. For now, the planners agreed the Mogadishu airport was sufficient for the AC-130s to land and refuel, even during the daytime.

The strikes had three objectives—first, weapons storage sites; second, Aideed's command and control infrastructure (to include Radio Aideed); followed by a ground sweep from coalition forces through Aideed's enclave to clear out SNA material, arms, and forces. Aideed's enclave area was affectionately termed "Mr. Rogers' Neighborhood." Each target engagement was to be preceded by a warning shot from one 40 mm round fired through the top of the targeted building's roof; five minutes later the gunship would get to work and destroy or neutralize the target. The results of the mission planning analyses, desired target list, and guidance from higher were then sent on to the aircrews in Djibouti, using a courier from the SOCCENT staff. The first engagements were planned for the early morning on June 12th.[1]

Before June 11th, the JSOTF, SOCCE, AFSOC Special Tactics Squadron personnel, and Ground Liaison Officers linked up with the UNOSOM II headquarters and 10th Mountain Division units; the same occurred for the coalition units. (The previous day, a firefight broke out near the Embassy compound with shots fired from fifteen or twenty armed Somalis as some sort of welcome to the new arrivals.) The SOF liaisons would provide communication connectivity and call-for-fires capability. One JSOTF liaison officer, with an attached radio operator, personally served Major General Montgomery. Most importantly, only Major General Montgomery could clear the AC-130 gunships to go hot and commence firing.

That same day, the AFSOD commander, LTC James (Doug) B. Conners, flew one AC-130 into Mogadishu with the intent to conduct a helicopter reconnaissance over the preplanned targets with key members of his staff and the JSOTF J2 targeteers. This action would help him refine planning between his staff and the JSOTF J2 and J3 staff. The next day, AC-130s began to fly UNOSOM II directed combat missions off the U3-generated Air Tasking Order (ATO).[2]

On June 12th, three AC-130Hs (call-signs Reach 67, Reach 68, and Reach 69) flew missions against Radio Mogadishu and the weapons storage site in the immediate vicinity of the radio station building. The Cigarette Factory weapons cache was hit, followed by a smaller weapons cache site. The mission resulted in the destruction of two weapons storage sites (one site had tanks stored in it) with a total of nineteen buildings hit.

During this operation, Reach 68 was assigned the destruction of Radio Mogadishu and during the set-up for the attack received a threat missile warning and began defensive maneuvers. After confirming a false report, the crew continued on with their mission. Reach 68 fired a total of ninety six rounds of 105 mm and successfully destroyed four power generators, the radio equipment itself, and inflicted further damage to the Radio Mogadishu building and others within the complex. Reach 69 destroyed the one tank which was operable at one of the storage sites with a direct hit.[3]

The JSOTF staff took to their roof around 0400 hours to watch the show and got their first taste of combat. CPT Al Glover, the SOCCENT J2, recalls the event, recorded in his log.

> . . . Up on the roof at 0400 to watch the show. Couldn't see much of the gunships, but had a great view of Cobras assaulting the radio repeater station. Tracers, rockets, it was quite a show. We began taking fire from a crew-served weapon after fifteen or twenty minutes and had to evacuate the roof.[4]

The JSOTF J2 flew the BDA (Battlefield Damage Assessment) mission during the day, took photographs, and brought it all back to the J3 air cell to ascertain target destruction levels. Using the information from prior inspections on Authorized Weapons Storage Sites #A1 and #A4, the SF B-team briefed AH-1 pilots, AC-130 crews, and Quick Reaction Force (QRF) commanders and staff on all targets prior to the seizure and destruction operations carried out on this day. Various members of ODA523 participated in 1/22nd Infantry's raids on these sites, guiding each of the infantry companies onto the various objectives and assisting in the destruction of the captured ordnance. Although the ground column was ambushed along the way, there were no casualties. QRF AH-1 Cobras provided fire support during the assault. ODA18E Engineers helped destroy over a dozen artillery tubes and disabled or destroyed numerous other weapons, including 106-millimeter recoilless rifles, mortars, and tanks. Other B-team operators were involved in combat that day, with company snipers claiming several hits.

On June 13th, two AC-130H gunships (one as spare) launched from Djibouti and conducted sensor alignments, wet bore-sighting, and a landing at Mogadishu airfield for refueling. Even though the airfield had previously been under sporadic sniper fire, no incidents occurred while the aircraft were on the ground. AC-130H Reach 67 had the primary mission for the night and early morning of the 13th to destroy Osman Atto's garage and weapons cache. Atto was using his compound as a primary storage site for his Technicals and various amounts of ammunition. Targeting officers aboard the gunships noticed the windows in the garage area taped up so no light could emit and also observed the glow coming from someone apparently arc-light welding in the garage. After Reach 67 rolled in and located the target, they fired one 40 mm round to mark the area for friendly troops and serve as a warning round for noncombatants in the area. The crew then went hot and fired the 105 mm and 40 mm Bofors cannon, destroying over fifty vehicles in the compound. The adjoining ammunition storage buildings were also targeted, resulting in their destruction.[5]

LTC Brian P. Cotts was one of the AC-130 pilots and reflected on the activities that night.

> Prior to the attacks, our planning teams and crews went to Mogadishu to get a first-hand look at the objectives. On the first night, I flew some of the interdiction missions. On the second night, I remember there were two targets scheduled: Aideed's house and Atto's garage. By the time we arrived, Aideed's house was off limits, so we worked over Atto's garage for about 45 minutes. I heard Chief Ward [SF GLO] on the radio call us after we fired to confirm friendly locations and he told us ". . . think you are in the wrong place," then followed by ". . . can't really tell."
>
> But we thought we had it right so we shot. We received a call from the GLO about a semi-tractor trailer that rolled up into the front of the garage being tactically parked, I guess to block the exit. We looked for it and noticed a few guys running out of the compound who jumped in and drove it off. We did not engage and landed at the Mogadishu airfield to refuel. I asked to go back to the target area with 105 mm but was refused—the ground party mentioned there were still secondary explosions in the area.[6]

Following this attack, the Italian contingent flew an aerial mission over the battle area and dropped PSYOP leaflets. The JSOTF close air support team positioned on K-7 and the JSOTF staff on the roof of their compound watched as thousands of paper leaflets floated down in the area. At K-7, they suddenly noticed engineer equipment and some bulldozers attempting to clear up the damage from the strike near Atto's house. The observation was passed to the JSOTF J2 to add to the targeting for the next round of sorties from the gunships.

On June 14th, two AC-130Hs launched, with one as spare, arriving over Mogadishu the early morning hours of the 15th. The AC-130 destroyed Aideed's garage and re-struck Atto's garage, destroying some of the heavy engineer equipment sighted earlier being used to repair roadblocks. Brigadier General Tangney, the SOCCENT commander, was present on the 10th of June to monitor JSOTF activities and serve as the senior SOF commander to Major General Montgomery during the accelerated combat period. After the attack, he rode with the JSOTF J2 targeteer, Captain Glover, to conduct the Battlefield Damage Assessment (BDA), post-strike. Al recorded the event in his log for that day.

> Took an 1100 PUMA [French Helicopter] ride as guide for BG Tangney. The French aircrew seemed genuinely impressed to be flying a U.S. general around. We flew all of our previous targets and some we hadn't hit yet. The Cigarette Factory was trashed, and several tanks at AWSS#3 were in pretty bad shape including a poor little M42. On our flight we also carried along the general's exec and a fully equipped SF operator in the event that we would have gone down. We didn't go down but the pilots did get a little disoriented a couple of times even with the map and planned course marked.
>
> We flew a loop around from the compound over Mr. Rogers' neighborhood, AWSS#1 and #4, way north into the country around AWSS#3, the Cigarette Factory, stadium, south through the middle of town, the Palace, Gen Aideed's radio station, through the Beirut area [nickname for one of the Aideed neighborhoods], down to the coast, over New Port, the airport and back to the compound. The weather was bad but it was a nice trip. We were at 200'–300' [feet]. We also flew by K4 where the Pakistanis had shot several demonstrators a couple of days before.[7]

On the night of June 15th, three AC-130s were scheduled for the mission but were cancelled for a 24-hour delay to allow coordination with the UNOSOM II's ground plan. Once the gunships were on station, they were to execute the following: destroy the roadblocks around the Aideed enclave, followed by illuminating the Aideed compound for ten minutes in concert with a five minute PSYOPS announcement and warning. Once the warning was complete, the AC-130 gunship would destroy or damage the compound with 40 mm and a couple of 105 mm rounds, fired into the roof of the building. This action was repeated for the Jess and Atto houses.

Additionally, the gunships were requested to destroy a guard house in the vicinity, as well as a large weapons cache in a house. Again, all targets were first illuminated in concert with a PSYOPS broadcast and warning, in order to mitigate civilian casualties. Once the targets were serviced, the gunships remained in orbit as long as fuel lasted (before declaring BINGO) to provide armed reconnaissance and deterrence.

On June 16th, three AC-130 gunship sorties were planned, but weather in the target area forced the cancellation of the June 15th Air Tasking Order.

Operation CASABLANCA

On June 15th, UNOSOM II issued FRAGORD 43 to OPLAN 1 for the conduct of Operation Casablanca, the third objective in neutralizing Aideed's fighting capacity and disrupting his operations. It would be a multi-national operation designed to sweep through the Aideed enclave area of Mogadishu ("Mr. Rogers' Neighborhood").

The enclave consisted of dense, urban neighborhoods with main streets running east to west and north to south. Each sector of city blocks was coded for simplicity (A, B, C, and so forth), and the streets were given the names of various animals for east-west running streets and colors assigned for north-south running streets (see map). Afgooye Street was labeled Phase Line (PL) COW.

Pakistani forces were assigned as the primary sweeping force, with the Moroccans 3rd Battalion, 1st Brigade providing a screening force to the northwest, north, and east, for about 2,100 meters of screen positions in total, resembling a horseshoe or U-shape along the perimeter of the battlespace. On the south, the Italian armored contingent from a company in the 3rd Battalion, 1st Brigade would provide a screen and blocking force along Afgooye Road. The French were designated as the emergency contingency force.

The Moroccan mission consisted of: (1) establishing a cordon, vicinity the northern portion of the USC/SNA-controlled area, no later than 170430C June 1993, in order to prevent large crowds and militia from entering or escaping the search area; (2) deny crowds of people and militia access to the area of operations; (3) be prepared to engage hostile forces in the vicinity of the Abdi House; (4) guard the Atto Villa (NT 33502557) to prevent hostile forces from moving against Pakistani forces; and (5) control access routes into the area.

The Moroccan task force utilized mobile infantry supported by wheeled armored guns (as their Quick Reaction Force) and heavy mortars, backed up with jeep mounted TOW missiles. The total force consisted of approximately 300 personnel. Major Knigge and the SF CST would be collocated with COL Abdullah Bin Namus for the operation in order to provide centralized fire control. The MK-19 mounted on the Special Forces Desert Mobility Vehicle (DMV) was added to the Moroccan fire support plan.[8]

Five liaison support teams from the QRF were provided by the 10th Mountain QRF (the Warrior Brigade, commanded by COL James L. Campbell); two with U.S. Army personnel, consisting of an officer, driver, and an additional NCO and radio operator; and three made up from Special Forces operators out of the SF Ground Liaison Officer element, the SOCCE, and A-teams. An ODA523 split team was placed with the Moroccans on the north edge of the battle area, manned by Master Sergeant Bell (ODA523 Team Sergeant), Sergeant First Class Bower, and Staff Sergeant Mason E. Fail. USAF MSgt Kim Dobson rode with Bell's Combat Controller as a GLO. Their DMV was armed with an MK-19 40 mm automatic grenade launcher. Additional snipers and M-203 grenadiers from the 1-22nd Infantry also supported the ground element, all backed up with snipers provided from operators assigned to SOCCE520.[9]

The SF coalition support team for liaison with the Italian contingent was led by the ODA523 commander, CPT Mike Hurst, along with two of his NCOs, McCoy and Shakeenab. The 10th Mountain Division (Light) LNO team with the Italians was accompanied by MSG Steve English (Army Special Forces) as the GLO from AFSOC to assist in aerial calls for fire. The Pakistani liaison team was supported by a split team of ODA525 and MSG Olen F. Kelly and Master Chief Larry Barret, both AFSOC Special Tactics personnel.

Other Green Berets participating in the attack included CW3 Vernon Ward (Army GLO) located at the U.S. Embassy Compound during the operations and MSG Mike England (Army GLO) located atop K-7 in support of the Italians. Major Jesmer, the SOCCE commander, and a couple of his company SF snipers were also atop K-7. Captain Bonk was deployed to UNOSOM II headquarters to serve as the SFLE (SF Liaison Element) with Major General Montgomery.

The operation was supported with aerial fires from three AC-130H gunships from the 16th SOW's AFSOD, using their GLOs and CCTs for the ground-to-air coordination. The PHOENIX aviation element (Company C, 5th Battalion, 101st Aviation) and the DIAMONDBACK aviation element from C Company, 3rd Battalion, 325th Aviation assigned to the QRF would provide OH-58 Kiowa observation helicopters and AH-1 Cobra attack helicopters.

The Pakistanis had the lead for the operation and were using the 7th Frontier Force Battalion and the 6th Punjab Infantry Battalion, with the 6th Punjab in the lead.[10]

By early evening on the 16th of June, all liaison teams departed the Embassy and University compounds and linked up with their respective coalition partners. Major Jesmer spent that evening attending a dinner with the Pakistani Commander and his officers.

In the early morning hours of the 17th, four AC-130H gunships departed Djibouti headed to Mogadishu; one diverted back to Djibouti due to a computer malfunction aboard. En-route, they conducted their communications check with the JSOTF air staff and the GLOs and spent a short time firing their weapons systems to confirm they were operable. The aircraft landed at Mogadishu airport under the aerial protective cover of the 325th Aviation and refueled—two were airborne to set up their targeting orbits by 0120 hours that morning.

One AC-130H, Reach 66, experienced an engine fire immediately after takeoff from the Mogadishu airfield once its refueling operation was complete. The aircraft subsequently experienced an additional engine problem, forcing the crew to conduct emergency procedures in preparation for an emergency landing at Mogadishu. After orbiting and dumping fuel, the bird was skillfully landed on the power of its two remaining engines. Reach 67 landed during the mission to pick up the crew and on-load ammunition from the damaged aircraft. The two remaining AC-130Hs serviced targets for up to four hours preceding the friendly forces ground movement to sweep the Aideed enclave, flying off station at approximately 0515C. The primary target for this night was the Jess compound. During their attacks, ten buildings, four roadblocks, and numerous targets of opportunity, such as technicals and construction equipment, were destroyed (Reach 67 targeted two front end-loaders).[11]

From the ground, the firing weapons and explosions appeared to the troops like a 4th of July show.

Meanwhile, the Pakistanis, Moroccans, and Italians were in their jump off areas ready to begin the operation at a quarter past five. Within a half hour of the operation beginning, the Moroccans and Italians reported small arms fires on their positions. A crowd of over 100 Somalis, armed with AK-47s, attacked the Italians in the vicinity of the Phase Line GREEN and LEOPARD intersection.

Within an hour into the operation, the Moroccans and Italians reported their cordons completed. About that time, around 300 Somalis were gathering near the Moroccans near Phase Lines ORANGE and DOG. Fifty Somalis gathered to their south. At both of these locations, the Somalis opened fire and began to erect barricades. One Moroccan soldier was wounded in this exchange. Soon, a larger crowd began to develop from the north. The liaison teams with the Moroccans requested the use of CS gas to disperse the crowds;

Major General Montgomery approved. They were assisted in dispersing the irritant from a PHOENIX unit helicopter, using its rotor wash to spread the irritant over the crowd.

The situation in the north began to look dire; the 1-22nd Infantry was alerted for possible reinforcement while QRF helicopters responded with 20 mm cannon fires and TOW rockets. One Army liaison team led by Major Knigge, the author of *Operation Casablanca: Nine Hours in Hell*, was near the Abdi house on Phase Line DOG along with Master Sergeant Bell's SF Coalition Support Team. They both added their fire to the Moroccan response. By now, fires were coming in from all along the north of Phase Line DOG, including sniper fires from the Digfer Hospital. Bell moved his vehicle east to the northeast quadrant of the operation (PL DOG and PL YELLOW intersection) where hundreds of Somalis were forming. (This position remained under constant fire for the rest of the battle. The Moroccans were finally relieved of this position by French armored forces arriving after 1500 hours.) Throughout the intense fight, Somali men used woman and children to mask their moves or fire their weapons from behind them, convinced that UNOSOM II peacekeepers would not fire on unarmed civilians.

Near 0630 hours, Master Sergeant Bell and a member of Major Knigge's team rolled softball-style tear gas grenades at the mob before them (estimated at 300 men, women, and children). The grenades were duds; Bell and his partner retrieved the grenades, fixed the problem, and threw them back at the crowd. Master Sergeant Bell was sprayed with CS gas ejecting from the grenade but doused his face with water and was back in the fight in under ten minutes. (He and his team did not have gas masks, deferring to the fact the Moroccans did not have gas masks to wear either.)[12]

Near 0700 in the morning, the 6th Punjab Pakistan Battalion attacked east down Phase Line BIRD to begin suppressing and clearing Atto's house and his weapons storage garage, then moving to an SNA support building and Omar Jess's house to the north of the road. To the south of the phase line, the same actions were taken to clear Aideed's compound and an adjacent SNA guard house.

By 0800 hours, the 6th Punjab Infantry had cleared the second block of their assigned buildings. They had two of their own M60 tanks and additional support from four Armored Personnel Carriers (Italian Panhards) in the Italian cordon. The 7th Frontier Force moved on line near Atto's house, then veered to the northeast to advance to Phase Line BLUE. Both forces received small arms and sniper fire along with heavy resistance from SNA fighters. The 6th Punjab maneuvered straight ahead and seized the intersection at PL BLUE and ORANGE. By now, the response from the Somalis had intensified with enemy gunmen deploying crew-served weapons. Near DOG and YELLOW, the Moroccan QRF liaison team (Major Knigge) and Master Sergeant Bell's CST were positioned together; Knigge's crew were returning fires to the north at Somalis gathering near the cemetery and Bell's crew firing the MK-19 to the southeast (SSG Michael D. Bower operated as gunner). All the while, attack helicopters from PHOENIX added their fires at this location (given the circumstances, the Moroccans now had priority for aviation fires).

UH-60 Black Hawk helicopters overhead reported armed militia moving in and around all the buildings surrounding Knigge's and Bell's position. After engaging the Somali gunmen with M-60 machine gun fire from their helicopter, the pilot reported to Major Knigge that there were possibly twelve Somalis killed or wounded with their weapons dropped on the ground. The two liaison teams were now just north of the Moroccan perimeter and felt it was about time to withdraw to a better position. Before leaving, Bell and Staff Sergeant Fail volunteered to recover some of the enemy weapons and soon dismounted under covering fire from Bower on the MK-19 and M-16 fire from Knigge's team. It was here the SF took one

12.7-millimeter round in their armored windshield, apparently an attempt by the Somalis to suppress the MK-19.[13]

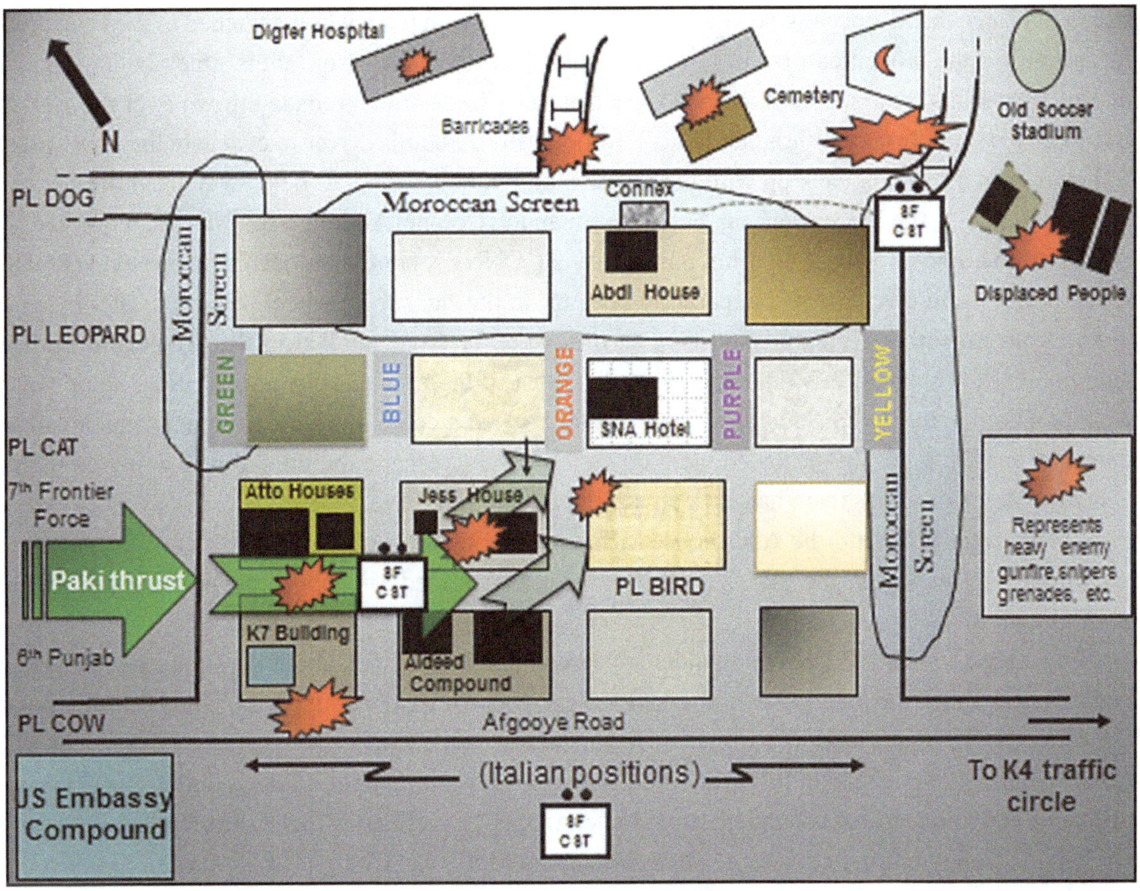

Operation Casablanca scheme of maneuver (adapted from maps in Major Tim Knigge's work, *Operation CASABLANCA: Nine Hours in Hell*. Chapel Hill, NC: Professional Press, 1995).

Tim Knigge remembers the following.

> We began to take heavier fire in this area. I directed the SOF team to engage a known target, the building immediately to our southeast from which numerous small-arms were firing. The target continued to snipe us all day. I had Bower engage at least six different targets with the MK-19 automatic grenade launcher. After getting a few rounds off at each target, I ordered the SOF team to pull up closer to my location for possible engagement of additional targets. Bower provided accurate fires, taking out numerous firing positions and Aideed militia members. He continued doing this while he himself was the recipient of heavy enemy return fire.[14]

Fearing the capture of the two SF NCOs (or their subsequent killing and mutilation), Sergeant First Class Bower and Major Knigge's crew maneuvered to provide covering fires after the two NCOs low-crawled to seek protection by a wall. Bell and Sergeant First Class English from Major Knigge's crew were picked up and extracted.[15]

As the fighting continued and more Somali gunmen joined the fight, the French were alerted to prepare to reinforce the Moroccans. A Quick Reaction Force liaison team was dispatched to the French forces in preparation for the movement. By now, the Italians added their fires from their attack helicopters. Then Major Knigge and his crew survived an explosion from a Chinese stick grenade thrown over the wall at his vehicle. With a wound from the shrapnel in his left foot, he ordered his driver to evacuate the now untenable area. They moved off to link back up with the SF team (about half a football field away from their position).

Around 0800, the Moroccan Task Force Commander, Colonel Abdullah Bin Namus, requested Major Knigge report to his position near the Abdi house, near a CONNEX container located in the street (PL DOG). However, the SF team's vehicle had apparently been hit in the right front wheel bearing socket by an RPG or a round from a recoilless rifle; most of its other tires were flat from small arms fire. The team had been in constant combat for hours at this time and expended most of their small arms ammunition; some weapons were malfunctioning due to constant use. The SF team hooked a cable to Major Knigge's vehicle, all under constant fire, and Knigge's driver towed them down PL DOG to escape the fires. It was going on 0900; the SF team and Major Knigge's team had been in constant combat for five hours.[16]

TOWs were fired into the Abdi house (a few mysteriously acting erratically in flight). These fires landed within forty meters of the SF CST vehicle. Bower continued to support this position (alongside the CONNEX) with MK-19 fires.

To Knigge, the Task Force commander appeared highly concerned about the strength of attacks from the Somalis near DOG and YELLOW and asked Knigge for AH-1 Cobra support. COL Namus then moved to the YELLOW/DOG intersection. Knigge moved at the same time to a position to the southeast along Phase Line DOG. Fighting intensified in this area, and the Task Force commander, COL Abdullah Bin Namus, was killed when a 106 mm recoilless round hit his vehicle. An AH-1F Cobra destroyed the recoilless rifle position with M-60 machine gun fire. In this same engagement, the Moroccan Company commander at this location, along with the Pakistani LNO, were also killed. The Moroccan command and control began to break down.

MAJ Al Glover was on the JSOTF J2 staff assisting in the targeting process for the SOF contingents during the month of June. He captured the highlights of this day's fighting in his personal log entry for Thursday, June 17, 1993.

> . . . Up at 0130 with the first 105 mm round impacting Mr. Rogers' Neighborhood. They worked the whole area all night with two gunships. It's 0915 right now and our power is out. No recce this morning. Helicopters are taking fire all morning. I saw an Agusta Mongoose [Italian attack helicopter] flying CAP. At about 0815, I was coming from the UNOSOM building when I heard the sound of heavy weapons fire. I turned around and there was a Cobra about two hundred yards away pumping out 20 mm. He fired a long burst once of about ten seconds. The whole neighborhood area was hot all day. I heard gunfire all day as the Pakistanis, Moroccans, and Italians swept the area. The Moroccans took pretty heavy casualties. We had an SF CAS team with them. They had a 106 mm recoilless rifle and an RPG fired at them. MSGT Kelly (Army) went into Aideed's house and Jess's house. He found a 60 mm mortar tube in Jess' yard by the front wall. Later in the day, I saw an Alouette III make a "Flak bait" run through the neighborhood. He seemed to draw fire the whole

way. Then a Cobra rolled in. Sometime during the day a Cobra nearly hit one of our SF teams with rockets and 20 mm. They had to call "knock it off." We worked up our targets for the night, sniper positions on the SNA Hotel and ABDI's house, but at 2030 hours, the missions were all cancelled for the night.[17]

At 0900 hours, the French were ordered to move from their positions at the old Somali military compound located at the intersection of Via Lenin Road and 21 October Road. Their mission was to reinforce the Moroccans, clearing Somali fighters as they advanced.

As the day grew on, many of the fires called for became "danger close." Somalis maneuvered to within fifty meters of Moroccan positions as the Moroccans continued to take casualties. At 1122 hours, the Pakistanis secured the Aideed compound and raised a UN flag. By 1400 hours, the Pakistanis withdrew west of Phase Line GREEN as their mission was now complete.

As the French arrived to reinforce the Moroccans, they engaged a large crowd of gunmen near the cemetery north of PL DOG and YELLOW. By 1500 hours, the fighting ended. Major Knigge moved back to the CONNEX on Phase Line DOG and helped tow Bell's damaged HMMWV back to the Embassy compound (around 1500 hours).

SFC Channing C. Bell, SSG Michael D. Bower, and SSG Mason E. Fail all received the Bronze Star with "V" for valor device; the attached Ground Liaison Officer, USAF MSgt Kim Dobson, received a Bronze Star medal. Major Knigge received the Bronze Star with "V" device and the Purple Heart medal.

It is hard to estimate the number of casualties among the SNA gunmen due to the nature of the fight, the confusion, and inability to accurately count the dead and wounded. A figure of as many as 150 killed appeared in some estimates. There must have been just as many or more wounded. The Moroccans had five KIA; over sixty peacekeepers were wounded among the coalition during the action, with the Moroccans suffering over forty casualties alone.[18]

After this fight, no further targets were authorized by CENTCOM or the National Command Authority, even though Major General Montgomery continued to request additional targets for servicing. The AFSOD remained with its aircraft in Djibouti as a ready standby force to deter Aideed from further attacks on UNOSOM II forces.

The SF Company snipers also scored hits during this fight, three of them shooting off the K-7 rooftop position.[19]

The failure to capture Aideed resulted in a new plan by UNOSOM II to hunt him down. ODB520 conducted mission analysis on their "asset recovery" concept and quickly taskorganized SF reconnaissance and security teams (R&S) to assist in this effort.

The Hunt for the "Rabbit"

As a result of the failed attempt to apprehend Aideed during the June 17th operations, UNOSOM planners developed a concept of operation to create a standing crisis reaction force with the capability to launch on very short notice in the event Aideed's whereabouts could be confirmed. In-place forces capable of the speed and agility to conduct this mission included elements of the 10th Mountain Brigade, Marines from the MEU-SOC, or the SOF assets in Mogadishu. As it was extremely difficult to garnish detailed intelligence on Aideed's projected activities, combined with his elusiveness, the concept hinged on an aggressive reconnaissance and tracking capability often deployed daily to augment existing intelligence mechanisms.

The first order of business was to set in place a system to track Aideed throughout Mogadishu, as he was known to move frequently or adopt disguises. LTC Moe Elmore, the Deputy Commander of the JSOTF, assisted the staff with preliminary planning, even though he considered the intelligence on hand "iffy" at best. This operation would be called Caustic Brimstone, but the staff colloquially called it "The Hunt for the Rabbit."

A concept was developed based on the constant flight of 10th Mountain helicopters over the city. They would now be incorporated as reconnaissance and surveillance aircraft (R&S), while a standby flight on the ground at the airfield would serve as a quick reaction capability once Aideed was spotted. The task organization would also consist of Special Forces sniper/observers and a dedicated platoon of infantry.

The aerial surveillance asset was named "Eyes Over Mogadishu." There would also be a classified network of ground assets managed by the intelligence community. All of the nodes of this operation were equipped with enough radios to provide redundant communications. Further, the existing UNOSOM II assets on the ground were incorporated into the search—checkpoints, guards, strong-points, and snipers.

Lieutenant Colonel Elmore brought the concept to Major Jesmer to look over. They were clear the QRF was not really "quick." As Elmore said,

> One must understand that the 10th Mountain Division rapid reaction force was never that. They could not assemble quickly (within minutes) and they were not trained to respond to any specific mission like a snatch. Anyway, Jesmer, myself, the SAS CPT from the UNOSOM II staff, and another SF NCO or two, sat down one evening and analyzed the mission and support available.

After analyzing the mission and ROE, SF capabilities on hand, the inability to train (rehearse) or move unobserved, and the insufficient or suspect intelligence on hand, Elmore decided it was in the "no good" box and refused the mission.

> I was not willing to take an inexperienced in Direct Action and under-strength element of the SOCCE (I was going to lead the operation because I was the only one with experience in planning and executing such types of DA) on such a mission given the circumstances. I remember the coalition SOF CPT saying, "Whew! I'm sure glad you decided that!" He had been on a unit with that type of experience and knew what we were up against. Afterwards he agreed that the mission was a bad one. I also did a risk assessment and did not think the reason was good enough to risk twenty plus soldiers on a doubtful mission.

When informing the U3 of their decision, LTC Elmore said, "The U3 was a U.S. Armor Colonel. I was told it did not seem to be such a hard mission and he could not understand why we were so 'chicken.' I just said he did not understand the job and left."[20]

> UNOSOM II passed the mission to the Italians; after a day or two, they also refused it and gave it back to the staff. It was deemed too complex and difficult to pull off in Mogadishu's environment.

Then the Egyptians took it but did the same after looking hard at the requirements. The Pakistanis looked at it but never took the mission. The mission concept was also shared with the Marines who felt they could conduct this type of "special operation" with the assets from the MEU (SOC). They briefed the chain of command at UNOSOM II and accepted the task. More experienced personnel in the room were suspect, though they admired the Marine's enthusiasm.

Lieutenant Colonel Elmore continued:

Anyway, the next day we were actually doing a wet rehearsal on how we would track "Elvis" if he was spotted, but we did not have the comms with the Marines fully worked out. I was on the command chopper and we got the word that Aideed was on the move. We tracked the vehicle in question and the Marines launched their CH-46s and in they came. The target was seen going into a building and the Marines came in and fast-roped into several places. They broke into several buildings, threw flash bangs, found no one and left. That was the end of their first and final attempt.[21]

It was clear the operation would need to be conducted by the "first team" of SOF. The Joint Special Operations Command (JSOC) sent three separate survey teams to Somalia over the next few weeks. The response from each survey team was that the mission was fraught with political risk, and the congested urban nature of the Mogadishu terrain nullified any tactical advantage which would be required for the mission to be successful. Somalis had a propensity to run to the sound of guns and could overwhelm any small ground force isolated among the clutter of buildings making up most neighborhoods in the city.

The UNOSOM II Crisis Reaction Cell could not operate on a 24-hour basis, nor could troops on hand stand continual alert, so the plan was modified to take advantage of the congruence of intelligence platforms available to UNOSOM II. If sufficient assets were on hand, the Crisis Reaction Cell stood up its staff and alerted the various units tasked with Reconnaissance and Surveillance (R&S) and placed the reaction force on standby. The force would be alerted through a series of intelligence reports, named "OXBOW," ranging from a target confirmation to other reports indicating SIGINT or HUMINT related factors which could lead to a target confirmation.

Operational requirements for the R&S team included the capabilities to land aboard shipping and on rooftop LZs. Aerial snipers were qualified to perform interdiction on stationary or mobile targets, while the ride-along observers were qualified in the variety of communications needed for the mission and the command and control procedures for target handoff to the reaction force.

Once the target was identified, the reaction force moved in for the apprehension. Forces in Mogadishu already understood the need to be fully qualified in urban operations and crowd control before conducting this mission. The reaction force would be backed by firepower from attack helicopters. The reaction force generally consisted of a platoon-sized element with enough helicopters to conduct air-mobile raids.

One idea did emerge from the brainstorming sessions at the SOCCE and JSOTF: taking Aideed when he was riding in a convoy. In that concept, about twenty operators would be loaded into two local five-ton trucks lined with armor once Aideed was spotted. There was an idea to use local truck drivers in order to blend into the population, with the understanding that the idea could fail if the trucks had to negotiate too many UN checkpoints (as well as Somali checkpoints).

An intelligence and command helicopter would position overhead, followed by two helicopters carrying support elements. They would be followed by a MEDEVAC helicopter and two AH-1 Cobras for additional firepower. The overhead helicopter would guide the trucks toward the convoy, where they could ram Aideed's convoy, which normally consisted of two vehicles—one with Aideed and one with his bodyguards. Assuming that battering the two Somali vehicles created some form of immobility, the two helicopters with the support unit would position themselves and form a perimeter while the assault teams in the trucks apprehended Aideed and dealt with his bodyguards (pepper spray was the preferred method to disable the guards; the assault team would only shoot if necessary).

Once Aideed was apprehended, along with any of his assistants, and any documents found after a hasty search were secured, all would be loaded into the trucks for a return to base covered by the Cobra's firepower to isolate the position from any Somalis approaching the site.

If a truck, or both trucks, were disabled from the battering, the operators would move to a quick pick-up zone for a rotary-wing extraction. The use of an AC-130 gunship was ruled out; the aircraft could not meet the timelines and be in position for such a hasty action.

Once again, this plan could probably have been executed, but the JSOTF and the SOCCE were unable to procure a local truck and driver and did not have any place to practice and rehearse the operation; also, no one thought the intelligence needed to pinpoint Aideed's location would ever be forthcoming. However, the concept was briefed to the Joint Special Operations Command (JSOC) survey teams. Interestingly, Task Force-Ranger's operation to interdict a vehicle convoy and later capture Osman Atto mirrored the intent of this plan with the additional use of a heavy sniper weapon to take out a vehicle's engine block.

The SOCCE received the task to develop standing R&S teams for use in this plan. The SOCCE was not tasked with providing the ground or an aerial reaction force, however, they continued to support the "Eyes Over Mogadishu" assignment with aerial sniper support. A rehearsal was conducted using the aerial snipers and a six-man team from the Marines to validate the R&S concept. The JSOTF J2 designed a scenario to employ some DMVs and a member from the staff in a ground convoy to serve as the "rabbit" on the move.[22]

The execution of this standing plan consisted of five elements:[23]

a. If intelligence indicated the target's whereabouts, the R&S team launched via UH-60s to the target area (the standing requirement was ten minutes for launch). A backup UH-60 was also alerted.

b. The aerial R&S confirmed or denied the target once on station.

c. If target confirmed, the aerial R&S maintained contact with the target and fed constant updates to the crisis action cell and the incoming reaction force. If the surveillance extended over two hours, the secondary helicopter and R&S team launched as a replacement.

d. The reaction force was alerted (SOF, USMC, 10th Mountain helicopter gunships) and launched (planning required this force capability within two hours of target acquisition). Upon arrival to the mission area, the reaction force commander assumed OPCON of the R&S and overall control of mission execution.

e. In the final portion, the reaction force either successfully interdicted and captured the target or aborted the mission.

On June 22nd, five members of ODA523, led by CPT Hurst, deployed on an R&S flight in support of a USMC led reaction force to capture Aideed, but the mission was not successful, even though the Marines bagged one suspect. This mission was conducted to the northeast of the Pasta Factory. This was the same day Italian forces were mobbed by a Somali crowd near the Pasta Factory, presumably to throw off UNOSOM II forces from capturing Aideed. Additional attacks occurred in the late afternoon on the University compound with fifteen to twenty Somalis firing RPGs and small arms. U.S. forces estimated four Somali gunmen were killed by return fire in this attack.

The Security Assessment

In the latter part of July, CENTCOM deployed a physical security team to ascertain the force protection methodologies employed in and around U.S. bases. Two Lieutenant Colonels from the Marines and USAF, respectively, arrived accompanied by a USMC Intelligence Warrant Officer. Lieutenant Colonel Elmore was tasked to serve as a ground component Subject Matter Expert, and the SOCCE supported the physical security and force protection assessment team with vehicles and security escort during its activities around Mogadishu.

The team conducted both daytime and nighttime visits to the airfield, Embassy compound, SWORD Base, HUNTER Base, and the New Port area from the 19th through the 21st of July, assessing both U.S. and coalition security measures. The team prepared detailed reports on the vulnerabilities of each of the various compounds. Generally, all sites required enhanced security lighting, night vision equipment, increased physical security structures (for example, bunkers, sandbagging of critical points, trenches, and so on), an integrated alarm and reporting system, and additional forces to conduct perimeter and area patrolling. The need for well-trained convoy escort forces was also identified and a recommendation made to reestablish vehicle check points.

When the assessment team completed their mission, they gave a detailed briefing to Lieutenant General Bir, Major General Montgomery, Brigadier General Ellison, and Colonel Campbell on their way out of Mogadishu. Overall, their assessment and visit was well received by all the contingents. A number of on-site corrections were made based on their recommendations and immediately incorporated into local security plans. Over time, the equipment recommendations in the areas of security lighting and night-vision gear were supplied into the theater, resulting in the mitigation of friendly casualties as the operation continued.[24]

The Remainder of July

On the 22nd through the 23rd, Colonel Bowra, the 5th SFG(A) commander, and his Command Sergeant Major conducted a visit to the troops in Mogadishu, departing just prior to an ambush on SOCCE members. Colonel Bowra visited the various sites occupied by the SF teams and conducted an office call with Major General Montgomery. Major General Montgomery was pleased with the performance of the SF to date but told Colonel Bowra he did not feel the need to expand the SOF capability in Somalia with additional assets. Colonel Bowra reported this item to the commander of USASFC(A) upon his return to Fort Campbell.

On this same day, the ODB was tasked to assist with enhancing force protection measures at New Port and the Mogadishu airfield and deployed two sniper teams to those locations.[25]

The Aerial Sniper Missions of CPT Hurst

(The following is taken from an Awards narrative for Captain Hurst, the ODA523 Detachment Commander.) CPT Michael R. Hurst began operating as an aerial sniper on June 19, 1993 as a result of military aggression toward United Nation Forces deployed in support of Operation Continue Hope. During the period of June 19th to August 20, 1993, CPT Hurst logged more than eighty hours in a UH-60L while operating in a combat environment. The Award narrative read:

> For exceptionally meritorious achievement while participating in aerial flight in support of United Nations Operations in Somalia on 22 June 1993. As team leader of surveillance and reconnaissance flights aboard a UH-60L for a personnel seizure mission, CPT Hurst provided the aerial command and control during a joint special operation on 22 June 1993.
>
> After confirming the target moving through the twisted streets of Mogadishu, CPT Hurst maintained surveillance and vectored in an aerial assault force in an attempt to seize the targeted Somali political figure. After the operation, CPT Hurst led an effort to assist the Italian forces in the area who were threatened by a belligerent Somali mob.
>
> Directing the placement of riot control agents in the crowd and utilizing the helicopter's rotor-wash, CPT Hurst's aircraft and crew enabled the Italians to maintain control of the situation until armored reinforcements arrived. At one point, CPT Hurst directed the aircraft to land near the Italians so that he could personally check on the medical condition of an injured Italian soldier. He then instructed the pilot to use rotor-wash to open a path through the mob and the hastily built roadblocks for a U.S. light vehicle convoy which needed to transit the dangerously congested route. His outstanding initiative and that of the crew under his command undoubtedly prevented further United Nations casualties.

The July 24th Ambush

Friendly forces were experiencing increased attacks on their positions and personnel. The Somalis were now combining and increasing their use of small-arms, machine guns, and RPGs. With the immediate responses from UNOSOM II and QRF soldiers anytime the Somalis attempted to attack big military vehicle convoys, the gunmen soon shifted their harassment attacks to smaller convoys (most driving with the three vehicle minimum for protection).

On July 24th, members of the SOCCE and ODA523 departed their compound with two vehicles headed out for the day to conduct some small-arms fire training at a local range (re-zero their weapons). Afterwards, they planned to perform a counter-sniper mission at one of the compounds. Lieutenant Colonel Elmore had asked to go along, hoping for a ride to the airport to catch his flight home. He had just missed a UN Russian helicopter at Jaybird pad and did not want to wait for the next helicopter; he did have time to stop off with the team after finishing up at the firing range. Sergeant Major Ballog was in the lead vehicle with Sergeant First Class Hoepner driving. Sergeant First Class Bower manned the gun. A company mechanic was driving vehicle number two with Lieutenant Colonel Elmore in the front seat, Sergeant First Class Berry on the gun, and Sergeant First Class Beebe in the back along with two intelligence personnel from the Military Intelligence Detachment (MID).

The lead vehicle carried the MK-19 grenade launcher, while the second vehicle mounted the .50-caliber machine gun. Sergeant First Class Berry carried a Squad Automatic Weapon light machine gun. Each member was armed with their M-16 rifle, pistol, and a basic load of ammunition. All had on protective armor. The vehicles chosen had the upgraded Kevlar soft kits (armored blankets to line the floors) and bullet-proof, armored front windshields. Upon departure, each member of the team was assigned a sector to cover, with their weapons facing out.

It was the team's practice to vary their routes each time they departed the Embassy compound in order to not set a pattern for Somalis watching their activity. This day, they chose a route north on a thoroughfare in an area cluttered with low buildings. The drivers kept their spacing at a fifty to seventy-five meter interval. About two miles into the trip, they noticed the absence of any activity, except for light traffic and a Somali woman grabbing her child to pull him back into a house. Both Lieutenant Colonel Elmore and Sergeant Major Ballog sensed danger ahead; Elmore warned his vehicle crew to stay alert. Suddenly, the convoy was ambushed.

Sergeant Major Ballog, who was wounded in the ambush, recounts the incident that day.

> [We] Left the compound about 1000 hours driving on side streets. Rounded a corner onto a straight dirt road of about 300 meters. The buildings on both sides were generally two story, flat fronts with just enough room for two vehicles to pass. The street was deserted (unusual, but at our driving speed we were on the road and moving before it sunk in) with a ruined blue bus sitting on our left front. About 200 feet down the road I saw a side street/alley; as that registered in my mind, our vehicle (lead DMV) was hit by two rounds center mass of each armored windshield.
>
> The shooters were in the blue bus. Simultaneously their assault element opened up from our right in the alley. The Somali squad (8 to 12 gunmen?) had AKs and threw grenades or fired RPGs. I heard explosions but don't know what caused them. The enemy was only about fourteen feet away so this all happened in a second. I was hit in the right shoulder and right hand by an AK round. The windshield round came partially thru on my side and I took shrapnel in the mouth. SFC Mike Bowers was hit above the right knee blowing out four inches of his femur.
>
> The next few seconds were beautiful. The individuals (especially SFC Beebe) in the trail vehicle "got busy" and immediately engaged with effective fire killing the Somalis attempting to sweep forward to our vehicle. I could still use my M-16 and returned fire as SFC Lance Hoepner (driving our vehicle) fired his 9 mm pistol over my shoulder as he drove thru the kill zone. The distance between us and the Somalis was about five feet during this period. The trail vehicle backed and fired as we moved forward. SFC Bower was combat ineffective due to the pain of his wound—the Zimmer vest "locked" him in the turret and he could not get out or down and had to stand on his shot leg. We drove as fast as we could to the airfield (on flat tires)."[26]

Lieutenant Colonel (Retired) Elmore added more details he experienced from his position during the ambush.

> The first shots came from a side street on the right and hit the guys in the first vehicle. They sped up to get clear of the area and we saw a guy with a G3 come out to continue shooting at the lead vehicle.

> Our driver stopped just as the guy came out. I shot him and he spun and went back into the alley. A second guy stepped forward and I shot him but he did not go down, I shot again and MSG Beebe (he had an M-21) shot the guy at the same time and he fell away from us.
>
> The lead vehicle kept going and about this time we could see they were receiving more fire. I think they had a couple of RPGs or rifle grenades fired at them but they missed (there were a couple of explosions in the area about 100–150 meters ahead at another intersecting road that came from the left). I remember thinking that the rifle I had borrowed had a nice trigger on the first several shots, but then the job became one of covering the lead vehicle and to suppress more than shoot for point accuracy and I switched to auto. From then on it was two round bursts. Beebe started to engage those he could see near the corner to our front. He shot one guy out of a tree and probably hit several more. I know I hit a guy that came out of somewhere on the right side of the street; he jerked and then limped off the street. There was a little mini-bus parked on the left (north side of the street) and a guy was in it shooting at Ballog's vehicle as it sped away. I engaged him with a couple of bursts and he dropped his rifle and scuttled into the shelter of the alley/road on the left.[27]

Soon the lead vehicle cleared the ambush zone after they topped a small rise in the road. The remaining Somalis kept trying to come around the corner and fire at Elmore's vehicle with small arms and RPGs. Sergeant First Class Hoepner could be heard on the radio calling that they were clear and headed to the airport to get medical assistance (Hoepner was an SF Medic and knew the severity of the wounds would take more than immediate battlefield care). Lieutenant Colonel Elmore had his driver back up out of the kill zone while he and Sergeant First Class Beebe continued to engage through the dust and the smoke, hitting more Somali gunmen near a corner where they turned, scattering a few gunmen who ran away to seek cover. They then sped to the airfield to linkup with the first vehicle, now at the hospital helicopter pad as the trauma team waited to provide medical care prior to evacuating the patients back to the Embassy compound. The two vehicles drove back on a much safer route.

Sergeant Major Ballog and Sergeant Frist Class Bowers both received the Purple Heart Medal as a result of their combat wounds.

SOCCE520 provided sniper teams throughout the remainder of July to enhance compound security during periods of heightened vulnerability. More important than the physical presence of a few snipers was the psychological impact of having the Special Forces soldiers alongside the support soldiers manning the guard-posts at night.

SWORD Base responded positively to several suggestions made by ODA523, including turning off the exterior compound lights at night to decrease their target profile. The presence of SF snipers contributed to a decrease in enemy attacks. The same pattern was seen elsewhere where the emplacement of sniper teams seemed to have a deterrent effect on the militiamen. ODA523 also continued to provide an aerial sniper team to the QRF's "Eyes Over Mogadishu" and expanded their operations to provide both daytime and nighttime coverage.

The aggressive response by ODA523 snipers working at SWORD base to a night attack in the latter days of July also resulted in a noticeable decrease in the frequency and boldness of the Somali attacks. While supporting the guard post positions along the perimeter, the team received fire around 2030 hours from Somali gunmen who fired three recoilless rifle rounds and multiple AK rounds. The detachment returned fire with their M-16s, M-24 sniper rifles, and M2 .50 caliber and MK-19 vehicular mounted weapons. The same pattern was seen elsewhere on other UN positions where the emplacement of SF sniper teams seemed to provide a deterrent.[28]

Activities in August

ODA525 conducted a Foreign Internal Defense mission with the Malaysian contingent to train them on their new sniper weapons and equipment. The Malaysians received the new gear just prior to arriving in Mogadishu but did not have time to incorporate the new tactics and techniques into their operations prior to deployment. ODA523 continued their support to the "Eyes Over Mogadishu" mission by providing aerial snipers in the ongoing reconnaissance and surveillance mission to locate Aideed.

The SOCCE provided support to a diplomatic trip conducted by the U.S. Special Envoy during the period of August 6th–9th. One SF 18E operator assisted Mr. Gosende's entourage with a communications package as the Special Envoy and his delegation visited Djibouti and three cities in Somaliland.[29]

During the period of August 9th–10th, SF 18E's (engineer specialty) from the teams assisted a QRF Explosive Ordnance Detachment with the destruction of a large weapons cache.

During the night of August 10th, an "Eyes Over Mogadishu" aerial recon was conducted by CW2 Joel W. Gordon flying one of QRF's UH-60 Blackhawks, PHEONIX 73. Along with the pilots and crew, Captain Hurst was assigned as the recon team leader over four other sniper/observers positioned aboard. Two of the team members were from his detachment—SFC Channing C. Bell and SFC Lance P. Hoepner. The flight was asked to check out a suspected mortar pit in the area which had been firing on UNOSOM II positions. Chief Warrant Officer Two Gordon had just completed a perimeter sweep of HUNTER Base and turned to the northeast traveling along October 21 Road. Below, they spotted a column of eight Malaysian vehicles near the junction of October 21 and Via Lenin roads. A little after nine o'clock in the evening, they observed a mortar round leaving the tube from a spot located 700 meters east of the Dairy Plant.

Turning to the east, the helicopter was immediately engaged by small arms fire from the Dairy Plant. The pilots also saw a large caliber anti-aircraft weapon firing at the second QRF helicopter (call sign DIAMONDBACK) from about 500 to 800 meters southeast of the Dairy Plant. The crew-chief, manning an M-60 machine gun, engaged the building where the small arms fire was coming from. The Malaysians began engaging both threats. As the DIAMONDBACK element broke left to evade the anti-aircraft gun, the Somali gunners turned their fire toward Chief Gordon's aircraft.

About that time, the helicopter flew over the suspected mortar pit. It was not a mortar; the Somalis had mounted a 57 mm rocket pod on the back of a Technical-style truck. All aboard the helicopter observed three or four rockets being shot in the direction of UNOSOM II positions at the airfield. Captain Hurst yelled for all aboard with weapons to immediately engage the rocket pod. In Gordon's witness statement written after the flight, he described what then transpired following the rocket firing sighting.

> About that time we flew over the suspected mortar site. As I banked left, we confirmed that there was a "technical" with a mounted rocket-launcher in the back, and approximately seven individuals in and around the vehicle. The enemy managed to fire another rocket as we passed over the "technical," while some of the dismounted troops fired their small arms at us. Immediately SPC Klein, CPT

> Hurst, SFC Hoepner, and SSG McCoy (who were on the left side of the aircraft) suppressed the enemy with their individual weapons. The enemy tried to disperse, but the heavy spray of fire coming from above quickly halted their flight. SFC Hoepner engaged the "Technical" with the Barrett .50 cal, quickly disabling it from use. While the left side needed to reload, I transferred the controls to 2LT Fortner who pivoted the aircraft to the left, exposing the right side to the "Technical." At that time SGT Clemons, along with SFC Bell, and SSG Angell, continued to suppress the vehicle with their weapons.
>
> I forwarded a SPOT report to Falcon 11 and then asked DIAMONDBACK to direct its 20 mm gun at the "technical" to destroy it for good. We repositioned to an over watch position and observed the DIAMONDBACK aircraft shoot three, short bursts into the vehicle before its gun jammed. Fearing that the rocket-launcher itself was not damaged, the order was given to direct our fire into the launcher. SGT Clemons fired his H-60D which in turn caused two secondary explosions from rockets still in the launcher. We were relieved by another aircrew at approximately 22:30. During post-flight, it was noted that our aircraft suffered damage from small-arms fire.[30]

CPT Hurst estimated at least two Somali gunmen were killed while the rocket pod had been disabled.[31]

The SOCCE continued its support to the UN plan to conduct activities in northern Somalia. ODA525 conducted some minor civil affairs projects in Bossaso and assisted the Ceasefire and Disarmament Committee as they oversaw the collecting and destruction of heavy weapons at various cantonments (for instance, Galciao during the period of August 22nd–24th). While on these trips, ODA525 conducted special reconnaissance to update area assessments, identify and meet with any NGOs operating in the area, and to check the status of policing, schools, and medical facilities. In one town, SF 18E Engineers, along with other members of the team, assisted local police to rebuild a roof on their station. Thereafter, they conducted a MEDCAP for the police and their families.

It was about this time the UNOSOM II intelligence staff began receiving reports of a suspected transshipment of SA-7 surface-to-air missiles in the coastal town of Hobyo (north up the coastline from Mogadishu). Major Jesmer began preliminary planning for a Special Reconnaissance mission, followed by a Direct Action mission to destroy the missiles if sighted, but he was not able to put all the assets in place before his company's departure. He turned this mission over to Major Robinson's B-team as they arrived in September. The mission would require a heavy lift helicopter to transport the DMVs, but no such helicopter was available in the U.S. inventory at Mogadishu.

On August 22nd, six American soldiers were wounded when their vehicle was ambushed by another command-detonated mine. This was the final straw for President Clinton and his National Security Council advisor, Tony Lake. He directed the deployment of the Joint Special Operations Command (JSOC) to Mogadishu. Generals Powell and Hoar agreed.

In another indication of weapons transshipments near Belet Weyne, ODA523 flew the aerial reconnaissance mission to locate the source on August 26th, but no activity was found.[32]

On August 26th–27th, JSOC deployed with a beefed up Squadron (Squadron C) commanded by COL Gary Harrell, a Ranger element commanded by LTC Danny McKnight, and little birds (AH-6s and MH-6s) from the 160th Special Operations Aviation Regiment (SOAR). Operation Gothic Serpentto capture Aideed was now on.

The final mission for the SF teams in the "Hunt for Aideed" was assisting Task Force Ranger with the handover of Somali detainees for transfer to a remote island off the coast of Somalia.

One of the final tasks for Major Jesmer was to provide input to a force augmentation concept for the upcoming fall period. As a result of the force protection analysis conducted by Lieutenant Colonel Elmore in late July, Major General Montgomery requested from all U.S. components the equipment and force augmentation packages required to enhance security. The 10th Mountain Division QRF submitted its concept to deploy an additional battalion task force while Major Jesmer conducted a staff analysis to outline the requirements for additional SOF.[33]

Major Jesmer remembered his thoughts while he drafted his input.

> Before I left, I heard of a concept which called for an expansion of U.S. forces to enhance force protection. The concept I heard discussed called for the deployment of an additional battalion task force, to include M2 Bradley IFVs. It seems to me that this approach treats the symptoms (and not too effectively at that), but ignores the cause of the strife. If we must remain in Somalia until stability returns and the nation is on the course to recovery, then perhaps Special Forces should have a much larger role. Perhaps Mogadishu should be treated like the rest of the country, with Special Forces detachments deployed into the various clan-based neighborhoods to establish contact with the indigenous leadership, to provide intelligence and to coordinate the reconstruction efforts. As personal ties are strengthened, the intelligence network will also strengthen and will assist in the detection and interdiction of intruders from other clans.
>
> This deployment of unconventional forces into the city would enable the Somalis in Mogadishu to see something positive develop from the UN military intervention. Thus, Aideed's strongholds would be isolated and perhaps co-opted at some point.[34]

The concept was titled the Protection Augmentation Force (PAF) Package. The intent of the proposal provided for measures to increase stability in the surrounding communities in Mogadishu by working with community leadership and reinforcing civil authority. Once this measure was completed, it was to be followed with a continuation of international relief efforts to increasingly build and develop the civil infrastructure needed by the populace. Presumably, UNOSOM II operations could then be conducted with minimal external disruptions. Over time, these measures would lead to the encouragement of Somali self-determination, setting the stage for the subsequent United States and UN disengagement.

In the event more SOF was needed, Phase II of the plan called for the deployment of a SF Group headquarters (SFOB) to command and control an additional Civil Affairs battalion, an SF Forward Operational Base (FOB), and a PSYOPs detachment of undetermined size, all supported by the 5th Special Operations Support Command's support element. The planned augmentation was never approved but would serve as a basis for the employment of additional forces when JTF-Somalia arrived after the Battle of Mogadishu to provide a show of force.[35]

The company conducted preparations for redeployment near the end of August and minimized any further tactical operations. On September 3rd, Major Jesmer redeployed, leaving command of the company to CPT Mike Hurst. Between September 10th–13th, ODB520 elements conducted transition and handover activities to the incoming ODB590 and redeployed home on September 13th.

During its six months in Somalia, SOCCE520 successfully planned and conducted a wide variety of missions throughout the strife-torn country, often in combat, usually with limited support and far away from friendly forces, and always in a hostile environment requiring all aspects of Special Forces individual skills. The experience gained and the lessons learned and recorded by Company B, 1/5th SFG(A) personnel in Somalia in areas such as sniper operations, mounted operations, urban combat, and peacekeeping would benefit the entire Special Forces community. The value of their experiences was reflected in 1994 by the awarding of their debriefing report as the USSOCOM Intelligence Report of the Year, 1994.[36]

Endnotes

1. Conners, James B. (USAF, Lt Col). AAR, Operation Continue Hope JSOTF Somalia, dated 15 Jul 1993, pp. 1–3.
2. Ibid, pp. 3–5.
3. Celeski, Joseph D. "Operation CONTINUE HOPE, April–August 1993." Draft Research Paper, 9 November, 2006, p. 13.
4. Personal journal of Mr. Al Glover, the then SOCCENT J2 targeteer, provided to COL (Ret.) Joseph D. Celeski in 2005.
5. Recommendation for Awards to the crew of Reach 67 written by Lt Col James B. Conner upon his return as the mission commander during the June–July 1993 deployment of the AFSOD, 7 Oct 1993.
6. Interview with LTC Brian P. Cotts on the 18th of April, 2002 at Hurlburt Field, FL.
7. Glover journal entry for that day.
8. Witness statement provided by then MAJ Timothy M. Knigge to the SOCCE as supporting document for Silver Star awards submission for SF CST team who accompanied him that day as the Moroccan LNO. MAJ Knigge described the fight and actions of the SF team members in order to provide a record of the events for later awards submissions. Copy of the 5-page statement provided to COL (Ret.) Joseph D. Celeski by Mike Hurst in late November, 2004.
9. Knigge, Timothy M. Operation Casablanca: Nine Hours in Hell. Professional Press, Chapel Hill, NC: 1995, pp. 1–29.
10. The events of June 17, 1993, are extracted from the mission timelines published in Timothy M. Knigge's book, Operation Casablanca: Nine Hours in Hell, published by Professional Press in Chapel Hill, NC, 1995.
11. Recommendation for Awards to the crew of Reach66 written by Lt Col James B. Conner upon his return as the mission commander during the June–July 1993 deployment of the AFSOD, 7 October 1993.
12. Knigge, p. 38.
13. Ibid, pp. 41–43.
14. Ibid, pp. 40–41.
15. Witness statement provided by then MAJ Timothy M. Knigge to the SOCCE as supporting document for Silver Star awards submission for the SF CST team who accompanied him that day as the Moroccan LNO.
16. Ibid.
17. Log entry notes of Al Glover.
18. Knigge, p. 35.
19. Memorandum dated 18 September 1993, Subject: AAR-SF ODB520. Deployment to Somalia, 24 Mar–19 Sep 1993, signed by MAJ David G. Jesmer, p. 2.

20. Moe Elmore's notes to author titled "Hunt for the Rabbit."

21. Ibid.

22. Daily Journal of ODA523 activities in Somalia during Operation CONTINUE HOPE, provided by Mike Hurst,

June 22nd entry.

23. UNOSOM II staff planning documents for the Crisis Reaction Cell and SRC SOP.

24. Elmore, paper to author on his role on the Force Protection team.

25. Celeski, p. 32.

26. E-mail between COL Joseph D. Celeski and CSM Patrick Ballog, Monday, February 4, 2002, Subject: Somalia.

27. E-mail from LTC (Ret.) Moe Elmore on the subject of the ambush from his perspective, March 17, 2002.

28. Detachment daily journal from Mike Hurst provided to COL (Ret.) Joseph D. Celeski in 2005. Also notes from executive summary and AAR in the Memorandum dated 18 September 1993, Subject: AAR-SF ODB520. Deployment to Somalia, 24 March–19 September 1993, signed by MAJ David G. Jesmer.

29. Memorandum dated 18 Sep 1993, Subject: AAR-SF ODB520. Deployment to Somalia, 24 March–19 September 1993, signed by MAJ David G. Jesmer, p. 3.

30. CW2 Joel W. Gordon. WITNESS STATEMENT DATE: 10 AUGUST, 1993, TIME: 21: 15, MOGADISHU(VCTY NH 3470 2691), Mission: CONDUCT AERIAL SECURITY FOR UNOSOM INSTALLATIONS/SUPPRESS KNOWN ENEMY TARGETS WITH PRECISION SNIPER FIRE.

31. Notes from citation to accompany award for CPT Mike Hurst for the incident on 10 August 1993. Copy provided to COL (Ret.) Joseph D. Celeski by Mike Hurst.

32. Memorandum dated 18 September 1993, Subject: AAR-SF ODB520. Deployment to Somalia, 24 March–19 September 1993, signed by MAJ David G. Jesmer, p. 2.

33. Celeski, p. 37.

34. Jesmer Memorandum dated September 18, 1993, p. 2.

35. Memorandum for COL Noe, 5th SFG(A) Commander, dated 26 August 1993, Subject: Proposal for U.S. Forces–Somalia Augmentation. Memo written by B-1/5 commander, MAJ David G. Jesmer, deployed at U.S. Embassy compound, Mogadishu, Somalia. Memo with supporting slides outlining commander's intent and a two phased deployment of augmentation SOF.

36. Executive Summary (Draft)–Company B, 1st Battalion, 5th Special Forces Group (Airborne) in Somalia, prepared by MAJ David G. Jesmer for input to official memorandum AAR. Copy provided to COL (Ret.) Joseph D. Celeski by LTC Jesmer in McLean, VA, 2002.

9

Operation Gothic Serpent: 25 August – 7 October 1993

Achieve an advance that cannot be withstood by rushing their weak sectors; achieve a withdrawal that cannot be entrapped by departing with unsurpassable rapidity.

— Sun Tzu, Book 6, *Superiority and Inferiority*

The Battle of Mogadishu marked the turning point for American and UN involvement in Somalia. The grand notion that a peace reconciliation could be accomplished through diplomacy with the warring clan factions disappeared. Title VII Peace Enforcement under the UN Charter was essentially neutralized when the media captured images of American soldiers being dragged naked through the streets after the battle. The "CNN effect" would now impact directly on American foreign policy; President Clinton, pressured by Congress and the mood of the American public, announced the United States would withdraw from Somalia by March of 1994.

The hunt for Aidid continued into August. It was clear UNOSOM II and the 10th Mountain QRF in Mogadishu were in an urban guerrilla war with Aidid's Habar Gidir (also spelled Habr Gidr) clan. Not a week went by in August and September without an outbreak of skirmishes and sporadic indirect fire between the opposing sides (at times, every night).

On 2 August, the *Golden Dragons* of the 2/14th Infantry officially replaced the 2/22nd Infantry. The incoming soldiers, even though having participated in Operation *Restore Hope* earlier, were untested in urban combat. There was no doubt, however, that on this tour they would become as battle hardened as the 2/22nd appeared to look to their arriving replacements.

After the tragic killing of four MPs by a command detonated mine on 8 August, work began on a bypass route around the city, to be named Main Supply Route (MSR) TIGER. In a very short time, the 2/14th soon got their wish for combat; the Somalis dropped mortar rounds on them nightly since their arrival. In a countermove, a Q-36 mortar direction-finding radar was deployed near the airfield.

As a result of increased hostilities from Aidid's warring clan faction, CENTCOM upped the ante and was approved to deploy USSOCOM's Rangers, Delta force, and 160th Special Operations Aviation Regiment's (SOAR) Night Stalkers, considered highly trained and skilled to hunt down and capture enemy leadership and/or their key supporters.

Arriving on the 25th, Task Force Ranger declared they were mission ready on the 28th of August. The task force consisted of around 450 personnel, with the bulk of the force in Company B of the 3/75th Rangers. CPT Mike Steele commanded the company; LTC Danny McKnight was in overall charge of the Rangers, including the logistic and support elements. Around forty-five Delta operators deployed as Squadron C of the 1st Special Forces Operational Detachment-Delta, commanded by LTC Gary Harrell (later to become the SOCCENT Commander during OEF and OIF). The U.S. Army Delta force was overall commanded by COL William G. "Jerry" Boykin (who would later command the U.S. Army Special Forces Command as a two-star general).

LTC Tom Matthews commanded the specialized helicopters of the 160th Special Operations Aviation Regiment. The force included MH-60Ls, AH-6Js, and MH-6s. (The MH-6s were referred to as "Little Birds.") The AH-6Js were armed with a six-barreled 7.62mm mini-gun and a 2.75-inch rocket pod.

The task force was supported by the addition of USAF combat controllers and PJs (para-rescue personnel), SEAL snipers, and an intelligence support cell. MG William F. "Bill" Garrison commanded the task force, overall.

TF Ranger settled into "The Hangar" on Mogadishu airfield. Their mission was to apprehend Aidid per UN Resolution 837 (issued 6 June, immediately after the Pakistani ambush, which intimated Aidid was the suspect perpetrator and called for his arrest). Gaining intelligence on Aidid, who was hiding in the city and constantly on the move, proved to be the most challenging task for Garrison. When the first few raid attempts came up empty, Major General Garrison expanded the Task Force's role to arrest Aidid's lieutenants, hoping one of them would talk and inform them as to Aidid's whereabouts. Task Force Ranger trained for two scenarios: one on a fixed target (inside a building) and the other on a mobile target (moving in a vehicle). During the month of September, the Task Force executed six takedowns, the most successful being the capture of Osman Atto, Aidid's financier, on September 21st.

There were some challenges. By now, everyone knew the task force was in town and looking for the fugitive warlord Aidid. Garrison could not achieve operational security during the hunt; the focus would have to be on tactical surprise and deception. The movement of helicopters and vehicles from the airport could not be concealed to the Somalis who immediately executed their early warning system during any activity on the part of the Task Force. The hangar area was constantly mortared so all were aware the Somalis knew who were in town.

Both the QRF and the task force noticed the pattern of Somali response during raids or sweeps: crowds would immediately gather, sniper and AK fires began, and more and more RPG fires were being used by the Habar Gidir. This included the firing of RPGs at low-flying helicopters. The Somali militia was successful in downing Black Hawk *Courage 53* on September 25th, killing three out of five crewmen when it crashed.

The Somalis were also constantly erecting roadblocks to impede maneuver and block reinforcements. Major General Montgomery requested American tanks and Bradley fighting vehicles to overcome this tactic, knowing the coalition had armor but would probably be reluctant to employ it on the crowded streets of Mogadishu. General Montgomery's request was denied by the administration; President Clinton and Secretary Aspin were hesitant to expand conflict in Somalia and were simultaneously exploring options to negotiate with the clan leaders and ratchet down the violence. Why TF Ranger was not told to stand down if this was the new policy is not known.

ODB590 – August 1993

Company C, 3/5th was alerted to replace ODB520 mid-August. Although scheduled to board a C-5A on the 15th of August, the cargo aircraft was diverted to pick up members of Task Force Ranger deploying into Mogadishu. ODB590 would not get a window to deploy into Somalia until the Labor Day weekend. They were also ordered to drive two of their up-armored Humvees to Fort Bragg, North Carolina, to sign over to the Rangers. ODA592 drove the two vehicles for delivery and would not see their vehicles back until after the October 3rd Battle of Mogadishu, albeit full of bullet holes and RPG shrapnel damage.

ODB590 deployed with three A-teams, ODAs 591, 592, and 593. To support the "Eyes Over Mogadishu" program, a small section of 5th Group snipers were attached to the company. Attachments were also provided to the SF company in maintenance personnel, an intelligence cell, and ministry support (Chaplain Ken Brown). The company was led by MAJ Bill Robinson and SGM Mike Williams. The company closed in on the Embassy compound, relieved ODB520, and reported themselves operational on the 8th of September. Like Jesmer, Major Robinson's company was placed TACON to UNOSOM II; the B-team was SOCCENT's SOCCE to the 10th Mountain QRF, but they were not under their command and control. Instead, ODB590 performed somewhat in the role of SOCCE to the commander, USFORSOM, Major General Montgomery and his staff. During their tenure in Somalia, ODB590 would conduct Special Reconnaissance (SR), Direct

Action (DA), area assessment reconnaissance, coalition support, and Humanitarian Activities (HA). The SF company also assisted the 10th Mountain G3 with force protection missions within the environs of Mogadishu (a supporting command relationship). These missions included the provision of SF aerial snipers for the "Eyes Over Mogadishu" program and UN-compound force protection snipers (armed presence). The Operational Detachment – Bravo (ODB) supported the program with aerial snipers three nights on, and then three nights off, rotating with the 10th Mountain snipers.

Outside of Mogadishu, UNOSOM II viewed the SF as deep sector reconnaissance assets, to be tasked by the U3 with Major General Montgomery's approval. UNOSOM II was still eager to implement their campaign plan to expand into central and northern Somalia, even amidst the rise in violence in Mogadishu. The Special Forces mounted teams were the only asset remotely capable of and willing to conduct this form of wide-ranging reconnaissance.

The ODB staff and SF teams readily assisted TF Ranger with whatever they could provide in intelligence and maps, situation reports, and assistance with crypto communications. One special item was requested by Task Force Ranger; they were not configured nor had the assets to guard, escort or provide security on detainees captured during their raids and needed a hand over procedure to another unit. Seeing the SF company as the only viable asset with capability to perform the task, ODB590 performed this role when needed. The most high-profile mission for the B-team assisting the Rangers was to escort the detainees from the Osman Atto raid during the prisoner transfer to a holding facility administered by UNOSOM II, a remote island off the coast of Somalia (dubbed "Gilligan's Island" by the troops).[1]

September 1993

September began with an increase of skirmishes and ambushes between Aidid's Somali National Alliance and UNOSOM II peacekeepers, which should have indicated the SNA was more than willing to take on their nemesis in the streets of Mogadishu. On September 5th, Nigerian soldiers were attacked when replacing Italian positions at Strongpoints #19 and #42. The Nigerian commander at SP#42 was told by a Somali they had no permission to occupy the area. Fighting broke out. When Nigerians from SP#19 attempted to help their brethren at SP#42, one of their columns was ambushed on Balad Road. An APC was badly shot up; the Nigerians sustained seventeen wounded in the altercation. Mysteriously, the Italians did nothing to help, adding to the long list of things occurring in the Italian sector creating distrust by the UNOSOM II leadership over the apparent siding of the Italians with the SNA.

On 9 September, an armored column of Pakistanis with tanks and APCs was attacked by small arms and RPGs while attempting to clear a roadblock on 21 October Road. The Pakistanis were swarmed by crowds and SNA militia fighters. It took the arrival of Cobra helicopters firing 20-millimeter cannons and 2.75-inch rockets to scatter the SNA. As the fighting intensified, Cobras fired TOW (Tube-launched, Optically-tracked, Wire-guided) missiles to allow for the extraction of the Pakistani column. It became clear to the Force Command (UNOSOM II) that Somalis were willing to fight and were increasing their tactic to mix unarmed civilians with their gunmen, knowing the peacekeepers were hindered by the ROE from shooting on civilians.[2]

Special Forces Reconnaissance to Hobyo

Major Robinson reflected on the first mission he was handed:

> I reported in to the 10th MTN Aviation Brigade commander [COL Casper] and informed him of our presence. I then went and reported into the U-3 of the UN forces (an American 0-6 [COL Ward]). Officially we were the SOCCE for the Aviation Brigade. But the UN HQ had used our predecessor as a personal recon force and had tasked them directly to do missions. The colonel in the U-3 shop wanted to maintain this relationship and the Aviation Brigade CO had no issue with it. The U-3 immediately gave us a warning order for us to conduct some type of recon. His initial mission to us was very vague in that he only waved his hand over a map of Somalia and asked if we could find where weapons were coming into the country along the Ethiopian border.
>
> The border with Somalia was over 1,000 miles long and he had no intelligence on where to even start to search. The only intelligence was that the border was mined. He did not want to put the mission in writing until I pushed the issue then he backed off the mission completely. The next day we had a new mission to go to the town of Hobyo and recon for possible shipments of weapons coming in by sea. [3]

Hobyo was a small to moderate-sized village located approximately 300 miles north of Mogadishu. Situated along the shoreline, it consisted of a harbor with a wharf for small- to medium-sized fishing vessels. The beach was wide enough to support boat landings, although not optimal due to the 12- to 15-meter-high sand dunes cropping up past the beach line. The town itself was run down, as most towns in Somalia were, but contained a market, a central mosque, and enough generators to provide a sporadic electric supply for the populace and to light the jetty at night. The town's roads resembled a spider network emanating outwards into sparse, sand dune-like terrain, spotted with camel thorn bushes. At the northeast edge of town stood an observation tower providing a view over the main dirt roads leading in and out of the town. Hobyo was a known area for Somali National Alliance activities, mostly used for rest and refit of their fighters.

Nomads surrounded the town, but the bulk of the population preferred to remain in its small buildings and houses situated along the beach to the bareness of the desert and its hot, high winds constantly blowing sand. Most of the population aligned themselves with Aidid's SNA faction and were considered not friendly to UNOSOM II forces. Hobyo was an ideal, out-of-the way location for smuggling and illicit activities, and it came under the scrutiny of the U2 as a potential site used by Aidid to move illegal weapons from aboard boats for transshipment through the town onward to Mogadishu.[4]

(The remaining activities during the mission were described below by Bill Robinson in detail, written in a letter to the author in 2002.)

> With the increase in helicopter activity over the city, and now with the introduction of TF Ranger's helicopter assets, it was predictable by those in the intelligence community the SNA would seek methods to rid themselves of what they feared most. Already hints were floating within intelligence channels of Osama bin Laden assisting Aidid with funds for weapons procurement.

There was no doubt an RPG could bring down a low-flying helicopter along with heavy machine guns mounted on Technicals. It was purported that Aidid was offering $10,000 to anyone who could bring down a coalition helicopter. Somewhere in the fog of reports the possible smuggling of SA-7 Grails [shoulder-fired heat-seeking missiles] appeared, with the small coastal town of Hobyo as a trans-shipment point. Although the UN had earlier passed a ban on the import of weapons into Somalia, in and of itself it was ineffective with no way to monitor or intercept a weapons shipment. The Special Reconnaissance (SR) mission was tasked to ODB590.[5]

One of the problems plaguing UNOSOM II at this stage of the operation was lack of specialized aerial platforms with a capability to conduct long-range infiltration and exfiltration of reconnaissance teams (TF Ranger's assets were off-limits). In what would be a poor substitute to infiltrate ODB590's SR assets into positions near Hobyo, the U3 coordinated for a Russian-made MI-26 HALO cargo helicopter, currently under contract to the UN. The HALO was one of the largest cargo-carrying helicopters in the world, with a five-man crew, and could lift up to 44,000 pounds of cargo. Its internal loading bay was similar in size to a C-130. The downside was the Russian contract crew was not trained to fly during periods of darkness (nor was the helicopter equipped for low-level, night flying). Major Robinson accepted the risk of a daytime insertion, hoping to choose a landing zone far enough away from Hobyo so the teams could maneuver undetected after unloading.

The intelligence package for the mission was sketchy—basically aerial photos from the daily P3 Orion sorties. One thing was evident in the photos: there was a lot of nomadic (herders) activity around Hobyo. The SR teams would face the challenge of unwanted encounters with locals, often accompanied with children and dogs, as they tended their flocks of goats and camels.

Major Robinson task organized for the mission into a C2 node, one surveillance team, and two security teams. The trip was expected to last up to the 15th of September, beginning with an insertion on the 9th. With the potential OPSEC compromise of a daytime infiltration, the B-team coordinated with the 10th Mountain QRF for an emergency extraction if required, with the plan to leave the vehicles behind. Even so, the flight time for a response force would be hours.

After the concept was approved and mission execution given, Major Robinson took one of his senior NCOs who spoke Russian to the airfield to continue the coordination with the MI-26 crew. A gift of vodka moved the conversation along when the pilots came to the conclusion the mission would involve the installment of long-range fuel tanks, but soon were happily installing them. Major Robinson and Sergeant First Class Dreller (Russian speaker) spent the night sleeping under the helicopters to the tune of small arms fires impacting the airfield.[6]

The company moved to the Mogadishu airfield with their vehicles on the morning of the 9th. The Russian crew, dressed in short pants and sandals, loaded the ODB's vehicles and personnel and departed for the 315-mile trip northeast to Hobyo (about two hours of flight time). The insertion went off without any problem, landing 15 kilometers north of Hobyo, although conducted in full daylight.[7]

Sergeant Mike Horan was the airborne imagery analyst for the B-team intelligence section at the time. He remembers what occurred upon Major Robinson's arrival, when the SF operators were seen by herders from nearby tents:

> Sure enough, the first thing that the guys started reporting back to us was the presence of several groups of nomads in the distance, several hundred meters away from their LZ [landing zone], mostly camels and tents, but also a couple of vehicles. So much for going undetected and moving into their observation sites, as it was only a matter of time until their presence was reported to the Haber Gedir in Hobyo, who would not necessarily be thrilled with this unsolicited, heavily armed visit.[8]

In a matter of hours, a motorized column of militia wound its way from the town to Major Robinson's position. The SF had enough assets on the ground to easily defend themselves, but if the situation got worse, there would be no assets available from UNOSOM II as a quick reaction force. Horan continued:

> The Somali interpreters and MAJ Robinson went out to meet them, since trying to hide was futile, and they figured they should find out what their situation was, so they could set up some type of bivouac/defense before darkness fell upon them. The group's leader was a real straightforward guy who did not mince words, and his message to MAJ Robinson was pretty simple, "leave the area now, or die."[9]

Per their Standard Operating Procedures (SOPs), the entire unit moved off about ten kilometers from the infiltration point, established a listening post in a small depression, and awaited darkness. In the night, the unit split and went into various directions. The surveillance team moved to establish a position in overwatch south of Hobyo itself. The two security teams moved to establish surveillance positions on what they determined to be key roads leading out of Hobyo, in hopes of interdicting any potential weapons shipments: one to the north of the town and one due west of the town.

Major Robinson established his command and control (C^2) and support site with the surveillance team in the western position, giving him the ability to maneuver his additional B-team vehicles in support of either of the two sites. The unit manned these sites for almost two days without any significant activity noted but were unable to escape detection by the nomads and populace roaming around the desert outside of town. Futile attempts were made to move the vehicles off to new positions, but the nomads always appeared wherever the SF were located. Although never approaching the teams, the nomads built signal fires and used pen flares at night in what appeared to be signals announcing the unit's activities.

The village elders were more than aware of the unit's presence, having been tipped off by the nomads, and contacted ODA593 on the 10th to establish the reason for UNOSOM II presence and to schedule a formal meeting with elders on the next day to conduct talks. The SF soldiers of ODA593 remembered the exchange between them and the villagers as friendly and courteous. Major Robinson wrote, "The UN policy was to inform all local chiefs that we were there to look for bandits and smugglers. They laughed at this and stated that they were all from Aidid's clan and that the UN had already said that they were all bandits. So much for that line."[10]

The company was resupplied the next day with serviceable tires (camel thorn bushes played havoc on the 2-ply sidewalls of the DMV tires) and water; however, a mission exfiltration helicopter could not be arranged for several more days. The company moved south of the town and established a laager to monitor the situation.

During the surveillance mission, the residents of Hobyo learned about the engagement between the

10th Mountain QRF and clan militia in Mogadishu, with the reports of several hundred people being killed by Cobra helicopters. Angry villagers with placards demonstrated against this action in front of ODA593 the next day. As the situation grew more intense, the meeting with the elders was called off.

Later, clan militia led by the military chief of operations for 11 provincial areas, Mohahadin Moalima, attempted to bluster the SF by approaching their position in a truck flying a Somali flag loaded with approximately seven armed personnel. The angry Somalis did not approach the team directly and maintained their distance at about 200–300 meters from the team's DMVs. The C2 element of ODB590 maneuvered to the team's position, and the armed gunmen from Hobyo retreated back into the town. Attempts at any further negotiations failed.[11]

Fortunately, an exfiltration helicopter was arranged for the 12th of September. ODB590 returned safely to Mogadishu. During the company After Action Review (AAR) that evening, the Somalis greeted the returnees with a five-round mortar barrage on the Embassy compound, a few landing only forty meters away from the B-team location.

While the Hobyo reconnaissance mission was being conducted, SFC Harvey Sills added the 5th SF Group sniper cell personnel to the force protection mission around Mogadishu, primarily conducting aerial flights for the "Eyes Over Mogadishu" helicopters. The snipers rigged a mess-hall wooden bench inside the helicopter to sit on and strung a rest strap across the doors. Sniper rifles, both the M-21 and the Barrett .50 caliber, were affixed by bungee cords to stabilize the barrels during flight. Out of the few engagements in September, the most memorable was the "killing" of a Somali bucket loader, being used to erect barricades. The bucket loader was previously owned by U.S. Engineers and had been captured by the militia when it was abandoned under fire during the Engineers assistance to the Pakistanis during the June 5th ambush. The "interdiction" mission was a success with the SF snipers putting it out of commission.

Now that Major Robinson was back from Hobyo, there remained the pending mission to conduct an area assessment in the Bossaso region. UNOSOM II was highly desirous of beginning the training of local and regional police, based on Ambassador Glaspie's earlier efforts to push this program during her visit there. ODA593 was tasked for the escort mission, to begin the latter part of September.

September 13th – The Battle of the Benadir Hospital Complex

Attacks on American vehicle convoys became more violent. Daily and nightly indirect fires did not cease. With spotters to identify the origin of militia indirect fires, and with the help of the Q-36 radar, it was determined the Habar Gidir was firing a mortar from the vicinity of the Banadir hospital (just to the east of the Embassy wall). On 13 September, the 2/14th Infantry was assigned the task to find the mortar and seize the crew. They would soon be walking into the next major firefight for the month. (See LTC Michael Whetstone's *Madness in Mogadishu*, Stackpole Books: 2015.)

Company B, 2/14th commanded by CPT Mark Suich and Company C, 2/14th commanded by CPT Mike Whetstone, exited the Embassy gate around 0500 hours to conduct a cordon and search of the Banadir Hospital compound (an SNA enclave). Company B would perform an inner cordon and search on buildings to the south; Company C's responsibility was the buildings to the north of the objective. The battalion headquarters element, supported with HMMWV-mounted .50-caliber machine guns and MK-19 40mm grenade launchers, provided C2 in over-watch from an outer cordon.

The first two hours of the operation were routine to the troops: surround the objective, warn the occupants to come out (using PSYOP), search and segregate people, seize any arms, and then search the buildings. As Company C began its search, they received fire from the area of the Banadir hospital. At the same time, LTC Bill David, the 2/14th commander, saw smoke from tire fires being lit (the SNA alarm system). Word was passed through channels the SNA may be gathering fighters over the next half hour or so. Soon, Company C located the 20mm mortar and crew during their search (in the Blood Bank building of the hospital complex). As the detainees were boarded onto trucks, and Companies B and C were prepared to withdraw, a barrage of enemy fires from small arms and RPGs hit the force.[12]

A heavy volume of fire was coming from militia hidden in the Banadir hospital. Whetstone described the fires in his book, *Madness in Mogadishu*:

> An RPG came screaming down from the hospital, all sparks and whoosh, and with an ear-shattering explosion blew up against the back wall of the Blood Bank. The roar enveloped us but caused no casualties. Simultaneously, intermittent fire directed at the command groups and the left flank of Bravo Company erupted from the hospital wing facing us. Evidently, we were now officially invited to a firefight.[13]

The company fired a volley of AT-4 rockets to demolish one section of the hospital to remove the threat. Knowing he could not "assault" a hospital, CPT Whetstone then had a hole blown in the compound wall to extricate his platoons to the vicinity of the Embassy wall, all the while still under tremendous enemy fires.

It took reinforcing fires from the battalion's heavy weapons, QRF snipers, Special Forces snipers at the Embassy, and Cobra gunships to get both companies through a dangerous, open area and back into the safety of the Embassy compound, seven hours later.

ODB590 Late September – Early October 1993

Meanwhile, the sniper program began to expand with the deployment of 5th SFG(A) snipers to positions at New Port on the 20th and SWORD Base on the 25th. SFC Harvey Sills established his position in a CONEX container at the New Port to provide counter-sniper support. The UN and U.S. forces stationed there continued to take fire from the surrounding houses every night. Earlier, two Italian soldiers were killed by a Somali sniper in that area. SFC Sills made a number of shots against suspected Somali gunmen during his posting there, but it is unclear if they were hits or kills due to the difficulty of verification at that location.[14]

CPT Weber, the company XO, developed a robust plan to use snipers outside of the compounds in the outlying strongpoint positions. He conducted liaison and coordination with the Pakistani, Malaysian and UAE forces on the potential use of snipers to provide force protection at these locations. The various UN contingents were either not trained for sniper operations, or if they did have snipers, unequipped or poorly equipped to employ them. The idea sold, and soon more SF snipers were deployed throughout the city. The initial coalition sniper support was allocated solely to the Pakistani contingent.

The SF snipers were broken down to provide support down to the individual Pakistani battalion level—snipers were sent to the 15th Frontier Force battalion manning checkpoints #1 and #69 in the southern half of Mogadishu. Shortly thereafter, the mission expanded to support the Pakistanis at CP#9, across from the Embassy Compound. The Pakistani contingent used the abandoned, seven story building as their headquarters (the building named K-7 on map graphics) and had established an observation post on top to provide overwatch of the Afgooye Road. The SF sniper detachment soon built an additional sandbagged bunker position on the top next to the Pakistani OP and began operations. To support and reinforce the SF snipers on K-7, the ODB established a sniper position atop the U.S. Embassy (now being used by UNOSOM II as its headquarters), across the street.[15]

On September 20th, ODA593 deployed with their vehicles via aircraft to Bosaso to conduct their assessment. This mission would last until their return on October 17th. During this journey, the team logged over 2500 kilometers of road travel and was able to assess nine out of the eleven UN-nominated towns to receive potential support. The team initially remained in Bosaso until the 5th of October, conducting numerous area recons, MEDCAPs, and interviews with local police officials and village elders. All of their activities were coordinated by local UN officials. They met with most of the senior leaders in the region (the King, the Governor, officials of the SSDF, and so on) as well as talking again with COL Yusuf (COL Abdullahi Yusuf Mohammed, the chairman and leader of the Somali Salvation Democratic Front - SSDF). In their conversations with the team, the SSDF representatives expressed their resentment of the lack of humanitarian activities in the region and warned the team that UNOSOM II should be careful to not choose sides with Aidid.

The team assisted the Bosaso police with a construction project to replace the roof on their police station, led by the 18Es (SF Engineers). Later, on September 23rd, they visited a German-run refugee camp and conducted a MEDCAP, assisting in the treatment of approximately 600 Somalis. Before they left Bosaso, Colonel Yusuf asked them to evaluate some potential storage areas for his future use as weapons cantonments, but the team did not bring up the subject of disarmament (the unpopular UN-pushed program) while conducting the site survey.

The detachment departed Bosaso on October 6th headed for Iskushuban and the remaining towns on their list requiring assessment. All the while, the team gathered information on the SNA, Islamic fundamentalists, road conditions (particularly the China Highway), location of existing schools and hospitals, police stations, water sources, bridges, minefields, weapon storage areas, airfields, and the mood of the local populace for their support for UNOSOM II operations. Generally working along the northern Somali and Eastern Somali coastlines, the team encountered some decent highways but, more often, encountered rough, dirt tracks or no roads at all, requiring them to move overland with their vehicles to reach some of the towns and villages.

As they stopped in each town, the team met with local police forces and officials, within the charter given them to assess the area for possible future training of police forces. The information gathered as a result of ODA593's work would go a long way in updating the UN files in Mogadishu and a final report was prepared and turned over to the senior police authority at the UN for his use. ODA593 returned to Bosaso on the 16th of October, remained overnight, and linked up with C-130 cargo aircraft for movement back to Mogadishu on the 17th of October.[16]

Upon the capture of Osman Atto on the 21st of September, TF Ranger coordinated the handover of Atto and other detainees to ODB590's control. The company provided security during his detention, and dropped him off in a UN-controlled detention facility. When the SF detachment (ODA592) noticed the unprofessionalism of the guards at the facility (a seventy-man guard force), they remained for a short time to provide training and advisory assistance to the guard force. There was also no medical facility or medical capability at the site; the SF medics performed this function until the team had to leave. The entire mission lasted seven days, after which, the team returned to Mogadishu.

ODA592 was then assigned to perform duties at the soccer stadium as a coalition support team with the Pakistani 19th Lancer (Armored) unit. The unit was being freshly deployed from the airfield to the soccer stadium (the teams called it the "Paki Stadium"). The Pakistani Sind Regiment (now known as the Sindh Regiment) was also stationed at the stadium; ODA592 split its operators to perform liaison and advisory duties for both units. A sniper position was established atop the Green Hotel, located nearby, which gave the snipers a clear view overlooking 21 October Road. Shortly thereafter, the team established enough rapport with the Pakistanis to allow them to place a sniper position in the old Parliament building.[17]

In late September, the company reacted to one of the SNA's nighttime attacks on the Embassy compound. As enemy fire impacted throughout the compound, the lights of the compound were extinguished and all of the forces inside occupied their defensive positions. Incoming RPGs, mortars, and small arms fire flew everywhere. Explosions impacted throughout the compound, all lit up by the flares fired by 10th Mountain mortars. Upon being notified that infiltrators were suspected on the compound, a small team of SF went searching for them, led by the ODA592 team sergeant (MSG Dave Bruner). He was subsequently lightly wounded in the face and eye from a piece of shrapnel from an explosion.[18]

The First "Black Hawk Down"

In the early morning hours of September 25th, a 10th Mountain UH-60 Black Hawk helicopter was performing its sortie in support of the "Eyes Over Mogadishu" operation. As the mission over the northern portion of Mogadishu was complete, the pilot chose to head back to the airfield for refueling, passing south over the Villa d'Italia area. The helicopter was hit by an RPG. The pilots struggled with the out of control bird, crashing on the street in the vicinity of New Port, and catching fire. Company C, 2/14th moved to their rescue.

Only two of the crew crawled from the wreckage and survived. A friendly Somali man kept them safe in his house until the arrival of soldiers from the UAE contingent. The Quick Reaction Company did their best to douse the flames at the wreck and recover bodies, but they quickly came under fire from armed militia, firing AKs and RPGs, and who were later reinforced with Technicals mounting heavy machine guns. Company C successfully extracted from a heavy gunfight, with three of their soldiers wounded. It was now apparent the Somalis could down helicopters with RPGs. The "Eyes Over Mogadishu" aerial flight program was terminated.

Later in the month, ODA591 was asked to focus their humanitarian efforts in the Medina district. Several of the American and UN compounds, as well as the airport, fell within the boundaries of the district. The populace seemed friendly overall, but a mistrust of the inhabitants began after the July 24th ambush of the ODB520 elements. Most peacekeepers, including the SF, began using the new bypass road which completely cut off the Medina citizens from any further humanitarian aid.

Both Major Jesmer's company and Major Robinson's company were supported by generous Americans on the home front who sent medical supplies, children's school materials, shoes and clothing and other necessities. Now with nowhere to distribute the humanitarian supplies, a large stockpile of materials began to build-up at the B-team compound.

ODB590 was asked to reconnect the citizens with the "hearts and minds" programs of UNOSOM II. Keeping the district pacified was the key to the safety and security of many of the peacekeepers. Major Robinson chose to build on the work of a Swedish humanitarian organization still operating in Medina. Major Robinson reflected on some of the generous donations they received from home:[19]

> I had a cousin teaching junior high in Iredell County, NC. She wrote asking if there was anything they could do to help us out. I asked for donations of clothing to use as a HA [Humanitarian Assistance] project at the orphanages and during MEDCAPS. She collected 70 boxes of clothing from the entire student population and my Mom mailed them to us. We used part at an orphanage and part during a MEDCAP. I sent a 5th Group Certificate of Appreciation to the students of the school. From all indications, it was a big hit with the kids.[20]

One of the tasks UNOSOM II wanted assistance with was trying to incorporate the Medina district to assist with local information to enhance security. If possible, it was felt the locals could also provide armed security and patrols to cooperate in keeping their area of town safe. Very few UNOSOM II contingents widely patrolled the streets of Mogadishu, so the Special Forces were asked if they could facilitate this initiative. Major Robinson used the venue of the medical assistance visits to refugee centers and various police posts to discuss the concept and elicit the help of the elders in the Medina district. Although the medical assistance was appreciated, it was not enough to sell the elders on overt cooperation with UN forces. The Special Forces operators did not give up their efforts to convince the elders and scheduled several meetings to talk over the subject. All of these efforts became moot after the October 3rd battle of TF Ranger with Aidid's militia. The initiative was dropped.[21]

While all of these activities were ongoing—the CA and MEDCAPs in Medina district, the recon to Bosaso, and the sniper and counter-sniper support to various locations—Major Robinson continued to monitor the activities of TF Ranger and the various operations of the 10th Mountain QRF for potential SF missions. Near the end of September, he discussed the role of his ODAs supporting TF Ranger activities with Major General Garrison. They discussed potentially using the SF mounted assets as security for the task force trucks going in to pick up the Rangers and any detainees after completion of their raids. Major General Garrison informed Major Robinson he would pass those recommendations of support on to his J3. ODB590 was never contacted on any involvement under this concept prior to the October 3rd raid. The SF company did not fall under the command and control of the Task Force nor the QRF, being TACON to UNOSOM II, and was probably not considered for use as an available asset during TF Ranger mission planning.[22]

By the end of September, Task Force Ranger was very successful on the line of effort to capture Aidid's lieutenants, with a little over 50% in custody off the Task Force's Tier 1 list.[23]

October 3 – 4: The Battle of Mogadishu (*Maalinta Ranger* – The Day of the Ranger)

On the morning of October 3rd, four Marine counterintelligence personnel were severely injured when their Humvee was hit by a remote-controlled detonation near the New Port area. The estimated 200 pounds of explosives tore the vehicle apart and killed the Somali interpreter.

Meanwhile, Sunday morning was a routine day for TF Ranger with physical training, maintenance, breakfast and finding something to occupy the time as the sun warmed the air. Around midday, an intelligence source gave the Task Force a tip that Aidid's lieutenants were going to have a meeting in the afternoon near the Olympic Hotel. Two of those at the meeting were Omar Salad Elmi, one of Aidid's personal aides and political advisor, and Abdi Hassan Awale, Aidid's Minister of Interior, along with others.

The target building was a few buildings north of the Olympic Hotel, along Hawlwadig Road, in the area of the Bakaara Market. It was considered one of the most dangerous spots in Mogadishu, deep in the heart of SNA and Habar Gidir territory. Intelligence felt it would certainly be defended by thousands of militiamen armed with light arms, RPGs, 12.7mms and probably some heavy weapons mounted on Technicals. Given the pattern of the gunfight between the SNA and the 2/14th Infantry on September 13th near the Banadir Hospital, it was also pretty clear that, once provoked, the SNA would rush to the scene of any gunfire, along with civilian crowds. A raid in this area had to be accomplished in under a half hour to beat the SNA's reaction time.

Task Force Ranger went on alert and prepared for the mission. 2/14th Infantry was put on notice an operation was going down.

ODB590 was holding a baptism ceremony for one of their NCO operators during the morning of October 3rd, down at the Mogadishu airport. Not everyone could attend; ODA592 was conducting a UNOSOM II-ordered reconnaissance flight aboard a UN MI-17 to assess the locations of police stations in central Somalia. LTC Leslie Fuller, the 3/5th SFG(A) battalion commander was present along with the battalion Command Sergeant Major. Lieutenant Colonel Fuller had deployed to prepare the unit for the upcoming visit of Colonel Bowra, the 5th Group commander.

With the baptism ceremony complete, a few of the members of the company stopped by the TF Ranger hangar to say hello to their friends in the unit before they boarded their vehicles and headed back to the B-team building at the Embassy compound. Suddenly, they were all asked to leave, told by their friends of an impending operation.

At 1400 hours, TF Ranger put the Bakaara Market area (the "Black Sea") off limits to any other UNOSOM II forces and the 10th Mountain QRF, including restricting the air space over the suspected target area. Around 1500 hours, Major General Garrison received confirmation from the intelligence source that the meeting was occurring and confirmed the location of the building.[24*]

This day's Quick Reaction Company was Company C, 2/14th, commanded by Captain Whetstone. His unit was placed on high alert just after 1530 hours. At 1537 hours, Task Force Ranger issued the code word LUCY, announcing the start of the operation. MH-60Ls with Rangers, AH-6Js and MH-6s with Delta operators, and a Combat Search and Rescue/Personnel Recovery along with the C2 Black Hawk lifted off the Mogadishu airfield. (There were 16 helicopters from 160th SOAR conducting the raid.) The target take-down would be performed by Squadron C of the 1st SFOD-Delta and two platoons of Company B, 3/75th Rangers, under the control of CPT Mike Steele and his headquarters element.

The code word IRENE was transmitted, indicating the raid was on at the objective area. Around 1,543 hours, four chalks of Rangers (about 12–15 Rangers in each chalk) fast-roped down to form the four corners of a square perimeter around the objective (a cordon) while Delta operators were delivered by the AH-6s and MH-6s. Brownouts from the dust blown by the helicopter blades hindered the operation, but all of the force inserted; the Rangers suffered one very serious injury when a Ranger fell off the fast rope. A few blocks away, LTC Danny McKnight stood by with the Ground Reaction Force (nine HMMWVs and three 5-ton trucks, with approximately 60 personnel) awaiting to be called forward to transport everyone and the detainees back to the airfield. The vehicles were lightly protected with sandbags, some armor plating and blast blankets, and some with bulletproof windshields.

Although every element on the raid experienced small arms fire and RPG fires, the operation was a success and had been completed in just under a half hour. As the detainees were being loaded on the trucks, Black Hawk *Super 61* piloted by Warrant Officer Clifton Wolcott was shot down by an RPG and crashed in an alley east of the objective, along Marehan street (1620 hours). The two pilots were killed, but six others survived and needed rescue. All Delta operators and Rangers on the ground began to race to the scene of the crash to secure the wreck and form a perimeter. The CSAR helicopter landed additional operators and snipers, along with highly trained medics. (Additional medics were USAF MSgt Scott Fales and TSgt Tim Wilkerson, both 23rd Special Tactics Squadron para-rescue men, and Combat Controller, Pat Rogers.) The CSAR bird, *Super 68*, was also hit by an RPG but managed to limp back to the airfield where the pilots conducted a controlled crash. As time passed at the objective, heavy enemy fires forced the operators to defend the area in isolated pockets stretched along three blocks of Marehan Road (running north-south) near the wreckage.

After sending back three HMMWVs to the airport loaded with dead and wounded, Lieutenant Colonel McKnight redirected the remaining ground convoy (now down to five HMMWVs and two 5-tons) to assist at the *Super 61* crash site. His force was constantly fired upon as he attempted a torturous route towards the scene. As he took more casualties and extreme battle damage to his vehicles, all the while under brutal fire, he made the decision to return to the airfield with the survivors of the column as he could not find a direct route to reinforce the Rangers and Delta operators at the wreck. The twists and turns to get back to a route towards the K4 traffic circle and then close in on the airfield became one of the most dangerous passages of the force. (It was during this same period the men in the three HMMWVs that had made it back to the airfield gathered more vehicles and men to form Ground Reaction Force #2 and moved back out to assist; they too were ambushed and their route blocked. Thwarted, they returned to the airfield meeting Ground Reaction Force #1 at the K4 traffic circle.)

Upon return to his headquarters, Major Robinson began to monitor the command net, listening for any information about the raid. Up to this time, he had not been tasked to support any UN or TF Ranger operation ongoing that day. He took it in stride, as the B-team was not part of the UNOSOM II QRF if they were needed. Frustrated, all anyone from the SF company could do was to climb atop the headquarters and living area roofs and watch the parade of Black Hawks and Little Birds flying towards the middle of Mogadishu. Just about a half hour into the operation, word came over the radio net of the first Black Hawk downed, *Super 61*.

The battalion commander, B-team staff, and ODA leadership assembled to begin expanding their situational awareness, sure they would be part of the 10th Mountain QRF activities, if alerted. Various courses of action were developed and discussed, but strangely, word came from the Aviation Brigade commander (COL Casper, commander of the Falcon Brigade) for the ODB to stand down in their planning efforts, even as the 10th Mountain QRF was being launched. Unfortunately, the ODB remained in the Embassy compound throughout the night, constantly monitoring the fight and awaiting instructions.[25]

Mike Horan remembered in disbelief his meeting with COL Casper, after he was sent by Major Robinson to tell the 10th Mountain QRF commander the B-team was "ready to go" if any assistance was needed:

> I rushed over to the QRF TOC running right up to COL Casper, "Sir, MAJ Robinson wants me to tell you that we are ready to head down to the airfield to hook up with the Rangers." What I would hear next shocks me to this day, I would have never believed in a million years that he would respond in the way he did. He looked me right in the eye, "You tell MAJ Robinson that you can stand down, and 10th Mountain will handle this."[26]

(In fairness to his decision, COL Casper did not command the ODB and knew they were tasked for other missions by UNOSOM II. He already had a motorized infantry company with heavy arms headed to the scene, and an additional two companies to back that up, if needed.)

With the shoot-down of *Super 61*, the 10th Mountain Quick Reaction Company (QRC) was notified to stand by. Shortly thereafter, they received the execute order. With updates on the battle, Company C, 2/14th chose to use the bypass road to get to the Ranger's hangar at the airport, instead of moving directly to the scene of the battle and likely suffering ambushes and casualties.

At 1641 hours, *Super 64* was downed by an RPG, south of National Street and the original target objective. Its four crewmen were presumed to be alive. MSG Gary Gordon and SFC Randy Shugart, both Delta snipers, received permission to fast rope down to the wreck and help protect the crew; Michael Durant and his copilot were observed moving, but there was no status on the two crewmen in the back.

Company C, 2/14th was ordered to move to the *Super 64* crash site. The column arrived to the Ranger hangar at around 1724 hours. The QRC rolled with ten HMMWVs and six 5-ton trucks (1735 hours), anticipating having to transport the Rangers out of the danger zone (accompanied with some reconstituted Ranger forces and their vehicles). Captain Whetstone's column was also blocked, forcing him to drive further and further north to find a road to turn east. Near the abandoned milk factory, the unit was ambushed. A massive firefight ensued, but Whetstone prevailed, extricated the force, and got them turned back around to the south. The drive southward was a gauntlet of snipers, ambushes and firefights; when it was clear he could not reach access to *Super 64*, Captain Whetstone also decided to preserve his remaining vehicles and force and moved back to the airfield, now covered during his movement with 10th Mountain AH-1 Cobra fires.

The operators at the scene of *Super 61* would defend bravely through the night, reinforced by incredible acts of flying courage by the 160th SOAR pilots to keep the SNA gunmen under fire and to also resupply the isolated contingents of Rangers and Delta operators. It was clear to all a ground armored force would be necessary to reach the isolated pocket, as well as the scene of the wreck of *Super 64*. It was now UNOSOM II's task to assist TF Ranger and the 10th Mountain QRF, conduct a link-up, secure and recover crewmen from both helicopters, evacuate the wounded, and then withdraw the force to safety.

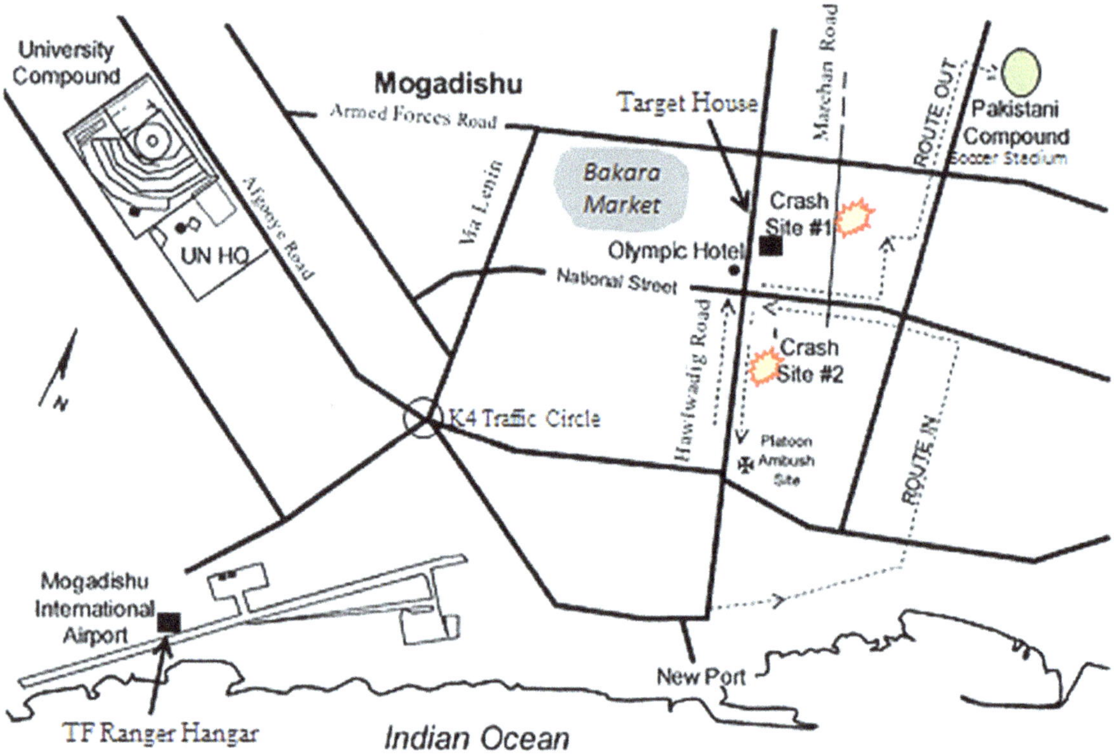

Battle of Mogadishu, 3 – 4 October 1993, adapted and modified from the map in Robert F. Baumann's and Lawrence A. Yates' work *My Clan Against the World: U.S. and Coalition Forces in Somalia, 1992 – 1994*, Fort Leavenworth, KS, Combat Studies Institute, 2004, 149.

A Coalition Quick Reaction Force Assembles

UNOSOM II, the entire 10th Mountain QRF, and elements of Delta and the Rangers all assembled at the New Port Area, en route from the airfield at 2130 hours. Four Pakistani tanks would lead the column, with 10th Mountain Infantry and Rangers inside Malaysian Condor APCs. Lighter vehicles were mixed among the column or were in trail. LTC Bill David, 10th Mountain QRF commander, led the division's infantry. The Deputy Commander of Delta, LTC Van Arsdale, led the Rangers and positioned himself at the head of the column. (Lieutenant Colonel David was the overall ground commander.)

After quick planning and a brief to the participants, and then sorting out the friction of coalition operations (language, operating styles, and so on), the column departed at 2,323 hours. They took a route east of the target area traveling north up Soccer Stadium Road to the Pakistani checkpoint at National Street, then turned west. It was around this time the Pakistani tanks fell out citing the dangers from RPGs and impassable roadblocks (they did support the movement returning RPG fires with tank rounds, however). The rest of the column pressed on up National Street, all the while under fire. AH-6s and AH-1 Cobras supported the movement with aerial fires.

Managing to reach the designated point on National Street, Team A (Company A, 2/14th Infantry) maneuvered north to the scene of *Super 61* and gained contact with the Rangers and Delta operators. Team

C (Company C) turned left with a few Malaysian APCs and braved a strong hail of bullets and RPGs to maneuver towards *Super 64*. Unfortunately, no pilots or crew were found at the wreck. After the body of Chief Cliff Wolcott was finally removed from the crumpled wreckage at *Super 61*, both helicopters were destroyed with thermite grenades.

All met back on National Street to begin the ten minute movement towards the Pakistani Stadium, around 0620 a.m. The vehicles in the convoy left in staggered sections, forcing those who could not ride, or missed a ride, to jog alongside the APCs back down National Street. Fortunately, the light-skinned vehicles of the Rangers and 10th Mountain CP awaited them for the rest of the ride to the stadium. At 0916 hours, all personnel were accounted for less six Americans missing.

A few hundred intrepid warriors had taken on thousands of Somalis. When the battle ended, 18 U.S. servicemen were dead and over 70 wounded and injured. An estimate from Somalis who participated in the battle of their losses ranges in the low hundreds killed and wounded to maybe over 500, with twice as many wounded, but an exact count has never been determined. (In the UNOSOM II force, two Malay were killed and seven wounded; two Pakistanis were wounded.)

Immediately after the battle ended, ODB590 sent as many 18D SF Medics as they could spare to the hospital at the airfield to assist with the wounded and injured; most, if not all, donated blood.

LTC John Holcombe (U.S. Army) was the Surgeon General of the 46th Combat Support Hospital in Mogadishu. He later on remarked about the work of the ODB590 SF medics at an urban warfare conference sponsored by RAND in FY2000 that "The SOF medics treated every casualty correctly. They made tough medical decisions. They ran out of supplies. They did an outstanding job. They were very, very good."

Lieutenant Colonel Fuller and his Command Sergeant Major, scheduled to depart back to Fort Campbell in a few days, were directed to remain in place by the COMUSFORSOM until further notice; Colonel Bowra's upcoming command visit was cancelled.[27]

As the week rolled on, ODB590 resumed their everyday activities, primarily operating sniper posts, as the tension remained high in the city. It was soon apparent something would change in Somalia; the U.S. would deploy more forces in the aftermath of the battle. As word came that part of the SOF reinforcement would be four AC-130 gunships, Lieutenant Colonel Fuller met with the Pakistani leadership to arrange for the acceptance of coalition support teams to assist with calls-for-fire from the gunships.

Aidid's SNA continued fighting, launching daily mortar and rocket fires and firing on UNOSOM II contingents with small arms. Tragically, on October 6th, two SNA-launched mortar rounds impacted in the vicinity of the TF Ranger hangar, killing one Delta Force operator and wounding 14 Rangers of the Task Force.

Endnotes

1. AAR dtd 1 Jan 1994, C Company, 3/5th SFG(A) titled "Operation Continue Hope". Summary prepared by MAJ William G. Robinson, 1-2.
2. Baumann, Robert F. and Lawrence A. Yates. *My Clan Against the World: U.S. and Coalition Forces in Somalia, 1992–1994*. Fort Leavenworth, KS, Combat Studies Institute, 2004, 122 – 123.
3. Letter provided by Bill Robinson to COL (Ret.) Joseph D. Celeski outlining the role of ODB590 during his command, 8 Feb 02.
4. Celeski, Joseph D. "Operation CONTINUE HOPE II JTF-Somalia: *ARSOF Operations September 93 – March 1994,"* Draft Research, Section I, dtd 1 Jun 2007, prepared for USAOC Historian, Fort Bragg, NC, pg. 4 and various internet sources.
5. Bill Robinson letter.
6. Ibid.
7. Horan, Mike. *Eyes Over Mogadishu*. USA: Xlibris Corporation (www.Xlibris.com), 2003, 49.
8. Ibid, 49 – 50.
9. Ibid, 50.
10. Bill Robinson letter.
11. Remarks derived from Mission After Action Review by ODB590 and ODAs 591,592, and 593, 3/5th SFG(A), conducted on 20 Jan 94.
12. Ibid, 126.
13. Whetstone, Michael (LTC, USA, Retired). *Madness in Mogadishu*. Mechanicsburg, PA: Stackpole, 2015, 84.
14. Bill Robinson letter.
15. Company C, 3/5th SFG(A) Unit History Report 1993, Executive Summary Fiscal Year 1993, dtd 24 May 04. This summary contained the unit's activities in Somalia for 4th QTR CY1993, 2.
16. Celeski, pp. 8–9.
17. Ibid, 2.
18. Ibid, 10.
19. Company C, 3/5th SFG(A) Unit History Report 1993.
20. Bill Robinson letter.
21. Company C, 3/5th SFG(A) Unit History Report 1993, 2.
22. Celeski, 10.
23. Neville, Leigh. *Day of the Rangers: The Battle of Mogadishu 25 Years On*. Oxford, UK: Osprey, 2018, 78. Forty-nine Tier 1 targets were on the list; after the first six raids, 26 had been captured.
24. * All mission times extracted from TF Ranger's Joint Operational Center operations log.
25. Interview between the author and Bill Robinson on B-team activities the day of the battle, conducted at

Fort Campbell, KY.

26. Horan, 92.

27. Phone discussions with Bill Robinson and COL (Ret.) Joseph D. Celeski during February-March 2002.

10

JTF Somalia: October 1993 - March 1994

Peace Operations may require years to achieve the desired effects because the underlying causes of confrontation and conflict rarely have a clear beginning or a decisive resolution. Although this is a principle often tied to debates about U.S. long-term commitments, its operational application is that commanders must balance their desire to attain objectives quickly with sensitivity for the long-term strategic aims that may impose some limitations on operations. (Joint Pub 3-0, Principle of Perseverance)

— Kenneth Allard, *Somalia Operations: Lessons Learned*

The October 3rd–4th Battle of Mogadishu was not well received by the American public and would be characterized by CNN and in other media as Clinton's Tet (1968 offensive by NVA, considered the Vietnam War changing point). The public watched in horror when mutilated U.S. service-members' bodies were dragged through the streets of Mogadishu. How could people just rescued from starvation by the good graces and big hearts of America now be killing Americans? Memorably, Senator Phil Gramm remarked, "The people who are dragging around bodies of Americans don't look very hungry to the people of Texas."

The shock of October 3rd unleashed the so-called media effect on America's foreign policy decision-making concerning Somalia and the United States partnering with the UN. Steven Livingston, in his work "Clarifying the CNN Effect," found that in this instance the news media coverage had an agenda-setting effect that drove the decision by President Clinton to abandon the Somalia mission and cast the UN adrift (albeit with some limited support to their efforts in order to save the reputation of the UN and hopefully have any success in future humanitarian interventions the United States may participate in).

There was also risk aversion on the part of policy makers who could not explain high American casualties to the American public during a time of relative peace across the world. The hunt for Aideed and the commensurate Battle of Mogadishu exposed the façade of the UN's Title VII peacemaking capabilities, when most of the nation-states still supporting the mission in Somalia were hesitant to use force to enforce the mandate.

Clinton's changed policy for Somalia was to pull out and not expose American forces to any further casualties. The posture of American forces in Somalia changed to one of force protection while accomplishing a total withdrawal of American assets by March of 1994. This would be a crushing blow to the UN-led venture and started the slow decline of ambitions by the UN in Somalia, resulting in their eventual withdrawal in early 1995. Some world problems were just too intractable and could not be solved by any New World Order philosophy.

In other areas of the world, events seemed to be on track for a new era in foreign policy. In September, Yitzhak Rabin (Israel) and Yasser Arafat (PLO) signed the peace agreement on the White House lawn. On the 4th of October, President Boris Yeltsin survived an uprising against him when his armed forces quashed rioters and armed dissidents at the Russian Parliament, ensuring some semblance of stability. Elsewhere, there was some concern when China exploded a nuclear weapon, violating the international moratorium. In another victory for global collectivism, the Europeans were successfully negotiating for the formation of the European Union, which became effective on November 1st with the Maastricht Treaty.

Other foreign policy challenges ongoing for Clinton during this time included the war in Bosnia and Operation Southern Watch (enforcement of no-fly zones over Iraq).

As a further decline in the UN's authority, the president of Haiti and his forces blocked the arrival of the UN Military Mission in Haiti (UNMIH) between October 11th and 28th, perhaps feeding off what was occurring in Somalia as Aideed and the SNA held off UN intervention in their country. Increased sanctions were applied to the country and President Clinton obliged salvaging the mission by sending six warships to enforce the mission.

In Somalia, security was heightened amid growing tensions between the Somali National Alliance and the UN. For all intents and purposes, the U.S. would remain on the defensive for their remaining time in Somalia.

U.S. forces continued to receive sporadic small arms and mortar fires in the days after the battle. Sadly, on October 6th, a mortar shell hit outside the hangar housing the Ranger Task Force, wounding several members of the Task Force (one later dying of his wounds).

President Clinton spoke to the nation on the 7th of October 1993. After praising the performance of the troops and pinning the blame for the tragedy on Somali problem makers, he promised the nation to complete the mission within six months: "All American troops will be out of Somalia no later than March the 31st, except for a few hundred support personnel in noncombat roles." Presumably, he meant the 50 or so Marine Security guards left to protect the USLO and the Ambassador to Somalia. The President also announced an increase in troop strength, sending an aircraft carrier, and the deployment of two Amphibious Ready Groups (ARGs) loaded with Marines.

Clinton's new show of force defined his objectives: protect the force, provide security for the continued flow of aid relief, and attempt to work with the Somalis towards reconciliation and forming of a transitional government by maintaining pressure on the clans. When asked by the President for help, Ambassador Oakley once again put on his desert boots and returned on October 10th to Mogadishu as the President's special representative for Somalia. One of his first achievements was to convince Aideed to release Warrant Officer Durant and a Nigerian soldier from captivity, no questions asked and no bargains made.

With the pending American withdrawal, UNOSOM II required a new Quick Reaction Force and additional troop strength to fill the hole left by the Americans. The UN looked to Egypt, Pakistan, and South Korea. Another task to beef up security required the completion of the National Police Force project. UNOSOM II would need a new logistics component when the Americans departed—whatever its replacement was would be protected by the Bangladeshis as its QRF.

President Clinton appointed General John Shalikashvili the new Chairman of the Joint Chiefs of Staff.

Aideed's response was to offer a unilateral cease-fire on October 10th after his losses in the battle, with no one but Aideed believing he could make it happen. On October 20th, Joint Task Force-Somalia was activated. The 10th Mountain Division (LI) Commander, Major General Carl F. Ernst (USA) was selected to command the JTF. He had previously served as the Commandant of the U.S. Army Infantry Center and School at Ft. Benning, GA. The parameters of the JTF mission included protecting the force per the President's guidance (including being the QRF for UNOSOM II), securing the Lines of Communication (LOCs), and preparing for the withdrawal. The JTF would serve under the tactical control (TACON) of the Commander, USFORSOM, Major General Montgomery. USCENTCOM was clear in its guidance—minimize coalition and American casualties.

Major General Ernst's staff developed a four-phased plan for the U.S. departure of land forces by March 31st:

- **Phase I**: Tactical Offense and Operational Defense
- **Phase II**: Tactical Defense and Operational Defense
- **Phase III**: Tactical and Operational Defense (achieved early since the JTF was not allowed to conduct any major tactical or operational offensives)
- **Phase IV**: Withdrawal

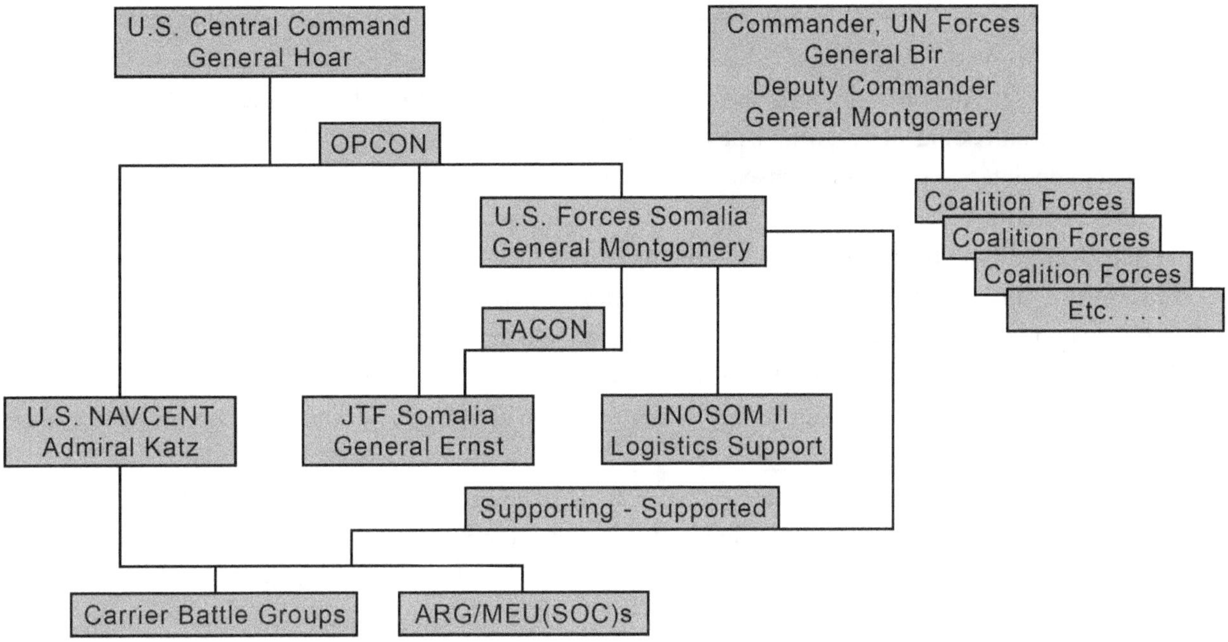

[10.1] The Somalia Theater of Organization in mid-October 1993. Courtesy of Diagram found in Gary J. Ohls' work, "Somalia. . . From the Sea." *Newport Papers* #34. New Port, RI: U.S. Naval War College, Jul 2009, 158.

On October 19th, Task Force Ranger departed Mogadishu. SOCCENT staff flowing into the theater spent a brief time with them at the Cairo West airfield in Egypt, awaiting overnight for a morning flight into Mogadishu. Many stories were shared. Although tired, the Rangers were proud of their accomplishment against overwhelming odds. Al Glover, the SOCCENT J2, described the somberness:

> It was 22 October, on a Friday. We remained at the terminal pax and saw the out-coming Ranger force of two groups from Mogadishu flowing through. I saw LTC Adams who was with COL McKnight when we were in Mogadishu before. Apparently they had quite a rough time of it and were involved in the 3 October fighting. His men were different from other troops. They looked like unmade beds with their blood types written on their boots, shirts, or pants pocket flap. They were looking forward to getting back home.[1]

In the last weeks of October, additional military reinforcements flowed into Mogadishu to make good on the President's promise. For the maritime component, the USS *America* Battle Group replaced the USS *Abraham Lincoln* Battle Group, and the ARGs replaced one another with the incoming 13th MEU (SOC) aboard the *New Orleans* and the 22nd MEU (SOC) aboard the USS *Guadalcanal,* for a total of 3,600 Marines. The switchover occurred on October 27th. The USS *America* carried AV-8 and F-14 jet attack fighters which, along with USMC AH-1 Cobras, would provide the primary air support for JTF-Somalia. With the carrier and the ARGs were SEAL Platoons: SEAL Team Five Golf Platoon aboard the USS *New Orleans* and SEAL Team Eight Delta Platoon (Amphibious Platoon). Also with the carrier was a Special Boat Unit (SBU). A Naval Special Warfare Task Group commander controlled both SEAL platoons (NSWTG-A) from his station on the USS *Guadalcanal*. NAVCENT assets were assigned in a supporting/supported relationship with the JTF and USFORSOM.

The USAF maintained the responsibility for the running of Mogadishu airport, traffic control, and the loading and unloading of cargo and passengers. There would not be a Joint Forces Air Component Commander appointed in the classic sense of Joint Doctrine.

For the Army, the 24th Infantry Division (Mechanized) deployed Task Force 1-64 armor, along with a mechanized infantry battalion (2/14th Infantry) and a 155 millimeter artillery battery equipped with COPPERHEAD artillery shells (for a total of 104 armored vehicles). The 43rd Engineer Combat Battalion (Heavy), now on their second deployment to Somalia, built a new base for them from the ground up. Situated north outside of Mogadishu, across the 21 October Road, VICTORY Base was a monument to rapid vertical construction. The 43rd chose the old military site previously used by the Somalis to manufacture SA-2/3 warheads plus other missiles.

There was big excitement one day (October 29th) on the Embassy compound when a huge explosion occurred, followed by a large mushroom cloud. Many in the compound thought they had come under some type of mortar or rocket attack and hit the ground. However, with little notification, an Explosives Ordnance Detachment exploded several thousand pounds of munitions, creating a shock wave and a strong boom, followed by a mushroom cloud rising several thousand feet. The destroyed materials were estimated at 80,000 to 90,000 pounds of TNT.

SOCCENT deployed a lean staff to form JSOTF-Somalia, commanded by COL (USAF) Richard R. Stimer, Jr. The JSOTF Deputy Commandeer, Lieutenant Colonel Darrell G. "Moe" Elmore (USA), double-hatted as the SOF liaison officer to UNOSOM II Headquarters. Within the staff was a three-man Special Operations Coordination Element (SOCOORD) for attachment to JTF-Somalia. SOCCENT's JSOTF became operational on the 26th of October.

The Joint Special Operations Air Component Commander (JSOACC) with elements from the 16th Special Operations Wing at Hurlburt Field, Florida, was bedded down in Mombasa with four AC-130 *Spectre* gunships and two KC-135s, along with four C-130 cargo aircraft which were added in early December. LTC Jim Conners commanded the JSOACC (Joint Special Operations Air Component Command).

While still a part of the Quick Reaction Force with the 10th Mountain Division (LI), ODA593 supported Ambassador Howe's trip to Gardo on the 9th through the 10th of October by providing security and communications support. ODA593 continued to expand support for attachment of Coalition Support Teams (CSTs) in discussions with the Pakistanis and Malaysians on the 11th. Between the SF Team and Lieutenant Colonel Fuller's earlier efforts, training would begin with the Pakistanis on the 16th of October along with the approval to put SF snipers into SP#1 (from ODB590) and SP#69 (from ODA593). SF Snipers would expand and include another position at Strongpoint #207 on the 23rd of October, with ODA592.

MAJ Bill Robinson's company moved to be under the OPCON of the JSOTF (along with his three SF mounted teams). COL (Ret.) Lawrence E. Casper, the commander of the Falcon Brigade, later reflected on the work of the SF B-team and Operational Detachment-Alphas (ODAs) when they were under his command (in correspondence with Major Robinson):

> I viewed the Special Forces soldiers working within the Brigade as a combat multiplier—from PSYOPs to the sniper teams, they played a pivotal role in our day-to-day operations. Although I

viewed many of the SF troops as cowboys, I could always rely upon them to tell me the ground truth—even when I did not want to hear it. I initially had a difficult time accepting the unorthodox dress and independent behavior of the SF troops supporting the brigade. Unfortunately, it took me longer than most to see the value they brought to the team and how important they were to our overall success.

On October 10th, the first "show of force" message was sent to Aideed's SNA; AC-130 gunships pounded an abandoned weapons storage site located on the north of the city.

Fort Bragg deployed a Joint Psychological Operations Task Force and a Joint Civil Affairs Task Force (JPOTF and JCMOTF) which would serve under the command of the JTF, not SOCCENT.

Special Assistant to UNOSOM II Staff

In late October 1993, Major James Realini was selected from the XVIII Airborne SOCOORD to serve with JTF SOMALIA. Major Realini was one of three Special Forces graduates of the Advanced School of Military Studies (SAMS school at Fort Leavenworth, Kansas). Upon arrival in Mogadishu, Major Realini was then assigned to Major General Montgomery's staff to assist the U5 (plans staff) with plans for withdrawing all U.S. forces from Somalia. The planning eventually included the ferrying by sea transport of all troops to Mombasa, Kenya for eventual air transport back to the United States. Once the planning for the withdrawal was underway in February 1994, Major Realini served as Major General Montgomery's personal representative to Rear Admiral Spencer's III Amphibious Task Force on board the USS *Peleliu.*

Major Realini completed his tour with Major General Montgomery's JTF-Somalia by serving as his Tactical Communications Officer on the last two days of the American deployment in Somalia and was the next to last soldier to leave Somalia preceding Major General Montgomery.

Sniper Reinforcements

Major Robinson received some needed reinforcements for the sniper, counter-sniper force protection mission in a novel way: "from the sea." Upon arrival offshore of Mogadishu, the NSWTU-Alpha Commander sought permission through Combatant Command channels to send the SEAL Platoon ashore in support of the Joint Special Operations Task Force (JSOTF). Colonel Stimer advocated for the SEALS; the JTF advocated for USMC MEU(SOC)'s Force Recon snipers. Both were quite capable of performing the mission, with the Marines a bit younger and less experienced than the SEALs.

NSWTU-Alpha's commander and selected staff flew ahead to Mogadishu and arrived on the 8th of October. Upon his arrival, the Commander of NSWTU-A attended daily briefings and met with the various contingents to gain an appreciation of the overall situation facing his unit upon their arrival. He conducted limited ground reconnaissance, supported by ODB590 security and transportation. The remainder of his unit arrived on the 18th of October as the fleet arrived offshore of Mogadishu.

Once permission was received to use the SEALs to support the JSOTF's sniper operation, the SEAL Team-8 Delta platoon moved ashore and set up sniper, counter-sniper, ROE enforcement, and special reconnaissance operations in conjunction with ODB590, TACON to the SF Company. CMDR Michael

McGuire commanded the 16-man platoon. About eight SEALs at a time rotated ashore, for up to 12 days. The SEALs lodged in tent city on the Embassy Compound and received their vehicle and logistical support from the JSOTF and the SF Company. (This platoon was from the USS *America*; at first the amphibious SEAL Platoon with the MEU was not considered, perhaps held back to support missions tasked by Admiral of the carrier battle group.)

The teams came equipped with the Model 700 Remington sniper rifle, with attached Leopold scope. There were also match grade M-14s available. For longer ranges, the SEALS used the McMillan TAC-50, .50 caliber; they preferred this weapon over the Barrett due to its accuracy, even though it was slower.

McGuire reflected on his mission:

> The SEAL contribution that received the greatest attention was scout/sniper teams augmentation to the JSOTF ashore. These scout/sniper teams, five in all, were tasked to provide surveillance of designated areas in Mogadishu for the purpose of enforcing UNOSOM II Rules of Engagement (ROE). These ROE were less restrictive than the previous UNOSOM I ROE and the standard peacetime ROE.
>
> Under these guidelines, U.S. military personnel were authorized to use appropriate measures, up to and including lethal force, upon detection of any crew-served weapon, Rocket Propelled Grenade (RPG), "technical" vehicle, or scoped weapon. The multiple SO [sniper/observer] teams that were rotated from NSWTU-A into three different fortified locations, received hostile fire on a daily basis. These teams were involved in four significant engagements which resulted in shooting thirteen ROE violators with no friendly casualties.[2]

To expand coalition liaison operations, one SF Coalition Support Team (CST) deployed for operations with the Bangladeshi contingent which was located at the Soap Factory (northeast of Mogadishu airport); they also established a sniper post to overlook the K4 traffic circle. On October 25th, ODB590 sent a Special Forces CST to the Nigerians. Discussions began with the United Arab Emirate contingent to discuss potential new sniper equipment training. That same day, SF snipers on K-7 engaged an RPG gunner at 400 meters and killed him. Clan fighting intensified between factions of Aideed and Ali Mahdi outside of the Embassy and University compounds; SOCCENT staff members emplacing antennas on the roof of the JSOTF headquarters were caught in the crossfire of bullets spilling into the compound; they immediately evacuated the roof.

The 5th group sniper team on K-7 engaged an RPG gunner, making a 400-meter shot. The following morning, Army Special Forces snipers on the UNOSOM headquarters building dropped a Somali carrying an RPG. They were incredulous as they watched a couple of men come out to pick up the body, followed by a young boy who retrieved the RPG.

On the 26th, SF snipers atop the Embassy killed a Somali gunman carrying an RPG. On the 27th, the members of the JTF SOCOORD supported Ambassador Howe's trip to El Bur, serving as his security detail and as Ground Liaison Officers (GLOs) for the AC-130 orbiting overhead.

Major Robinson gained more help with the sniper mission that day when a 13-man USMC sniper detail (Force Recon), led by Sergeant Weston, was CHOP'ed TACON to him. The Marines were in awe when Colonel Stimer, the JSOTF Commander, had them built an air-conditioned Temper-tent, with a refrigerator and TV, along with cots. The tents were stocked with sodas, bottled water, and rations. The Marines also now had access to the 10th Mountain Division Dining Facility located just across the street and shopping at the Post Exchange on the University Compound.

The Marines initially served on the U.S. Embassy sniper post and on top of K-7, across the street. Whenever USMC snipers were in an engagement, they met the criteria for awarding of the U.S. Navy Combat Action Ribbon (as was the same for the SEALs). The competition by the remaining Marines on board the ARGs to rotate in and out to serve Major Robinson was stiff.

On the 29th, another RPG gunner was engaged from K-7 (Aideed apparently did not believe the JTF and UNOSOM ROE).

In what was to be a sign of coming times, snipers atop K-7 and the Embassy reported to the AC-130 orbiting on the night of the 30th/31st period that RPG and heavy machine gun fire (potentially a 23 millimeter anti-aircraft gun) was shooting up at the gunship, originating from the Olympic Hotel. The crew of the gunship prepared to fire back in response, in accordance with the ROE (self-defense). The JTF staff intervened quickly with an order to hold operations, while they woke Major General Ernst or Major General Montgomery for approval. The JSOTF readily made the case no approval was needed, as this was an act of self-defense per the ROE.

The decision came after a brief wait: ". . . no fire authorized." Although potentially setting a bad precedent, it was later learned the command did not feel it prudent to conduct aggressive fires in light of the announced Aideed cease-fire. There was a desire for a cooling-off period to stimulate negotiations, thus no need to exacerbate the situation. After all, the gunship was not hit. Plus, all agreed there was no way Somali weapons had the range to hit the gunship. With this new declaratory posture, the AC-130s would never shoot against enemy forces for the remainder of their time in Somalia. In the following months, the endless droning over the night skies of Mogadishu, "cutting donut holes in the sky," would wear on the crews and airframes of the 16th SOW contingent.

The following day, Ambassador Oakley met with Aideed and his SNA.

November

November began with a bang. JSOTF snipers at the Pakistani Stadium used their Barrett .50 caliber to engage Somalis who were off-loading RPG-7s and ammo into a building. The range was 1600 meters. Although they fired three times before the Pakistanis asked them to cease firing, the snipers missed. That same day, the Commanding General from the MEU-ARGs visited his Marine snipers and toured the observation post on K-7.

By early November, ODB590 employed snipers in ten positions: Strongpoint #1, Strongpoint #69 "The Alamo," and Strongpoint #207 (with the Pakistanis); a sniper position atop the Green Hotel near the Paki stadium (Pakistani position); K-7 building (Pakistani position) reinforced by snipers atop the Embassy building; a sniper position at New Port created by hiding the team inside a CONNEX on the wharf with observation and shooting slits cut out of the side; the Soap Factory (Bangladeshi position); atop a water tower north of town to support SWORD and VICTORY base, exclusively manned by SEALs; and as needed, snipers atop the B-team building or within the University compound to support the 10th Mountain Division sniper teams, or for any arrangements made with the coalition partners for sniper support.

The strongpoints in Mogadishu were established by UNOSOM II contingents responsible for a designated sector. Strongpoint #1 (SP#1), was the most desired position as it overlooked a suspected clan command and control building and an arms transfer and cache node. Strong clan activity always occurred in this area. SP-1 and SP-69 were in the vicinity of the old Parliament building.

When not firing, SOF snipers provided the JTF with the most relevant situational awareness of the sector they were employed in, having eyes on the sector 24-hours daily. SOF snipers were beneficial in vectoring in AC-130 orbits whenever suspicious activities began to occur; the mere droning of the aircraft engines overhead were often enough to quell the Khat-inspired enthusiasm of malicious clansmen. The ROE were clear: any Somali gunmen armed with crew-served weapons (RPKs, MGs, RPGs, and armed technicals), or a sniper rifle with scope, could be engaged without provocation.

Clan fighting intensified the evening of November 5th; USMC snipers on K-7 and the Embassy building reported heavy gunfire and what appeared to be a firefight between warring factions, lasting almost 30 minutes. The sniper team at Strongpoint #1 also reported gunfire and an explosion. These types of incidents had been increasing in Mogadishu over the last week. The following morning, Marine snipers on K-7 once again engaged a Toyota pickup, focusing on a gunman carrying an RPD (light Soviet-style machine gun). The Marines used their Barrett .50 caliber to make a long-range shot and appeared to hit the gunman in the chest. SEALs at Strong Point #1 engaged a Somali riding in a pickup carrying an RPG, but they missed (it was later confirmed the gunman was hit in the shoulder). And on the 9th, USMC snipers on K-7 engaged an RPK gunman around mid-afternoon, with one round from the Barrett and one round from their M40.

Radio Aideed followed the K-7 sniper shooting with its typical propaganda broadcast, claiming that the Marines had killed and wounded twelve Somalis in indiscriminate shootings. (In this same period, the Marines from the Fleet Anti-Terrorism Security Team Company had separate engagements against RPG-7 and RPK gunners, one near the Benadir hospital.) A half hour after the incident on the 9th, a Somali approached the Embassy Compound to claim the snipers shot a woman. When a member of the JSOTF staff drove over to K-7 to investigate (in the now mandated three vehicle minimum for convoys), the curious media congregated as one helicopter was seen overflying the area. The incident was subsequently reported by the Associated Press as a shooting of a pregnant woman. (Later, Aideed tried to reduce tensions with U.S. forces by now claiming on Radio Mogadishu that it was the Pakistanis who had made the engagement. It was later found out the pregnant woman was in an accident and had been hit by a truck.)

COL John Noe, the 5th SFG(A) commander, conducted a quick visit to Mogadishu in early November, given his proximity in Egypt for the *Bright Star* annual exercise.

On the 6th, the USS *America* carrier battle group was scheduled for departure. Earlier, SEAL Team Eight snipers were pulled off their positions for return to the ship. Incoming SEALs from ST-5, Golf Platoon aboard the new carrier battle group replaced them, awaiting word from the JTF for authorization from the Pakistanis as to when snipers could reoccupy positions at SP#1, SP#69, and at the Paki stadium. This occurred on December 11th for SP#1; SOF snipers would not reoccupy SP#69.

Initially, it was Army SF snipers who established positions at K-7 and the Embassy building. K-7 had the best overview of the Aideed enclave, the Atto garage compound, Afgooye Road, the Olympic Hotel, and over-watch of the U.S. compounds (it was one of the highest buildings in Mogadishu, with a view all the way to the airfield, the ocean and the ships from the U.S. Navy). The position atop the Embassy gave excellent observation of the Benadir hospital and the surrounding areas. The USMC sniper detachment assumed responsibility from the SF snipers for both of these positions in early November. On November 11th,

USMC snipers engaged a Somali with an RPG and also engaged a Somali rifleman setting up in a window to take a shot at coalition forces. They fired on the gunman as he was aiming his rifle out of a window; they saw the rifle fall from the window but were unable to ascertain the fate of the gunman.

13 November 1993

Some of the SOF sniper engagements made the news, with the fact that special operators were called "American soldiers" or "sharpshooters" to hide their identity.

"U.S. soldier kills Somali; bandits attack Arab troops"
The Stars and Stripes

MOGADISHU, Somalia (AP) – An American soldier has killed a Somali carrying a rocket-propelled grenade launcher near the U.N. headquarters compound, officials said Friday.

The incident Thursday was the second time in two days that U.S. sharpshooters had wounded or killed men carrying heavy weapons. Later, bandits attacked a position held by about 25 United Arab Emirates soldiers near Afgoi, just outside Mogadishu, U.N. officials said.

U.N. military spokesman Capt. Tim McDavitt of New Zealand said the shooting early Friday continued sporadically for an hour. None of the soldiers were wounded. Zimbabwean and UAE forces sent reinforcements in armored personnel carriers, but they were not needed,

In Mogadishu, Army Col. Steve Rausch, the U.S. military spokesman, said the American soldiers who shot the Somalis were acting under the rules of engagement in the country. The rules allow them to fire if they see heavy weapons. If they see someone with a small-caliber weapon, they will respond only if threatened.

Rausch said a U.S. soldier atop a building in the U.N. military headquarters compound spotted a Somali carrying the rocket-propelled grenade launcher within 500 yards of the compound. The grenade launcher has an accurate range of 500 yards. The Somali was killed by the first shot, Rausch said. The U.S. sharpshooter missed a second Somali who picked up the weapon and fled.

Soon afterward, the American position came under fire from an AK47 assault rifle from a second-floor window in the same general vicinity. After return fire, the gun fell from the window.

Ambassador Oakley arrived to Mogadishu in preparation for his continued talks with Aideed and the SNA. He met first with the JTF commander. Strongpoint #69 was re-occupied by UNOSOM II. The 24th Infantry Division Mechanized Brigade off-loaded at New Port and convoyed to the now completed VICTORY Base. Fifty-nine detainees captured during the October 3rd battle were released on November 6th as an indication to Aideed the fighting was over.

To continue to support UNOSOM II, Pakistan observed the lessons of Task Force Ranger, and on November 9th asked the U.S. for M113 armored personnel carriers. The request also included tanks and trucks and armored wheeled vehicles (around 150). The U.S. loaned the Pakistanis 70 M113s and 30 M60A-1 model tanks. ODB590 assisted the Military Training Team (MTT) with instructors for the .50- caliber machine guns provided with the armored vehicles.

COMSOCCENT BG William Tangney conducted a visit to the JSOTF and the ODB on the 9th. A picture of the activities of the JSOTF for this period included:

- ODA591 was at the airport with the SEALs at the Bangladeshi outpost in the Soap Factory
- ODA592 was positioned with the 1st SIND Pakistani battalion
- ODA593 (split-team) was performing CST mission with the 15th Frontier Force (FF) at the airfield
- USMC snipers were positioned in K-7 and the Embassy sniper posts
- Two snipers from ODA593 were on duty at Strongpoint #1
- Two additional snipers from ODA593 were on duty at Strongpoint #69
- The ODB headquarters was supporting the JSOTF with armed convoy escort and also providing snipers to train coalition contingents
- The AC-130s refueled by the KC-135s performed their armed reconnaissance mission nightly

Brigadier General Tangney assessed the plans and progress of the JSOTF, met with the senior leadership of the JTF and UNOSOM II, and approved of the direction of the operational set for SOF. He departed on the 13th of November. By the end of the month, the JSOTF was fully employed in current operations, set for the operation to retake 21 October Road, and begin the transition to the airfield in December. Initial plans also began at the ODB for the CST mission to Kismayo, in support of the arriving Indian contingent.

Requirements Assessment

With the sniper and coalition support teams fully deployed and the nightly coverage with the AC-130s in place at the beginning of November, SOCCENT turned its efforts to the operational level planning details for the tasks still remaining: the analysis of sufficient SOF in theater, the transition plan for the withdrawal, and supporting plans to the JTF objectives. Major General Ernst and the JTF staff set the pace to ensure sufficient combat power was employed throughout Mogadishu to prevent any further major attacks on the United Sates and UN forces in the time still remaining before withdrawal. Another objective was to alleviate the degradation of delivery of humanitarian relief supplies by opening back up access to Lines of Communication in the city so relief supplies could flow again, unhampered. The arrival of the AC-130s, armor, and artillery solved the combat power dilemma.The JTF staff now began the plans process to develop Courses of Action (COAs) to open back up the 21 October Road, the roads in the area of the Digfer triangle, and the Balad Road.

Looking at the remaining operations ahead, the JSOTF's first task became one of conducting a troop-to-task analysis to determine if the force on hand was sufficient to execute the tasks. At the UNOSOM II level, critical intelligence gathering outside of Mogadishu was best served by the mounted SF teams who could deploy for long-range, long-duration missions and provide "eyes on" reporting and situational awareness. The U2 (UNOSOM II Intelligence Officer) was still concerned with the myriad of reports regarding weapons shipments in the vicinity of the Ethiopian border and the disposition of Ethiopian forces along the northern border region. The U3 in conjunction with the UNOSOM Justice Division developed the final concept for the training of a 300-man Somali Security Element for employment in the city as a protective force around hospitals and medical clinics, but the plan hinged on Foreign Internal Defense (FID) trainers for the force (with the SF as the force of choice if a separate security assistance Military Training Team could not be activated).

The Joint Task Force - Somalia had the benefit of American armor units from the 24th Mech Division. Shown here is an M1 tank at a checkpoint. Other armored vehicles included M2 Bradleys, M113s, and self-propelled 155-millimeter howitzers. (*Courtesy of Al Glover.*)

To provide effective reconnaissance and area assessment coverage of the border region alone would require two-to-three ODAs rotating in for weeks at a time; subsequently, this unilateral mission would then need effective Command and Control in the form of an additional Special Forces B-team (company leadership and support staff).

The sniper operation, although highly effective, had never achieved an optimal manning level whereby there were sufficient four-man sniper teams with capability to serve appropriate duty shifts. The SOF snipers were stretched thin across the city, often in two-man teams in some cases, serving a full 24-hour shift before replacement. While the introduction of the additional SEAL and USMC assets went a long way to fix this, the Army SF sniper teams still lacked sufficient strength to cover their missions without incurring risk to the sniper-observer team becoming highly fatigued (with vision degradation after staring long hours through scopes).

The Coalition Support Teams' efforts would be key to the transition and withdrawal of U.S. forces as coalition contingents assumed the duties of major U.S. combat formations, once they extracted from outlying positions in Mogadishu. Using additional SF teams to conduct the liaison, provide the communications connectivity, and for access to AC-130 and other fire support assets, would be a significant multiplier during this phase of the JTF's withdrawal OPLAN. The three ODAs on hand were already at split-team operations to handle the Pakistani units and the Bangladeshi position, and then further divided to provide sniper teams at these locations. In December, the French and Belgians were leaving UNOSOM II, to be replaced by the Indian contingent in Kismayo, which might require an SF CST. The Nigerian, Egyptian and UAE contingents at the airfield would need to be considered for placement of CSTs, as well.

Adding it all up, more teams, more snipers, another B-team, and so on, would require more robust staffs at the JSOTF level to handle the expanded operations (mission planning, targeting, tracking) and two ODBs (SF company headquarters) on the ground would now need an effective command and control system between the ground element and the JSOTF—the addition of an SF Battalion headquarters (FOB). Without a doubt, a force of this size would require dedicated rotary-wing lift support, currently not available to the JSOTF in November.

The JSOTF provided their assessment to Brigadier General Tangney for review and approval. On the 5th of November, COMJSOTF-Somalia sent the request for additional forces to Major General Ernst's headquarters: five additional SF teams (four for coalition support and one for the FID mission with the Somali Security Force), six additional snipers, one ODB with three mounted ODAs for border recon, and additional staff augmentation for the JSOTF and ODB590. Absent from the message was any discussion of additional SOF air assets and the desire for an FOB to C^2 the ground SOF elements. "Black Hawk Down" might have dampened enthusiasm for any further deployment of SOF aviation to the theater.

After mission analysis by the JTF staff, much of the request for additional SOF was cut; the staff augmentation could be handled internally by SOCCENT and the 5th SFG(A), so this did not need to be in the purview of a Request for Forces (RFF) to CENTCOM. Sticking with the intent of the JCS to not expand operations in Somalia, and to focus on protecting the force until it withdrew, external operations outside of Mogadishu were out. The additional sniper teams now readily available to the JSOTF from the SEALs and the USMC would takeover and replace the SF snipers, allowing them to consolidate in fewer positions. Major General Montgomery did accede to the need for more SF teams to conduct the tasks needed in Mogadishu, and on the 9th sent on the request for five additional ODAs, specifically mentioning the Foreign Internal Defense trainer requirement for the Somali Security Force.

Back at Ft. Bragg, USASOC sent out the commensurate warning order for the five ODAs, but staff discussion still hinged on the need for the SF battalion to C^2 the expanded force. In time, astute planners, working with SOCCENT and the JSOTF, looked ahead to what the force requirement would be in just a month, as forces transitioned to the airfield, and the magic number was only five ODAs (not eight). Two additional ODAs from ODB550 were alerted for deployment to Mogadishu, arriving on the 1st of December, to give Major Robinson and the JSOTF the five ODAs needed, presumably, when they would be moving to and located on the airfield. Any discussion on increased roles for SOF, or augmentation forces, ceased. The withdrawal date of March 31, 1994 became firmly fixed as the strategic end state.

The CENTCOM Force Protection Assessment

LTC Elmore, the JSOTF Deputy Commander and Liaison Officer to UNOSOM II was appointed to serve on the Force Protection Assessment mission as part of meeting one of the Commander CINCCENT's objectives during the withdrawal. His memory of the task included:

> USCENTCOM sent a team to review general physical security of U.S. forces and their bases. The team had a USMC Warrant Officer (general intelligence matters) and two LTCs, one Marine and one from the USAF. I volunteered to stay as a ground tactical guy and the SF company there provided transportation and security for movement between elements.

> Obviously the physical plant of each organization was different. Some bases were totally U.S. controlled and others (Mogadishu airfield for example) involved elements from several UN countries. Our USAF activities and later SMU's [Special Mission Units - TF Ranger] were based there, along with the Egyptians, some Bulgarians and other elements whose nationalities escape me. Each U.S. organization had different missions, and they were geographically separated by up to several kilometers. The SF Company, SOCCENT, JTF, some 10th Mtn Div elements, the U.S. diplomatic presence, UN HQ and many UN forces were located within a large walled and fenced facility.
>
> During the inspection a number of areas surfaced that I believe should have had an effect on U.S. Army doctrine, unit structures and training policies.

One day, some of the team in a separate vehicle were ambushed by gunfire; the rounds thankfully did not penetrate the bulletproof glass in their HMMWV.

On November 12th, after granted a Security Council extension for continued operations in Somalia, the Secretary General provided his own assessment as to where the United Nations effort stood. His review first covered the progress made in reducing starvation and improving public health. The number of hospitals, medical staff, and clinics had grown in the last year. Additionally, clean water projects (120 rehabilitation projects) improved the drinking water in Mogadishu and other cities.

The number of schools reopened was very positive. The agricultural industry and the cattle industry were revived. Commercial trade also returned to Somalia's ports, allowing exports of some goods for the first time.

All a rosy outlook. The next challenge for the UN in its "nation building" rested on switching to long-term reconstruction efforts for the country's social and economic infrastructure. With another round of support from donors, the next priorities were refugee return and resettlement, the building of political institutions through local and regional councils, a judicial system, national reconciliation efforts, and building a security apparatus, both armed forces and police.

The Secretary General reiterated that the problem in Somalia was a lack of disarmament progress. With the announcement of the United States leaving, he urged the member states for another round of increased effort if UNOSOM II was to succeed. Therefore, he concluded, they were faced with three options: leave the UN mandate unchanged (peacemaking); eschew coercive measures and try to get along with the Somalis, with UNOSOM II only exercising self-defense measures but keeping the flow of humanitarian goods coming into the country; and the third option, a minimal approach to keep the main airfields and ports open for deliveries of humanitarian aid to the outlying regions.

The Security Council chose to continue the mandate of Peacemaking under Title VII, requiring an additional brigade of peacekeepers. The mandate was renewed until May 31, 1994.

ODB590 Activities

Two reconnaissance missions were launched in the October–November time frame. ODA592 went to Kismayo to recon an airfield for use by UN forces to support the Indian contingent (for the SF, these are known as Special Reconnaissance—SR—when conducting the mission in semi- or non-permissive environments). After applying appropriate camouflage to their vehicles and switching to woodland pattern uniforms to match the terrain around Kismayo, ODA592 was flown in with USAF C-130s. ODA591 conducted an overland movement to Baidoa to support the removal of communications equipment during the evacuation of that HRS.

Because of C company's liaison with the Pakistanis, Major Robinson was asked by the JTF to escort two of their generals out to the USS *Abraham Lincoln* (CVBG) before its departure. This move by the JTF hopefully would help to establish partnering for emplacing SF assets with other UNOSOM II elements in positions around Mogadishu.

The ODAs began as Coalition Support Teams to the Pakistani units at the airfield and at the stadium. ODA592 occupied the stadium position and ODA593 was at the airfield. This was the start of the relationships with the other UN forces: Egyptians, UAE, and Malaysians. After the arrival of the JSOTF, Major Robinson expanded his sniper positions to include the Soap Factory overlooking the K4 traffic circle and the airfield. (The Soap Factory was under the control of the Bangladeshis.)

SEAL Team Five, Golf Platoon emplaced its first sniper team atop the water tower west of HUNTER Base. The *Raven* Aviation Task Force from the 10th Mountain Division assisted with the lift of sandbags to fortify the position using UH-60 helicopters. The sniper position downtown near the Parliament building remained in operation.

On one occasion, an SF Sniper Observer team was receiving effective machine gun fire and called for assistance. A group of off-duty SEALs (three) mounted an emergency Quick Reaction Force (QRF), arriving at the scene two miles from their compound in less than 18 minutes. The emergency QRF quelled all fire on the position, searched buildings, and disarmed a number of Somalis in the surrounding area a full two hours before the designated QRF arrived on scene.

ODB590 also conducted a number of MEDCAPS and some civic action activities. One day they were asked by an NGO Doctor, who was also an Army Reservist, to support one of his visits to an outlying village. The company coordinated for medical personnel and medical supplies to support the trip. It was a tremendous success; the ODB conducted a second visit, this time taking staff members of the JSOTF. (MEDCAPs and DENCAPs were one of the few means for U.S. personnel to get outside the stifling confinement of Mogadishu and conduct activities which felt like a contribution was being made to assist the Somali people.)

There were two other civic action trips made by the company, unilaterally. Of interest during this period, Major Robinson and his men escorted a journalist who was along to record these humanitarian activities. Ironically, they went to the home of Musa Sudi Yasalow to conduct the MEDCAP. He was the current warlord in that area, and pro-American. This visit was captured on video and aired on two of the national networks back in the United States. Like most of the units on the ground, the family support network back home readily assisted the teams with gifts and supplies to hand out to the Somali people and children. Much of this was coordinated through the unit's chaplain and religious support detachments.

From the Green Hotel position at the Paki stadium, snipers had a commanding view of the 21 October Road, east to the Pasta Factory, and west to the Cigarette Factory. A "Technical" production factory was located immediately to the west, a target of constant concern to the JTF, but constantly under surveillance by the sniper team. It was from this position SF snipers shot a Barrett .50 caliber over 900 meters to hit a technical mounted with a machine gun, scaring off the crew. The vehicle was not disabled, and the driver immediately left the area. On November 29th, the sniper team engaged a Somali carrying an RPG and killed him; they then engaged another Somali with a machine gun, wounding him. A crowd of about 150 Somalis gathered to protest but, after some time, melted away.

Naval Special Warfare Activities

The attached SEALs conducted dozens of insertion, extraction, and resupply convoys through heavily-contested areas in Mogadishu and the surrounding countryside. These convoys routinely received direct harassment to include verbal taunting, rock throwing, small arms fire, and attempts to overwhelm the vehicles with crowds. Internal weaponry, fighting skills, and reliable radio communications were the only sources of security for these convoys.

The first SEAL Scout/Sniper rotation was collocated with a Bangladeshi infantry brigade located adjacent to the airport at the abandoned Soap Factory. The sniper position at the Soap Factory was situated about four-to-five stories high out in a wing of the building (no windows remained). This position gave excellent fields of fire over the airport perimeter and the K4 traffic circle. The SEALs replaced the SF team assigned to the position on the 9th through 10th of November. The structure of the Soap Factory provided a natural vantage point that overlooked the K4 circle. There were known arms caches and drop off points in the area, so any surveillance promised to produce actionable intelligence. Throughout the first day, the SEAL Sniper Observer (SO) team observed and logged a variety of small arms being transported and noticed a concentration of activity in a gas station approximately 550 yards from their position. Groups of men arrived by vehicle and on foot. Within the false roof of the gas station, there appeared to be a pile of debris or possibly a person laying in the structure.

In the afternoon of the 10th, a white Toyota pickup arrived at the side of the gas station, and a Somali began to off-load a crew-served weapon. The weapon was clearly identified as a heavy machine gun with a butterfly trigger. Per the ROE, the on-duty SEAL spotter directed the sniper to engage the individual handling the crew-served weapon. The first shot struck the Somali and killed him instantly, causing several Somalis in the area to scatter. A second Somali bravely attempted to retrieve the weapon to bring it to the gas station building but was shot by the sniper and wounded in the shoulder area. Yet a third individual went for the crew-served weapon but was also shot in the abdomen by the SEAL sniper. As the situation escalated, a second sniper-observer (SO) team moved into position. The second spotter observed that the mass of debris in the false roof was actually a person draped in camouflage netting. The person in the false roof moved forward and produced an RPG that he was wielding in the direction of the Soap Factory and preparing to engage. When the RPG gunner fired, the snipers dropped him.

The Bangladeshis appeared furious that the calm around their position had just changed. A Bangladeshi NCO attempted to stop the SEALs from engaging. The SOCCENT J2 Targeteer, Al Glover, noted in his personal diary, "The SEAL not shooting 'physically restrained' him."

This engagement created a buzz of activity from the clan militia at the K4 traffic circle area, who now began to return fire on the SEALs. The JSOTF requested support and soon Cobras arrived on the scene to add their muscle while the ODB590 launched a QRF to the location. (In an earlier engagement by ODA593 from this position, Major Robinson had also requested Cobras as he launched his company internal QRF, but no Cobras appeared.) All in all, at least eight gunmen were hit and wounded as the

SEALs began to take return fires. This firefight received quite a bit of interest and scrutiny from stateside with higher commands second-guessing why U.S. forces were in combat in Somalia when there was a cease-fire and upcoming withdrawal.

This "incident" rose to the levels of White House interest, which then called CENTCOM. Unbelievably, someone on the CENTCOM staff drilled all the way down, bypassing the chain of command, and spoke personally to Major Robinson. The first story as usual was wrong and was soon straightened out when Colonel Stimer relayed the facts back up the chain of command. Even at that, the story made it on to a CNN television broadcast. The SEALs were removed from the Soap Factory. This was not a good omen for the future of ROE enforcement and the assigned force protection mission for the snipers.

In the afternoon of that same day, Marine Snipers at the Embassy engaged an RPG gunner with the Barrett .50 caliber. A half hour later, the Marine Corporal observed a weapon sticking out of the window of a building and engaged the target. The weapon fell. It could not be ascertained if the Somali gunman was hit, but shortly thereafter the Marines observed people running out of the building in panic.

The second SEAL sniper assignment during the period of the 14th through the 17th of November was at a location known as Strongpoint #69, also nicknamed "Fort Apache." This observation point was actually located on top of a bell tower in a densely-populated section of Mogadishu. This position gathered key intelligence and observed activity in a highly contested area on the dividing line between two rival clans (Habr Gedir and Abgal). Throughout the four-day rotation, there was near-constant gunfire coming in from all directions.

The third SEAL sniper-observer assignment was on the 21 October Road water tower (22nd through 25th of November 1993). This site was on the border of the Habr Gedir-controlled territory and provided over-watch of UNOSOM forces in the adjoining SWORD Base and adjacent area known as Gotham City.

The water tower north of Mogadishu was evaluated as a potential commanding position for snipers; it overlooked the 21 October Road and the various JTF compounds located along the road (HUNTER Base). However, access to the top was only possible by climbing a single ladder on the inside of the structure. Barrier materials were needed atop the structure to build a fortified position, and the 10th Mountain Aviation Task Force, *Raven*, solved the problem, sling-loading materials onto the top.

As the SEALs arrived, they were assigned this position and remained exclusively there until it was shut down on the 30th of November. During the month, this position came under attack from armed factions occupying a warehouse complex to the east. The SF company was alerted and rolled out the gate with loaded Desert Mobility Vehicles. MAJ Robinson directed the attack against the warehouse complex, but the suspected gunmen had fled.

The SEAL SO teams provided over-watch for hundreds of convoys transiting 21 October Road and actively prevented theft of government property from the coalition forces compound. The SO teams received numerous taunts and threats by groups of Somalis; the SO team received effective weapons fire six times. On one such occasion, the team engaged a group of armed belligerents in a vehicle at a distance of 770 yards. The extent of enemy casualties was undetermined, but with the assistance of the JSOTF QRF, the vehicle was detained and several Somalis were temporarily disarmed.

Another final SEAL sniper-observer team rotation was sent to Fort Apache and stayed for several days (27th of November through the 1st of December). Like previous rotations at this location, there were constant gunfire exchanges in the immediate surrounding area, coming from all directions. The SO team was threatened a number of times and taunted by Arabic-speaking belligerents who used the phrase "Americano you die!" The team came under harassing automatic weapons fire from a location 150 yards away. The SO team detected and engaged armed belligerents moving a heavy machine gun near a position 300 yards away, resulting in one wounded and two killed Somali personnel.

Strongpoint #69 was dubbed "Fort Apache" by the SOF snipers. There were numerous engagements fired at Habr Gedir gunmen from this location. The snipers worked out of the tower in center of picture. (*Courtesy of Bill Robinson.*)

Lieutenant Commander McGuire was proud of his unit's achievements and noted the great respect they all had for Major Robinson's leadership and the support from the SF teams, the company, and the JSOTF:

> During the period spent in Mogadishu, Somalia, the SEAL contingent benefited greatly from the technical and tactical guidance of the 5th SFG Company. The SF ODB was extremely supportive at all times to the SEALs deployed under their tactical control. Many important lessons were learned by the Navy SEALs during the deployment to Somalia. Recommendations were carried back for improvements to training and equipment. Intimate familiarity with standing and theater specific ROE was flagged as a critical concern. Most importantly, careful precise planning and professional behavior and performance are imperative to gain the confidence of the military leadership and to get employed in any theater.

Aideed provided testimony on the effectiveness of the various SOF snipers when he announced around the 13th of the month on Radio Mogadishu a warning to his people with their crew-served weapons to stay away from the compounds, because if they were seen carrying those type weapons, the UN snipers would kill them.

On November 16th, the first report came in from the snipers having seen what appeared to be a Somali with a sniper weapon. They were pre-cleared by the JSOTF commander to engage him if they saw him again. If any of the SOF snipers had become complacent because they were never engaged by Somali counter-sniper fire, this certainly got their attention.

During the same period, the UN Security Council established a commission of inquiry to investigate the attack on Pakistani peacekeepers back in May (UN Resolution 886). The hidden agenda looked to be absolving Aideed personally and disposing of the writ for his arrest. With the Americans leaving and no longer pursuing him, UNOSOM II had no wherewithal to accomplish his apprehension, so why be embarrassed maintaining the façade? (The real slap in the face came later when U.S. troops watched Aideed escorted to the Mogadishu airport by U.S. officials on December 2nd to board an American provided C-12 for his flight to Addas Ababa in Ethiopia to attend peace talks.)

Lieutenant General Ellis, the commanding general of ARCENT, and General Mundy, Commandant of the Marine Corps, visited the troops and various commands in Mogadishu. On November 18th, Ambassador Oakley met with Aideed and then departed to meet with Yusef in Somaliland. Lieutenant General Shelton, Commanding General of the XVIII Airborne Corps, conducted a visit to Mogadishu.

The 21 October Road Operation

To flex the offensive spirit of the JTF, Major General Ernst was urged by Ambassador Oakley to reopen and clear the 21 October Road, in accordance with his task to keep open major lines of communications. 21 October Road, running along the northern edge of Mogadishu's city limits, was named after the date Siad Barre seized power in a military coup (October 21, 1969). The road was no longer used that much since the completion of the bypass road. Not all sections of the road were impassable, and some of the roadblocks thrown up by the Somalis were temporary, based on food convoy traffic appearing. Most of these incidents occurred between the Via Lenin Street intersection and down east to the Pasta Factory.

In the plan, two American battalions would approach the road from the north, coming out of the desert (1/64th Armor and 2/22nd Infantry, with 2/14th Infantry battalion in reserve at VICTORY base). The 22nd MEU (SOC) was scheduled to occupy the American left flank at Balad Road intersection with 21 October Road, after coming ashore and conducting a circular, roundabout road march.

One would have thought this amount of force by the JTF would easily crush any opposition along the road. Strangely, the clearing of the road was given to the Pakistanis driving with M113s and tanks from the west to the east. Major General Montgomery was not a fan of the raid. Why do it when 21 October Road access was no longer needed, now that the bypass road was available? Was it an opportunity for Pakistani payback after the 5 June deaths near the Pasta Factory?

The JSOTF portion of the plan was providing support overhead with AC-130 gunships and the provision of Coalition Support Teams scattered throughout the Pakistani armored battalion.

The rehearsals, adjustments to planning, and arguments about the plan were incessant. Finally, the mission was called off. CPT Al Glover was attending a Sunday briefing on November 21st (ironic date!) at the JTF when the word was first announced of the cancellation: "At 1300 hours we attended an intelligence briefing for the JTF's planned upcoming operation. In the middle of the briefing, after MAJ Smart was finished with her part, LTC Sardo walked in and said that the UN (Lt Gen Bir) did not bless it."

Throughout the months of November and December, the JTF executed a number of civil projects and medical and dental visits outside of Mogadishu and within the city (called MEDCAPs and DENCAPs—medical and dental capabilities demonstrations). These occurred in Merca, Qoryooley, and the Old Port area of Mogadishu (the Old Port operation was conducted on December 2nd). The humanitarian operations were named Operation SHOW CARE and MORE CARE.

The Mogadishu Star

The JSOTF-Somalia's JSOACC (SOF air component) was given the responsibility of the *Mogadishu Star* intra-theater airlift, eventually growing to four USAF C-130s collocated with the AC-130s at Mombasa. As the JTF operation expanded logistic requirements, there was a need to provide additional intra-theater fixed-wing lift to move passengers and cargo around. It made a perfect fit to assign the C-130s to Lieutenant Colonel Conners: same airframe, same engines, same types of maintenance, and so forth, as the AC-130s. The unit could also share the products of Conners' air planning staff and weather personnel. An airfield assessment team of Combat Controller Teams (CCTs) were part of the package. The CCTs ensured the serviceability of any outlying airstrip being used by a Mogadishu Star aircraft prior to landing. The Mogadishu Star's call-sign was *Rhino*.

The C-130 detachment reported in to the JSOTF on December 1st. Three of their air mission planners were placed on the JSOTF staff at the Embassy Compound. This would prove beneficial to support the JSOTF's Rest & Recreation plan when sending personnel to Mombasa for three days of leave, being able to catch a ride on the C-130s.

Some of the flights were allocated to UNOSOM II, giving the JSOTF credit for also supporting UN operations. Daily, the unit received its missions from the air tasking order published by the LOGSUPCOM and the UNOSOM II U3. The routes flown included Mombasa, Mogadishu, and some of the outlying airfields (Baidoa, Belet Weyne, and Kismayo). The flights also included conducting MEDEVACs and resupplying NGOs with fuel bladders.

As was done earlier during Operation Provide Relief, CCT personnel, an SF medic, a communications NCO, and up to three additional SF operators (from both ODA554 and ODA555) accompanied each flight (calling themselves the Joint Security Team). To assess the airfields prior to its use by the Mogadishu Star, the ODB and JSOTF personnel, accompanied by a USAF Special Tactics Team or CCT airmen, flew from Mogadishu to the location using the UN's Russian-contracted helicopters (generally the MI-8; to move vehicles along with the assessment team, the MI-26 HALO was employed).

Absent MH-47s and MC-130s to move the SF teams and SEAL teams around for the conduct of their missions, the C-130s of the Mogadishu Star provided a valuable 'pseudo infiltration platform' for the JSOTF. For example, C-130s infiltrated and exfiltrated ODA592 for their mission to Kismayo, along with their vehicles.

The Mogadishu Star provided a much-needed service during its tenure with the JSOTF; thousands of tons of supplies were delivered and hundreds of personnel were transported throughout the theater. The most memorable transport occurred when flying the singer Larry Gatlin into Mogadishu where he entertained the troops just prior to Christmas. The unit also flew General Shalikashvili and his entourage from Kenya to Mogadishu.

On November 21st, the Marine Corps Commandant personally visited with the JSOTF Marine snipers. He also spent a few moments to meet and talk with the JSOTF J3 Operations Sergeant, Gunnery Sergeant Tim Hess (USMC). On the 23rd, USMC snipers at the K-7 position spotted a Somali gunman leaving the Benadir Hospital around 1015 hours with a 7.62mm machine gun; at around 1430 hours, the same snipers on K-7 spotted a lone gunman near Atto's garage with an M-60 machine gun and about 200 rounds. The Somali was engaged and fell to the ground. Al Glover recorded in his log: "Later an elder came by K-7 and said the gunman was in the Benadir Hospital. After he was shot a crowd gathered and put him in a Toyota pickup and the M-60 machine gun disappeared."

Marine and SEAL snipers had engagements the following two days; K-7 took one round on their position around 1015 hours, then the snipers on the embassy took two rounds. No one was hit, but the snipers did not return fire due to a lack of a clean shot. The following day (November 24th, Wednesday), the SEALs on the water tower across from SWORD Base returned fire when they were shot at but, due to the extreme range, could not confirm any hits. Thanksgiving was quiet for the JSOTF; Major Robinson, after eating with his own troops, spent the rest of the day delivering Thanksgiving meals to all the sniper positions he could reach.

On November 22nd, USMC snipers on K-7 wounded a light machine gunner; on the 27th, they killed one RPG gunner and wounded another light machine gunner. Again, from K-7, snipers shot at Somalis carrying a light machine gun, wounding one and killing two. (Aideed and the SNA had still not got the message about carrying crew-served weapons; the snipers atop K-7 were doing their best to send the message.)

On November 29th, the JTF's use of snipers to enforce the ROE heated up further. The SF snipers at the Paki Stadium saw a Somali with an RPG-7 and one Somali with an AK47; the RPG gunner was killed in the engagement and the other gunman wounded. In another engagement, SEALs on top of the water tower were shot at in the afternoon by a drive-by shooting. Minutes later, the Marines at K-7 spotted three Somalis loading an M-60 machine gun into a white car. They shot the Somali with the M-60 and killed him. Then the driver came out and swung an AK-47 into action; he was engaged and killed. The passenger took off running and shooting with an AK-47; he was engaged and wounded.

Back at the water tower, one round ricocheted off the structure. That night, a mysterious gentleman in civilian clothes visited the JSOTF night staff to relay the snipers atop the Embassy building had shot an Ethiopian dignitary. He appeared furious and demanded the removal of the Marine snipers. Colonel Stimer was sleeping; Major Celeski, the J3, was on the night shift. Celeski replied, "I'm not taking any snipers off position in the middle of the night, and you are not going up there. I'm certainly not removing young Marines for doing the job they were asked to do by the JTF and the ROE." It was later proven the claims of the mystery man were false.

Major General Ernst commended the sniper effort:

> There was more work for Special Forces to do than we had Special Forces to do it, so we cut a deal. Rear Admiral (RADM) Jack Dantone started it. When he left, he was replaced on station with the America Group (RADM Art Cebrowski commanded America Group) and the arrangement continued. In this arrangement the SEALs were chopped on a handshake to the JSOTF. . . The SEALs came ashore and reinforced the JSOTF . . . Our special operations units were worth their weight in gold and a great combat multiplier.

A synopsis of the SOF sniper activities for the remainder of November included:

- On the 16th of November, snipers at the embassy position report indications of a Somalia sniper operating very professionally in the area
- On November 22nd, snipers from K-7 observed many armed Somalis in the vicinity of Atto's garage, engaging them and killing one carrying an RPK
- Between the 23rd and 29th, snipers at K-7 and the water tower reported taking small-arms fire from Somali gunmen or drive-by shooters in vehicles
- On the 29th, Army SF snipers report engagements on RPK gunner and RPG gunner; USMC snipers on K-7 report two kills on RPK gunners, and one WIA

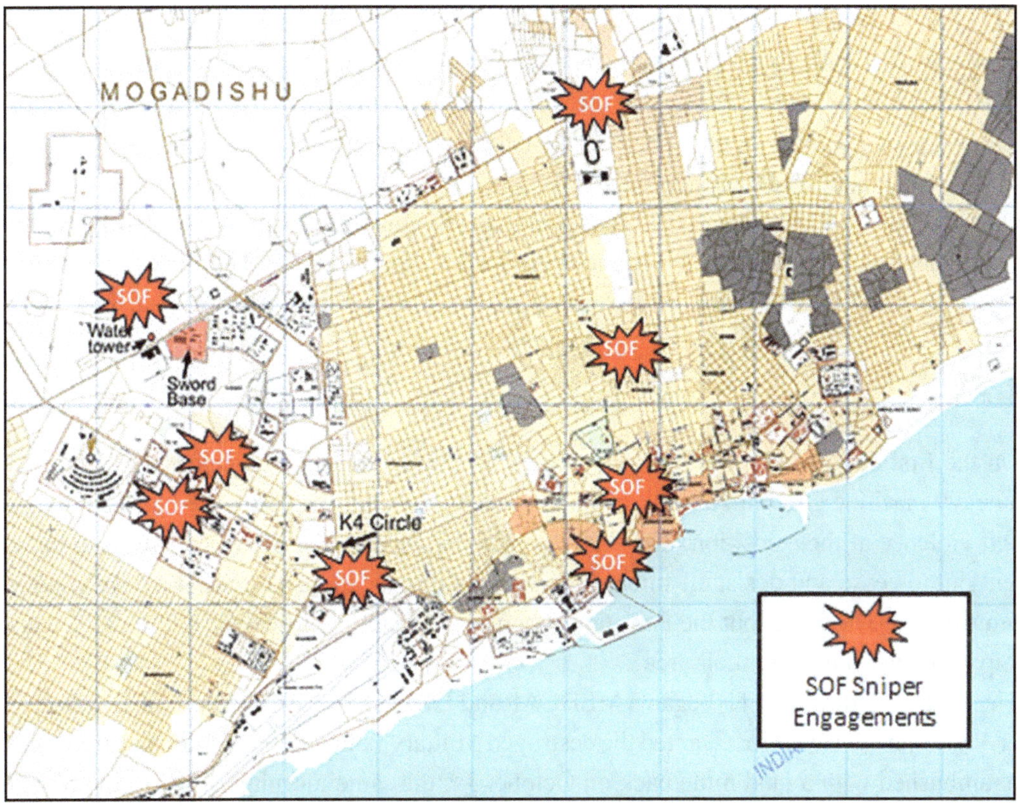

SOF (Special Operations Forces: Army, Navy, Marines, Air Force) Sniper Engagement Map

Between October and into mid-December, SOF snipers accounted for over 26 successful engagements on armed clansmen. SOF snipers were in constant danger and frequently were under fire from Somali gunmen. The situation improved dramatically when all of the positions were evacuated in late December to support the phased withdrawal to the airfield and the Redeployment Support Area (RSA). From January onward, SOF snipers operated only in the vicinity of the airport, New Port, and from the Soap Factory position.

Sniper operations degraded near the end of the month due to the Pakistani contingent rotation of its battalions inside the city, shutting down operations at the Strongpoints and at the Paki stadium until they reset their units. The Pakistanis were further concerned about the more-than-expected firing and kill rates of the snipers, which caused either return fires or hostile crowds to form in protest near their positions. U.S. Army SF snipers reporting on events and incidents in their locations perceived the reports were not passed along by the Pakistanis to higher; weapons interdictions had ceased. (Knowledge of the snipers caused the Somalis to hide their weapons when in the vicinity of checkpoints or Strongpoints—only the infantry on the ground at these positions were the solution for stopping and searching personnel and vehicles to find the weapons; this appeared to the SF snipers as not being conducted by the Pakistanis.) Snipers also complained about occasions when Pakistani soldiers even went to some lengths to walk in front of the snipers to prevent their shots.

The JSOTF staff took repeated steps to get authorization to reoccupy these positions; the loss of situational awareness and observation in these locales hampered all the staffs in collecting information. Bad blood began to develop, requiring the intervention of the JTF J3, Colonel Bedard, and the JSOTF J3, Major Celeski, to visit with all the Pakistani leadership and explain the value of the positions to the JTF and UNOSOM II. Rounds of tea were drunk in efforts to reach a conciliatory agreement amenable to all, and soon the snipers would be back in position after the Pakistani realignment in the first week of December was complete (with the exception of Strongpoint #207).

On November 30th, the JSOTF J3 began efforts to begin coordination with the Egyptians and the Malaysians at the airfield for attachment of the CSTs for later in December. The Malaysians would become the standing UNOSOM II QRF upon the withdrawal of the 2/14th Infantry from Mogadishu.

December

On the first day of December, SEAL snipers on Strongpoint #69 engaged two Somalis who were loading a heavy machine gun onto a truck, wounding them both. At this point, the Pakistani contingent had had enough violence at their position. The Paki commander ordered the SEALs down from the tower, and orders went out to cease and desist all sniper operations at that location. The SEALs were conducting their engagements per the JTF ROE, but the Pakistanis operated on making those decisions locally, based on their relationship with the Somalis in their area.

The following day, the JTF began a MEDCAP and DENCAP at the Old Port area. Following that, on December 3rd, Major General Ernst wanted the destroyed Military Police HMMWV in that area recovered (the MPs were ambushed with a land mine back on October 3rd, the same morning of the Battle of Mogadishu). Planning a complex operation, which included flying an AC-130 in the daylight (very high operational risk to the aircraft, in plain view of enemy air defense fires), the matter was solved when the UAE contingent in the area merely attached a chain and towed the wreck back into the airfield perimeter.

ODAs 554 and 555 arrived into Mogadishu the 1st of December to bring the ODB strength up to five ODAs. The requirement for the Somali Security Force FID training remained in limbo, convoluted in the legalese of getting the weapons and equipment to them from a donor nation, and still hinging on the grey line of authority for SF to train foreign police forces.

On December 9th, ODB590 conducted a MEDCAP and DENCAP to Kismayo. On December 14th, Major Robinson sent ODA592 to Kismayo. ODA592 loaded the Mogadishu Star C-130s and deployed to the Kismayo airfield, conducting link-up with the Belgians and the Indians. After settling in, the team moved to the countryside to conduct preliminary area assessments in preparation for the JTF scheduled MEDCAP/

DENCAP operation later in the month. One of their directed tasks was to ascertain the serviceability of an unimproved dirt airstrip out in the countryside for potential use by UN forces (it was not feasible upon inspection as the dirt strip was too grown over with vegetation). ODA592 met the UNOSOM II leadership when they arrived for the transfer of responsibility ceremony between the Belgians and the Indians on the 16th. (The members of the detachment were in awe at the formal military dress and obligatory dining which occurred—similar to the British protocol for formal ceremonies. Needless to say, the ODA felt underdressed and a bit worn out in their field fatigues.)

The ODA also conducted joint combat patrols with the Indians, enjoying excellent rapport they had not previously experienced with the Belgians. With the capacity now on hand to call for AC-130 fires, the JSOTF integrated the nightly flights of the AC-130s to include passes over Kismayo on the aircraft's way in and out of Mogadishu airspace. The AC-130s responded to a call for fire during a night attack on Indian position, but by the time they arrived, after looking for targets, they did not find any enemy worthy of engaging.

The ODA provided valuable planning data to the JTF for the upcoming MEDCAP/DENCAP operation during their reconnaissance and stay with the Indians. On the 23rd, they were replaced with ODA554. Unfortunately, the ODA received word the operation was delayed until January and then informed it was cancelled altogether. The ODA redeployed via C-130 back to Mogadishu on December 30th.

In early December, SOF snipers on K-7 killed an RPG gunner. On December 12th, USMC snipers at K-7 engaged an armed Somali (carrying an RPK) riding atop a bus on the Afgooye Road. The same type incident occurred again on the 14th. On December 15th, the QRF responsibility was turned over to the Malaysians. By the end of December, only 5,000 to 6,000 U.S. troops remained ashore. Brown & Root contractors took charge of logistics for UNOSOM II.

For the remainder of December, the JSOTF focused its operations on activities in support of the build-up of the Redeployment Support Area (RSA). A new mission had emerged for the JTF earlier in the month: be prepared to conduct a Noncombatant Evacuation Operation (NEO), if directed by CENTCOM, during any of the remaining time on the ground. JSOTF J3 mission planners took the task at hand and began developing Courses of Action with Joint SOF in them.

CSTs were attached to the Egyptians and Nigerians and one to the Malaysian QRF (ODA555 split team) when the Malaysians formally assumed the duty for UNOSOM II (and the JTF) on December 15th. ODAs 554 and 555 split themselves three ways to cover the other CST missions at the airfield, with the third team now reported as the Joint Coalition Support Team (JCST) providing support to the Egyptians.

Over the 20th and 21st of December, General Shalikashvili visited Mogadishu. SOF snipers from ODB590 supported the arrival of General Shalikashvili; the snipers arrayed themselves in several over-watch positions at the Mogadishu airfield and at the Embassy compound to provide VIP protection support. General Shalikashvili spent time talking with the troops at the Embassy compound and was personally briefed by Major Robinson on Special Forces capabilities inherent within ODB590. As part of the briefing, a selection of the equipment and weapons used by the SF teams were on display, with team NCOs explaining the use and purpose of each item to the Chief. During his visit, the musician and singer Larry Gatlin performed a concert for the troops. Chaplain Wylie Johnson from ODB590 was assigned as his personal escort. On the 21st, an AC-130 performed a live-fire demonstration for the Indian contingent.

Also in December, the renewed interest in resurrecting the stalled Somali Police Program floated its way to the JSOTF to conduct an FID (Foreign Internal Defense) mission and provide trainers, sponsored by the Department of State. There was one small problem: the JSOTF staff once again reminded all involved it was against U.S. law for American troops to use Title 10 military funds to train foreign police. The job to train Somali police went elsewhere.

On December 23rd, ODB590 conducted a MEDCAP/DENCAP in the vicinity of Afgooye.

After General Shalikashvili's visit, the remainder of the month of December got interesting as clan violence increased. Snipers on K-7 shot at a technical with an RPD gunman. About a half hour later, snipers at the Embassy position received one round of enemy gunfire aimed at their position.On December 26th, the Olympic Stadium snipers observed over 500 Somalis making roadblocks on 21 October Road. Three Somalis were engaged to enforce ROE. The following day, a bomb detonated at the New Port, destroying nine CONNEXs. This same day, Aideed departed Addis Ababa (having been there for discussions since December 2nd) and went to Nairobi, Kenya. Sporadic clan fighting in Mogadishu continued.

On the 27th of December, the sniper position at the Green Hotel, Paki stadium, was closed out. ODB590 prepared for redeployment with the impending arrival of ODB550. The JSOTF command was relinquished by Colonel Stimer; COL Dave Plumer took the command. SOF operations in January through March of 1994 would primarily all be conducted within the vicinity of the Mogadishu airfield, with only the AC-130s still flying nightly patrols over the rest of the city.

On December 31st, USMC snipers atop the U.S. Embassy building fired on a pickup truck with a gunman carrying an RPK. The shot missed; it was a 500-meter shot on a moving target traveling around 25 to 30 miles per hour.

LTC Elmore, now retired, reflected on the activities and role of the sniper program:

> We used snipers to deny access to specific locations, to eliminate some crew-served weapons and RPG gunners, and to create a *cordon sanitaire* around designated facilities. At times we employed SF snipers, a USMC sniper platoon and SEALS from the fleet. Overall we had excellent results from all elements and we never took a casualty from enemy fire when the snipers were covering a site.
>
> There were several aspects of sniper employment that bear comment. First the negative points. In view of the lack of a valid counter-sniper capability by the Somalis some of our snipers did not worry about their background and would take positions that were easily spotted. They ignored some tactical measures and would tend to engage people in place of systems. (Our forces did capture a scope-mounted hunting rifle from a building within 100 meters of the airfield but no valid sniper threat ever materialized on the Somali side.) The priority of targets was as follows: Technicals (SUV/pickup trucks with crew served weapons mounted on board), Crew served weapons (recoilless rifles and heavy machine-guns), light machine-guns and RPGs). I told them that the vehicle and weapon on a technical were the priority part of the system and the driver and gunners were secondary. The rationale being that it was easier to replace a crew than the vehicle and weapon. Still they consistently shot off the crews and did not destroy the technicals.
>
> Good aspects were the overall discrimination and discipline they used. All elements acted in a mature disciplined fashion with good target identification/confirmation and excellent results. They adopted and easily integrated new systems into operations with some excellent results. We found that the Somalis had an armored vehicle park a few miles outside of Mogadishu. The vehicles included a number of medium tanks (M47 or M48s I think) and some personnel carriers.

We did not have any antitank weapons that we could use effectively and we were not allowed to land and destroy them on the ground with explosives. We had some intel that some of the vehicles were operational and if so, they would pose a very credible threat. So we flew out in a helicopter and used the .50 cal to engage the tanks and APCs from the top through the grates over the engine compartment. We effectively destroyed the fuel and wiring as well as damaging the engines themselves.

On another occasion a team with a .50 and two 7.62 rifles (one an SR25) spotted a group of armed Somalis approaching the UN compound. The group had a machinegun and at least one RPG. They got fairly close (within 200 meters) and took cover behind a concrete block wall. The team engaged the wall with the .50 using a Raufuss round (a kind of chemical round that creates an enhanced effect on a solid target). The impact knocked a large hole in the wall and showered the area with sparks. The Somalis scattered like quail and the smaller rifles took out several of them. No more problems that night.

SF Special Assistant to Admiral Howe (December 1993 – March 1994)

In the fall of 1993, COL "Russ" Howard was posted to Ft. Lewis, Washington awaiting assumption of command of the 1st Special Forces Group (Airborne) when he was asked to serve as a special assistant to Admiral Howe, the UN Special Advisorto Somalia. He deployed to assume those duties in Somalia in late December of 1993, partnering with a Navy SEAL Captain.

After deploying by commercial aircraft to Mogadishu, he linked up with General Sheehan, Admiral Howe's Deputy, at the University complex where Ambassador Howe was living in a trailer. Now retired Brigadier General Howard remembers his duties: "I took my tasks and missions from daily discussions with Admiral Howe and General Sheehan. The tasks were formulated from attempts to move the governmental and UN process for rebuilding some form of Somali governance."

Of highest importance was the resurrection and reset of operational training for the Somali National Police program. His cohort, the Navy SEAL Captain, served as the project director, with Howard assisting in the securing of the donated police program equipment, such as 2-1/2 ton trucks, jeeps from Malaysia, police uniforms, and light weapons. These duties often took the two special assistants to Baidoa, Kismayo, and Bale Dogle to find old buildings or warehouses they could use to run the police training. Other requirements emerged for the assistants to handle, as a course of their duties. Once, Howard was asked to go to Bale Dogle to talk to tribal elders when it was known they had used their people to raid a food warehouse.

Since U.S. law prohibited military forces from training police forces of another country, various problems emerged to plague the program and for finding sponsors for the trainers. Lieutenant Colonel Elmore was involved in the first stages of finding a U.S. sponsor. Major Robinson remembered the earlier attempt in September: "LTC Elmore (UNOSOM II LNO) was trying to get a local police force established and we went to a couple meetings at the Nigerian compound with many of the tribal elders from the local clans. There appeared to be some promise here but not real intention on the part of the UN to move this forward."

Upon the announcement of the withdrawal of the JTF by March of 1994, the SEAL special assistant went to great effort to secure the donated Police Program equipment and get it under U.S. control for its removal.

Brigadier General (Retired) Howard completed his assignment in the latter part of March 1994 and redeployed home. Some highlights of his three-month tour include his participation in key meetings and visits, defusing tribal tensions, and having the opportunity to observe high-level diplomacy within the State Department and the UN.

As the drawdown of U.S. forces began in mid-December, the JSOTF-Somalia began a phased withdrawal plan to the environs of the airfield. Earlier in December, the JSOTF staff and the ODB staff conducted site surveys on potential locations on the airfield to move their operations. Planning and coordination for the NEO mission continued.

By the end of the year, France, Belgium, Italy, Sweden, and Tunisia had all withdrawn their force commitments to Somalia, questioning why they should stay if the Americans would not stay.

C-3/5th SFG(A) with its five ODAs transitioned to sniper positions surrounding the airfield and continued CST missions to UNOSOM II forces located on the airfield perimeter.

The Withdrawal and Last SF Company Rotation – January to March 1994

By January 1994, ODB590 was nearing its six-month deployment mark and required replacing. With the amount of total military forces now assembled in the Mogadishu "pocket," it was a moot point, doctrinally, that any further role for Special Forces was required. It was also clear the intent for the JTF was an increasing drawdown, not a one-for-one replacement of units as they gradually departed Somalia. The claim Special Forces snipers were essential for the remaining two months of the task force was nebulous at best, given that SEAL and USMC snipers were available offshore with the naval Task Groups and the 10th Mountain Division and selected coalition partners having received sniper training from ODB590. For some reason, this did not enter the decision to just send the SF company home; a new SF company would replace ODB590, albeit with only two teams vice five teams.

It was once again 2nd Battalion, 5th Special Forces Group's time in the deployment cycle. Lieutenant Colonel Faistenhammer selected Company B (ODB550), commanded by MAJ Kerry Allen. Since two of his teams were already in Somalia (ODA554 and ODA555, with ten operators apiece) and only had about a month in Mogadishu under their belt, they would remain in theater and be transferred from ODB590 back to their parent company upon Major Allen's arrival with his staff.

On January 2, 1994, the World Food Program compound in Baidoa was raided by Somali gunmen. Guards were able to kill one Somali and wound another, but the entire foodstuffs stored were gone. Secretary of Defense Les Aspin resigned in January; he had been preceded by the Assistant Secretary of State's resigning in the fall.

On January 3rd, SEAL Team Two afloat with PHIBRON 4 relieved SEAL Team Five. The JSOTF headquarters and staff moved to final sites at the airfield on the 4th of February. The parade of dignitaries visiting Somalia carried over into the New Year. On January 4th, the Commander of USCENTCOM visited ODB590's sniper display (it was becoming apparent the sniper program was one of the most active and effective force protection measures implemented by the force).

Clan fighting intensified the evening of January 5th; USMC snipers on K-7 and the Embassy building reported heavy gunfire and what appeared to be a firefight between warring factions, lasting almost 30 minutes. The sniper team at Strongpoint #1 also reported gunfire and an explosion. These types of incidents had been increasing in Mogadishu over the last week. The following morning, Marine snipers on K-7 once again engaged a Toyota pickup, focusing on a gunman carrying an RPD. The Marines used their Barrett .50

caliber to make a long-range shot and appeared to hit the gunman in the chest.

Although a relatively quiet period, security for convoys remained in force due to the anticipation of an ambush, more often for the eventuality of being mobbed by Somalis attempting to steal goods from the rear of trucks loaded with supplies. Non-lethal methods were used to disperse Somalis from climbing aboard, but sometimes the crowd got out of control. On the 6th, Brigadier General Tangney, COMSOCCENT visited the JSOTF and the other SOF elements in country. While COMSOCCENT was visiting the AFSOD in Mombasa, USMC snipers on K-7 shot a Somali riding atop a Toyota truck, carrying an RPD, hitting him in the chest with their Barrett .50- caliber round. At Strongpoint #1, SEALs engaged an RPD gunner in a Toyota truck but could not ascertain the effects of their shot (it was later learned he had been wounded). Brigadier General Tangney flew to Mogadishu with Colonel Stimer to receive the briefing on the incident. There were expectations of reporting to higher in infinite detail each time a sniper engagement occurred (and the staff found themselves preparing reports on the incidents for the following three days). General Tangney departed the following day after attending the handover of the JTF to Major General Montgomery. Quite an introduction for COMSOCCENT of what to expect during the remaining time for SOF in Mogadishu.

USFORMSOM SF Contingent
8 Jan 1994 – 24 March 1994

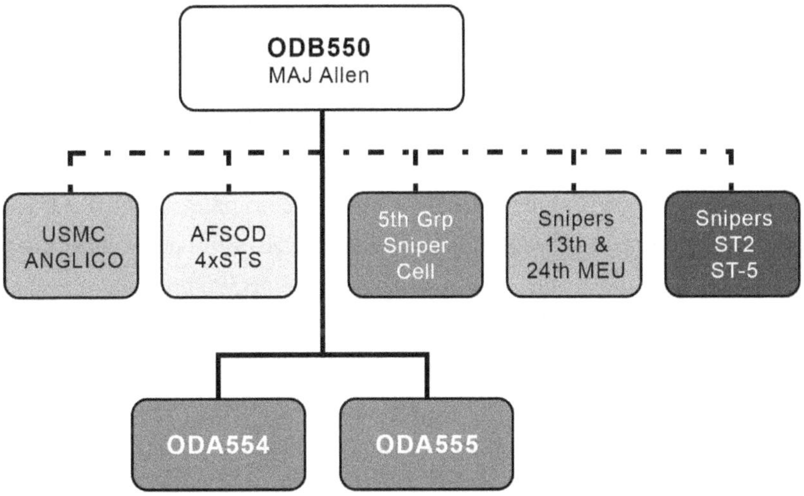

On 7 January 1994, Major General Montgomery departed his duties at UNOSOM II and assumed control of the JTF as General Hoar's Commander of both the JTF and USFORSOM. General Montgomery wryly said, "It's time to get out. At some point in time you've got to stand up and take responsibility, and Somalis will not take responsibility." Major General Ernst departed country.

Company B, 2/5th deployed to Mogadishu on January 6, 2004 and arrived on January 8th in Mogadishu to replace ODB590's operation with JSOTF-Somalia but did not arrive in time to conduct a full transition hand-off with the commander of ODB590, who had departed early in January with key elements of his staff due to air flow constraints. Chaplain Wylie Johnson and the Chaplain's Assistant for ODB590, SGT Speckert, remained in Mogadishu to assist with the arrival of ODB550, and then they both deployed home on the 15th of January, 1994.

Two of the company's ODAs working for ODB590, ODA554 and ODA555 (ten operators each), remained in-country to receive the B-team, with the detachment commander of ODA555 assuming the role as senior commander of the SF contingent until the arrival of Major Allen. Upon link-up with the teams at the airfield, the B-team moved to the ODB headquarters at the Embassy compound and assumed C^2 over its two ODAs, with TACON over the 13 attached snipers from the 13th MEU (SOC), the USAF STS personnel, and the snipers deployed from SEAL Team 5. (An ANGLICO was attached to the company in February to assist in support to the Egyptians. Major Allen signed for two armored Humvees from the 10th Mountain to provide the ANGLICO mobility and protection.) It was a short stay in the 'luxury accommodations' at the Embassy compound; on January 8th, the ODB moved to the airfield to occupy its final position (now living in tents).

ODB550 assumed responsibility for the two remaining mission areas: sniper and counter-sniper operations around JTF and UN facilities in Mogadishu and providing coalition support teams to the Egyptian and Malaysian contingents. A tertiary mission task was to be prepared to conduct humanitarian activities in support of MEDCAP and DENCAPs. Although not in receipt of the NEO mission, Major Allen understood it to be an implied mission and dedicated some of his staff for the planning.

On the 8th, new SEAL snipers rotated in from the carrier. Sniper and counter-sniper missions were performed at the Embassy Compound, K-7, the Soap Factory, New Port, and in the Redeployment Support Area (RSA). Army SF snipers performed their duties at New Port and the RSA. USMC snipers still held posts at the Embassy compound and K-7, with occasional duty at the ammunition supply point. SEAL snipers still remained in place at the Soap Factory. For the most part, January was relatively quiet with very little need for sniper engagements, but the few that occurred had a major impact on the use of snipers for further engagements.

On January 9th, COL Dave Plumer arrived to assume command of the JSOTF. That day, the SOF snipers on K-7 engaged a Somali machine-gunner. Per the ROE allowing fires on any Somalis carrying crew-served weapons, USMC snipers on K-7 still assigned to ODB590 engaged a Somali carrying an RPK light machine gun. A spokesman for the Joint Information Bureau told reporters the shooting was within the ROE: "It's a judgment call. Unless they have a good shot they are not required to take it. They don't just blaze away."

Due to the previous day's sniper engagement, the Pakistanis insisted all U.S. sniper operations cease at that location.

On the 10th through the 19th of January, the B-team assisted the task force with medics to conduct a MEDCAP in the village of Jasiira, located south of Mogadishu along the coast, assisting with the treatment of over 300 Somalis.

On January 14th, when snipers reported hearing gunfire near the K4 traffic circle and seeing men moving a mortar and mines, they queried the command for any latest ROE guidance, given the recent uproars about sniper engagements. General Hoar reportedly amended the ROE to limit enforcement to only engage mounted heavy machine guns. Clan fighting continued near the K4 the following day, all watched by SEAL snipers at the Soap Factory who did not engage anything.

Back on the 9th of December, K-7 snipers shot at an RPK gunner and missed. An uproar occurred when a Somali representative reported the stray round hit a woman. Although it was later found out upon examination of the woman at the hospital she had been injured when a truck hit her, the "black propaganda" of the day spun the JSOTF staff up once again to answer queries from higher. The news story the next day magnified the event by reporting a pregnant woman had been hit, based on their Somali sources. This event almost neutralized the JSOTF's ability to employ the snipers for force protection. On the 14th of January, when JSOTF snipers reported observing Somalis moving a mortar and mines and requested ROE guidance, the order came not to fire.

The Marines at the U.S. Embassy compound received enemy fires on their position throughout the 11th to the 14th of January. The USMC and SEAL snipers performed missions at their assigned locations from January 10th through February 2nd, relieved in place by the incoming 24th MEU (SOC) and SEAL Team Two snipers. (SEAL Team Two snipers soon became the 10th Mountain Division's reserve sniper team.) The 24th MEU (SOC) snipers would go on to serve the remaining time in Somalia in the Embassy, Mogadishu airfield, Soap Factory, and New Port positions.

The JSOTF staff deployed out to the USS *New Orleans* on the 16th of January aboard one of the USMC's CH-46 to coordinate the desert fueling needed to support the concept derived for the NEO plan. The plan was still in the coordination stage.

ODB590 snipers supported the visit of the Chief of Staff of the Army on the 22nd by providing VIP force protection.

ODA554 continued to perform as the CST with the Egyptian contingent until the 13th of February, when they were relieved by elements of the 24th MEU (SOC). During their tenure with the Egyptians, they assisted in providing training for base defense matters and Close Air Support training. ODA555 remained with the Malaysians through 15 March 1994, conducting several combat patrols with them. The attached CAT I sniper on the team (SFC Dwight Comer from ODA551) responded to the request of the Malaysian Ranger Battalion (Malbatt II) to set up a sniper selection and training course. The Malaysians had seen the effect achieved by snipers for force protection and wanted to ensure they had the same capability as the U.S. after the March 31st withdrawal. SFC Comer's efforts were highly successful—assisted by the team, his FID efforts paid off with the Malaysian Ranger battalion adding six CAT II sniper teams to their force. (A CAT I sniper is the highest trained and experienced; CAT II level means well-trained.)

The Malaysian contingent assumed the duties of the UNOSOM II Quick Reaction Force (QRF) in mid-December. ODA555 was assigned as their CST, operating with a split team, and would remain with this mission until redeployment (once again together as a full team). Their most interesting experience was supporting the Malaysian Ranger Battalion (Malbatt II) operation to test the Malaysians' performance as the QRF in February. ODA554 covered the Egyptian CST mission.

The CST mission was fairly static, and Mogadishu was quiet; Major Allen's AOB also participated where he could in any of the JTF's convoy escort missions and local MEDCAPs and DENCAPs.

With the release of Osman Atto on January 18th, all SNA detainees from the Battle of Mogadishu (*Malinta Ranger* in Somali language—The Day of the Rangers) were finally free. At the close of the month, most NGOs departed the country after seeing the looting of WFP and Save the Children warehouses (over 250 tons of foodstuffs lost, at the least).

There were other foreign policy challenges to digest in the New Year. A potential intervention in Rwanda to halt the genocide was being looked at in American policy circles. Certainly, as a minimum, the conduct of a NEO there would probably be looming over the horizon. Bosnia continued to simmer. Meanwhile in Somalia, the countries of Germany, Italy, and Turkey joined the exodus from UNSOSOM II.

The LOGSUPCOM was turned over to a civilian contract, bid on by Brown & Root Company. Much of the logistic equipment and base was transferred or loaned to the company. They would now serve UNOSOM II.

The remaining three months for U.S. forces were not peaceful, even with Aideed's professed "cease-fire." Lest anyone thought the urban guerrilla war was over, on February 1st, Marines escorting American diplomats to meet with Aideed were fired on by Somali snipers; a firefight ensued. Eight Somalis were killed, and 24 were wounded. There were no casualties on the friendly side, except for bullet holes in one vehicle. During the firefight, Somali gunmen attempted to reinforce their brethren, but the Marines fought their way out of the engagement and completed their escort mission.

February 1994

On the 3rd of February, ODA555 participated with the Malaysians for the UNOSOM II QRF call out exercise; one Malaysian infantry company supported with Pakistani tanks convoyed out to HUNTER Base. The exercise to test the QRF concept was successful, only hampered by a large crowd of Somalis gathered at HUNTER Base in protest. They were quickly driven off after the QRF fired warning shots. On the 13th, the sniper positions at K-7 and the Embassy compound were finally closed out with the snipers moving to new positions at the New Port area.

That same day, sniper replacements for the JSOTF came from SEAL Team Two on PHIBRON 4. The USS *Peleliu* arrived with the 11th MEU (SOC) and with the 24th MEU (SOC) aboard the USS *Inchon* in early February. An additional mission added to the ARGs was to be prepared to conduct a NEO in Somalia, if required.

The JSOTF conducted their final move to the Mogadishu airfield on the 3rd and 4th of February, closing out the Embassy compound position.

There were some minor combat activities in the month of February with ODA555 ambushed when they were departing the airfield on February 10th by gunmen employing small arms fire from a rooftop. The team was attempting to negotiate an obstacle-type road block outside the gate when the gunmen opened fire, but no casualties occurred, and the team drove away from the engagement area. In the New Port area, the company snipers were often shot at with harassing fires, increasing in intensity during the period of the 17th and 18th of February. One of the SOF snipers returned fire on a threatening Somali gunman and killed him.

The SEALs departed the JSOTF and returned back to ARG on February 14th. By February 15th, the SEALs from ST-8 and the 13th MEU snipers were gone. For the Army land components, most had departed via air and their heavy equipment loaded aboard shipping.

The Somalia Noncombatant Evacuation Operation Plan

USCENTCOM anticipated a non-combatant evacuation (NEO) of Department of State, U.S. civilian, and other foreign diplomatic citizens stationed in Somalia as the final withdrawal date approached. One of the last tasks for USFORSOM was to assemble a NEO task force, cobbled together from forces remaining until March 31st. The warning order in early February from Major General Montgomery's headquarters included the JSOTF as one of the NEO forces listed to support a MEU (SOC).

The JSOTF staff deployed out to the USS *New Orleans* on the 16th of January aboard one USMC CH-46 to coordinate the fueling for their aircraft needed to support the NEO concept.

ODB550—along with one infantry company (L Company) from the 24th MEU (SOC); their CH-53 helicopters; a C-130 with an internal, long-range fuel bladder; and support from AC-130s—conducted reconnaissance of potential high-risk evacuation sites (Kismayo, Baidoa, Bardera; 12 members of ODB550 returned to Kismayo on the 14th of March to retake photos for the NEO plan files) followed by a full task force rehearsal in Mogadishu on February 18, 1994. The rehearsal completed, the NEO plan was updated and the task force stood ready. Fortunately, no NEO was required.

On the 22nd of March, all USMC snipers and the ANGLICO terminated their attachment with ODB550 and returned to the PHIBRON 4. One of the USMC snipers from the 24th MEU (SOC), then a LCPL, remembered his time supporting the SF company:

> My overall impression from those times was that we were all pretty young (compared to SF folks) and were truly impressed by the professionalism and actions of the team. My very first impression upon getting into country was that we were staying in the SF hootches and had to stand the typical fire watch. As a *gung ho* young Marine, I volunteered for the 2 – 4 a.m. shift. As I was making my way to the post – a sandbagged bunker looking out at one of the roads – I walked up to the guy standing at the M-60 MG and when he turned around it was Major Allen. That always impressed the hell out of me. An SF officer doing duty like everyone else . . . Interesting days in Somalia.
>
> During this time we were working for the SF folks, there were only two or three shots out of our total of thirteen confirmed that were taken when we worked for the SF. The first confirmed shot was mine and it was taken overlooking the road that ran in front of the port. I think another Marine friend of mine got the other, and I believe there was one more before we had to CHOP back to the 24th MEU. I remember MAJ Allen came out to my position to talk to me about the shot. I was a young CPL that had never taken a shot and I thought it was something else to see an SF Major take an interest in my well-being and treat me like one of his team. I had another shot from the Soap Factory that was around 500 yards or so.

To further refine the NEO contingency planning, the B-team employed its SEALs and ODAs to conduct another reconnaissance of Kismayo (Feb 23rd via UN MI-17), Baidoa, and Bardera in order to ascertain airfield data and build NEO evacuation folders with pictures and diagrams for use in further planning.

March 1994

In early March, UNOSOM II assumed the role as QRF and the coalition began to unravel. The JSOTF staff withdrew to Mombasa.

On the 14th of March, a final tragedy befell the Special Operations Forces. In the early morning hours of March 14th, AC-130H call-sign Jockey 14 completed its sortie over Mogadishu and headed back to Mombasa. As it neared the coast of Kenya near the seaport town of Malindi, a 105mm round in the gun tube prematurely exploded, sending shrapnel through the number one engine and piercing the fuel tank. The engine caught fire while the aircraft was at 9,000-foot altitude. The fire grew into an immense flame. The pilot immediately cut power to the engine and dove the aircraft in an attempt to ditch in the shallow waters along the coastline. Three crewmen bailed out before the AC-130 hit the water at 140 knots, breaking the hull and separating the tail section from the rest of the aircraft.

The plane came to rest about 200 meters from the shoreline and sank in 15 feet of water. Three crewmen inside the aircraft survived; the remaining seven crewmen still inside perished during the crash and sinking of the plane.

The JSOTF and the JFACC were notified of the emergency as it occurred. The USS *Duluth* joined the search party, along with the Kenyan civil aviation authorities, and conducted a recovery of the survivors. One of the crew who made it out by parachute went missing during the recovery mission. Although an extensive search was made, the missing crewman could not be found. After an exhaustive search, the recovery terminated. The Kenyan government posted a reward in hopes local fishermen might find the missing crewman later.

The aircraft was recovered by a team deployed from the 16th SOW in Hurlburt Field, Florida. Cables were attached to allow the recovery team to drag the frame onto the shoreline, where it was dismantled as much as possible, then loaded onto USAF cargo aircraft for a flight back to the United States.

A memorial service was conducted at Mombasa the following day, attended by local dignitaries and members of the JSOTF. Back in Mogadishu, ODB550 completed its CST mission with the Malaysian QRF. The company staff wrapped up any final tasks and withdrew to the Navy fleet offshore, using Marine amphibious assets and helicopters. They then steamed to Kismayo. At the airfield in Kismayo, C-130s transported ODB550 to Mombasa to connect with C-5A's for the ride home on March 19th. ODA's 554 and 555 completed their missions as CSTs and on March 24th, redeployed to CONUS.

The Marines were the last to go, coming ashore from the ARGs to guard the final perimeter of American holdings in Somalia as the Falcon Brigade and TF 1-64th departed. Called Operation *Quick Draw*, all U.S. Forces were gone from Somalia by March 25, 1994. Consistent with their behavior, Somalis quickly swarmed and looted any positions left unprotected and vacant. Not knowing it then, but the Marines, Army Green Berets, and AC-130s would return to Somalia from January through March of 1995 to assist UNOSOM II forces with their final withdrawal when the UN military mission finally collapsed.

Only Ambassador Richard Bogosian and his 50 Marine security guards at the USLO compound remained to serve American interests, along with American civilian contractors working for Brown & Root. UNOSOM II mustered approximately 20,000 remaining troops (mostly Pakistanis, Egyptians and Indians making up the bulk of the force).

It would not be the end of American involvement in Somalia, however. As the clans rearmed and violent incidents continued, the need to conduct a NEO for U.S. and foreign diplomats, the USLO and Marine security guards, and the hundreds of U.S. citizens working for Brown & Root and various NGOs still remained a possibility. The MEU-ARGs rotating in and out of CENTCOM's AOR and SOCCENT with the 5th Special Forces Group, Task Force 160th SOAR, the 16th SOW, and the 75th Rangers (if the MEU was absent) took turns in 1994 to serve three months at a time as the on-call NEO Task Force.

When UN efforts in Somalia were doomed to fail, the day would predictably come for the withdrawal of UNOSOM II. That day came, and the United States conducted Operation United Shield January through March of 1995 to ensure UNOSOM II's safe exit.

A drift to unilateralism vice multilateralism began to steer Clinton's foreign policy.

Endnotes

1. Al Glover diary notes.
2. McGuire interview.

11

Somalia Noncombatant Evacuation Planning: 1 April - 24 July 1994

An operation whereby noncombatant evacuees are evacuated from a threatened area abroad, which includes areas facing actual or potential danger from natural or manmade disaster, civil unrest, imminent or actual terrorist activities, hostilities, and similar circumstances, that is carried out with the assistance of the Department of Defense.

— JCS Pub 3-68, *Noncombatant Evacuation Operations*

As the New Year began, President William Jefferson Clinton neared his first year as commander-in-chief. There still remained a number of military contingency operations around the world for his administration to deal with. He did in fact achieve his goal in Somalia: a total withdrawal of U.S. military forces by March 31st. Now only a small contingent of State Department employees remained (about 27) along with a platoon of 54 Marines from the USMC Fleet Anti-terrorism Support Team (FAST) to protect the U.S. Liaison Office (USLO). Additionally, a Marine Expeditionary Unit–Amphibious Ready Group (MEU-ARG), was placed in the vicinity of the Indian Ocean. The FAST was funded up to June 30th in anticipation of a decision by the UN to either continue the Chapter VII mandate or abandon the mission.

U.S. contract personnel remained to administer the security assistance equipment programs provided to UNOSOM II–about $44 million worth of equipment sold to date and $4 million worth of military equipment leased to the various contingents attached to UNOSOM II. Other contractors worked for Brown & Root providing purely logistic services. Additionally, there were several U.S. noncombatant citizens who were members of the various NGOs working throughout Somalia, and not considered in the GAO account and budget.[1]

The Clinton administration turned its eyes to the conflict in Bosnia, with NATO seemingly mired down in its mission. In Bosnia, Clinton would turn to diplomacy (not another troop deployment) in an attempt to broker a cease-fire, followed by some kind of peace arrangement. But for now, the United States kept its role to a minimum by only providing air assets to NATO for Operation Deny Flight. Elsewhere, the enforcement of the no-fly zones in Iraq continued.

Somalia 1994

After President Clinton's announcement on October 7, 1993 terminating U.S. military involvement in Somalia, Congress responded to ensure no more funds would be designated for military operations in that country by passing the 1994 Defense Appropriations Act (P.L. 103-139) in November 1993. Included in this legislation was the "Byrd Amendment" (section 8151), limiting Department of Defense monies spent in Somalia to only those operations up to the March 31, 1994 deadline; the President could not request additional funds past the deadline without the approval of Congress. Congress was aware of the residual requirement for limited military operations in support of the Department of State (DOS) and agreed to provide funds supporting a limited number of troops to protect the diplomatic facilities and U.S. citizens remaining in Somalia after March 31st. The provision also provided for the U.S. government employees and civilian personnel needed to assist and advise the UN commander.[2]

As the final aircraft and ship carrying U.S. military forces departed Somalia in March, UNOSOM II still had the unenviable task of providing security and stability in Mogadishu and out in the Humanitarian Relief Sectors, now without the significant military backing and power of the United States. The United States government mildly desired to remain active in the process of bringing the warring factions to the peace table in concert with UN attempts to form and develop a transitional Somali government.

During the month of the United States withdrawal, UN diplomatic efforts were reinforced at the Somali Peace Conference talks held in Nairobi, Kenya. Disagreements by the factions led to a breakdown, then failure of the talks. There was a second attempt to reconvene the conference in Mogadishu and get the factions to move toward reconciliation and security cooperation as a first step toward ultimately forming a transitional government for Somalia, but that effort also failed. The only success for some form of reconciliation came in the Jubba region conference (Kismayo).

The same old disagreement persisted—would it be Aideed or Ali Mahdi as the head of the transitional government? With a weaker military presence from UNOSOM II, the Nairobi Peace Conference's failure, and the United States absent, it only took Aideed a few months to once again stir the pot with an increase in violence against the peacekeepers and NGOs.[3]

In April, Somali National Alliance militia forces (Habr Gedir) seized the town of Merka, southwest of Mogadishu. Clashes between the Habr Gedir and the Hawadle clans erupted in Mogadishu. Between May and June, over 17 peacekeepers along with a handful of humanitarian NGO personnel were killed. Additional attempts for peace and reconciliation conferences in Mogadishu were for naught.

The viability of the UN Chapter VII Peace Enforcement mission hinged on a minimum UN troop level of around 16,000 to 22,000 soldiers in order to achieve several objectives consistent with the UN campaign: to maintain a secure environment (the level of security based on the conditions created by earlier UNITAF operations); to encourage reconciliation amongst the warring factions for government creation; to continue the developing infrastructure; and to put a system in place for administering justice (effective police, laws, courts, judicial appointments, and so on). The UN's earlier attempt to coerce the factions to disarm was dropped. As the United States departed, UNOSOM II troop strength approximated 19,000 boots on the ground, with most of the troops provided by India, Egypt, and Pakistan. The 1994 extended mandate for the UN peace enforcement operation was due to expire in May.[4]

There was also a deadline set for the Pakistani equipment under lend-lease, as shown below.

> Items leased include 80 M-113A2 armored personnel carriers from stock in Europe, 30 M-60A3 tanks from stock in the United States, 8 AH-1S Cobra attack helicopters from the Hawaii National Guard, and 5 OH-58C Scout helicopters from stock in Europe. Also included are associated guns, grenade launchers, radios, radar sets, and test equipment. The leases run through December 1994 and require that the items be returned to the United States in the same condition in which they were received by the United Nations and that the items are to be paid for if destroyed.[5]

There always loomed the specter of a UN withdrawal contested by the clans, or at least accompanied by a rise in violence as UN peacekeepers departed. CENTCOM planners began contingency planning for a noncombatant evacuation operation (NEO) as early as the fall of 1993, not knowing exactly for how long the UNOSOM II mandate would be extended, or worse, if it would be abandoned. In the fall of 1993, CENTCOM planners were already looking ahead to foresee this contingency and began the process for preliminary NEO planning and rehearsals with Joint Task Force-Somalia. While JTF-Somalia was still ashore, Special Operations Forces (SOF) operators conducted the first surveys for NEO evacuation and pick-up zones covering Kismayo, Mogadishu, Bale Dogle, and Baidoa. After the American withdrawal in March of 1994, the Marine FAST platoon with the USLO and a MEU-ARG in the region would serve as CENTCOM's response force, at least until June 30th before funds ran out.

Aircraft carrier deployment and MEU-ARGs in the CENTCOM Area of Responsibility generally followed a schedule of six months' duration to support the requirements of the regional combatant commander. If the UNOSOM II mandate ended by summer, the United States government (USG) could cover a NEO if needed with current naval and USMC assets in place. The issue for the CENTCOM planners was the gap created when the Peleliu ARG was scheduled for departure—the new replacement (Tripoli ARG) would not be on station until later in July. Contingency plans were also needed in the event UNOSOM II remained for another extension period, which would take them through the remainder of 1994.

In March 1994, CENTCOM began crisis-action planning procedures and afterwards requested component commander estimates on courses of action to solve the "gapping" problem—specifically coverage for a NEO during the June–July period. With the Navy and Marines absent during the gap, CENTCOM looked to the Air Force, Army, and Special Operations as potential leads for conducting a NEO. Any NEO in Somalia would require a ground component, so the Air Force would have to assume a supporting role. When the U.S. Army outlined its course of action, it was deemed far too heavy for the CENTCOM planners. General Peay chose his Special Operations Command—CENTCOM's (SOCCENT)–course of action to meet the requirement.

In April 1994, the Commander USCENTCOM tasked the Commander SOCCENT to be prepared to conduct a Noncombatant Evacuation Operation of U.S. government personnel and U.S. citizens remaining in Somalia, with June 1994 as the anticipated time frame to be prepared for the event.

The Conduct of a Noncombatant Evacuation

The responsibility to conduct noncombatant evacuations from any Embassy, Consulate, or Liaison office rests with the U.S. Department of State. The Ambassador has the lead responsibility to ensure the evacuation of American citizens, host nation nationals, and other foreign personnel the United States is allied with or has diplomatic relations with. The priority of evacuation is to the Embassy Team first, then American citizens, and then any others chosen by the Ambassador (the senior diplomatic officer at any state department facility in a foreign country is called the Chief of Mission). U.S. Embassies and consulates are considered United States sovereign territory.

In the absence of the Ambassador, the Chief of Station or the highest-ranking member of the diplomatic service assumes the responsibility for the evacuation.

Evacuations may be peaceful; if a situation in a foreign country becomes tense enough that the Ambassador wants American citizens out of the country, it is at that time the Embassy will issue an alert. Generally, an evacuation of this type is for withdrawal of non-essential personnel and their families. The Ambassador may choose to keep critical members of the Country Team—key diplomatic staff—until the situation becomes calm and everyone can return.

Embassies and consulates maintain a list of potential evacuees known as the F-77 list. It contains each evacuee's personal information, location, and contact procedures. The F-77 report is critical during a NEO to serve as an accountability roster for all noncombatants being evacuated.

On a day-to-day basis, the State Department oversees the security of any diplomatic post. Overall, security requirements are managed by the State Department's Diplomatic Security Services department. The regional Security Officer (RSO) on the Embassy staff is responsible within the country team for matters of security on the ground. The RSO oversees the force protection requirements for the Embassy based on the level of declared threat in the host country. The RSO is assisted by a USMC Security Detachment which is armed and provides immediate protection to the Ambassador. Outside the grounds of the Embassy, the host nation is responsible for guards and security.

A noncombatant evacuation may be required when the threat level is raised. A civil war in the host nation or an increase in violence during anti-American protests can create an environment of uncertainty and make it more difficult to evacuate Americans. When semi-permissive or non-permissive environments are created, the State Department will call on the Department of Defense (DOD) for assistance to prevent any harm to Americans. If the situation becomes dire, everyone evacuates under military protection and abandons the Embassy or Consulate.

All embassies and consulates prepare an Emergency Action Plan (EAP). The EAP contains essential information to perform the planning and execution of an evacuation. The EAP lists assembly areas for citizens and other noncombatants, evacuation sites, maps, communication plans, host nation support, and medical needs as identified on the F-77 rosters. Any other details which might enhance the evacuation and ensure its safety are included within the plan.

In conjunction with the EAP, the U.S. State Department coordinates with the Department of Defense to prepare classified details of the facilities used by the U.S. government (USG) to help prepare military NEO forces in their plan for barrier removal and entry techniques in order to infiltrate facilities and secure noncombatants and the Ambassador for evacuation. These details are known as the Integrated Survey Program (ISP).

All geographic combatant commanders (GCCs) have the responsibility to conduct NEOs in their area of responsibility. In this role, they work directly for the Secretary of Defense. The GCC chooses the force option based on the threat and level of hostilities in the environment. Prior to the evacuation, some form of an advanced party of military liaison personnel conduct a link-up with key personnel of the Embassy or Consulate, such as the Chief of Station and the RSO, to coordinate the details of the actual NEO. These teams can be designated by a variety of names, but they all perform similar duties. For example, for a Somalia NEO, the military liaison team was named the Military Support and Assistance Liaison Team (MSALT). The team may adopt other names; the SOF team, when organized for SOCCENT's NEO plan, was named the Joint Special Operations Assessment Team–JSOAT.

Part of the military force planning is selecting a staging base for the NEO force. This space is known as the Intermediate Support Base (ISB). The NEO force will use this site to station assets and control the NEO under execution. If distance is a factor in reaching the evacuation zones, the NEO force commander can also designate a forward support base (FSB).

The second locale designated is the Safe Haven, the location where the evacuees are being transported for return to U.S. government control. This location is normally in an adjacent country, or one nearby, using a major airport or maritime port facility. Rarely is a NEO conducted using vehicle transport cross-country into an adjacent friendly country, but it could be planned for if there are no other means to evacuate noncombatants. If the situation is dangerous, the Joint Task Force NEO commander has the authority to designate a temporary safe haven.

Once the evacuation begins, the assembled NEO force conducts movement to the Embassy (or Consulate), as well as to major assembly locations designated in the Emergency Action Plans. At each site, the senior military commander designates a check-in center (the Evacuation Control Center—ECCTR). The evacuees are signed in and accounted for and separated into evacuation chalks (groups) for airlift. For security reasons, all baggage and carry-on items are inspected prior to loading any transportation. Excess baggage is removed to save weight. Also, any medical problems are seen to within the capability of the NEO force.

The Ambassador is often the last evacuee. Once all noncombatants have been successfully evacuated and turned back over to the State Department's control, the NEO force is dissolved.

Special Operations Command–Central Command and the Special Operations Forces Option

SOCCENT would serve as the Somalia NEO force formed as a Joint Special Operations Task Force (JSOTF) commanded by BG William Tangney, the SOCCENT commander. The 5th Special Forces Group (Airborne) would serve as the major ground component. COL John Noe, the 5th SFG(A) commander, was designated as the Army Special Operations Task Force commander (ARSOTF). A Joint Special Operations Air Component Command (JSOACC) was task-organized from the 16th SOW. Aerial fire support was provided by MH-6s from the 160th Special Operations Aviation Regiment (SOAR) and AC-130 gunships from the 16th Special Operations Wing (SOW). Special Tactics Squadron (STS) personnel were attached to the SF NEO teams to provide air control. Additionally, four MH-47 helicopters from the 160th SOAR and four MC-130s from the Air Force Special Operations Command (AFSOC) were provided for tactical airlift of the civilian evacuees.[6]

In mid-April, SOCCENT sent its NEO warning order to all of its components. The 5th SFG(A) commander, COL John Noe, chose LTC Charles B. Paxton's 3/5th battalion to serve as the NEO ground force. The 3rd battalion briefed back their concept to Colonel Noe on April 25th. Colonel Noe organized the force with two companies of Special Forces, one company of Rangers, a detachment of PSYOP (4th PSYOP battalion), a Civil Affairs section (96th CA battalion), and a handful of STS personnel. He also requested slices from other Army SOF—the 112th Signal Battalion and the 528th Logistics Battalion, both located at Fort Bragg, North Carolina. The 5th SFG(A) staff prepared their requests for information (RFIs) pending the launch of an assessment team to Somalia.[7]

At the end of April, SOCCENT then held a NEO planning conference in Tampa with delegates from the Army SOF components. Colonel Noe settled on the assumption that the force would bed down in Mombasa (the interim support base) and establish a Forward Support Base at Bale Dogle in Somalia. Colonel Noe organized the force with a NEO ground component (eight Special Forces teams), a coalition support team cell with three five-man teams (three SF and two STS members per team), the Rangers, and a Combat Search and Rescue/Personnel Recovery element (CSAR/PR).

He then gave his guidance. He wanted the NEO pick-up zone teams trained thoroughly, able to conduct operations at night replicating the four Mogadishu PZs (pick-up zones) and the three outlying PZs. He envisioned it would require multiple lifts from each PZ to the Forward Support Base, and wanted that built into any rehearsal. He also wanted to exercise the fire support which would be provided from the AC-130s and the AH-6 Little Birds. His last directive for the planners was to conduct a recon to Somalia to assess all sites and answer requests for information. Most important would be ascertaining the number of AMCITs and others listed on the F-77s and how many could be counted on for each pick-up zone.

The results of the planning conference were delivered to SOCCENT in a Commander's Estimate. SOCCENT sent its Commander's Estimate to CENTCOM on May 10th. CENTCOM forwarded their concept to the Joint Chiefs of Staff in mid-May; the concept was approved, and Task Orders were sent throughout SOCCENT's components with a "be prepared" date as of June 4th (the ARG's withdrawal from the area). There was one caveat; the NEO Task Force would not be positioned forward in theater awaiting the call. SOF components would be kept on a short string inside the United States; if a NEO needed to happen within a period of days, the Peleliu ARG would be ordered to steam back into the waters off Mogadishu.[8]

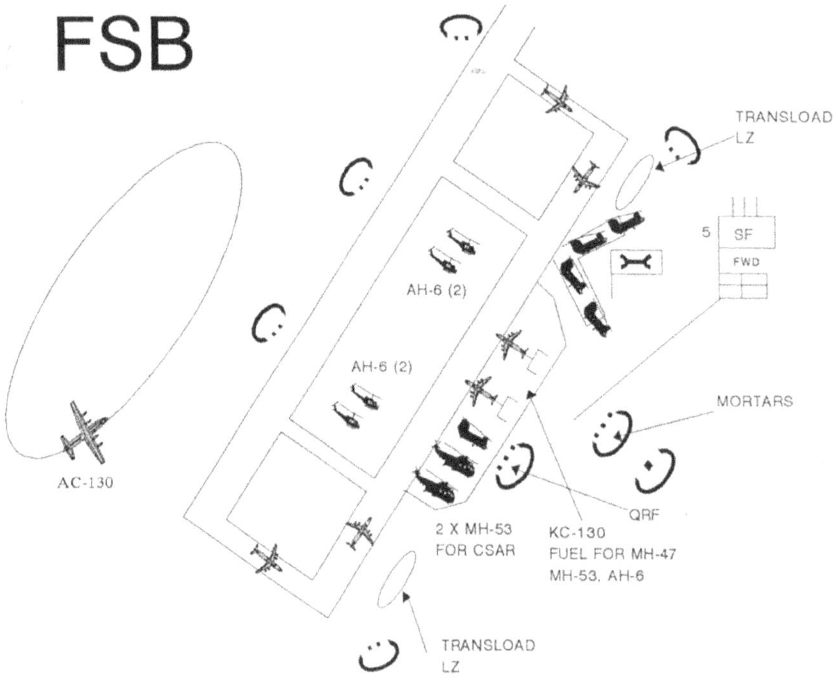

Planning template of course of action at the Forward Support Base for the NEO force (from author's notes).

In the event of a NEO called by the Ambassador, the NEO Task Force would assemble the force at Mombasa in Kenya. After the Ambassador and his staff issued the execute order for American citizens (AMCITs) to move to their pick-up zones, the Ranger company would then insert into Bale Dogle, while AC-130 gunships circled overhead for fire support. Once the airfield was secure, additional aircraft (MC-130s) would deliver the Little Birds (AH-6s) and SF teams and also establish parking and refueling areas. The MH-47s and MC-130s followed afterwards with the SF NEO teams. Once the airfield was secure and the force was in place, the evacuation would then begin with helicopters flying to Mogadishu and the outlying zones. AC-130 gunships provided overhead aerial fire support for Mogadishu, along with the AH-6s positioned with a Forward Air Refueling Point (FARP) established by the Rangers (also serving as the CSAR/PR force in vicinity near Mogadishu).

All noncombatants would then be flown to Bale Dogle for trans-loading onto MC-130s and the flight to Mombasa in Kenya. (The C-141s and the MH-53s, along with the Marine KC-130s, depicted in the diagram had since been withdrawn.)

The final phase would be the complete withdrawal of the NEO Task Force from Somalia and then eventually its return to home station.

The Joint Special Operations Assessment Team (JSOAT)

It was the desire of all components to have a reconnaissance and updated assessment conducted in Somalia to answer key questions the planners had on the concept. SOCCENT formed a Joint Special

Operations Assessment Team (JSOAT), similar to the function of an MSALT, headed by MAJ Joseph D. Celeski (SOCCENT J3 Ground Operations Officer) and assisted with MAJ "Hawk" Holloway from the 5th SFG(A), an Army SF Senior Warrant Officer, Chief Townsend, and an SF Senior NCO. The JSOAT had four key tasks.

1. Meet with all the key players in Kenya and Somalia (DOS, UNOSOM II, USMC FAST, and so forth)
2. Survey the NEO evacuation sites (four in Mogadishu, three in outlying areas)
3. Review all existing plans for NEOs and their associated F-77 rosters
4. Answer all Requests for Information (RFIs) from the components[9]

The team departed for Kenya on May 13, 1994. In the first week on the ground in Kenya, the team conducted coordination with the U.S. Embassy Military Group (MILGRP) and representatives from the Department of State, reviewing NEO plans and the F-77 rosters. They met with Ambassador Brazeal. Contact was established with the KUSLO (Kenya USLO), which requested a liaison team be attached to them during the NEO. The MILGRP and Embassy contractors were helpful to ascertain the needs for lodging the force at Mombasa, basing needs, and housing for the Force. Although the team worked to get the NEO Task Force bedded down at a Kenyan military base, the Kenyan government was insistent the force use Mombasa in order to keep the operation low-key and out of sight of the Kenyan populace; with this decision, the DOS immediately selected Mombasa as the safe haven for the NEO plan. Coordination was made to arrange for tents, water, food, and lodging for the noncombatants. The list of evacuees still held at around 400 AMCITs, along with some additional contractors, American NGO workers, and foreign nationals working for the United States.

The Kenyan military was keen to help in any other way. The team coordinated for aerial and small arms ranges so the task force operators and gunships could practice weapons firing while awaiting a call to conduct the NEO. An alternate airstrip for emergency use was needed. Wajir was ruled out due to the political sensibilities with all the Somali refugee camps in that location, but Malindi was found to be suitable after the team conducted a day recon to assess the site. Surprisingly, while on the ground, the team observed USMC CH-53E Super Stallion helicopters and HC-130s conducting landing, refueling, and take-offs from the small airstrip; the base had been used several times by the Marine Corps and would be added to the NEO plan as an emergency or refueling airstrip.

After a productive week, the team took a private, small twin-engine aircraft to Mogadishu, flying up along the coastline. They were met by Marines from the FAST upon their arrival. During their two day period on the ground, they met with Ambassador Bogosian and with the USLO staff, coordinated with UNOSOM II for the coalition support team (Pakistan, India, and Egypt), and surveyed the four pick-up zone sites around Mogadishu. The Pakistani and Egyptian contingents agreed to help secure two of the pick-up zones in the city and near the airport once the NEO operation commenced. Other UNOSOM II contingents manning the outlying locations agreed to provide up to two companies of infantry to secure pick-up sites during the operation.

Next, the JSOAT met with the Ambassador Richard W. Bogosian (called the Ambassador, but was Chief of Mission) to seek his guidance for the NEO. The Ambassador informed the team he would be the last to go in the evacuation. If conditions permitted, he might also remain in Mogadishu with the FAST at the USLO compound, to maintain a diplomatic presence in Somalia. The team and Ambassador both agreed evacuees could carry only one bag of personal belongings, and no pets or contraband would be allowed aboard the evacuation aircraft. If an evacuee failed to report to a pick-up zone, the NEO teams would not go searching for them; those who missed the flight would have to make their way to Mogadishu or Bale Dogle. The Ambassador requested the NEO teams be prepared to provide food, water, and medical assistance at the evacuation zones.

Ambassador Bogosian did his best to run down the threats he anticipated for the summer. Even though the factions had been told to move their military forces north of the 21 October Road, there was still fighting among the clans in the city streets. There was also an increase in sightings of armed Technicals roaming the streets. Relevant for what time of day to conduct the NEO, the Ambassador passed on that the fighting among clans generally occurred between the hours of 1400 and 2000 daily. This observation was an important piece of information for the Task Force to consider conducting the NEO evacuations during the late night to early morning hours (the safest option).

The Ambassador closed the meeting with one contingency: if Mogadishu was too dangerous for the evacuees to assemble, he desired a string on the NEO Task Force for up to one week in order to move civilians and AMCITs to Bale Dogle as the assembly point. He also passed on that his replacement in June would be Ambassador Dan Simpson and requested the team return during the handover to update the new Ambassador.[10]

While the two Army SF Majors conducted key meetings, the SF Warrant Officer and NCO not only conducted the four Mogadishu pick-up zone surveys but also were assisted by the FAST with trips to Baidoa and Bale Dogle to update their surveys.

One item was identified which needed resolving. The UNOSOM II ROE was more restrictive than when the Americans were in Somalia earlier. Most UNOSOM II contingents reduced their altercations with the clans and were ordered to use lethal measures only for self-defense. The NEO Task Force would need broader ROE to engage apparent threats in a semi-permissive or non-permissive environment, plus additional ROE for the AH-6 gunships and the AC-130s. This predicament could not be solved by the JSOAT during the trip. The team took the information back to the States with them for higher command to resolve. (It was during this same time period the C-141s, KC-135s, MH-53s, USMC HC-130s, and Military Police requests for the NEO Contingency Plan were disapproved by the JCS.)[11]

Brigadier General Tangney ordered a full mission rehearsal before reporting to General Peay that the NEO Task Force was ready to go on standby for the mission. All components on the force list assembled at Fort Campbell, Kentucky using the 5th SFG(A) facilities to conduct mission planning. Assembly areas and remote airfields in the Eglin Air Force Base range complex were chosen to replicate the flight distance from Mogadishu to Bale Dogle.

SOCCENT and the 5th SFG(A) coordinated for a rehearsal planning meeting during May 25th and 26th.

The JSOAT departed theater on May 27th. Upon their return, they conducted an update briefing to the planning staffs. The rehearsal period was scheduled to begin on May 31st.

Exercise INTREPID WARRIOR

SOCCENT received approval and funding to conduct a full mission rehearsal from CENTCOM, scheduled at Fort Campbell, Kentucky, from May 31st through early June. The NEO Exercise was named Intrepid Warrior. Fort Campbell was chosen as the rehearsal base replicating Mombasa for its facilities and the fact that both the 5th SFG(A) and the 160th SOAR were already based there. All forces on the task list were chopped TACON to SOCCENT for the rehearsal.

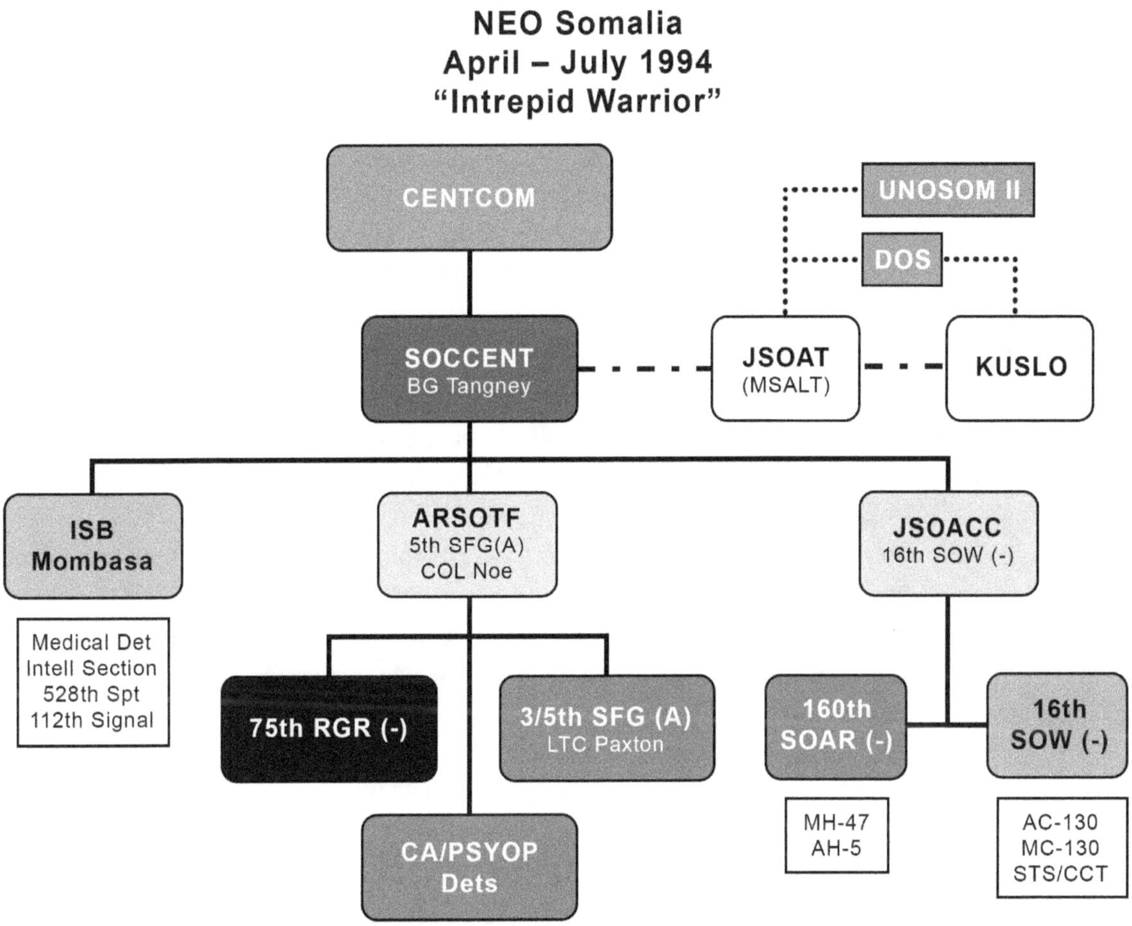

Once all forces for the exercise were assembled, SOCCENT delivered the mission brief: "When directed, SOCCENT deploys a JSOTF to Kenya and in coordination with UNOSOM II and DOS, prepares to conduct a NEO in southern Somalia to secure the safety and protection of U.S. citizens; on order, conduct NEO operations."[12]

The NEO Task Force knew they would be operating from a forward support base at Bale Dogle once the NEO began. Duke Field and the Eglin AFB ranges would replicate locations in Somalia to mimic the outlying NEO pick-up zones and areas around Mogadishu, and the range of flight times anticipated from Bale Dogle. Other airstrips between Fort Campbell and Nashville were looked at to get a more remote location, but the idea, proposed by Lieutenant Colonel Paxton based on his use of them during previous, local exercises, was not considered in order to preserve OPSEC and to not excite local civilians with "black helicopters landing during the night."

Using Eglin Range allowed the AC-130s to operate from their home station at Hurlburt Field in Florida, saving some wear and tear. To support the effort, ground force NEO teams and other support (5th Group members dressed in civilian clothes to replicate the noncombatants) convoyed down to Eglin with their vehicles. Part of their mission was to prepare mock-ups and targets for the plan scenarios.

The 3/5th SFG(A) would perform as the NEO ground component with two SF companies of four Operational Detachment-Alphas (ODAs) each. Other ODAs from the other two battalions, along with Special Operations Team-Alphas from the Military Intelligence Detachment (MID), augmented the force along with attachments from the Civil Affairs and PSYOPs. The Rangers would be attached to the ARSOTF under Colonel Noe. The SF teams would serve as the NEO evacuation zone teams, under the command of MAJ Pat Higgins, the Company B commander.

An evacuation zone team averaged twenty personnel. For mobility, the teams would use their modified HMMWVs (the Desert Mobility Vehicle—DMV) and motorcycles. Although this level of mobility was not needed in Mogadishu, the outlying stations needed the DMVs for firepower and maneuver if a threat arose. Mounted teams as far away as Baidoa and Kismayo also planned for ground Escape and Evasion (E&E) with the vehicles driving to the coast or the Kenyan border if they came under duress during the evacuation.

The 75th Rangers' task was to seize and secure the forward operating airfield, help to establish a forward refueling point (FARP), and serve as a CSAR/PR or quick reaction force. Some Rangers were added to the helicopter crews flying in the NEO teams in order to secure the helicopters at their landing and extraction zones.

The SOF air component had several missions: AC-130s for armed reconnaissance and direct fire support, AH-6s to assist the Rangers in the event of airfield seizure, MH-47s to fly to each NEO pick-up zone and transport evacuees, and MC-130s to deliver the ground force, conduct FARP operations, then be used as back-haul for AMCITS transported to Mombasa. Additionally, the wet-wing refueling capability of the MC-130s would provide refueling to support the force.[13]

Brigadier General Tangney set up his JSOTF headquarters' operations in 5th Group's Isolation Facility (ISOFAC), Rowe Hall. The facility was used to replicate the USLO warehouses at Mombasa airfield in Kenya, used many times during named operations into Somalia.

Tangney's guidance to the force was simple: conduct a night operation to mitigate risk, leave securing Mogadishu airport to UNOSOM II, Command and Control (C2) the operation from Bale Dogle, and incorporate PSYOP and disinformation to enhance the operation of the mission and protect the force. His goal was clear. Evacuate all identified American citizens safely and turn them over to Department of State control. The operation would not be complete until the NEO Task Force was withdrawn from Somalia. Brigadier General Tangney also did not want the force to get tied up in a hostage situation. If one occurred, he would take guidance from the commander of CENTCOM on any further course of action.[14]

Colonel Noe then issued his mission and intent for the ARSOTF ground force during the operation. His staff saw it as a five-phased operation:

1. Pre-mission preparations 18–31 May
2. Deployment – On Order from CENTCOM

3. Lodgement and build-up in Mombasa; conduct of rehearsals between 4 June and a date to be determined (TBD)

4. Execution of the NEO

5. Extraction of the NEO force from Somalia and redeployment[15]

One task was added by Colonel Noe for the NEO ground force. He wanted to prepare some of the SF to serve as coalition support teams for elements of UNOSOM II, based on input from the JSOAT reconnaissance. (Two additional SF teams were added to the force, for a total of ten teams.) The 5th SFG(A) would also add a KUSLO liaison cell for coordination.

The COL James N. "Nick" Rowe Isolation Facility at 5th Special Forces Group (Airborne), Fort Campbell, Kentucky. James Rowe was held prisoner by communist forces in South Vietnam for five years before escaping captivity (one of only thirty-four POWs to escape during the Vietnam War). He later taught survival, escape, rescue, and evasion courses to the Special Operations community. He was murdered by a Philippine communist hit squad in 1989 (Author's photo).

Lieutenant Colonel Paxton was charged with the NEO evacuation zones mission. In Mogadishu, these zones were located in the U.S. Embassy compound's vicinity, the north ramp at the airfield, the HUNTER base compound, and an area in the Old Port area. Lieutenant Colonel Paxton also considered requirements for outlying evacuation operations in Kismayo, Baidoa, and Bardera. He reiterated what had been said previously on the concept, adding a caveat: be prepared to go any time and prepare for an uncertain environment (mostly urban). The key to success would be a well-rehearsed and understood plan. He had already increased the force from eight SF teams to ten based on the JSOAT assessment and the additional missions from Colonel Noe.[16]

MAJ Pat Higgins would command and control the NEO pick-up zone ODAs from the FSB by establishing an Advanced Operations Base (AOB) once the airfield was secured by the Rangers. Additionally, he would also be in charge of the trans-load site at the FSB to process the incoming evacuees and board them on aircraft for their flight to the safe haven. Prior to the exercise Task Force's arrival, Major Higgins prepared his company for the mission by conducting day and night rehearsals of their key tasks. He explained preparations for his mission as shown below.

> We had clear tasks to accomplish in the pre-exercise period: the first was to gather any and all information and TTPs [Tactics, Techniques, and Procedures] on how to conduct NEOs. The most useful product we found was the 82nd Airborne Division's NEO handbook, which we later used and rewrote as a 5th SFG(A) NEO planning guide. The second task was to set up and rehearse NEO assembly zones and flesh out the team SOPs [Standard Operating Procedures] for the operation. We did this by a crawl, walk, run method, utilizing local training areas—particularly the MOUT [Military Operations in Urban Terrain] site at Fort Campbell. The last task was to conduct rehearsals, day and night, of the sequence of aircraft loading and unloading. We accomplished this once the exercise aircraft arrived to the airfield. Finally, while not possible during the actual exercise and rehearsal, we were able on a later date to practice calls for fire with the AC-130 gunships.[17]

As the planning progressed, it was clear the entire mission could not be conducted in one period of darkness; there were too many evacuees to extract in the limited period of one night. The 3/5th S3 Operations Officer, MAJ Matt McGuiness, discovered this "glitch" while preparing the detailed Master Events List and the ground force H-hour list. The force would be required to lay-up during the day following the first night's activities and then finish the mission on the second night. Remaining at Bale Dogle during one whole day could possibly compromise and endanger the force. The only solution would be to beef up the Ranger contingent to handle this additional security contingency. However, the Air Force Special Operations Command mission planners, limited to the aircraft on hand in the task force, were quick to crunch the numbers and report something would have to give to increase the Ranger assets; the SF ODAs were reduced and now fixed at four teams.

The Ranger staff planners adjusted their Master Events List, focused on their security mission at the airfield, but would not fold their time lines into the overall 3/5th H-hour timelines, causing friction. A red star cluster (euphemism for "something is wrong") was raised to the SOCCENT staff on a potential glitch in overall coordination for the mission, but this issue was not fixed in time for the rehearsal.[18]

The period for the full rehearsal was chosen as the 8th through the 9th of June. Unfortunately, a sudden and severe storm blocked the air contingent as they neared the Alabama/Florida state line; for safety, the aircraft returned back to Fort Campbell. (The storm had been predicted, but the air team felt they could fly around the worst parts.) The Airborne Mission Commander for the 8th SOS (Combat Talon I), then LTC Ken "Redman" Poole, remembered the attempt to fly on that day.

> When participating SOF airmen (Combat Talon I and II, Army/Air Force helicopter, and Spectre gunships) recall this SOCCENT NEO rehearsal, they all remember one thing—the WEATHER! The weather 100 miles south of Fort Campbell, KY, through the state of Alabama all the way to the Gulf Coast of Florida was the worst flying weather observed or heard of, covering this region, for any five-day consecutive period that I can remember in the past thirty-four years. Marginal visibility, 100-foot ceilings with persistent thunderstorms and with expansive feeder bands, lighting, hail, and wind gusts—in places up to forty knots—made it impossible to fly safely to the planned training objective areas in the Eglin Range.[19]

Brigadier General Tangney pushed to have the exercise period extended and retain Tactical Control of the force (TACON); it was important that all key elements of the plan were rehearsed before everyone returned to home station. Without that, Brigadier General Tangney could not validate the force. As an alternative, the SOCCENT and 5th Group plan staffs met to consider their options, realizing the essential, critical tasks of securing Bale Dogle and conducting NEO pick-up zones could be accomplished at their location on the Campbell Army Airfield (CAAF) and throughout the local Fort Campbell training areas.

This maneuver would require dousing all the lights on the CAAF in order to replicate the airfield seizure portion of the plan needed to complete the Ranger portion of the mission. The staff soon met with Ft. Campbell post officials, who readily agreed to assist, and the lights were turned off.[20]

Brigadier General Tangney agreed with the staff proposal as long as the critical tasks of the plan were rehearsed. Unfortunately, the air-to-air refueling, AC-130 fires, and FARP operation would have to be rehearsed at a later date. CENTCOM agreed. The rehearsal was executed over a fourteen hour period with no flaws.

The SF NEO teams benefited greatly from experienced Rangers who shared their internal SOPs and handbooks on conducting a NEO. Some of the teams stranded at the Eglin ranges were even able to conduct calls-for-fire with the AC-130s from Hurlburt Field in Florida. General Tangney validated the force, reported to the SOCCENT commander he was ready, and released the components back to their home stations between June 9th and 11th.

During the build-up to the exercise, the Joint Search and Survival Agency (JSSA) provided a training team to the Task Force to cover aspects of escape, evasion, and recovery. The team remained till June 10th and was critical to preparing the force for the operation. Of importance, they assisted the SOCCENT J3 PR/CSAR/E&R cell, led by CWO3 Mike Dozier, on the development of the real-world Search and Rescue response in the event of any aircraft going down.[21]

As summer began, the NEO Task Force remained on standby. Any key leaders replaced during the summer were briefed thoroughly on their anticipated role. Fighting increased in Mogadishu as the summer wore on, and intelligence reported the first appearance of numerous Technicals near the K4 traffic circle. More ominous was a purported threat from shoulder-fired Surface to Air missiles (SAMs). The 5th Group intelligence analysts kept a keen eye on their old enemies, the SNA in Mogadishu and the al-Itihad al-Islamiya (AIAI) throughout southern Somalia. Colonel Jess and Colonel Morgan were still at each other near Kismayo.

On June 26th, Brigadier General Tangney, Colonel Noe, and Lieutenant Colonel Paxton conducted a visit to Kenya where they assessed the Mombasa facilities and met with the new Ambassador in Nairobi. It was then they learned the amount of noncombatant evacuees might total up to 800 people (the change mainly coming from additional NGOs now working in Somalia). The plan was adjusted to request more helicopters and C-141s for this eventuality, and sent higher.

Somalia – Summer of 1994

In July 1994, the Secretary General sent a special mission to Somalia to ascertain force requirements for an extended mandate going into the fall of 1994, albeit with a requirement to downsize. After discussions with the force commander, it was agreed upon that the mission could effectively continue with a reduced force strength of around 17,200 in the ranks (it was around 22,000 at the time), with this drawdown level to be completed during September. The special mission also noted any remaining force strength which dropped below 15,000 would render UNOSOM II ineffective; it was envisioned this troop level could only be achieved later in October or November if the situation in Somalia was more stable.[22]

Seeing the writing on the wall, a clan militia overran a UNOSOM II contingent at Belet Weyne on July 24th.

By late August, there was little to show for the efforts to achieve national reconciliation among the clans. Harassment attacks against UNOSOM II increased, primarily fomented by the Hawiya sub-clans (primarily the Habr Gedir of Aideed and Ali Mahdi's Abgal). Several NGOs were also reporting deterioration in security. The UN Security Council had agreed on the reduction of forces but were now concerned it could not be achieved without a solid path on how national reconciliation could be achieved, requesting a plan to achieve this goal from the Secretary General. The primary importance for any initiative concerning Somalia was the safety and security of UNOSOM II forces as well as other international agencies and NGOs.

June passed without the need for a NEO. As the first weeks of July also passed without incident, the Tripoli ARG arrived on station and SOCCENT was relieved of the mission for a standing NEO force.

The United States Liaison Office Evacuation – Operation Quick Draw

In light of the eventual withdrawal of UNOSOM II, the waning interest in Somalia by Americans, and a noted increase in tension throughout Mogadishu, the State Department decided to relocate its USLO delegation in Mogadishu back to Kenya. This move was urged by the new Ambassador in Mogadishu, Dan Simpson, who felt there were no more diplomatic or humanitarian relief reasons for the United States to remain in Somalia; their job could be conducted from the Embassy in Kenya. Only a handful of U.S. State Department employees along with the Marine security detachment remained at the CONOCO compound in the old city area representing American interests.[23]

On August 13, 1994, the *Somalia News Update*, Volume 3 Number 21 published by Bernard Helander, spoke on the impact of the FAST team's potential withdrawal:

> One factor that was expected to play a role in the debate on UNOSOM was the disposition of U.S. Marines currently guarding the U.S. Liaison Office (USLO) compound in Mogadishu. Department of Defense was attempting to have them withdrawn from Somalia by August 15th, on the grounds that they were needed elsewhere (the 50 man unit is one of the elite 'FAST' platoons [Fleet Anti-terrorism Security Team]). The State Department argued it could not keep the Liaison Office open without FAST protection. Somalia News Update has learned that, in reality, the issue was over who would foot the bill–the Defense department was paying $150,000 a month for the team to stay in Mogadishu, and has lately been protesting strongly against peacekeeping operations being paid for out of the defense budget. There was a real concern within the UN that if the U.S. pulled its Liaison office out, other contributing countries would withdraw their troops. A compromise was worked out–the Marines will stay on until September 30, with the Department of Defense paying; thereafter, a Congressional amendment calls for their withdrawal by the new fiscal year, October 31.

On Friday, September 16, 1994, the *Washington Post's* journalist Julia Preston wrote the following section in the article, "U.S. Troops May Aid in U.N. Withdrawal from Somalia."

> The United States would like to see the Somalia mission closed down by the end of this year, U.S. officials said. The Security Council is scheduled to review the mandate for the mission by Sept. 30.
>
> In a meeting this morning with the five nonaligned nations on the council, U.S. Ambassador Madeleine Albright argued that because Somali leaders have made no progress toward a settlement, the mission is not producing results that justify the huge international commitment, U.S. officials said . . . Because of the risks, the United States today finished closing down its Somali embassy in the heart of Aideed-controlled southern neighborhoods of Mogadishu. U.S. Ambassador Daniel Simpson and the last of about 80 U.S. diplomatic employees were expected to leave Mogadishu today.

On August 22nd, the 3/5th SFG(A) placed a liaison officer in the U.S. Embassy in Nairobi to assist with the evacuation operation. His duties included: liaising between the U.S. Embassy and UN forces in Somalia, providing updates to CENTCOM situation reports, and coordinating diplomatic travel for members of CENTCOM, as required.

After his arrival to Nairobi, the Company C, 3/5th SFG Liaison Officer deployed to Mogadishu with an Embassy team via the State Department Beechcraft aircraft. Upon their arrival in Mogadishu, the team was transported by helicopter to the U.S. Embassy compound and linked up with the USMC and USLO operations staff to assist with preparations for the upcoming withdrawal.

The Tripoli ARG (11th MEU-SOC) was ordered to depart the Arabian Gulf to support the USLO's withdrawal to Kenya. The USS *Cleveland* (LPD-7) departed Jebel Ali, UAE on September 2nd after completing scheduled repairs, and arrived off the coast of Mogadishu on September 8th.

Steven M. Sullivan captured the MEU-SOC's operational concept in his article for the *Marine Gazette*:

> The ARG was directed to proceed to a new ModLoc [modified location] off Somalia no later than 8 September, prepared to conduct NEOs, provide security elements, and furnish a QRF as needed. Evaluating the situation ashore and the progress of the USLO drawdown, the MEU determined that it would face two major requirements. The first phase involved protecting the USLO compound and the route to the airport or, alternatively, conducting an immediate NEO. The second phase involved protecting the USLO assets that had been moved to the airport awaiting transportation via air to Nairobi, Kenya. It was estimated that the first phase would require a sizable mechanized combined arms team. The designated helicopter company and the designated mechanized company were selected for the second phase because of their ability to withdraw unassisted (without landing craft) from the airport upon completion of the security mission. Finally, a series of helicopter-borne force packages, including the MEUs Maritime Special Purpose Force, were designated as the QRF and held in specific alert conditions. The efforts of the Marines and Sailors of the ARG during this tense and sensitive operation were exemplary. For eight days, 24 hours a day, they provided vital support

to the USLO relocation, safeguarding the lives of U.S. citizens. The deterrent effect of the MEU offshore was evidenced by the relative calm that overcame Mogadishu during the period.[24]

In early September, SOCCENT established a small JSOTF in Mombasa with AC-130s and a couple of SF teams (mounted) from the 5th SFG(A). The SF teams were transported forward to Mogadishu Airport and the U.S. Embassy compound to assist during the evacuation. They remained for three days until the operation was complete, then returned to Mombasa.

The USLO evacuation was conducted in a semi-permissive environment. In the first stages of the operation, the USLO contingent at the CONOCO compound was relocated to the U.S. Embassy compound (about 120 to 150 personnel). Once there, U.S. Embassy and attached military personnel and equipment were loaded into the armored HMMWVs of the 5th SFG(A) teams and armored Land Cruisers of the State Department, forming a twenty-six vehicle convoy. The column drove on the bypass road to reach Mogadishu Airport. Other Americans were already assembled in their pick-up zones awaiting the convoy's arrival. American civilians and contractors took commercial flights out of Mogadishu. A skeleton crew of Embassy personnel was transferred to UNOSOM II to assist in its upcoming withdrawal from Somalia.

Excess equipment and materials from the failed Somali Police Program were loaded out on contracted Russian cargo aircraft. State Department personnel and their employees flew to Nairobi; the Marines of the FAST platoon, along with elements of the SF, loaded their equipment aboard the USS *Cleveland* for transport afloat to Mombasa. The USS *Cleveland* arrived in Mombasa on September 15th; USMC Security Force Battalion's FAST platoon redeployed three days later back to their home station. Elements from the 5th SFG(A) flew military transport back to Fort Campbell, Kentucky.

The 3/5th SFG(A) USLO LNO departed Mogadishu for Nairobi with members of the State Department on September 12th; the mission was a complete success, and the SF Liaison Officer (LNO) position was discontinued on September 16th.[25]

While the UN Special Representative (Gbeho) began discussions with clan and religious leaders inside Somalia, the UNOSOM II Force Commander concentrated his remaining military contingents into four areas, three of which were Mogadishu, Baidoa, and Kismayo. The Secretary General informed the Security Council of these measures on September 17th. Ahead lay the potential to end the UN mandate in Somalia. While all of these developments were being ruminated on, he also asked for an extension to the mandate for one month (up to October 31st). The Security Council approved the extension with UN Resolution 946 on September 30th, with an eye toward contingency planning for a UN withdrawal.[26]

Endnotes

1. Celeski, Joseph D. (COL, USA). "Somalia NEO Operations: ARSOF Operations March–September 1994." Section V, Draft Research for the USASOC History Office, May 2007, p. 4. Numbers extracted from GAO Report "Peace Operations: Withdrawal of U.S. Troops from Somalia," GAO/NSIAD-94-175, June 1994.
2. Davis, Richard (Director National Security Analysis). The United States General Accounting Office, Report to Congressional Requestors. "Peace Operations: Withdrawal of U.S. Troops from Somalia," GAO/NSIAD-94-175, June 1994, Washington, DC, p. 4.
3. Ibid, Fogarassy, Helen. Article sent to the author in 2007 titled, "A Civilian View of the U.S. Military in Somalia from the UN Organization." Helen Fogarassy was the Editor-in-Chief for the *UNOSOM Weekly Review* in Somalia during 1994.
4. Davis, p. 9.
5. Ibid, p. 7.
6. Celeski, Joseph D. (COL, USA). "Somalia NEO Operations: ARSOF Operations March–September 1994," Section V, Draft Research for the USASOC History Office, May 2007, p. 4.
7. Celeski, p. 6.
8. Ibid, p. 9.
9. The author was the team leader for the JSOAT while he served as the SOCCENT J3 Ground Operations Officer. This section was prepared from his notes and pictures.
10. Celeski, p. 11.
11. UNOSOM II Mogadishu, Somalia, March 19, 1994, Appendix 6 to Annex C to UNOSOM II OPLAN 1, Rules of Engagement (ROE), signed by LTG Aboo Samah, March 19, 1994.
12. SOCCENT Somalia NEO Briefing files, June 2004.
13. Celeski, Joseph D. (COL, USA). "Somalia NEO Operations: ARSOF Operations March–September 1994."Section V, Draft Research for the USASOC History Office, May 2007.
14. Ibid.
15. 5th SFG(A) Somalia NEO Briefing files, reviewed June 2004.
16. Celeski, p. 16–17.
17. Discussion and interview with COL(P) Pat Higgins on May 29, 2007 at Ft. Bragg, NC.
18. Celeski, p. 16.
19. Quote provided to author on March 21, 2007 at Joint Special Operations University, Hurlburt Field, FL. COL (Ret.) Ken Poole was serving as a SOF Senior Fellow in JSOU's Strategic Studies Division.
20. Celeski, p. 16.
21. Celeski, p. 19.
22. UNOSOM II: August 1994–March 1995, www.peacekeeping.un.org, May 25, 2020. "UNOSOM II Downsized."

23. Bogosian, Richard (Ambassador). Foreign Affairs Oral History Project, Diplomatic Studies and Training, conducted on April 1, 1998 (page 124). Interview conducted by Vladim Lehovich and published in 2000.
24. Sullivan, Steven M. "Forward Presence: MEU(SOC)s in Action Today and Tomorrow," *Marine Corps Gazette*. Quantico: August 1995. Vol. 79, Iss. 8; p. 38.
25. Ibid, p. 6.
26. UNOSOM II: August 1994–March 1995, www.peacekeeping.un.org, May 25, 2020. "Situation Deteriorates."

12

The United Nations Withdrawal from Somalia: Operation United Shield, January - March 1995

The world may have bitten off more than it could chew in terms of trying to bring the Somalis to a government . . . People who look at the Somali situation now with even a small amount of optimism operate from the premise that it may be that now the foreigners are actually leaving Somalia, the Somalis themselves may be able to get together, close the door inside the family and say, 'Okay, now it's time to wrap it up and make a government.'

— Dan Simpson, U.S. Special Envoy to Somalia

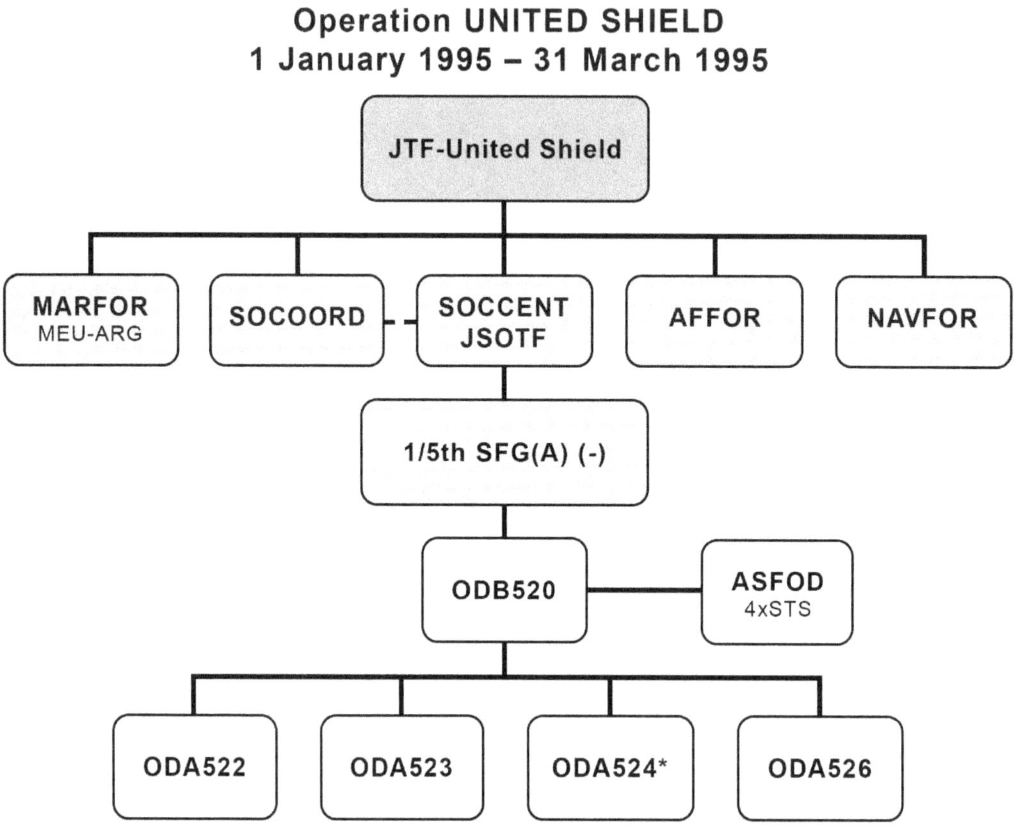

* Worked for and augmented the B-Team

Operation United Shield was conducted between January and March 1995 to withdraw the remaining UNOSOM II forces from Somalia, after all attempts by the UN to resolve major issues in forming a Somali government and bringing some semblance of stability to the country seemed for naught. It was ironic that the UN, with the largest military force in East Africa, could not design a safe withdrawal plan without asking the United States for military assistance. The organization had come a long way from its hubris of "peacemaking" on the African continent. The demise of Chapter VII Peace Enforcement operations by the UN came to a conclusion when mission creep changed a humanitarian intervention into a nation-building operation, without the commensurate UN leadership and participation by its most important member states.

The New World Disorder began. International terrorism was on the rise and was becoming the new national security challenge for America. The war in Bosnia had been raging for years, resulting in NATO—not the UN—intervening to stop what many feared as a genocide. American diplomats began negotiating some sort of peaceful arrangement between the warring states, and the CNN effect shifted to Europe.

It still remained a unipolar world in 1995. China was slowly developing its economy and brand of communism which would allow it to grow into a strategic competitor for the United States. With Russia occupied in a transition to a market economy, the old national security challenges in the Horn of Africa lessened as the area was now free of Cold War tensions and strife. In America, the administration looked forward to a "peace dividend" and began the process of cutting military budgets and downsizing the military.

The last act in Somalia for American and UN efforts was to salvage the reputation of the UN and begin learning the lessons of any further, ambitious UN ventures. Future humanitarian interventions would have to be limited to those which could be effective and affordable.

The American-led task force was commanded by LTG Anthony C. Zinni, the commanding general of the 1st Marine Expeditionary Force (I MEF). The U.S. components for the operation consisted of USMC, U.S. Navy, and Special Operations forces. Naval contingents from Pakistan (two ships) and Italy (two ships) would join the other six ships of the United States Navy to make up the air-ground task force. U.S. forces and remaining UNOSOM II forces would comprise a little over 4,000 troops for the operation.[1]

UN Peacekeeping in Somalia in the Fall of 1994

In early October, the Under-Secretary-General for Peacekeeping Operations, Kofi Annan, prepared a written report due October 14th to the Security Council after visiting Mogadishu. It was not pretty. It did not appear as though the goals of national reconciliation were being achieved by the clans. Commitments made under the Addis Ababa Agreement and the Nairobi Declaration were not carried out. Security for UNOSOM II forces in Somalia was tenuous, and soon support from the member states still participating in support efforts would become cost prohibitive. Even in the light of these dire predictions, the Secretary General advocated for a five-month extension to the mandate (till March 1995) in hopes the political process towards reconciliation would improve. This idea also rested on the notion Somali clan leadership would begin providing security for the various humanitarian organizations remaining—a lofty dream.

Regardless, the withdrawal of UNOSOM II became etched as one of the realities for Somalia's future. After a visit to Somalia by a seven-member consultation team, their report concluded the mandate should end on March 31, 1995. The Security Council quickly issued UN Resolution 954, confirming the recommended end date as March 31, 1995. Priority would be given to the security of remaining member state military forces and an assumption that member states would assist in the withdrawal. The Somali factions were "politely" asked to refrain from any further violence directed toward UNOSOM II during this period. A ceasefire was preferred.[2]

Helen Fogarassy was the Editor-in-Chief for the *UNOSOM Weekly Review* and served in Somalia during 1994. In that role, she attended daily briefings given to the UNOSOM II senior staff. She remembered the announcement of the UN withdrawal during one of these meetings.

> The announcement was made at a morning staff briefing that the October 31 decision by the UN Security Council to draw down the Somalia mission by March 31, 1995, was historic. It was the first time that the UN was leaving a country before its mission was accomplished and it was the first time the UN was abandoning a country.[3]

(Helen Fogarassy is the author of *Mission Improbable: The World Community on a UN Compound in Somalia,* printed by Lexington Books, March 25, 1999. It is an excellent work on understanding the daily workings of the UN and the UNOSOM II during a difficult year in Somalia.)

In a move which may have been wishful thinking, the UN Interagency Standing Committee (the IASC, a grouping of humanitarian and operational agencies) urged to continue to plan and implement humanitarian relief and diplomatic initiatives now that the military option was out. In addition, plans would need to be developed to establish humanitarian bases at key airfields and ports and transfer final end-use plans for UNOSOM II equipment—then look to establish standing committees to implement this initiative. Of course, key to any continued UN assistance to Somalia would be a reduction in violence and continued measures by Somali leadership to establish governance (to prevent the continuation of a failed state).[4]

Early Withdrawals

Peacekeepers had already been evacuated from Bardera, Hoddur, Wajid, and Balad. By September 12th, UNOSOM II force strength dropped from around 19,000 troops to fewer than 15,000, as some member states began early force withdrawals.[5]

In early October 1994, Saddam Hussein deployed major forces to his southern border with Kuwait for the purpose of intimidation and as a challenge to the American presence there. CENTCOM responded to the threat by deploying forces back into Kuwait (including sending elements of the Maritime Prepositioning Force to Saudi Arabia). The 5th SFG(A) deployed one SF battalion to support Operation Vigilant Warrior. The standoff soon became a staring contest, with neither side starting a conflict; however, some punitive bombing from the United States still occurred. Saddam Hussein backed off. In northern Iraq, a Joint Special Operations Task Force conducted Operation Provide Comfort to defend the Kurdish population and assist humanitarian efforts to feed them.

On December 7th, Somali clan militias attempted to prevent the Bangladeshi contingent in Afgoye from relocating to Mogadishu. A quick reaction force from UNOSOM II thwarted their attempt. By late December 1994 and early January 1995, remaining UNOSOM II forces were all consolidated in the Mogadishu area.[6]

The Joint Chiefs of Staff (JCS) anticipated there would at least be some American involvement to assist the UN. If nothing else, a large quantity of U.S. equipment still in Somalia would need to be removed. Although what the President would ultimately decide upon could not be ascertained, in prudence, the JCS directed course of action and force planning in early August 1994. In the event of an American role in any withdrawal operation of both equipment and UNOSOM II forces, CENTCOM would take the lead (now commanded by General Peay). If CENTCOM was going to require Special Operations Forces (SOF) in any course of action (COA), it would certainly involve the 5th SFG(A) as their Joint Strategic Capabilities Plan-apportioned SOF ground contingent. On August 15, 1994, the JCS sent a warning order to General Peay to develop the course of action for both equipment and supply removal and for assistance in withdrawing the remaining UNOSOM II forces (CJCS message 152218Z Aug. 94).[7]

To the planners, the requirement expansion would need a robust task force, operating from Mogadishu's airport and maritime port. A short duration operation with limited boots on the ground was desired. Clearly, an amphibious task force would best serve this purpose.

Without a doubt, the 5th SFG(A) was confident it would play a role in the operation, although it was still unknown what force size would deploy. Without an understanding of the current ground condition in Somalia, early thoughts in the SF Group envisioned deploying up to a battalion, with at least two to three companies, to travel out to the Humanitarian Relief Sectors still in existence and conduct link-ups with UNOSOM II contingents, then assist them with their withdrawals back to Mogadishu. The Group was quite comfortable with this course of action based on their previous experiences in Somalia. The only other foreign commitment for the Group was the standing SF Company deployed to Kuwait; with the Somali operation being seen as short duration, it would not pose any challenges for the Group to support.

In Somalia, with the United States' involvement almost guaranteed, the U.S. Liaison Office (USLO) felt many of the UNOSOM II contingents would rapidly begin to withdraw from Somalia. Anticipating a worsened security situation until a U.S.-sponsored task force arrived, the USLO departed the CONOCO compound on September 19th and repositioned into offices in Mombasa, Kenya, leaving behind the USMC guard and security contingent (about 50 Marines were assigned to the USLO to provide security while in Mogadishu).

USLO equipment and vehicles were packaged and moved to the airfield, and the compound's buildings were closed out. Other materials for extraction included equipment bought or donated to support the Somali Police Program. The Marines loaded their armored HMMWVs aboard Russian-made commercial cargo aircraft and departed by U.S. Navy ship.

During this period, SOCCENT positioned three AC-130s at Moie International Airport in Kenya to provide ongoing force protection for UNOSOM II. Also in September, planning staff from CENTCOM deployed to Mogadishu to begin planning for the withdrawal operation with the UNOSOM II staff.[8]

CENTCOM designated the 1st Marine Expeditionary Force (1st MEF) at 29 Palms, California as the lead contingent for planning and force sizing on September 21, 1994. The named operation was called United Shield. The task force commander was designated as COMUSMARCENT (LTG Anthony C. Zinni, I MEF commander).

SOCCENT was then tasked to provide a planning cell to the Marines on September 30, 1994, forming a combined team of SOCCENT J3 and J5 planners, led by MAJ Gary Danley, an SF officer assigned as one of the ground SOF staff in the J3 Directorate. After initial mission analysis and course of action analysis, SOCCENT issued a warning order to the 5th SFG(A) along with a warning order to its other SOF contingents. Major Danley and members of the planning staff conducted numerous trips throughout October and December to continue refining SOCCENT's role in the plan.

1st Battalion, 5th Special Forces Group (Airborne)

COL John Noe, the 5th SFG(A) commander, chose LTC Tom Csrnko and his 1/5th SFG(A) battalion to support SOCCENT. The 1/5th was the designated deployment battalion when the warning order was received (it was in its GREEN cycle). The 1/5th was already supporting the Kuwait mission with one SF company and was also scheduled for an upcoming ARTEP evaluation (Army Training and Evaluation Program) to be conducted at Ft. Polk, Louisiana. Fortunately, these activities provided an operational security cover (OPSEC) to mask the increased activities of the battalion as it prepared for Operation United Shield.

Lieutenant Colonel Csrnko, based on a course of action requiring only one company of mounted SF, then chose his B Company commanded by MAJ Bryan Whitman to deploy as an Operational Detachment Bravo (ODB). Since SOCCENT's JSOTF in Mombasa would serve as the higher operational headquarters for SOF, the battalion commander and his staff saw no requirement to deploy to provide command and control (C^2). ODB520 could serve as an Advanced Operating Base (AOB) to the Marines during the operation.

Two senior NCOs from the battalion were provided for SOCCENT's Air Force Special Operations Detachment (AFSOD) to assist with control of air assets (from 1st Battalion's A Company–Sergeants Jeff Jilson and Tom McKoy).

Lieutenant Colonel Csrnko kept his finger on the pulse of the operation out at I MEF Headquarters by also participating in the planning with members of his staff. He remembers:

> It was a great marriage from the get-go. We went out there with a kind of idea to brief them on anything we could possibly do. After subsequent planning visits transpired, the task organization gelled into the four ODAs as coalition support teams (CSTs) assigned to the UNOSOM II countries needing them, as designated by the USMC requirements for the operation. These were the Bangladeshis, Pakistanis, and Egyptians. We left *Desert Storm* with the moniker of Coalition Warfare teams, and there was great discussion on that term. We weren't sure the UN conducting a humanitarian operation would necessarily understand this type of role for the ODA. . . We devised a new name, which is in use today in our doctrine, the Coalition Support Team (CST).[9]

Ultimately, Lieutenant Colonel Csrnko would deploy on the operation when LTG Zinni saw the need for some form of a SOF liaison and command and control element aboard his command ship. Csrnko flew to Tampa to discuss the requirement and receive operational guidance from Brigadier General Tangney, the SOCCENT commander. Lieutenant Colonel Csrnko then formed a three-man Special Operations Coordinating Element (SOCOORD): himself as commander, MAJ Craig Brown (his S-3), and an Air Force Special Operations officer experienced in AC-130 gunship operations, appointed by SOCCENT, Captain Morrow.

The Introduction of Non-Lethal Measures

During the I MEF planning, the staff considered part of the threat would be from hostile crowds which could easily transition to riots, looting, and harm to the coalition. The treatment of deceased U.S. servicemen during the Battle of Mogadishu was a stark reminder of how angry mobs could beat, kill, and mutilate those that fell into their hands. The ROE (Rules of Engagement) always allowed for self-defense by lethal means. By the same token, American commanders did not want to create civilian casualties if use of force became lethal. There was a vague ground between verbal warnings and shots fired into the air, then the use of lethal force commanders might have to employ on the front lines. Marines were trained in riot control, but the question still remained as to what action should be taken when the riot, or angry mob, became violent and they were not armed or part of a militia. Army Special Forces, not viewing their doctrinal role as policing, had not trained for riot control, although on previous rotations to Somalia the teams carried bats and large sticks to deter civilian looters from boarding their vehicles. (The 10th Mountain Division and UNOSOM II employed riot control agents during some of their operations to disperse crowds and clan militia.)

Lieutenant General Zinni asked the staff to consider non-lethal means to fill the gap because he was sure the Task Force would face unanticipated swarms of civilian personnel. The Somalis had demonstrated a preference for this method, swarming with crowds of people to loot and pilfer.

"Non-lethal" is a Department of Defense term, but it applies to a range of activities when the use of deadly force is not required. For instance, computer attacks and military information operations could be considered as non-lethal military activities. The definition is not precisely focused on its use solely to control civilian crowds. This aspect of the definition would have to be refined to then determine the equipment and tactics, along with training, which would be required for Somalia.[10]

The military in the main eschewed the use of non-lethal agents out of concern that they would not be seen as *lethal*. As a result, the non-lethal technology industrial base produced items of equipment primarily for law enforcement, and in small numbers. It had never been tried for rapid expansion for equipping large military orders.

Any non-lethal item produced for the Marines required matching the item's use to a weapon system already carried by the military. The planning staff decided the delivery platforms best suited were the 12-gauge shotgun and the M203 grenade launcher (40-millimeter, mounted below the M-16 rifle).

I MEF was assisted by the USMC Combat Developments Center to identify non-lethal weapons on the market, delivery equipment, and the training required for their use. Each viable selection would require a fast legal review in order to meet the shortened schedule for the upcoming operation. I MEF collaborated with the Los Angeles Police Department, national laboratories, and vendors during this process. For the operation, one Marine infantry company was designated to go ashore with this capability (they already were equipped with riot shields, plastic face masks, and batons).[11]

After testing several proposals, the Marines found the best effects were achieved through using the following: rubber rounds and bean bag rounds, pepper spray, stinger and flash-bang grenades, and pyrotechnics. The addition of barriers also helped to provide a stand-off distance (or denial), and channel any crowds; but using any form of sticky foam to build a barrier (or apply to a person) was deemed problematic. However, this material was still available to the Combined Task Force (CTF) during the operation.

Road-spike strips and caltrops were common items used by law enforcement that had some form of utility as an area denial method, and a good old bat or large stick-like baton still had utility. (Tasers, eye lasers, and acoustic blasters were a bit over the horizon in development.)[12]

The use of these now-preferred systems was approved by the I MEF after a safety and legal review. To fill the ROE gap between a warning, then lethal use of force, a specific ROE was drawn up for front line use by troops focused on crowd control, and only against unarmed civilians. (It would not include apprehension of rioters or suspects–a riot control or law enforcement function.) The Joint Task Force United Shield, Rules of Engagement, Unclassified ROE Card Series #1, dated January 11, 1999, stated, "When U.S. forces are attacked by unarmed hostile elements, mobs, and/or rioters, U.S. forces should use the minimum force necessary under the circumstances and proportional to the threat."[13]

Training began for the Marines of the 13th MEU (SOC) and continued when they were aboard the Task Force flotilla steaming toward Mogadishu. The goal for employment of non-lethal force was to prevent hostile crowds from interfering, impeding, or harming any friendly personnel during the course of the operation, with no civilian casualties as a result.

MSG Frank McFadden (now a retired Command Sergeant Major) remembered the SF unit's first introduction to "non-lethal" training.

> There was some non-lethal training (pepper spray and sticky foam) that the teams took part in. You have to remember the time; Somalia was winding down and the State Department wanted a kinder, gentler removal of forces. Lieutenant General Zinni went along with the requirement for training, but

> I don't remember any of these items being deployed ashore other than maybe pepper spray (we had brought along our own from Ft. Campbell).[14]

To announce this capability to the Somalis, Psychological Operations products and a witting press helped spread the message. Ultimately, only sticky foam and caltrops were taken ashore, but they were not used. (Fortunately, no swarming mobs or riots on or against Marine positions erupted during the operation.)

ODB520 Mission Planning

After the company Team Leaders and Team Sergeants were notified of the upcoming mission, they set up a mission planning site inside the 5th SFG(A) Isolation Facility (ISOFAC) in October to begin mission planning. ODA524 took the lead for planning, being the most experienced Somalia team and with previous assignments alongside the Pakistani Army.

Much time was spent on course of action development, beginning with a link-up with designated coalition partners outside of Mogadishu, then movement back into the city. Lack of specific information on the status of UNOSOM II forces and the dearth of information on any potential clan threats plagued the teams initially. This situation would not be rectified until the unit neared their departure time to Mombasa, Kenya and the deployment process began. For instance, a concern arose on the current relationship between the Pakistani and Indian contingents, tense at best, and how this dynamic would affect moving with both units during passage of lines operations.

Other factors included team roles as Coalition Support Teams (CSTs) and command relationships between their coalition contingent and the U.S. command element. Ultimately, the teams focused on providing liaison support, the ability to call for aviation support, and conducting area assessments and gathering intelligence for the task force. All teams factored in casualty rates expected from clashes with the SNA faction (Somali National Alliance—Aideed and Ali Mahdi militias), particularly around the Mogadishu area.

As the planning became more refined, it was learned the teams would conduct link-ups with their coalition partners only when the remaining UNOSOM II contingents were gathered in final positions as they collapsed back into Mogadishu. As ODB520 neared its deployment, the Indian contingent conducted their withdrawal from Somalia, removing the concern of any friction between them and the Pakistanis. Some contingents did not want, or need, American assistance (for example, Saudi Arabia and the Egyptians). Also, many of the UNOSOM II forces had already left Somalia as the mission date approached, leaving only the Bangladeshis, Pakistanis, and Egyptians to consider. Only three or four SF teams were required to support the mission.

The company also took into consideration the mission task to remove U.S. equipment and munitions to prevent its loss by falling into Somali hands. It was an extensive amount and included arms and ammunition, American-lent M60A3 tanks and M113 Armored Personnel carriers, equipment stockpiled for the Somali Police Program, other wheeled vehicles (including ambulances)—it even included the recovery of OH-58 Kiowa helicopters. All of this gear had to be collected and recovered for transport to Mombasa. Ultimately, ODB520 would play no role in this task.

With more clarity for the perceived mission, Major Whitman adjusted his mission planning guidance to the teams, wrapped up his selected course of action and, after ten days of the unit in the ISOFAC, took brief-backs from the four teams. The unit then conducted a five-day stand-down to spend some time with their families, since the operation would not go until January 1995. Other activities during the pre-departure period consisted of preparation on weapons ranges and maintenance and load out list refinements for the vehicles (a trailer for each team was added for their assigned two HMMWVs—Humvees—and ammunition basic load was increased).

As the deployment date neared, the unit learned of the final UNOSOM II contingents needing support: Pakistan and Bangladesh. The Pakistani brigade with three battalions would occupy the southern and north-central portions of the Mogadishu airfield, with the Bangladeshis securing the New Port area. Major Whitman then selected four of his most Somalia-experienced Operational A-detachments (ODAs) to conduct the mission: ODAs 522, 523, 524, and 526 (ODA524 was designated as a reserve team at this time). They were numbered consecutively as Coalition Support Teams 1–4.

On January 5, 1995, the Pentagon announced that 2,600 Marines would be deployed to support Operation United Shield as part of a seventeen-nation Task Force. Ultimately, 1,800 Marines deployed, along with 350 Italian Marines attached. When adding in the thirty or forty 5th SFG(A) personnel, the sailors of the naval task force, and SOCCENT's JSOTF, the deployment number easily approached the original announcement of force strength.

ODB520 departed Ft. Campbell via C-5A Galaxy aircraft on January 27, 1995, and arrived on January 29, 1995, in Mombasa, Kenya. For this phase of the operation, U.S. personnel lodged for five days in contracted hotels at Mombasa's Shamuvillage. Just prior to transfer aboard shipping, the company moved to the USLO hangar at Moie Airfield and collocated with SOCCENT's JSOTF. It was at this time the STS personnel (Special Tactics Squadron–USAF Special Operations air controllers) from the Air Force Special Operations Detachment (AFSOD) were attached to their respective SF teams; their role was to coordinate air support. An STS team was equipped with a robust communication capability and was a welcome addition to the SF team's capabilities.

Major Whitman participated with the SOCCENT staff in the attendance of daily briefings with the Combined Task Force. ODB520 continued to refine their preparations for the mission and, where possible, conducted weapons training and communications procedures. One of the most important aspects of the operation involved the communication plan between SOCCENT, air support, and other CTF contingents. For the most part, none of the SF teams on the operation had any experience with calling for naval gunfire. Lieutenant General Zinni had designated all of his assigned air assets as the CTF Quick Reaction Force, and rehearsing the comms procedures between SOF and USMC air assets was critical.

While at Mombasa, the SF Company prudently set up a Modular Demonstration (MOD DEMO) to display their weapons, vehicles, communications gear, and other capabilities such as demolitions and medical. Lieutenant General Zinni and his staff visited the MOD DEMO to gain a better understanding of the role and capabilities of a mounted SF team. During his visit, and much to the chagrin of the teams, he informed them there would be no need for mounted reconnaissance or vehicular patrols outside the security perimeters established on Mogadishu airfield and the New Port area. The DMVs would now be consigned to the role as a 3,500-pound, static gun position platform.

Lieutenant Colonel Csrnko and his SOCOORD moved to the USS *Belleau Wood* to assume their duties. In final preparations for steaming, Major Whitman and selected members of his staff and SF teams flew out to link up with Lieutenant Colonel Csrnko to participate in refinements to the plan, over on the USS

Essex. Game time was nearing; upon Whitman's return to the company ashore, he ordered all made ready for immediate departure.

Four days prior to departure, Master Sergeant McFadden and CPT Galyn Cross from ODA524 moved aboard the USS *Belleau Wood* to serve as a reception team for the company. (The company had been scheduled to board the USS *Fort Fisher*, but a change was made just prior to embarkation day.)

LHA-3 (Landing Helicopter Assault) USS *Belleau Wood*, the "Devil Dog," was named after the famous battle of Belleau Wood in World War I when Marines earned their nickname from the Germans as "Devil Dogs." She was commissioned in September 1998 and home-ported for several years at San Diego, California, where she served PACOM and the Pacific fleet during exercises and contingencies throughout the region.

She was designed as a Tarawa-class amphibious assault ship, capable of carrying up to thirty helicopters and a complement of AV-8 Harriers. The USS *Belleau Wood* was capable of fitting one battalion of Marines and their support elements aboard.

In the fall of 1992, she was forward based in Japan. During Operation United Shield, the USS B*elleau Wood* served as the command ship for Lieutenant General Zinni. Later, in 2002, she served in support of Operation Enduring Freedom. She was decommissioned on October 28, 2005, and, a year later, was sunk off the coast of Hawaii as a target ship.[15]

On the 7th of February, the company departed the USLO warehouse by vehicle convoy (Chief Warrant Officer 2 Eggen from ODA524 and two SF soldiers from the 1/5th remained behind with the JSOTF). They crossed the ferry at the port of Mombasa and loaded aboard U.S. Navy LCUs (Landing Craft Utility) for transport to the USS *Belleau Wood*. Master Sergeant McFadden and Captain Cross received the unit and got them settled, followed by giving them a short tour of the vessel to familiarize themselves with how to navigate the passageways of the ship to reach their quarters, messing areas, and where to store arms and ammunition. Key rules and procedures were briefed, most notably the storage and accountability of ammunition. The most dangerous catastrophe a ship could face was a fire on board. The teams were admonished to have strict control over their explosives and thermite grenades.[16]

Mission planning continued daily. Rehearsals and immediate action drills, where possible, took place. The teams were allowed to fire off the fantail of the ship to ensure weapons were operable. New to U.S. Navy operational doctrine ". . . *From the Sea*," some operators were amazed at how fast weapons could rust in the salty air. Each day, the plan was talked over within the B-team and modified based on input from the planning staff and the SOCOORD.

While steaming toward Mogadishu, Major Whitman and the B-team developed a mission concept (MICON) for approval with Lieutenant Colonel Csrnko; once it was approved, there followed an OPORD brief-back to Lieutenant General Zinni on the SF role in the operation. Each ODA commander participated in the brief-back to explain his detachment's role. The restated mission was finally approved as ". . . Collect ground truth on Pakistanis and Bangladeshis; perform liaison duties for the passage of these forces through the Marine Corps ground elements; provide Close Air Support (CAS); and report on positions and activities in your area (to include the Egyptians)." Lieutenant General Zinni approved the concept of operation and gave additional guidance and intent to the ODB commander: "Gain the confidence of the coalition battalion commander, become a part of his battalion, and lead them safely back through the USMC."[17]

Additional activities for the teams involved non-lethal measures training and a review of the CTF's PSYOP products (all PSYOP products were U.S. Navy and USMC products focused on the upcoming withdrawal operation). A few Non-commissioned Officers (NCOs) from the company assisted the civilian techs and USMC instructors to impart non-lethal training to the Marines aboard.

In order, the ODAs, followed by the B-team, conducted amphibious landing operations from LCACs (Landing Craft Air Cushion) onto the GREEN Beach sector (south of the airfield) within the Pakistani sector, beginning on February 8, 1995.[18]

The Pakistani Brigade (*Tirwana*) had assumed overall security responsibility of the airfield perimeter on February 6, 1995.[19]

The Pakistani and SF positions could be seen from the neighborhood overlooking the main gate and from an encampment known as Halana village (near the Pakistani general hospital). One of the weaknesses of the perimeter was a gate for Halana village residents to visit the hospital, which remained a concern until the Pakis closed the hospital one week prior to their evacuation. Other measures were taken to limit Somali civilian movement about the perimeter: security checkpoints and barbed wire, all covered by elevated observation posts.

SF Coalition Support Team Alignments

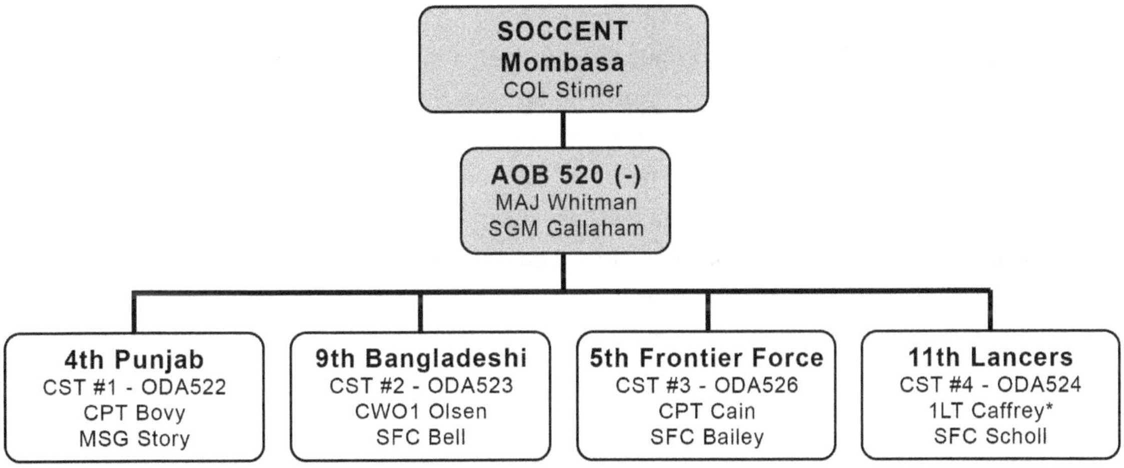

* SFC Lance Caffrey was "frocked" to 1st Lieutenant in order to conduct duties with the commander and officers of the 11th Lancers.

The B-team (AOB 520)

MAJ Bryan Whitman's B-team (ODB520) mission was to provide situational awareness of the SF teams and their operations to the JSOTF in Kenya (commanded by COL Rich Stimer) and to provide minimal support and Command and Control (C2) for the ODAs on the ground in Mogadishu. The B-team served in this role as an AOB (Advanced Operating Base). The task organization was as follows:

- Commander – MAJ Bryan Whitman
- XO – CPT Gaylen Cross (ODA524 augmentee)

- Company Sergeant Major – SGM Hank Gallahan
- AOB Operations – MSG Frank McFadden (from ODA524)
- AOB Operations/Air NCO – SFC Dave Asher
- Medic – SSG Eric Anderson (EOD background)
- Commo
  – 18E SSG John Caldera from ODA524 and attachment from Group signal shop
  – SGT Bobby Brown
- Two USAF STSs were attached at Ft. Campbell and remained throughout the operation

(CWO2 Chuck Eggen from ODA524 remained with the JSOTF in Mombasa to serve as MAJ Whitman's representative to the JSOTF.)[20]

The B-team was transported from LCUs (Landing Craft Utility) to LCACs (Landing Craft Air Cushioned) and landed at GREEN Beach along with the initial ODAs with their vehicles. The ODB had two DMVs loaded with .50-caliber machine guns. The B-team occupied two trailers, behind a hill, alongside the beach on the northern end of Mogadishu airfield. Unfortunately, this put them in close proximity with the international media, and they did not escape news coverage attention.

The B-team had no operational role in the evacuation operations; they merely monitored and provided over-watch. The USMC was in charge of all the evacuations. Prior to the evacuation, members from a SEAL team afloat came by once looking for work (to serve as snipers as they had done in previous Somali operations), but MAJ Whitman declined to have them attached.[21]

Prior to the amphibious operation to bring the Marines ashore, the B-team facilitated and coordinated with their advance parties; once the USMC was ashore, the B-team was responsible for attending all USMC staff calls and providing LNOs to their staff (Master Sergeant McFadden and Captain Cross served in this role). Other liaison duties included coordinating all passage points for the coalition forces.

Tremendous explosions on the airfield were a daily occurrence for all forces ashore due to USMC Explosive Ordnance Detachments (EOD) destroying excess and unwanted munitions in the area to prevent them from falling into the hands of clan militias. This daily blasting did much to raise the landing force's apprehension of danger.

Operational Detachment Alpha – 522

ODA522 was assigned as a Coalition Support Team #1 (CST) to the 4th Punjab Battalion, equipped with M113s. ODA522 was led by CPT Joe Bovy; his Team Sergeant was MSG Daniel Story. The 4th Punjab occupied the right-hand side of the northern Mogadishu Airport perimeter. Apparently, the 4th Punjab was living out of their vehicles and in some tents during the final days before withdrawal, so the team chose to bed down on the large sand escarpment on the southern ocean side of the airstrip (in the UN Village). This arrangement also put them in close proximity to the B-team. The team departed each morning to visit with and coordinate with the 4th Punjab, but they were not required to man any security positions or observation posts or to conduct dismounted patrolling. The 4th Punjab had more than sufficient firepower, so the additions of the team's .50-caliber machine gun and the MK-19 grenade launcher were not necessary. If air support was required, this task could easily be performed from atop the sand berm which afforded the team full visibility of the 4th's position and the airport's main gate. At night, a few members of the team, along with their STS support, did collocate with the unit to provide nighttime CAS support.

Operational Detachment Alpha – 523

The most depressing assignment for the SF Company was coalition support for the 9th Bangladeshis. ODA523 was commanded by CW2 Bryan Olson; the team sergeant was SFC Channing Bell. (Other team members assigned included SFC Lance Hoepner, SFC Anthony Pettingill, SFC Doug Sims, SSG Eric Tanner, SSG Franklin Brown, and SSG Karl Feisheim, but it was not known who among them were actually on the team in Mogadishu.) The arrival of the CST to the New Port area was met with confusion and reluctance on the part of the 9th. Why did they need a CST, and why was the attachment not coordinated prior to their arrival? Apparently, the Bangladeshis were confident in their capabilities to secure the New Port area and became somewhat miffed at the perception they needed assistance. Explaining the comms links to American air assets as one benefit smoothed some of the ruffled feathers. It would all be a moot point when the Bangladeshis were relieved in place by the USMC contingent and boarded ships to return home.

After providing some assistance during the handover and left with nothing to do, ODA523 split some of its operators to other coalition support teams and in support of the ODB located on the sand berm at the UN Village site.

Operational Detachment Alpha – 524

While in Mombasa preparing for their Coalition Support Team mission, ODA524 was still unaware whether or not the team would be assigned to an Egyptian or a Pakistani unit. With the lack of comprehensive intelligence on the situation in Mogadishu, and very little mission guidance, all the teams envisioned their roles to be similar to the previous time the unit operated with coalition forces in Somalia.

The team understood the basic role of liaison to their UN contingent—to wit, provide situational reports to their higher command on the activities of their UN partners, serve as a communication link between their assigned contingent between them and U.S. forces, and provide connectivity for CTF air support. Ultimately, it was the team's job to provide any assistance to help enhance the Paki or Egyptian's UN mission, and once under CTF control, continue the mission for withdrawal.

ODA524 conducted pre-mission training and rehearsals on the critical parts of their operation: vehicle drills, weapons firing, and communications—given the restrictions of operating on the Mombasa airfield. A new aspect of their mission required further training alongside their USMC infantry counterparts once they were aboard the USS *Belleau Wood*: non-lethal weapons.

Little was known of the capabilities and composition of either the Pakistani contingent or the Egyptians or, for that matter, any other remaining UNOSOM II forces positioned throughout Mogadishu (such as the New Port area). Most glaring were deficiencies in intelligence with respect to the current situation of Aideed's and Ali Mahdi's forces: size, location, capabilities, and so forth. UNOSOM II human intelligence (HUMINT) was still lacking. ODA524 anticipated some combat with clan forces, but to what extent it could not be ascertained.

ODA524, as with the other SF teams deploying, was equipped with two Desert Mobility Vehicles (DMV) and one trailer. One vehicle was armed with a .50-caliber machine gun and the second with a MK-19 grenade launcher. This, along with the organic mix of weaponry for the team, would allow for sufficient firepower to conduct aggressive mounted patrols outside the defensive boundaries of their assigned sector. The team was finally designated as Coalition Support Team #4 with the assignment to serve alongside the 11th Pakistani Lancer Battalion, from the 19th Lancer Brigade. Initially, the team boarded the USS *Belleau Wood* but, once arriving off Mogadishu, transferred to the USS *Fort Fisher* as their designated debarkation ship. From there, they would be transported to GREEN Beach by LCAC.

The exact date of their landing was based on tide, ground conditions, and illumination (both night and day). Since this requirement was constantly being refined, it became an on-again, off-again drill for the team. When the announcement finally came on February 11th, ODA524 was given only 90 minutes to board the LCAC. They were ready, thanks to constant drilling, and soon emptied their weapons locker, procured their ammunition, and loaded the DMVs in what was soon to be one of the first amphibious operations for Army Special Forces in modern military history. (There was confusion in the CTF as to how many CSTs were required to support UNOSOM II and what those remaining coalition commander's requirements were for a liaison element.)

Prior to their arrival on GREEN Beach, ODA524 split their team to provide for just enough personnel to man the two Desert Mobility Vehicles (DMVs). One USAF Special Operations STS NCO was attached to the team (Staff Sergeant Urenda, 23rd STS) to provide air control and calls for fire from the AC-130s and USMC attack aircraft. This formation gave the team six personnel: SFC Lance Caffrey (who was "frocked" to 1st Lieutenant after concern from Lieutenant Colonel Csrnko and Major Whitman about rapport to ensure an American officer was advising a Pakistani senior officer, not a sergeant, whose rank might insult the sensibilities of a coalition member), SFC David Scholl, SSG John Caldera, and SSG Brad Berry from ODA524. SSG Patrick Mitchell from ODA522 joined ODA524.

It was a clear day with blue skies; other than the ten foot waves, the movement to the shore was uneventful. The team soon arrived to their designated landing site on GREEN Beach, a strip of sand just northeast of their designated coalition partner—the Pakistani 11th Lancer (Cavalry) Battalion of the 19th Lancer Brigade. Like all the other SF teams, they were surprised at the amount of sand blown on and into their equipment by the LCAC fans, and required a short stay on the beach to clean everything.

The 19th Lancers were an armored regiment of the Pakistani Army (about a brigade-size unit with additional support elements). It had its origins in the 2nd Regiment of the Mahratta Horse (later re-designated as the 18th Regiment of Bengal Cavalry) during the late 1800s. They merged with the 19th Lancers (Fane's Horse) raised in 1860 for the Second Opium War (both units amalgamated in 1922). In World War I, they were named the 19th King George V's Lancers, and assumed the name of the 19th Lancers after Pakistan became a republic. (The 19th Lancers served in Somalia both in 1993 and in 1994 and were part of the rescue force to extract the Rangers during the Battle of Mogadishu in October 1993.) It was a well-performing unit that kept their armored vehicles and tanks in top-notch condition.

As one of the 19th's three battalions, the 11th Lancers were responsible for the southwestern tip (left flank of the Paki security perimeter) of the airfield and provided security from the shoreline, then north and northeast along the perimeter by the Pakistani hospital, culminating with its flank near the main gate (the MiG derelict aircraft monument fixing its location); they then were tied in on right flank with ODA526's coalition support position.

The team passed through the 19th Lancer Brigade's gate, the only access to the position, and joined the 11th Lancers to conduct an initial coordination meeting to introduce themselves to the Paki leadership, followed by settling into their campsite near the motor pool. Arrangements were made to issue the team security badges, set up their support, and receive their assigned mission from the battalion.

The Lancer Battalion was equipped with loaned, U.S.-model M60A1 tanks and M113 Armored Personnel Carriers (APCs). The headquarters was a small, white concrete building. Troops lived in tents when not on duty, while the armored vehicles were laagered together in one large motor pool. The battalion had taken care to improve the defenses of their position by emplacing barriers, triple-strand barbed wire, checkpoints, and observation towers, which also served as a position for snipers. Access to the camp required security badges. Armed guards conducted roving security patrols, backed up by a quick reaction force of a tank platoon, if needed. The SF team was impressed upon their first contact with the unit. SFC Lance Caffrey noted the following.

> My first impression of the 11th Lancers was very positive. All of their equipment was in good shape with weapons very clean, even though they had held this position on the beach, with blowing sand and dust. The troops were highly disciplined and well dressed–not the image we had seen of other contingents over the years we were in Somalia.[22]

The ODA lived in a tent alongside the unit and positioned their vehicles within the cantonment area, given the prohibition by the CTF that there would be no mounted patrols outside the defensive perimeter. ODA524 added the firepower of their .50-caliber machine gun and their MK-19 grenade launcher to provide defensive fires from protected positions assigned by the commander and the battalion's operations staff.

There was very little harassment from the Somalis at this position (occasionally, random gunfire). No clan was willing to challenge this heavily secured and defended portion of the airfield. Their team, however, did respond with three personnel to ODA526's request for assistance when clan Technicals fired upon one another near the main gate while random sniper fire was also occurring. ODA524 participated in this force protection action for three days, adding their numbers to serve as sniper observers, and a counter-sniper capability, if needed.

As with all combat deployments, a monotonous routine soon set in when danger was not present. The team met daily with the 11th Lancer's leadership, mostly to participate in the planning and withdrawal operation. Other tasks included vehicle and weapons maintenance and, when possible, practicing calls-for-fire with the AC-130 gunships.

Staff Sergeant Urenda, the assigned 23rd STS member, controlled close air support; fortunately, no live fire missions were needed during the deployment. Urenda remembered his role (in an article printed for *Air Commando* magazine in 1995) as shown below.

> I was there to provide control for airpower if needed. The shooting seemed to pick up between 11 a.m. and 3 p.m., then it would slack off at night because many Somalis were celebrating Ramadan. . . When the Somalis heard the tanks moving, they got braver, thought it was time to move in and would aim our way. The Pakistanis got a little nervous and returned fire lots of times, but we had a good button on what was going on.[23]

Operational Detachment Alpha – 526

ODA526 came ashore via LCAC on February 8th. They dusted off, watched, and listened to gain situational awareness before moving out to their assignment with the Pakistanis. As CST #3, they would position themselves with the 5th Frontier Force commanded by COL Rana. The 5th FF was a mechanized infantry battalion equipped with M113 armored personnel carriers (APCs). Their M113s were distinctive, equipped with two light machine gun positions atop the crew hatch along with protective metal shields for the gunners (gun turrets for the main, heavy machine gun). The vehicles were camouflaged in deep green and black foliage-patterned colors.

But first, a meeting was called for all American forces ashore by the senior USMC commander. He imparted his guidance for the operation and position of the forces; the teams once again were disappointed to hear from him there would be no vehicles conducting patrolling outside the airport perimeter. In fact, he emphasized, ". . . go where you meet up at your position and park them." They were also ordered to remain in full kit while on the assignment, notwithstanding taking in any provisions for the heat and humidity. Welcome to good old infantry dismounted patrolling.[24]

Strangely, the Pakistani contingent had no such strictures, with many even wearing civilian clothes. Perhaps they imagined soldiers wearing helmets and protective vests signaled a willingness to begin a fight with the Somalis. The Pakis eschewed joint foot patrolling with the Americans as long as they displayed this tougher image of a peacekeeper on patrol (no problem with Americans joining them at combined observation posts located in towers where there was no interaction with civilians). Throughout every previous deployment to Somalia, American forces noticed the ambivalence of coalition contingents to treat the Somalis as a potential threat but more as a nuisance—as if to say, "don't bother us and we won't bother you."

The Pakis adopted this posture to impart a show of friendliness to Somalis within their sector, which the team took for a lack of discipline. The team perceived signs of a laxness in security with the Pakistanis loosely fraternizing with Somalis along their perimeter and visiting food stalls and tea houses outside of the airfield. They found various holes in the concrete wall along the Airport Road and were perplexed when the Pakistanis neglected to block off these access points.

The 5th FF was responsible for the center of the northern perimeter, linked into the Paki armor battalion in the southeast near the Paki hospital. Their territory then encompassed the northern main gate and extended right over to the Soap Factory. The Soap Factory was located outside the airport fencing on a rise in terrain—it afforded a fantastic view of the entire airfield, the K4 traffic circle, and the villages adjacent to the airfield. (The CTF wanted an extended perimeter beyond the airport fence and out to 1,000 meters; this amount of terrain was unrealistic since American forces could not patrol in some of these environs.) The Soap Factory location was the furthest piece of territory which could be occupied. ODA526's right flank was covered by ODA522. One of the team's first tasks was to conduct a link-up with ODA524 with the armor battalion on the 5th's left flank.

In this assigned position, the SF team was lodged in a building along the airport fenceline. Since their vehicles were now relegated to static displays, they prepared parking positions to afford them the best siting for adding their firepower to the defenses. The area outside of their position was open; it would be necessary to conduct dismounted patrolling to provide a stand-off from any potential threat of small arms fire. To add to their security, a guard post was established atop the Paki hospital.

CSM (retired) Jose Baily reflected on what was a typical day for the team.

> It consisted of our collecting ground truth on the units in our area along with any other important activities, checking the status of the security perimeter and barrier materials, planning and liaising for the withdrawal, and patrolling. We patrolled in two-man teams to conduct our inspections of the perimeters and the guard towers. A lot of the perimeter was in bad shape with breaks in the concertina wire; in other places, sandbags were used to block and fill holes in the concrete wall. We pitched in with our coalition counterparts to assist in the repairs where we could. The remainder of the team assisted Captain Cain [Steve Cain, Detachment Commander] in the planning effort with the Pakistani staff to fine-tune the withdrawal scenario.[25]

CPT Steve Cain recalled other activities.

> We registered targets and talked to the AC-130s every night. Our SITREPs were due in the evening. I, or a senior member of the detachment, attended daily meetings held by the Pakistanis. The passage of lines instructions and coordination with the 2IC [second in command] and the operations staff took the whole time we were on the ground to get it right. We also conducted some cross-training and basically tried to interact with the soldiers, trying to also co-opt the NCOs.[26]

An essential task for the team was ensuring connectivity to aerial fires, particularly during the periods of darkness. They availed themselves of the nightly sorties from the AC-130s to practice calls-for-fire. To simulate live fires, and to illuminate suspected unfriendly positions, they had the AC-130 use its searchlight instead of its guns to target the areas pinpointed by team lasers. With the Somalis' ingrained fear of the AC-130, this practice was soon put to a halt after complaints from locals to the still functioning UNOSOM II headquarters.

Now retired, CSM Jose Baily described more about their physical surroundings at this position.

> During this time the detachment was under gunfire almost every day. Although the newspapers would call it "sniper fire" in their reports, the team was convinced much of it was aimed at us. The Habr Gedir and Hawiya [SNA] clans predominantly controlled the airport surroundings, with boundaries overlapping along the northern perimeter. The Darod clan controlled the New Port area with boundary conflicts with the Hawiya clan.

Even with assurances to Lieutenant General Zinni from Ali Mahdi and Aideed to not interfere with CTF operations, the word apparently did not trickle down to the streets and neighborhoods surrounding the airport (or some sub-clan leaders were impotent in enforcing security). Many Somalis were pre-positioning themselves for a day of lucrative looting once the airfield was empty of forces. The agreement to also form joint clan security elements to control the airport and the port area led to some inter-clan tensions as well—apparently due to jockeying for positions under this arrangement. Fortunately, no clan militias were bold enough to directly engage Pakistani or USMC forces. Much of the random gunfire between the clans inadvertently spilled over from the airport village and landed among friendly positions. Command Sergeant Major Baily continued:

> The team house had a small alley between its buildings, and often when team members crossed in front of it, we were shot at from what we called the "sniper courtyard." Other names we attached to the alley included "*CMH Alley*" and "*Purple Heart Lane.*" We had to adopt a technique to throw off their shooters called a "stutter step" for use when exposed to the alley, in an attempt to complicate the shooter's aim. At least 95% of the shots fired by Somalis emanated from this area, even though they were single shots. Most offensive to the team was the daily shooting full of holes of the plastic water storage container atop of our team building.[27]
>
> A wide-open field, similar to a soccer field, in front of the team's position became a convenient gathering place for Somali crowds to chant and yell at us Americans whenever they were operating in that location. For instance, on one occasion Fail and I were up in a CP tower when a huge, menacing crowd gathered at the gate near the airfield village. Fortunately, no incidents occurred.
>
> We did, however, receive incidental fire when a clan firefight broke out in a depression across from the northern edge of the airfield, but again, we did not return fire, as it was not aimed at us or other coalition forces.

A tragic accident occurred one day when Pakistani forces shot a Somali woman when the soldier's gun accidentally misfired. Captain Cain helped to defuse the situation with the growing crowd by negotiating with the local chief.

The largest incident involving gunfire between the locals and Pakistani forces occurred when an Italian news crew went outside the airfield perimeter main gate and headed towards the K4 traffic circle. They soon became surrounded by an angry mob followed by a local militia with a technical. Clearly, someone was attempting to capture a hostage that day. An argument arose between rival militias, and soon gunfire and one RPG filled the air. In the ensuing firefight, Carmen La Sorella, a famous journalist in Italy, fled on foot to seek safety; her cohort was not so lucky and was killed. Captain Cross and Master Sergeant McFadden were liaising that day with the Paki Brigade command and were situated atop the Soap Factory, witnessing it all.

> They attempted to get clearance to fire their weapons in assistance as well as to call in CAS, but were refused by higher. Initially, it was thought that the Italian female reporter was being captured and held. In reaction, the Italian naval contingent moved their frigates and destroyers within ½ mile of the shore as a show of force. The other Italian journalists were eventually freed within three to four days.[28]

When two SF sergeants conducting air support became pinned down, Captain Cain commandeered a Paki armored personnel carrier and drove to their rescue. LTC (retired) Steve Cain related the series of events when this incident occurred.

> Meanwhile, along the airfield perimeter line, the Pakistanis exchanged gunfire with the clans, growing into a heated gunfight. During this firing, many rounds were spilling over into our detachment's area and pelting the area around detachment members on the roof of our team building. The Pakistani commander immediately ordered all movement to stop and the airport gate was locked down.

> When this incident occurred, the two SF NCOs serving as AFSOD air liaisons for the JSOTF became pinned down atop the old Egyptian Embassy building (Jeff Jilson and Tom McKoy from A/1/5). They immediately requested extraction.
>
> I secured permission to attempt the recovery and commandeered a Paki M113 after stripping it of any UN markings. My team sergeant [MSG Jose Bailey], my commo sergeant, and our attached STS airman monitored this action from the team TO [Tactical Operations Center]. While I drove the APC, SFC Korenowski manned the .50-caliber.
>
> The SF air liaison team was pinned down in the corner of the courtyard of the Embassy building while DsHK fire was peppering the top of the roof. I drove into the courtyard, spun around, dropped the ramp, and recovered the team. We all returned back to the safety of the airfield without further incident.[29]

(That day, Osman Atto lived up to his agreement with Lieutenant General Zinni that he would do his best to enforce security around the airfield. He maneuvered one of his large Technicals to the vicinity of the clash and used his clan militia to take on the rogue element and remove the threat from the area.)

Ms. La Sorella was indeed captured and in the custody of rogue militia. The Italian government demanded her release, backed up with the threat of action. In a few days, the situation calmed, and she was handed back over to a combined special operations team.[30]

The Task Force Operation and UNOSOM II Evacuation

During the last week of December 1994 and through the first week of January 1995, all UNOSOM II positions outside of Mogadishu were evacuated. An Indian naval contingent assisted this effort in Kismayo while the other contingents conducted overland movements into the environs of Mogadishu, occupying the Embassy compound and other bases previously used by American forces. Indian, Malaysian, and Zimbabwean forces departed Somalia, leaving only the Egyptians, Pakistanis, and Bangladeshis and the remnants of the UNOSOM II staff, for a total of about 8,000 troops. Moving the remaining forces to the airport was accomplished on January 8, 1995. Predictably, the Embassy Compound and other vacated encampments were quickly overwhelmed by looters (not just men, they included women and children), stripping these areas clean of anything useful. Media film clips captured even the looting of broken chairs and empty 55-gallon barrels being hauled off. Outgoing peacekeepers noted armed militia, Technicals, and RPG-carrying Somalis occupying the now-vacated areas.

CTF United Shield set sail from the coast of Mombasa on February 1, 1995 after conducting a full rehearsal of the operation (less the beach landings) and arrived off the coast of Mogadishu on the February 7th.

In the face of American might assembled offshore, Ali Mahdi and Aideed, along with other clan factions (primarily the SNA and the SSA), met with Lieutenant General Zinni who came ashore with Special Envoy Daniel Simpson on February 19th to assure Lieutenant General Zinni that no harm would come from their sub-clans against UNOSOM II and the CTF during the withdrawal. There was a caveat: they both informed Lieutenant General Zinni they could not, however, control rogue elements of their clans not under their control. Surprisingly, they agreed to share power and seek elections in the future and work towards the goal of establishing a transitional government. All the clan factions present in the meeting assured the CTF commander they would take steps to not "militarize" the situation while the withdrawal took place. This commitment included the removal of their Technicals and roadblocks.

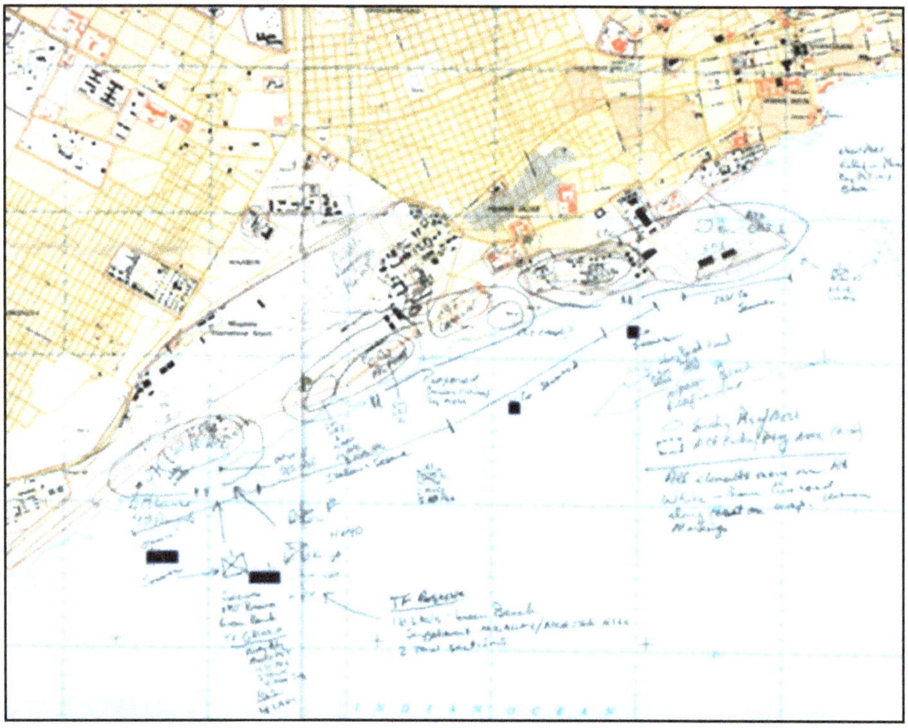

Planning map of MAJ Gary Danley, a member of the SOCCENT J3 Directorate, for the day of the Marines' amphibious landing of CTF United Shield. Actual unit designations have been concealed (*Courtesy of Gary Danley*).

In follow-on discussions, the rival factions agreed to a power-sharing over the airport and the port area through the use of joint militias to be provided with distinctive markings on their equipment. (Surprisingly, they followed through immediately after the UN withdrawal from the area, allowing markets and routes to and from the commercial areas in Mogadishu to be opened.)[31]

During February 12th–15th, around 7,000 Pakistani troops were loaded aboard shipping and departed. Between February 17th–22nd, the Egyptian contingent departed. In preparation for the final phase of the operation, a small contingent of Pakistani forces from the Lancer Brigade withdrew from their defensive positions at the Old Port area (the flank security position). This was completed on February 20, 1995.

Another contingent of Pakistani peacekeepers withdrew between February 23rd and 27th. All that now remained of UNOSOM II forces in Mogadishu was the Pakistani brigade with three battalions and the Bangladeshi brigade at New Port.

The Amphibious Operation

In the days prior to the final exfiltration of UN forces, Lieutenant General Zinni held one large coordination meeting for all involved in the operation. First, all coalition nations conducted their briefs, followed by a U.S. forces' brief. Although the SF Coalition Support Team operators were proud of their role in assisting their coalition partners in the planning for the briefs with the CTF commandeer, they remained silent and stood proudly as their coalition counterparts conducted the session. When finished, Lieutenant General Zinni approved all plans.

The Marines' movement to shore began on the morning of February 26th by sending in a small command element to link up with the UNOSOM II headquarters. Lieutenant General Zinni moved ashore the following day to assume command of UNOSOM II forces from General Aboo.

On February 28th, the amphibious landing portion of the operation began in the early hours of the morning with a battalion landing team from the 3/1 Marine Air Ground Task Force (Kilo and Lima companies, along with their support units), accompanied by Italian Marines from the San Marco Battalion. The force was commanded by Colonel Garrett, the 13th MEU (SOC) commander. On this day, General Aboo and his staff departed Somalia.

As the sun rose on the Mogadishu airfield and New Port area, a solid perimeter was established from GREEN Beach in the south to the northeast at New Port. Prior to the arrival of the landing parties, engineer and PSYOPs personnel had been inserted days before the landing to prepare positions and inform the Somali people of the operation. (And, of course, the SF A-teams were present since early February.) To prevent any instance of fratricide, the CSTs shared their sector sketches and updated information on their areas to the Marines as they occupied their own troop positions.[32]

Now that the Marines had arrived, about 800 Bangladeshi peacekeepers departed from New Port. Although the area was pretty secure, and lined with CONNEX steel containers erected as a wall barrier, the Bangladeshis spent each day running off crowds of Somalis loitering outside the main gate. This included charging with APCs as a deterrence to swarming crowds. A small contingent of Pakistani forces assumed security of the area as the Bangladeshis departed.

The Pakistani Brigade's units were all located at the airfield and would require forming up into convoys, then they would conduct a passage of lines through the Marine forces, and close on into New Port. Thus, the New Port area for UNOSOM II forces withdrawal and GREEN Beach for the final U.S. forces' withdrawal became key terrain for protection.

The ODAs prepared for the withdrawal by participating with their counterparts through a series of sand-table exercises. Only two ODAs (522 and 526) would provide vehicular escort for the two convoys, with one Desert Mobility Vehicle in the lead and one in trail. A key task for the SF teams was serving as the communications link between CTF forces with their air support and the withdrawing Pakistanis, lasting from the moment of initial movement until the Lancer Brigade closed at New Port.

It would be a four-phased operation:

1. Evacuation of the perimeter and observation post positions.
2. Simultaneously, pre-position and line up all vehicles using existing buildings as cover for the assembly area.
3. Assemble on airport aprons in order of sequence, establish a new security perimeter, and account for all forces. (Previously, Pakistani tanks were placed near USMC security checkpoints to over-watch the withdrawal).
4. Move units in sequence to southern end of the airfield and use both the north and south runways to move by convoy up to the USMC checkpoint at the New Port Road. All weapons and gun tubes

> were to be positioned at the rear until reaching this point. (The Pakis had responsibility for their security while on the New Port bypass road and while at New Port; weapons could be redirected during this portion of the move.)

SF Activities

During the phased withdrawal, ODA522 (CST #1) escorted the 4th Punjab down the airstrip over to New Port (the 4th Punjab was second in order for the evacuation of the airfield).

Coalition Support Team #4

Sergeant First Class Caffrey, commanding ODA524, received his guidance from Colonel Sikander as to the best role and positions for the team during the actual withdrawal, making it quite clear that ". . . we will not place the tanks into positions too early in order to not give away the plan to the Somalis, who will be watching."

The CTF conducted a rehearsal for the withdrawal of the final contingents from the airfield. The usual bevy of news reporters filmed the rehearsal, handing the Somalis a glimpse of what the pending operation would look like. (The team made it onto a CNN news clip.)

On the day of the withdrawal to New Port, Sergeant First Class Caffrey collocated with the 19th Brigade commander at the airport main gate. The remainder of the detachment assisted the unit during the assembly phase. ODA524 positioned themselves accordingly, fully kitted up, on a small sand dune at the southern end of the airfield alongside Pakistani M60A1 tanks in over-watch positions and provided any necessary communications coordination between the Pakistanis and the CTF.

As the movement commenced at midnight (March 1st), the Pakistanis received some small arms fire from crowds gathering, who had sensed the time for final withdrawal of UN forces was near. Two Pakistani M113 APC crews engaged them with warning fire from light machine guns and their .50-calibers. The STS controller with one of the teams watched the engagement.

> As we retreated, the Somalis came out of the woodwork. They were like cockroaches. The concertina wire didn't stop them. It was like a herd of cattle coming down a hill. They were all firing at each other.[33]

Sergeant First Class Caffrey then fell in formation in the trail position with the second team GMV as they closed out the camp with the 11th Lancers and collapsed the Lancers portion of the perimeter.

Coalition Support Team #3

ODA526 escorted the 5th Frontier Force element through Marine checkpoints to the New Port area. Once their mission was completed, the team assembled with the other ODAs and the B-team at GREEN Beach and awaited extraction (via LCAC) back to the USS *Belleau Wood*.

The Morning of March 1, 1995

As daylight approached, the assembled news media captured the Somalis gathering along the airport perimeter—some of the Somalis from vying clan militias—with at least two Technicals with heavy machine guns observed. The Technicals began firing at no one in particular and apparently as a sign of celebration. Swarms of people (described by one ODA member as "a stirred-up ant nest") began to slip over the wall and existing physical barriers for a chance to loot, particularly near the main gate. No building was left untouched. Within CTF ROE, soldiers and Marines from the Task Force shot warning fire.

One of the most bizarre events captured by the media was individual Somalis closely running a foot race behind the last elements of the Pakistani forces as they convoyed down the northern runway. ODA524 brought up the rear as overhead air cover provided by USMC AV-8 Harriers and AH-1 Sea Cobras began to vector into the area.

Lance Caffrey recalled the following.

> It was an unbelievable scene behind us–Somalis forging over the perimeter fence in a wave of humanity, pouring onto the airfield, all along the entire perimeter! I watched people fighting over anything they could get their hands on–wood, tin, etc.–but I did not see any arms of any kinds, I did not feel like this was a threatening gesture. They were just after subsistence items. We even saw some Somalis wearing what appeared to be U.S. military gear worn by the Rangers from the earlier battle in Mogadishu.[34]

Master Sergeant Bailey of ODA526 overheard one radio call from a member of the escorting teams aboard their Desert Mobility Vehicles while he was located near the B-team's commo operator at Lieutenant General Zinni's Command Post location who declared that, "the whole hill is moving like a black mass! Don't know how many there are but there must be a bazillion!" He watched Lieutenant General Zinni turn towards the commo operator on the hill and politely ask for clarification, inquiring, "How much is a bazillion?"[35]

When a few Technicals made it onto the airfield apron, they appeared to position themselves in order to drive towards the Command Post on the sand berm. Heavy machine gun fire from Marine Light Armored Vehicles (LAVs) drove them off, and they departed to apparently go exchange gunfire with another rival clan (including fire with a recoilless rifle). This action momentarily surprised the collected crowd near the CP, wondering why people were firing.

The withdrawal was executed professionally and completed by 0630 hours. As each vehicle passed the Marine checkpoint to enter the New Port bypass road, the vehicle crews reoriented their weapons forward to now provide for their own security. The speed of the withdrawing units did pose one challenge; it created traffic congestion at the New Port area. The ODAs were mission complete at this phase of the operation and were eager to rejoin the Marines, B-team, and other ODAs for assembly at GREEN Beach, but they waited a few hours until the road cleared to allow for passage on the New Port bypass road. The team was a bit nervous while they waited, since one of the requirements in the New Port area was clearing all weapons (a condition imposed on the Pakistanis by the contractors in order to board commercial shipping). Fortunately, even though the collection of forces was postured as a bunched-up lucrative target to the Somalis, no incidents occurred at this site. It would take two days to complete the maritime evacuation for the Pakistanis due to problems with the ships under contract.

The passage of lines was completed. The San Marco Marines extracted that night. Once all UNOSOM II units were reported as successfully evacuated, the Marine over-watch force, CTF staff, and the SF B-team withdrew to assemble at GREEN Beach. The SF CSTs and the AOB had previously extricated excess personnel, equipment (tents, radios, generators), and non-essential vehicles back to the MEU-ARG to store on the USS *Belleau Wood*.

Brigadier General Csrnko reflected on his unit's performance during these final days of the operation.

> General Zinni was very knowledgeable about Special Forces capabilities and welcomed our participation in the operation. As the B-Team conducted the operation, the SOCOORD monitored the evacuation from their station afloat to ensure interface of requirements were coordinated between the AOB, the JSOTF, and the CTF.
>
> We had not foreseen in the CTF the various flow of aircraft and shipping occurring by others during the evacuation: what type of ships were actually going to be used to conduct the evacuation? The port area was shallow water and became an issue. Also, a variety of international media appeared who had not been scheduled in the press pool. These problems were all worked out.
>
> General Zinni was cognizant of the fact to limit more personnel going ashore; he trusted his folks just as I trusted the B-Team to get the job done. I did go ashore with the CTF C-3 on only one occasion (25 February), but the AOB had things well in hand.
>
> The operation was extremely successful. General Zinni relayed to me that he found the SF teams highly beneficial in providing situational awareness and current dispositions and capabilities of the UN coalition military contingents ashore. He further thanked us for the critical communications capability we provided during the evacuation.[36]

Between March 2nd and the early morning hours of March 3rd, the remaining forces ashore (Marines, the AOB, and the CTF staff with Lieutenant General Zinni) collapsed their positions and assembled on GREEN Beach for further extraction back to their assigned ships.

At this time, perplexingly, the Somali clans and gunmen decided to conduct a last sendoff in a show of machismo. The elements at GREEN Beach began to receive gunfire, both from AKs and machine guns, as well as directed RPG fires. The CTF gave the Somalis a last wave-off in return. Marines returned Somali gunmen harassing fires. Most of the threat was neutralized and brushed aside with overwhelming firepower, including 25-millimeter from their LAVs, fire from snipers (USMC and SF), and threats from the air quick reaction forces—helicopter gunships, AV-8 Harriers, and AC-130Hs. At least five Somalis were killed in this exchange of gunfire; any Technicals in the area departed. There were no friendly casualties.[37]

The successful operation lasted a little over 72 hours. All remaining American units completed the backhaul to ships via LCACs and LCUs. Lieutenant General Zinni departed with the final amphibious movement on March 3rd. As far as America was concerned, this was goodbye forever to Somalia.[38]

Along with UNOSOM II military forces, about 150 civilians, mostly from UN staff and various NGOs, and over 100 combat vehicles were evacuated. All was complete by March 3rd. LTG Zinni was the last Marine off the beach. The 13th MEU (SOC) commander, Colonel Garrett remarked, "I think it was probably the closest to a perfect operation that I've ever been on." In the report Lieutenant General Zinni sent to CENTCOM, he stated, "Mission Accomplished." There were no casualties to the CTF.[39]

The Task Force steamed back to Mombasa, arriving in about three days. The 1/5th elements remained at Mombasa a little over a week conducting After Action Reviews, cleaning equipment, and palletizing their gear for redeployment. All SF forces arrived back to Fort Campbell by March 10, 1995, via military air transport.

On March 17, 1995, CTF United Shield was disestablished. The United Shield mission was a resounding success for not only CENTCOM and UNOSOM II, but also the professional efforts of the Special Forces CSTs and their assigned STS air controllers, directing AFSOD C-130 gunships. The special operators built the rapport necessary with Pakistani contingents to be instrumental in making United Shield happen.

Although the U.S. Army Special Forces support to the UN operation only achieved tactical objectives, vice operational and strategic objectives, the maturity level of the teams—with their combat and foreign internal defense experience in this region—and ability to converse and live with foreign partners–contributed in making the Green Berets a force multiplier in operating by, with, and through their coalition counterparts. (This organizational structure and capability is not inherent in an Army or USMC infantry squad, platoon, or company.) An added feature was the capabilities inherent within an ODA: robust communications, expert medical knowledge, weapons and mounted capabilities, and forward air control, just to name a few.

The 5th Special Forces Group (Airborne) closed out the 20th century by a return to their routine activities of conducting Middle East military exercises, Army Training and Evaluation Tests, and team training. In the five-year period post-Somalia, the Group continued to modernize and refine their mounted tactics. All of this progress went to good use when they were selected by the President and CENTCOM to perform their most important strategic assignment: the conduct of Unconventional Warfare with the Afghan Northern Alliance to defeat the Taliban immediately after 9/11.

Up to this day, there remains a deep skepticism by the American public on using the U.S. military in UN peacekeeping or peacemaking operations.

Endnotes

1. There would be 2,600 Marines in the Task Force. U.S. Army Special Forces consisted of one SF company as coalition support on the ground at the Mogadishu airfield and a small SF battalion staff contingent serving as a SOCCE afloat to Lieutenant General Zinni (approximately 50 personnel). SOCCENT provided a small JSOTF staff in Mombasa for the command and control of four AC-130 gunships from the 16th SOS. Remaining UN forces numbered around 2,500 (Pakistani and Bangladeshi). Supporting ships were also provided by Britain, France, and Malaysia.
2. UNOSOM II: August 1994–March 1995. www.peacekeeping.un.org, May 25, 2020. "Secretary-General Takes Stock."
3. Fogarassy article.
4. UNOSOM II: August 1994–March 1995. www.peacekeeping.un.org, May 25, 2020. "Inter-Agency Statement."
5. Fogarrasy article.
6. Ibid.
7. CENTCOM United Shield Chronology dtd, May 5,1995, Sub-series: Somalia, Title: United Shield Chronology, Unclassified (copy on file with USASOC Historian, Ft. Bragg, NC).
8. Ibid.
9. Phone discussion between Brigadier General Csrnko and COL (Ret.) Joseph D. Celeski conducted at Ft. Bragg, NC, August 3, 2006.
10. Lorenz, F. M. "Non-Lethal Force: The Slippery Slope to War?" *Parameters*, Autumn 1996, p. 52–53.
11. From The Sea.
12. Lorenz, p. 54.
13. Ibid, p. 55.
14. E-mail between CSM (Ret.) Frank McFadden and COL (Ret.) Joseph D. Celeski, October 13,2006.
15. Wikipedia. Accessed on July 13, 2020.
16. McFadden e-mail.
17. As told to COL (Ret.) Joseph D. Celeski by LTC Steve Cain during interview conducted at Ft. Bragg, NC on September 21, 2004. This is an excerpt from "ARSOF in Somalia, 1992–1995" draft research paper provided to the USASOC History Office in 2007.
18. Kliene, Gregory F. (MAJ USMC). "Operation United Shield: A Case Study. CSC 1999. Found on ISYY web (www.192.168.2.226/isysquery/irllff5/28/doc), March 23, 2006.
19. CENTCOM United Shield Chronology, dtd May 5, 1995, Sub-series: Somalia, Title: United Shield Chronology, Unclassified (copy on file with USASOC Historian, Ft. Bragg, NC).
20. E-mail on B-team operations during United Shield operation, sent from now-retired CSM Frank McFadden on October 17, 2006.
21. From interview conducted over the phone between COL (Ret.) Joseph D. Celeski at home in Buford,

GA, and Dave Asher from his location in western Iraq on Feb 5, 2006, 0930 hrs.

22. Interview with Lance Caffrey on October 25, 2006, 0900 hrs, conducted by COL (Ret.) Joseph D. Celeski at the Special Warfare Center ASOT committee building, Ft. Bragg, NC.
23. Rhodes, Phil (MSgt). "Exodus: Special Tactics Team effort critical during Operation United Shield," *Commando Magazine*, March 24, 1995, p. 4.
24. Interview between CSM Jose Bailey and COL (Ret.) Joseph D. Celeski at Ft. Bragg, NC, February 8, 2006.
25. Interview between CSM Jose Bailey and COL (Ret.) Joseph D. Celeski at Ft. Bragg, NC, February 8, 2006.
26. Interview with LTC Steve Cain by COL (ret.) Joseph D. Celeski conducted at Ft. Bragg, NC, September 21, 2004.
27. Interview between CSM Jose Bailey and COL (Ret.) Joseph D. Celeski at Ft. Bragg, NC, February 8, 2006.
28. E-mail between CSM (Ret.) Frank McFadden and COL (Ret.) Joseph D. Celeski, October 13, 2006.
29. Interview with LTC Steve Cain by COL (ret.) Joseph D. Celeski conducted at Ft. Bragg, NC, September 21, 2004.
30. Ohls, Gary J. "Somalia. . . From the Sea." p. 185.
31. UNOSOM II: August 1994–March 1995. www.peacekeeping.un.org, May 25, 2020. "SNA/SSA Agreements."
32. Ohls, Gary J. "Somalia. . . From the Sea." Newport Papers #34. New Port, RI: U.S. Naval War College, July 2009, Ibid. pp. 186–187.
33. Rhodes, Phil (MSgt). "Exodus: Special tactics team effort critical during Operation United Shield," *Commando Magazine*, March 24, 1995, p. 4.
34. Ibid.
35. Interview between CSM Jose Bailey and COL (Ret.) Joseph D. Celeski at Ft. Bragg, NC, February 8, 2006.
36. Csrnko interview.
37. Kliene, Gregory F. (MAJ USMC). "Operation United Shield: A Case Study." CSC 1999. Found on ISYY web (www.192.168.2.226/isysquery/irllff5/28/doc) March 23, 2006.
38. Ibid, p.188.
39. From Gary J. Ohls's interview with Garrett in preparation of his work, "Somalia. . . From the Sea." *Newport Papers* #34.

Epilogue

> The character of small wars and other acts of savage violence can vary as much from the character of contemporary 'regular' warfare as regular warfare in, say, 1914, varied from that of 1918, 1945, or even 1991. It should flow that if strategy and war do not change their nature among different historical periods, so small wars, other forms of organized (or semi-organized, given the relativity of the quality of the organization) violence, and regular warfare should be capable of analysis within the same conceptual framework . . . Armed forces that decline to take small wars seriously as a military art form with their own tactical, operational, and political–though not strategic–rules invite defeat.
>
> — Colin S. Gray, *Modern Strategy*

Armed violence did not end in Somalia with the departure of the UN in March of 1995, although American policy in the region became more subdued as a result of the experience. In December 1993, General Morgan took advantage of the departing Belgian peacekeepers in Kismayo and finally captured and held the city until 1999. In late July 1996, Aideed and his militia fought against Ali Mahdi and Osman Atto in Mogadishu. Aideed was wounded during the gunfire. On August 1, 1996, Mohammed Farah Aideed suffered a heart attack as a result of his surgery or injuries and died at 61. He was succeeded by his son, who was, ironically, an American Somali who had served in the U.S. Marine Corps.

General Warsame was arrested on April 19, 2011 by U.S. military forces in the Gulf of Aden. He was convicted in 2013 of conspiring and supporting both al-Qaeda in the Arabian Peninsula (AQAP) and al-Shabaab.

After Aideed's death, the power of Somali warlords waned. The country was exhausted by war and, without a functioning government, turned to their religion as a means to restore order. An attempt to bring security and stability to Mogadishu began in 1996 as an initiative by Aideed's Somali National Alliance (SNA), allowing clerics to reinstate Sharia law in the city. The idea spread into southern Somalia. To back up clerical rulings, a system of courts began to enforce Sharia law. Its enforcers came from clan militias with little else to do, and soon they became a quasi-military, religious police.

9/11 and the Rise of al-Shabaab

Events on September 11, 2001 would change the calculus for America's reintroduction into the Horn of Africa once again. No longer on a quest to stabilize the region or conduct peacekeeping missions, U.S. policy would now be to frame involvement in the region as part of the Global War on Terrorism (GWOT). If al-Qaeda decided to operate in the Horn, then the United States would pursue them in not only central Asia and the Persian Gulf but also Somalia, if need be. Finding and attacking al-Qaeda, its associates, and infrastructure would not be tied to nation-building or humanitarian gestures; it was a tightly-focused policy matching diplomatic efforts with military assets. America would indeed return to Somalia, in time.

While the fronts against al-Qaeda (AQ) ramped up in Afghanistan and Iraq, counter-terrorist experts in the American government were concerned the spread of terror cells would occur in the Horn, as militants and terrorists fled from American pressure elsewhere. It was known, or suspected, that the AQ cells responsible for the Kenya Embassy and Tanzania Embassy attacks had fled to Somalia and were receiving help from other radical jihadists (in 1998). In short order, Ahmed Nur Ali Jim'ale, a senior leader in the Islamic Courts Union (ICU), and Sheikh Hassan Dahir Aweys, the founder of al-Itihad al-Islamiya (AIAI) in Somalia, were both put on America's terrorist list. (AIAI was effectively neutered by the Ethiopian Army during their attacks in Somalia in late 1996. The AIAI fled into the mountainous region of the Gedo Province.)

A top priority for the U.S. Department of State became preventing the growth of AQ in Somalia. Due to America's absence from the country for over seven years, little was known on AQ's whereabouts and operations. A first step would be the need to increase situational awareness and increase intelligence on al-Qaeda within Somalia.[1]

To be effective, U.S. military forces had to be in the region. There was still apprehension about putting any boots on the ground in Somalia. The United States would look to fight the new war using proxies, unconventional warfare, covert operations, and indirect fires. To that end, Combined and Joint Task Force–

Horn of Africa (CJTF-HOA) was established in Djibouti at Camp Lemmonierin December of 2002. A small contingent of U.S. Army Special Forces operators were part of its force list.

As the movement to implement Sharia law in Somalia grew, the system of courts to enforce the edicts was formalized with the creation of the Islamic Courts Union (ICU) in 2003. At first, this movement was very popular among the Somali populace as justice and stability spread. However, its fundamentalist approach and possible support of extremists drew the attention of the Central Intelligence Agency. Without the political will to insert U.S. troops on the ground in Somalia, the CIA began a proxy program to hire warlords and militias to root out ICU jihadists and extremists, as part of the CIA's role in the GWOT to prevent the further spread of AQ in the region. Many ICU leaders were purportedly kidnapped or disappeared by Somali agents. In retaliation, the ICU senior leadership directed Hashi Aden Ayro of the al-Itihad al-Islamiya youth wing to conduct a counter program targeting the very same warlords and militias working indirectly for the United States. Ayro's organization was called the Hezbul Shabaab, or the "Youth" in Arabic.

The al-Shabaab organization was formed by returning Somali Afghan War veterans with the goal to create an Islamic state in Somalia as part of the greater Caliphate ideology preached by Osama bin Laden and the al-Qaeda.[2] There were several reasons for the rise of al-Shabaab as a militant Islamist organization. By 1988, many of the warlords were weakened in their influence over the populace, and some of the factions began to fracture. Disparate militias began to commit crimes as an economic means to restore lost pay from the warlords. This dynamic led to a failure on the part of the warlords or any fledgling temporary government arrangement to bring security or justice to the population—which found themselves in the middle lacking protection. The lack of any successful governing or political ideology (clan-based, communism, socialism, and so on) forced Somalis to turn to their religion, though still influenced by the clan system and its pressure.[3]

Another proxy counter-terror program created by the CIA was the Alliance for the Restoration of Peace and Counter-Terrorism (ARPCT), allegedly also responsible for targeting Islamic Courts Union leadership (and now, al-Shabaab). The Somalis who worked in this arena were characterized as a mafia, working for the highest bidder. (There were other financiers of their work, most notably Somali businessmen and elements of the Transitional Federal Government, both with the same goals to rid themselves of the terrorist problem.)

The al-Shabaab struck back in Mogadishu. Without a firm hand on the tiller of the ARPCT operations, many of those paid by the Agency went rogue and, ultimately, the program was deemed a failure. After successfully countering the program's operations in Mogadishu, the Islamists and al-Shabaab took control of the city; as they countered agents of the ARPCT elsewhere, al-Shabaab ended up controlling most of central and southern Somalia by June 2006.[4]

Looking to a Regional Solution

American foreign policy experts searched for a solution to address terrorists in Somalia without the need for U.S. boots on the ground. If America was not going to send troops, then someone else needed to provide them. However, American diplomats needed first to shed their reticence of interference in another state's affairs based on the international distaste of those experiences in the post-colonial period. As the GWOT progressed, the United States still challenged any notion of regional troops intervening in Somali affairs; on the other hand, American diplomats would not support another peacekeeping operation because the U.S. often got saddled with a disproportionate share of the cost. Other experts stuck to the opinion that the arrival of foreign troops in Somalia would only feed Somali resentment and bolster the rise of political Islam.

But East African nations were more than aware they had to do something, or the problem in Somalia would fester and potentially spread to their countries. They found the U.S. too noncommittal. The East African Intergovernmental Authority on Development (IGAD) began a push to establish an occupation force similar to the UN's Title VII operations (peace enforcement), almost like a UNOSOM II-redux. The UN passed a resolution to support the measure, but the U.S. objected and warned the UN it would not help fund any type of operation like this; American diplomats did not support the resolution. As if mimicking American foreign policy, the Somali National Alliance party also objected to foreigners occupying their country.

American counterterrorist experts may have perceived this type of plan by the UN because a regional force would disrupt their counterterrorism efforts against the Islamic Courts Union and al-Shabaab. The African Union (AU) was not to be rebuffed, however. IGAD authorized a peace support mission for Somalia in 2005 (named IGASOM and blessed by the AU) requiring 8,000 troops to conduct "peace enforcement." However, the effort failed based on reliance for frontline states for the troop commitment, which was sorely lacking.[5]

And then U.S. policy for the region changed to one of accepting a regional solution for Somalia. America would put no boots on the ground but would stand by, and even support, African nations that would. Queue up the Ethiopians, who saw it in their national security interest to support the Transitional Federal Government (TFG) residing in Baidoa (Mogadishu was too hazardous a location to govern from).

By 2006, intelligence reports indicated ICU Islamists with ties to al-Qaeda in the Arabian Peninsula (AQAP) were operating in Mogadishu. U.S. Navy SEALs and other forces from CJTF-HOA in Djibouti began combat actions against al-Shabaab. American Special Operations Forces (SOF) did return to Somalia, albeit indirectly. On December 6, 2006, the UN passed Resolution 1725, authorizing frontline states to intervene in Somalia. This was the green light for the Ethiopian Defence Forces to move south and capture Baidoa. Continuing its operation, the Ethiopians attacked and secured Mogadishu and expelled the ICU from the city, along with ICU elements in nearby Afgoye. The attack continued as the Ethiopians drove ICU forces as far south as Kismayo. At long last, the Transitional Federal Government (TFG) moved back into government buildings in Mogadishu.[6]

It would not be long before Somali insurgents lashed back at the Ethiopian occupation and the TFG. Once again, Somalis found something to unite them: their hatred for Ethiopia, their historical enemy. Feeling political pressure in Ethiopia and a rising casualty rate, Prime Minister Meles Zenawi was eager to withdraw back to Ethiopia and call the mission complete. However, many in the region felt departing too soon threatened to leave a vacuum not yet filled by the frontline states' troops. American diplomats went on a scramble to visit several African countries to get a pledge of support and an early deployment into Somalia, backing the deal with $40 million and promises of training and equipment for the new peacekeepers. Unsurprisingly, most African nations wanted nothing to do with Somali internal affairs. Some, like Ethiopia, were willing to train troops and policemen in their own countries as a show of support and an alternative to the African initiative.[7]

On January 19, 2007, the African Union adopted the mission for Somalia. Called AMISOM (African Union Mission in Somalia), the operation very much resembled the old UNOSOM mission—it was to be a humanitarian, peacekeeping, and development mission. AMISOM would be organized with nine battalions of 850 men each from the participating frontline states (around 8,000 peacekeepers).

One willing participant was the country of Uganda. With some funds and logistics from the U.S., they were a professional force ready to go. Their only stipulation was asking the U.S. to handle the maritime reconnaissance interdiction portion of the plan to prevent arms and drug smuggling into Somalia (generating funds needed by al-Shabaab).

True to their promise, the first Ugandan troops landed in Mogadishu on March 6, 2007, led by Major General Levi Karahanga (who would also become the first commander of AMISOM). They received the Mogadishu welcome from insurgents; the receiving ceremony was shelled by mortar fire. By their third day in country, they experienced their first improvised explosive detonation (IED) attack on one of their vehicles near the K4 traffic circle. Ugandan troops clashed with Somali insurgents at the traffic circle, suffering two wounded. Welcome to peacekeeping in Mogadishu. These early fits and starts were noticed by other African states pondering whether or not to support the AMISOM mission, making it a difficult task for the AU to recruit more battalions. [8]

While getting off to a rocky start amidst insurgents fighting the TFG and the Ethiopians, the Ugandans kept their professionalism and continued to practice "hearts and minds" among the populace.

The African Union Mission to Somalia Expands

Al-Shabaab joined the Osama bin Laden "Global Jihad" narrative and partnered with Hezb Islam to challenge both the TFG and AMISOM. The two terrorist organizations began to train, support, and house other al-Qaeda affiliates. Most dangerously, the tactic of suicide operations preferred by AQ began to appear in Somalia. Ugandan and Burundian forces became more aggressive to confront the threat.

On September 14, 2009, the U.S. began Operation Celestial Balance, ground attacks on AQ conducted using U.S. Navy SEALs and the 160th Special Operations Aviation Regiment. The raids marked the first use of American "boots on the ground" since 1995. In what appeared to be retaliation, al-Shabab conducted a suicide bomb attack on the AMISOM base in Mogadishu. Once again, the troops in AMISOM vowed to step up their activities against both the Hezbul Islam and the al-Shabaab. Fortunately a schism developed between the two enemy organizations, knocking Hezbul Islam to the sidelines. By 2010, al-Shabaab was the clear enemy confronting the forces of AMISOM and the Transitional Federal Government.[9]

AMISOM leadership wanted to do more and become an offensive-minded force, yet felt partially restricted by the rules of engagement. After al-Shabaab killed and captured tourists on small resort islands off the coast of Kenya, the government of Kenya took off the gloves and shed their reticence to participate in Somalia. They joined the fight, launching two battalions across the border to confront al-Shabaab forces near Kismayo and the Juba River (October 2011).

The small country of Djibouti then joined AMISOM. Troop strength rose to over 17,000 troops. AMISOM divided central and south Somalia into four sectors and assigned African countries' peacekeepers accordingly (for instance, Djibouti took the area around Belet Weyne, closest to its distance from Somalia). Belet Weyne was recaptured in December of 2011. With more troops and a loosening of UN restrictions on the peacekeepers, Mogadishu was totally in the hands of AMISOM and the Somali National Army (SNA) of the Transitional Federal Government in early 2012.[10]

A new form of "peace warriors" began to evolve to replace blue-helmeted peacekeepers. This development may have resulted from the realization that warriors in irregular warfare environments eschew civilized notions of peace, rules of war, and reconciliation. No better example existed than the decisions made by AMISOM to confront the al-Shabaab threat in light of this reality. Brigadier General Olum, the commander of AMISOM in 2014, changed his tactics to confront the al-Shabaab when traditional "peace-making" techniques continued to result in non-compliance by the militias and increasing casualties to friendly forces. Opiyo Oloya wrote in his book *Black Hawks Rising* of the General's thinking amid AMISOM operations against al-Shabaab during the operation conducted southwards along the Somalia coast, heading towards Kismayo.

> It was a different fight, in a different forum, one that AMISOM needed to get used to. Having demonstrated in the battlefield what peacekeeping looked like in the era of well-armed, well-resourced, well-educated and media-savvy insurgents, and how to confront, engage and contain them, AMISOM commanders like Brigadier Olum needed to learn to speak to the larger world audience about what really goes on in the battlefield, and why the old notion of peacekeeping where blue-hatted soldiers rattled around in white-painted vehicles no longer worked. The world needed to be educated that the new African peace warriors as abundantly demonstrated by AMISOM forces, unlike undisciplined militias and insurgents with no respect for authority and civilians' lives, are professionals whose approach to peacemaking is based solidly on deep regards for all humans to live in peace.

AMISOM had redefined peacekeeping.[11]

American Reentry into Somalia

The U.S. began drone strikes (MQ-9 Reaper) and airstrikes in 2007 targeting al-Qaeda and al-Shabaab as part of Operation Enduring Freedom – Horn of Africa. USAF AC-130 gunships struck terrorists locations at Ras Kamboni on the 7th and the 23rd of January. On May 1st, a cruise missile strike killed Aden Hashi Ayro in the Galguudud region. Another cruise missile strike was launched from the 5th Fleet against AQ targets in Puntland.

As mentioned earlier, Operation Celestial Balance targeted al-Qaeda operatives in 2009. The strikes continued to grow each year as part of the counterterrorism campaign in the Horn of Africa. Another example was a drone attack on al-Shabaab leadership suspected of working with Al-Qaeda on June 23, 2011.

In 2012, a forward support base for unmanned drone operations was established at the former Somali Air Force base at Bale Dogle. In 2013, U.S. Army Special Forces began the training of an elite Somali commando brigade, called the Danab ("Lightning") Brigade.

American SF trainers used the budget authority under the 127e program, allowing special operations forces to use local surrogates to conduct counterterrorism operations. The U.S. Army Special Forces ran a six-month course at Bale Dogle airfield, training 150 men in the first iteration. Somali commandos were recruited by merit, with diversity of clans among the ranks, and became some of the best-equipped and well-trained soldiers in the Somali Armed Forces. The mission goal for the Green Berets was to train at least 1,000 men for the Brigade. Ultimately, the plan was to train and man the Danab to a little over 3,000 commandos.

The military specialties of the men trained for the Danab focused on counterterrorist night missions and VIP protection. They also became capable of conducting counter-guerrilla warfare. The Danab is a light infantry unit. The unit's first mission was in helping the Somali National Army clear al-Shabaab from Mogadishu. U.S. Army SF operators accompanied the Danab during their missions as part of their "advise and assist" mission parameter but did not conduct the raids themselves. They arranged close air support or drone strikes to support the Danab. No U.S. Army SF operators were killed on these missions.

American troops returned to Somalia in 2014 in the form of a military coordination cell in Mogadishu to work with the Somali government's armed forces and AMISOM. The cell numbered in the low twenties. (The U.S. officially recognized the new Somali government in 2013, although no Embassy has been reestablished.)

That same year, the first American special operations death occurred in Somalia since the deployment of Task Force Ranger in 1993, when a U.S. Navy SEAL was killed during a raid on an al-Shabaab camp near Barii on May 5, 2014. An estimated six to eight al-Shabaab fighters were killed during the raid.

Feeling the effects of drone strikes and American advisors with the Danab Brigade, al-Shabaab attacked Bale Dogle with a car bomb to breach the gate, then sent its fighters in September 2014. The defenders fought off the attack without friendly casualties (there was one case of a concussion injury). The Americans conducted defensive airstrikes during the attack and killed an estimated 10 enemy fighters. A simultaneous attack was launched by al-Shabaab against an Italian convoy in Mogadishu.

Key leaders of both organizations continued to be targeted by American strikes, including the strike that took out Abdi Gadane near Kurtenwaary as he fled in his SUV during a Ugandan attack. American intelligence sources pinpointed his communications and launched a Hellfire missile on September 1, 2014.

One of the largest mixed drone and aircraft strikes occurred on March 5, 2016, near the town of Raso. Munitions hit the training facilities of al-Shabaab, destroying the camp and killing an estimated 150 fighters.

The U.S. Africa Command (AFRICOM) increased its role in Somalia in 2017 with Security Assistance (Foreign Internal Defense), sending additional forces into Somalia from the 101st Division to assist with the "train and equip" mission for the Somali Armed Forces.

Between 2018 and 2020, combined totals of unmanned drones and airstrikes were 47 in 2018, 63 in 2019 (killing an estimated 900–1,100 fighters), and 51 in 2020. After a pause in strikes by the U.S. in early 2021, the first bombings of the new year occurred in March.[12]

In January of 2021, American forces departed Somalia under President Trump's orders (an estimated 600 U.S. troops) and turned the Bale Dogle base over to the Somali government. After a national security review by the Biden administration, President Biden authorized the introduction of 500 U.S. troops for Somalia in May of 2022 (never say never again).

U.S. Army Special Forces conducted a range of special operations missions throughout their time in Somalia and contributed to the success of each Task Force they worked with. They uniquely supported UNOSOM II's objective to expand UN operations in north-central and northern Somalia when no other force was available with a capability for long-range ground reconnaissance. The U.S. Army Special Forces experiences in Somalia represent one of the better case studies in irregular warfare from the perspective of conflict in complex environments. No matter the task, the Green Berets displayed professionalism in all their encounters with partner forces and with the Somali population. Their experiences in the Somali irregular warfare environment helped them prepare for Operation Enduring Freedom and serves as a model for future clashes with irregulars in nontraditional warfare environments. America will surely experience another irregular war as it becomes more common as a means of conflict in the security environment of the 21st Century. U.S. Army Special Forces, the Green Berets, provide a viable capability as one of America's measured responses to irregular warfare adversaries.

Endnotes

1. Oloya, Opiyo. *Black Hawks Rising: The Story of AMISOM's Successful War Against Somali Insurgents, 2007–2014*. UK: Helion and Company, 2016, pp. 35–36.
2. Hansen, Stig Jarle. *Al-Shabab in Somalia*. New York: Oxford University Press, 2013, pp. 20–22.
3. Ibid, pp. 3–4.
4. Oloya, pp. 37–39.
5. Ibid, p. 47.
6. Ibid, p. 61.
7. Ibid, p. 63.
8. Ibid, pp. 68–77.
9. Ibid, p. 114.
10. Ibid, p. 184.
11. Oloya, p. 203.
12. AFRICOM.

Somalia Background, History, and Demographics

Left: Somalia is an Islamic country with an overwhelming religious majority of moderate, Sunni Muslims. There are only a few small pockets of non-Islamic population (like the Bantus in the Jubba Valley) as well as extremists like the al-Itihad al-Islamiya (AIAI) fundamentalists. Shown here is a large mosque in Mogadishu. (Courtesy of Ken Bowra.)

Right: Somalis derive their origin from the Samaal, their forbearers. The population is largely pastoral and nomadic, divided into clans and sub-clans. It is a patriarchal society. Outside of major towns and cities with their Italian-style architecture and villas, the average Somali lives in villages constructed from local materials. (Courtesy of Ken Bowra.)

Left: When peacekeepers arrived, there was a large refugee population due to the civil war and drought. Many Somali people subsisted from food provided by NGOs and lived in handmade shelters (often covered in blue tarp). (Courtesy of Tom Daze.)

The pre-war economy of Somalia was diverse, primarily consisting of agriculture centered in the Jubba River valley, and income derived from the raising of cattle (of such importance to their culture, a picture adorned the national twenty-Shilling bill) and camels. (Shilling courtesy of Chaplain Wylie.) There was also a mercantile economy in large towns and cities, particularly along the coast. The fishing industry was very small. The importation of Khat, an amphetamine-like leaf when chewed, was a lucrative business, more so during the war (and surprisingly, not illegal). As the crisis continued, a new economy emerged from stolen, pilfered, and looted humanitarian aid supplies.

Left: Somalis bought their daily food and household goods from a central market. Shown here is the Bakara market in Mogadishu. (Courtesy of Moe Elmore.)

Right: Food stalls at the street market in Mogadishu. (Courtesy of Paul Holthaus.)

Above: Somalia was a constitutional-based government prior to the civil war which included a prime minister, a president, and a Council of Ministers. Clan and tribal elders governed at the local level using the *Xeer* system, customary law. Shown here is the destroyed Parliament building in Mogadishu in 1993. (Courtesy of Tom Daze.)

Bottom Left: Typical countryside in Somalia – low scrub, trees where water exists, and turning into desert-like conditions as one travels north. There is a circular barrier of thorns and scrub brush called a *zariba*, used by herdsmen to protect their camels, goats, or cattle. (Courtesy of Tom Daze.)

Bottom Right: An example of the mountainous region in Somalia, in Bakool Province, near Oddur. (Courtesy of Jon Concheff.)

Top Left: The Jubba River in southern Somalia; one of two major rivers, the other the Shebelle. Note the agricultural areas along the river. (Courtesy of Robert Biller.)

Top Right: Typical dirt road found outside major towns and cities. Shown here is an SF mounted team conducting food convoy escort during Operation Restore Hope. (Courtesy of Gary Ramsey.)

Middle Left: Old Port area of Mogadishu Aideed's personal villa and the CONOCO compound were located in the area to the lower right. (DOD.)

Bottom Left: The Indian Ocean coastline south of Mogadishu. These waters were shark-infested. (Courtesy of Paul Holthaus.)

Above: The contested GREEN line avenue, separating Ali Mahdi's Abgal clan (to the east) from Aideed's Habr Gedir clan (to the south and west) in Mogadishu. (Paul Holthaus.)

Right: The K4 traffic intersection. Five major streets in Mogadishu converged here, located near the Mogadishu airport. This was the scene of both clan fighting and engagements by UNITAF, 10th Mountain Division, Rangers, and Navy SEALs. (DOD.)

Bottom Right: New Port. The major port used in Mogadishu to deliver seaborne humanitarian supplies and for the delivery of UNOSOM and U.S. forces heavy equipment. (Courtesy of Ken Bowra.)

Top Left: A cityscape of Mogadishu with the Radio Aideed building and antennas in the foreground. (Courtesy of Tom Daze.)

Top Right: The University Compound in Mogadishu (U.S. Embassy compound can be seen in upper left in photo.) The 10th Mountain Division (Light) used this area for their base. (Courtesy of Tom Daze.)

Above: The most notable landmark in Mogadishu was the U.S. Embassy and its compound along Afgoye Road. Evacuated in January of 1991, it was re-occupied by UNOSOM I and II and U.S. forces for operations in Somalia. (DOD.)

Bottom Left: The Mogadishu airport – U.S. forces Port of Embarkation/Debarkation (APOE looking south).

Clan Militia and Primitive Warfare

Left: Ali Mahdi and Mohammed Farah Aideed meet at one of the reconciliation conferences. Both vied as contenders for the office of Somali President. Ali Mahdi's Abgal sub-clan fought Aideed's Habr Gedir sub-clan for control of Mogadishu. Both were from the United Somali Congress's armed factions. (Courtesy of Moe Elmore.)

Below: Colonel Omar Jess (on right in pink shirt) led the Ogadeni-based Somalia Patriotic Movement (SPM) against former regime loyalists led by General Morgan near Kismayo (the SNF). (Courtesy of Tom Daze.)

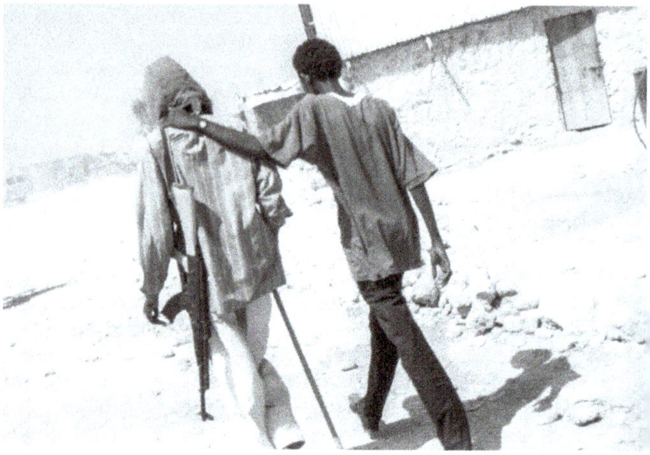

Top Left: Clan militias had opportunists and brigands throughout their ranks. Shown here, gunmen position themselves at a Somali airfield to extort humanitarian food relief agencies to pay exorbitant fees or pay to protect the NGOs and their warehouses and distribution centers by food and high wages exchange. Most NGOs paid; there was no alternative way to get the food into the hands of the starving. (Gary Ramsey.)

Top Right: Clan militia of the SSDF near Bossaso. (Courtesy of Moe Elmore.)

Right: A well-kitted out clan militia in the vicinity of the Baidoa/Bardera humanitarian relief sectors. Note attempt to wear camouflage uniforms. They utilize a Toyota truck as a "Technical," mounted with a DsHK heavy machine gun. Basic small arms were the AK-47 and its variants. (USMC.)

Left: The gunmen of the militias used a variety of Soviet-patterned weaponry and American, Italian, and British arms confiscated from regime forces during the civil war. Although the UN instituted an arms embargo for Somalia, it did not stop weapons sold on the market nor illicit gun smuggling over the borders. Shown here is a display at the Special Operations and Airborne Museum in downtown Fayetteville, NC, depicting an arms merchant and his wares in Mogadishu. (Author's photo.)

Top left: Confiscated weapons by SF teams near Belet Weyne. These are a variety of ex-government and Soviet-pattern weapons. (Courtesy of Ken Bowra.)

Top right: The massive amount of arms, mines, and demolitions found by the SF teams were collected and destroyed with explosives. (Gary Ramsey.)

Above: All militias utilized some form of Technicals to provide heavy weapons support to their militias, such as this SSDF one shown in Bossaso. (Courtesy of Tom Daze.)

Right: Technicals and other heavy weaponry were ordered to be placed into cantonments and kept under UN observation. Shown here is a variety of Technicals and heavy truck gun platforms consolidated near Baidoa. (Courtesy of Jon Concheff.)

Top Left: A Habr Gedir SNA Technical in action photographed by an SF team. It is firing on a rival clan faction near the Mogadishu airport during Operation United Shield in 1995. (Courtesy of Bryan Whitmann.)

Top Right: The remnants of Siad Barre's military littered the country. Shown here is a tank park in Mogadishu with both a Russian-supplied T-54 tank and an American-made M-47 tank. Both countries supported Siad Barre during the Cold War. (Courtesy of Al Glover.)

Right: SF teams found operable tanks used by the militia in the Belet Weyne Humanitarian Relief sector along with mortars, artillery, and recoilless rifles. (With permission of Charles B. Smith, SF ODA Commander – copyrighted.)

Left: The Somali Air Force at Mogadishu airfield. Russia supplied the Somali Air Force with various types of MIGs and jet trainers, all junk when peacekeeepers and U.S. forces arrived, due to the civil war. (Author's photo.)

Operation Provide Relief

Left: When USAF C-130s for Operation Provide Relief required security to operate on remote Somali airstrips, the 5th SFG(A) at Ft. Campbell, KY deployed their 2nd Battalion, with LTC Faistenhammer commanding. He chose MAJ Lelon Carroll's B-team to provide four SF mounted teams for the mission. Shown on the left is the 5th SFG(A) Headquarters circa early 1990's. (Author's photo.)

Right: Other nations provided cargo aircraft, shown in the far rear of this photo. (Courtesy of Mark Hamilton)

Left: A variety of USAF Reserve C-130 cargo aircraft on the ramp near the USLO warehouse on Mombasa airfield. Note temporary Red Crosses affixed to some of the tails to signify to those on the ground these are humanitarian relief cargo aircraft. (USAF – Staff Sgt Dean Wagner.)

Left: The USLO warehouse area on the Mombasa Airport served as the headquarters for JTF Provide Relief. The compound consisted of two long, parallel buildings. (Courtesy of Mark Hamilton.)

Below: A team photo of ODA542. SFC Wendell Greene is in back row, standing at far left in the picture. (Courtesy of Glenn Wharton.)

Right: Preparing for a day mission to deliver a humanitarian relief supplies of food and other staples to one of the four designated airfields in Somalia. Above, USAF cargomaster ensures USA-donated grain bags are secure before takeoff. (Courtesy of Mark Hamilton.)

Above and Right: While serving as the ASAT, detachment operators were rarely used for any emergency on the ground. It was an opportunity to catch up on reading, training, and sleep. HMMWVs were loaded and ready to roll off the aircraft once delivered to an airfield, if an emergency broke out. (Courtesy of Ken Barriger and Gary Ramsey.)

Left: Armed Somalis were constantly present; many were hired as security guards by the NGOs to protect the food. (Ken Barriger and Gary Ramsey.)

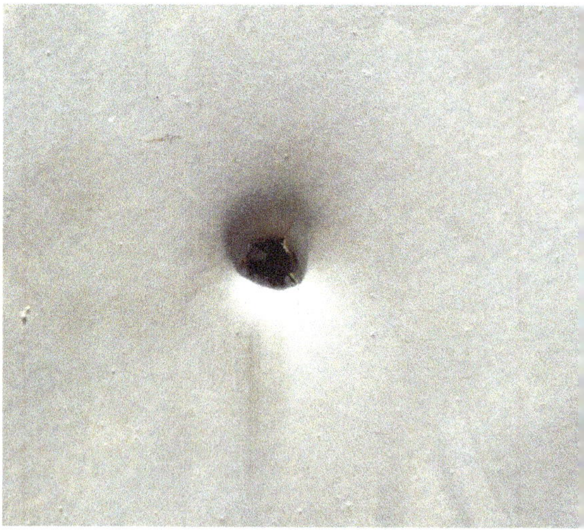

Above: An actual bullet hole in skin of a C-130, shot at while conducting a mission. There were a variety of threats which could endanger the food flights during Operation Provide Relief. The threat of anti-aircraft fire posed the most serious. (USAF.)

Below: Seen below are two "Technicals," armed trucks, used by the NGOs and paid for using "technical" funds, thus the name. SF detachment members look on. (Glenn Wharton.)

Right: MAJ Kent Listoe commanded Company C, 1/5th SFG(A) on the second rotation to Mombasa for Operation Provide Relief. He began with four SF teams, but as Operation Restore Hope began to form, two additional teams from the 2/5th SFG(A) battalion beefed up his numbers. (Ken Barriger and Gary Ramsey.)

Left: CWO3 Bruce Watts (ODA546) with C-130 pilot and USAF assessment team. In the early portion of Provide Relief, detachment members and CCTs spent the few moments on the ground to gather area assessments, take pictures, and survey the airfield. Later, as Operation Restore Hope deployment began, it was possible to conduct these missions in full military posture on the ground. (Glenn Wharton.)

Right: ODA and CCTs inspect runway surface.

Above: SFC Ken Barriger of ODA565 inspects destroyed Somali aircraft and bunkers to ensure munitions are identified for disposal and patrols the airfield to assess its condition. (Ken Barriger and Gary Ramsey.)

Below: SF Sergeant mans a guard position at Somali airfield (Bale Dogle). It would be important to own and control key airfields and facilities to support the upcoming Operation Restore Hope. The SF remained in these positions until met by arriving Marine or Army forces. (Ken Barriger and Gary Ramsey.)

Operation Restore Hope

Above: The Unified Task Force (UNITAF) for Operation Restore Hope, conducted December 1992 to March 1993, was led by Marine Lieutenant General Robert Johnston, shown above at the Mogadishu airport with Ambassador Oakley (center), the UN Special Representative. (DOD.)

Right: COL Thomas Smith (center), SOCCENT Operations Officer, commanded the JSOFOR for UNITAF. Shown here with MAJ Holthaus, the J2, on right in picture and unknown senior NCO on left. To their rear is a coalition helicopter. (Courtesy of Paul Holthaus.)

Left: LTC William Faistenhammer, Jr., the commander of 2/5th SFG(A), receives an update briefing from his staff. In January, he replaced COL Smith to command the downsized ARFOR, with his headquarters in Mogadishu. (Photo courtesy of Lee Carroll.)

Left: The first mission for Army SF on Operation Restore Hope was to occupy and assess airfields needed by UNITAF. Here SFC Barriger, ODA565 Detachment Commander, watches over his team and the CCTs and USAF REDHORSE Engineers as they conduct their assessment and clear Baledogle airfield of any danger. (Photo courtesy of Ken Barriger.)

Right: Mounted team of ODA565 clears the airfield of suspected threats from Somalis (weapons) and identifies unexploded ordnance prior to any landing of UNITAF aircraft. (Photo courtesy of Ken Barriger.)

Left: A remarkable still captured from some of the few videos recorded during Operation Restore Hope by Green Berets. This shows ODA562, commanded by CPT C.B. Stevens, conducting a reconnaissance of the streets of Mogadishu prior to any other coalition elements (24 December). (Video still courtesy of Charles B. Stevens.)

The scenes of Mogadishu when the first United States and coalition elements arrived reminded many of those pictures seen during the Battle of Stalingrad in World War II. Above: Appears to be National Road. Below: USMC and Italian Marines conduct a security patrol in the Green Line area Mogadishu using their armored LAVs and an Italian Fiat OTO Melara 6614, noting very few buildings intact or without some level of destruction. Phase I of the Restore Hope operation was UNITAF securing Mogadishu and its port and airfield. (Photos courtesy of DOD.)

Coalition Forces of UNITAF

Above Left and Above Right: Throughout December, coalition peacekeepers deployed to Mogadishu to reinforce the UNOSOM Pakistani peacekeepers and begin Phase I of the operation. (All photos courtesy of Ken Bowra, Moe Elmore, and Al Glover.)

Left: UN helicopters of UNOSOM operate out of Mogadishu airfield. (Al Glover.)

Below Left and Below Right: ODA562, commanded by CPT C.B. Stevens, was assigned as a CST to the Pakistani contingent. He established his team position in a small, fenced compound on the bluffs overlooking the airport. (Courtesy of Al Glover.)

The Coalition Support Teams

Right: ODA526 partnered with the 5th Royal Airborne Battalion from Saudi Arabia, shown here arriving at the port of Mogadishu. CPT Williams, the detachment commander, had made friends with Prince Fahd, one of their officers, during Desert Storm. (NARA.)

Left: A composite team from the 10th SFG(A), led by CW2 Bell, liaised with the Belgians at Kismayo. The team members were chosen for their French language ability. On right, Belgian military police run a vehicle checkpoint to search for weapons and contraband. (NARA.)

Right: ODB560 with four SF teams supported the Canadians at Belet Weyne. Pictured are both forces conducting a meeting with elders at a village. (Courtesy of Ken Bowra.)

Above Left: CPT Salimon leads the first 300-man French contingent of 3/2 R.E.P. to Baidoa; the French then moved to HRS Oddur.

Above Right: The 10th SFG(A) French-speaking composite team led by CW3 Jon Concheff. Standing left to right: SSG Pardoe, SSG Toth, SSG Tarhee, SFC Quentin, and CW3 Concheff; kneeling, SSG Sanchez. Two USMC communicators were attached to the team. (Courtesy of Jon Concheff.)

Left: Police fort used by the French at El Berde. (Photo courtesy of Jon Concheff.)

Right: SSG Toth sits with French medical personnel on 1 Jan 1993 at El Berde. (Photo courtesy of Jon Concheff.)

PSYOP & Civil Affairs

Right and Below: The 4th PSYOP Group deployed the 8th and 9th Battalions, led by LTC Charles P. Borchini, to support UNITAF in all phases of Operation Restore Hope. They were put under the operational control of the UNITAF J3. Other ARSOF personnel deployed to Somalia included soldiers from the 112th Signal Battalion and the 528th Support Battalion, also located at Ft. Bragg, NC. (Courtesy of Al Glover.)

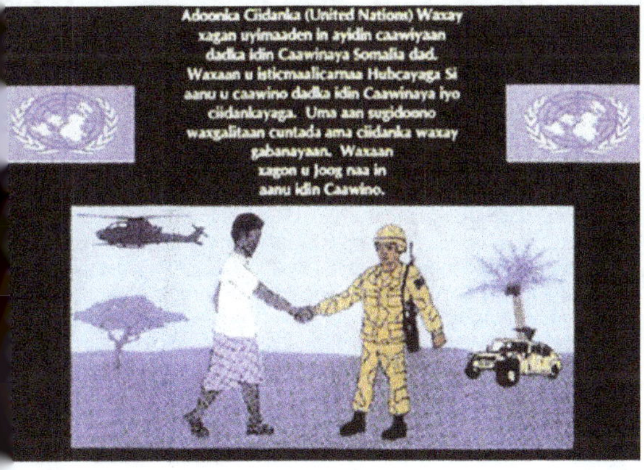

Right: A Civil Affairs team the team conducts civil military-operations southwest of Mogadishu in the Marka region. (Courtesy of Bob Biller.)

Left: One of the objectives after securing Mogadishu was to begin occupying the first four of the nine Humanitarian Relief Sectors (HRS): Bale Dogle, Belet Weyne, Kismayo, and Baidoa. Once secured, the lines of communication between them and Mogadishu were improved to get truck transport through. These efforts were led by the 10th Mountain Division (LI) and Marines from the 15th MEU (SOC). (DOD.)

Right: 2-87th Infantry from the 10th Mountain Division (LI) clears and secures a village in a nighttime operation. (DOD.)

Below Left: Belgian Scimitars (the "Curt") with 30mm RARDEN cannons move along an unimproved road between Mogadishu and Kismayo. (Courtesy of Bill Faistenhammer.)

Below Right: An SF ODA helps escort food supply trucks along one of the lines of communication near Belet Weyne. (Courtesy of Gary Ramsey.)

Restore Hope – HRS Merka, Bardera, Belet Weyne, and Kismayo

IV

Left: ODA526 assisted the 10th Mountain Division in opening HRS Merka, beginning on 1 January 1993. (Shown here outside Brava.) CPT Tim Williams is on the right, standing, in photo. SFC Bailey, the team sergeant, is on right kneeling. (Courtesy of Jose Bailey.)

Right: A-teams 314 and 526 helped to provide security to an aerial-delivered food flight at the airfield outside of Merka on the 5th of January 1993. (Courtesy of Jose Bailey.)

Left: BG Magruder led the 10th Mountain Task Force in Kismayo to reinforce the Belgians and to confront Colonel Jess and General Morgan on ceasing belligerent operations, or suffer the penalty from UNITAF. Shown here meeting with Colonel Omar Jess during January of 1993. (DOD.)

HRS Bardera

Above Left and Above Right: SFC Jose Baily, Team Sergeant for ODA526, in austere living quarters provided by the Marines at Bardera.

Left: Italian colonial-style architecture in this street scene.

Bottom Left: A camouflaged U.S. Marine AH-1 Seacobra patrols in the HRS Bardera sector.

Bottom Right: a typical village near the Jubba River. (Photos courtesy of Jose Baily.)

HRS Belet Weyne – AOB 560

Right: AOB 560 moved with its five SF teams to Belet Weyne to operate during Canada's Operation Deliverance for Phase II and III of Restore Hope. The SF Company established a camp at the Canadian Airborne Regiment's Battle Group headquarters. Important in this sector was surveillance of the Ethiopian border, border meetings with the Ethiopian Army at the post of Fer Fer, and assessments to monitor the activities of the three warring factions in the HRS. (Courtesy of Ken Bowra.)

Left: MOWAG Piranha armored vehicles. A Canadian Cougar six-wheeled (left) and an eight-wheeled Bison of A-Squadron, Royal Canadian Dragoons. (Courtesy of Ken Bowra.)

Right: LTC Faistenhammer, ARFOR commander, looks over a 427th Squadron Tactical Support CH-135 Twin Huey from the Canadian task force. (Courtesy of Ken Bowra.)

SF Detachment Activities – HRS Belet Weyne

Left: CPT Barber's ODA564 poses on a captured heavy weapon Technical on the 27th of January in the village of El Gaul. (Courtesy of Lee Carroll.)

All the SF teams identified abandoned or stockpiled ammunition throughout the area, to include locating and marking minefields.

Above Left, Above Right, and Left: Here, ODA565 secured a large stockpile, prepared it for destruction, and with assistance from EOD teams, destroyed it to prevent its use from warring clans in the Belet Weyne HRS. (Courtesy of Gary Ramsey.)

Right: C Company and ODA mission planning in the camp. (Courtesy of Lee Carroll.)

Left: All ODAs conducted limited humanitarian assistance activities throughout the sector, when possible. Here, they deliver relief supplies from one of the main NGOs. (Courtesy of Gary Ramsey.)

Right: COL Kenneth R. Bowra (located in center of the picture), the 5th Group Commander, visits with the Team leaders and Team sergeants of AOB 560 in early February. (Courtesy of Gary Ramsey.)

Left: This abandoned heavy weapon Technical was found in a cantonment 20 kilometers north of Belet Weyne. It is a ZPU-4, 12.7 mm used by the USC. (Courtesy of C.B. Stevens collection.)

Right: A tank and heavy armor cantonment inspected by USSF and Canadian forces, somewhere near Matabaan in the Belet Weyne HRS. (Courtesy of Canadian Land Forces Command.)

Left: MSG Al Beuscher, Team Sergeant for ODA562, prepares for a patrol out of Camp Holland near Matabaan. (Courtesy of C.B. Stevens.)

Left: A meeting with a district council in a village near Belet Weyne. (Courtesy of Ken Bowra.)

Below: SF mounted ODA conducts a route reconnaissance mission on Main Supply Route 'Orange' out of Belet Weyne headed towards Mogadishu. Teams noted the presence of any minefields, barricades, roadblocks, militia activity, and damaged sections of the road. This information was passed on to higher headquarters. (Courtesy of Gary Ramsey.)

Below Left: Routine HRS Belet Weyne activities. ODA562 prepares for overnight mission to ambush bandits setting up roadblocks.

Below Right: All Detachments conducted some form of civil affairs to deliver humanitarian relief supplies or donated items from the families back at Fort Campbell, KY. (Courtesy of Ken Bowra.)

Left: When going to a meeting at Balenbale on March 3, 1993 to talk to the SNF, ODA562 was ambushed by a mine laid on their route. SFC Deeks, the team medic, was killed by the explosion. Three other operators riding in the vehicle were wounded. (Courtesy of C.B. Smith.)

Below: SSG Jimmie Wilson, ODA562, checks the route for other mines with a mine detector after the explosion. (Courtesy of Gary Ramsey.)

Below: A portion of CPT Stevens's actual map reconnaissance denoting a minefield north of Belet Weyne.

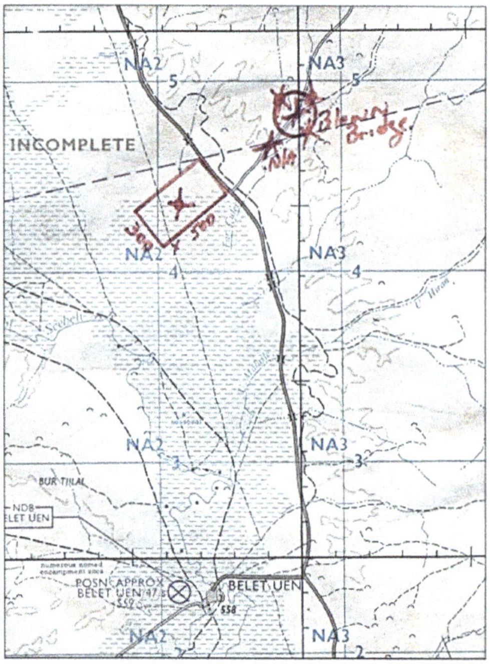

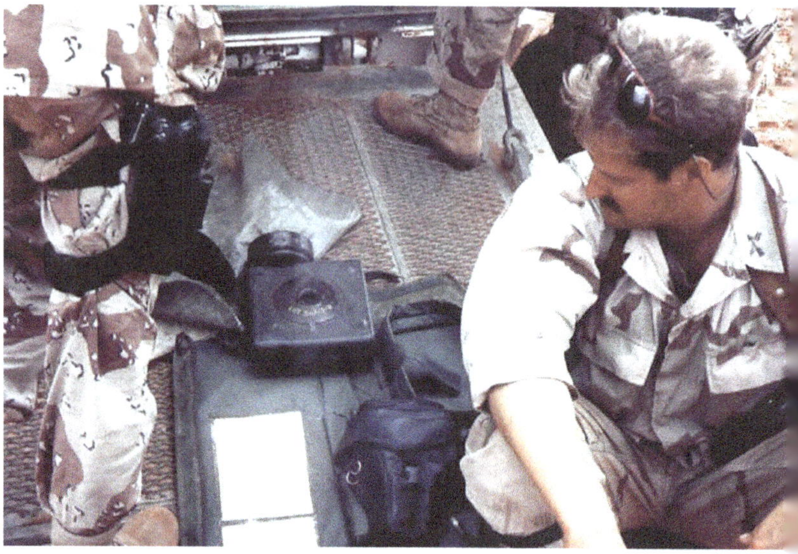

Right: CPT Stevens inspects a TM-46 anti-tank mine recovered on a routine route reconnaissance. (Courtesy of Charles B. Stevens.)

Kismayo HRS – The SOCCE Period (February to early March 1993)

Right: The Belgian 1st Parachute Battalion, led by Colonel Marc Jacqmin, stands parade at Kismayo. (DOD.)

Below: MAJ Lee Carroll led the SOCCE (-) to the 10th Mountain Task Force at Kismayo with a few personnel from his B-team staff. Along with him were ODAs 561 and 564. Shown here enroute to a meeting with General Morgan. (Courtesy of Lee Carroll.)

Below: The SOCCE (-) established its headquarters in a small building on the concrete quay in Kismayo. Shown here is CW3 Rick Detrich, the SOCCE Operations Tech. The SOCCE and the two A-teams erected ARFAB tents for their lodging. (Courtesy of Bill Faistenhammer.)

Above: CPT Barber, Detachment Commander for ODA564, poses on right rear door of his DMV along with his team members during a patrol to recon positions of COL Weyrah and General Morgan. SFC Kilcoyne was his Team Sergeant. (Courtesy of Jose Bailey.)

Below: ODA564 completed a route reconnaissance and area assessment south of Kismayo on a five day patrol. When finished, they moved to the coast and then loaded U.S. Naval transports to sail back to Kismayo. (Courtesy of Bill Faistenhammer.)

V

The Hunt for Aideed

Left: LTG Çevik Bir, a well-experienced Turkish general, was assigned as the UNOSOM II commander in Mogadishu. The 10th Mountain QRF and the Army Special Forces were under his tactical control; CENTCOM retained operational control of these units, through SOCCENT. (DOD.)

Left: MG Thomas Montgomery served as the Deputy UNOSOM II Commander, with a dual-hat as the Commander of U.S. Forces Somalia (USFORSOM). He is shown at right with a Congressional Delegation and soldiers from the 10th Mountain Division (LI) QRF. (U.S. Army photo.)

Right: MAJ Dave Jesmer commanded SOCCE 520. His first major mission was a deployment to Kismayo with his SF teams n support of the QRF. He is shown here in a meeting with clan leaders and Osman Atto. (Courtesy of Dave Jesmer.)

Left: Company B, 1/5th SFG(A) headquarters and team building on the Embassy compound. (Courtesy of Dave Jesmer.)

Right: ODA523 was commanded by CPT Mike Hurst (in glasses). He was recommended for several awards performing aerial sniper duties on the "Eyes Over Mogadishu" operation, logging over eighty hours of flight. The ODA was in heavy combat during Operation Casablanca on the 17th of June, 1993. (Courtesy of Moe Elmore.)

Left: ODA525, a SCUBA-specialty team, was commanded by CWO2 Warrington. Here they depart the Embassy compound on a mission. Note the shark emblem on the vehicle fender denoting them as a dive team. (Courtesy of Dave Jesmer.)

Right: AC-130H Spectre from the 16th SOW refuels at the Mogadishu airport. Four AC-130H gunships supported UNOSOM II strikes on Aideed's and SNA assets between 12 June and 17 June 1993. The gunships used 40mm Bofors cannons and the 105mm howitzer to fire on enemy targets. The aircraft were based out of Djibouti. (Courtesy of Al Glover.)

Left: The Radio Aideed facility after AC-130 gunship strikes flown on 12 June 1993. (Courtesy of Al Glover.)

Right: AC-130H gunship hits its target. (Dave Jesmer.)

Left: 10th Mountain Division QRF UH-60L over the Pakistani Stadium. OH-58s, AH-1 Cobras, and UH-60Ls worked the targets during daytime between the 12th through 17th June engagements. (Courtesy of Al Glover.)

Right: Post-strike photo of one of Aideed's guard houses and adjacent vehicle truck park. (Courtesy of Al Glover.)

Left: MAJ Jesmer, SOCCE 520 Commander, positioned himself atop the K-7 building during the engagements on Aideed's enclave, along with a CCT, the Army Ground Liaison Officer, CWO3 Ward, and SF snipers. (Courtesy of Dave Jesmer.)

Mission Targets 12 – 14 June 1993

Right: The Pasta Factory along October 21 Road. Note the truck park to its rear, where vehicles were converted into Technicals for Aideed's militia. (Courtesy of Al Glover.)

Left: Authorized Weapon Storage Site #3 (AWSS #3). This site was hit both day and night by coalition and U.S. air assets. Later, SF snipers engaged Somali militia atempting to repair the engines of M-47 tanks; the snipers destroyed all the engines with armor-piercing rounds from the Barrett .50 sniper caliber rifle. (Courtesy of Al Glover.)

Right: The Cigarette Factory alongside the 21 October Road. Note 105mm shell holes from the AC-130 gunships. (Courtesy of Al Glover.)

Left: A weapons storage site in an old Somali Army bunker complex. These sites were raided during the day by QRF forces along with SF 18E MOS Engineers. They were always fired upon by Somali gunmen. (Courtesy of Al Glover.)

Right: Nightime fires from Cobra helicopters and AC-130 gunship 40mm cannons. (Courtesy of Dave Jesmer.)

Left: MAJ Jessmer, SOCCE 520 Comander, attends UNOSOM II battle update during the strike period (standing in center of picture). BG Bill Tangney, SOCCENT Commander, stands or far left in photo. (Courtesy of Moe Elmore.)

Above: ODA523 CST, commanded by MSG Bell, fought in this area at the intersection of Phase Lines DOG and YELLOW during Operation Casablanca. It was here their vehicle took a hit in its windshield from a 12.7mm round, as Somali gunmen tried to suppress fires from their MK-19 grenade launcher. Later, the tire on the HMMWV was damaged from either a 106mm recoilless rifle round or an RPG. They were towed back to the left by a nearby 10th Mountain QRF liaison team along Phase Line DOG into the safety of the Moroccan perimeter. (Photo courtesy of Tim Knigge.)

Above: Abdi House where MSG Bell and his team positioned their vehicle during the fight, next to a CONNEX on 21 October Road (would be off top left of picture). (Courtesy of Al Glover.)

Right: With his radio dismantled, Aideed reverted to speaking at crowds of thousands in the old stadium. (Courtesy of Moe Elmore.)

Left: CPT Mike Hurst, ODA523 Detachment Commander, conducts Foreign Internal Defense (FID) training along with his team, to assist the Pakistani Peacekeepers with their newly issued U.S. M113 Armored Personnel Carriers. Here, the Pakis conduct M2 .50 caliber range training. CPT Al Glover, the SOCCENT J2 Targeteer, stands behind CPT Hurst wearing sunglasses. The SF teams also conducted FID to train the Malaysians on their newly-issued sniper weapons. (Photo courtesy of Al Glover.)

Right: On 24 July 1993, members of SOCCE 520 and SF team members from ODA523 were ambushed between the Embassy compound and the Mogadishu airport. In the intense gunfight, the SF Company Sergeant Major and one other NCO from ODA523 were wounded. Shown here is the vehicle they were riding in, with bullet holes in the windshields. Note blood on the passenger seat along with Kevlar liner from DARPA. (Courtesy of Dave Jesmer.)

Above Left: Ambassador Glaspie went to Bossaso to deliver UNOSOM II's ultimatums of intent to expand operations into Somaliland to COL Yusef representing the SSDF. (Photo courtesy of Moe Elmore.)

Above Right: CWO2 Warrington, ODA525 Detachment Commander, and one member of the JSOTF, were on the trip in order to conduct a survey and area assessment while the Ambassador visited. (Photo courtesy of Al Glover.)

Above: The main administrative building at the Bossaso "airport," a dirt strip with potholes. Note greetings from the local SSDF political party. (Courtesy of Al Glover.)

Right: The delegation for Ambassador Glaspie and the SF survey and assessment team were constantly shadowed by an SSDF Technical wherever they went in Bossaso. They were told it was their "courtesy" security escort, even though they had fifteen armed Marines with two HMMWVs and heavy machine guns to protect the Ambassador. (Courtesy of Al Glover.)

Above: SOCCE 520 could not complete their SR mission to Hobyo to look for a weapons transfer at this coastal city. When ODB590 arrived to replace them in early September, MAJ Robinson took the mission. Above, elements of ODB590 were inserted with their vehicles using a UN MI-26 Hook helicopter; no SOF rotary wing assets were yet in country. Unfortunately, using the UN asset meant they infiltrated in the daylight, spoiling their OPSEC. (Photo courtesy of Bill Robinson.)

Right: SF laager near Hobyo. As the mission progressed, the team's location was compromised by local Somalis aligned with the USC/SNA. No matter where they moved, this occurred. When the locals appeared to become hostile and warn them away, the team extracted back to Mogadishu. They did not see any evidence of the suspected weapons transfer. (Courtesy of Bill Robinson.)

VI

The October 3rd Battle of Mogadishu

Left: COL "Kip" Ward served as the Force Command U-3 (UNOSOM II). COL Ward generated the missions for Company C, 3/5th SFG(A) under their TACON relationship. (Courtesy of Tom Daze.)

Right: There was a sharp increase in firefights and indirect fire aimed against American and coalition forces during August and September of 1993. This precipitated the request in August for Task Force Ranger to apprehend Aideed. (Courtesy of Tom Daze.)

Left: The "Hangar" on Mogadishu airfield occupied by Task Force Ranger in late August. Six raids against Aideed's accomplices and his lieutenants were accomplished before the deadly raid near the Olympic Hotel on 3 October 1993. (Courtesy of Moe Elmore.)

Left: AH-6Js and MH-6 "Little Birds" of TF Ranger sit in front of the hangar. (Courtesy of Tom Daze.)

Right: ODA593 conducts area assessments and route reconnaissance in the provinces surrounding Bossaso, northern Somalia, 20 September thru 16 October 1993. They are shown here on a stop along the "Chinese Highway." (Courtesy of ODA593.)

Left: During the Bossaso area assessment by ODA593, the team visited several police stations and conducted medical assistance missions as they toured each province. Additional information gathered included the status of humanitarian relief, schools, medical clinics, and any known activity from the Islamic Fundamentalists – the AIAI. (Courtesy of ODA593.)

Right: On 13 September, the 10th Mountain QRF experienced a day of engagements with Somali gunmen in the Benadir Hospital compound during a weapons confiscation raid to its southeast. It was one of the large urban battles in Mogadishu. SOF snipers provided covering fires during the withdrawal. Picture taken atop K-7 sniper position; U.S. Embassy just to right off picture. (Courtesy of Moe Elmore.)

Left: The M-60 tanks of the 19th Pakistani Lancer battalion located at the soccer stadium. It would be here TF Ranger and the coalition QRF withdrew the morning of the 4th of October after the day and night battle on October 3, 1993. (Courtesy of Moe Elmore.)

Right: Picture taken atop the Green Hotel near the Pakistani stadium showing the visibility and observation ODA592 had in their sniper position. It was from here one of the longest sniper shots in Mogadishu was fired at an armed Technical later in the fall of 1993. (Courtesy of Al Glover.)

Left: In late September, ODB590 supported UNOSOM II's humanitarian relief efforts to expand support into the Medina district, which had suffered when the airport bypass road was built. The unit passed out clothing, school supplies, medical and other items during the latter part of September. (Courtesy of William Robinson.)

Above: The Olympic Hotel (white high-rise on left in photo) along Hawlwadig Road and the area of the target building. (Courtesy of Al Glover.)

Above: The only known combat photo taken during the raid shows Ranger chalks at the stone wall of the target building. The high-rise building is seen center-right in photo, just behind the trees. (U.S. Army photo.)

Below Left: Black Hawk Super 64 crashed in this dense neighborhood, about in the center of the photograph. Rescue efforts were hindered by gunmen, swarms of Somalis, and multiple barricades. Three crewmen perished; Warrant Officer Michael Durant was taken prisoner, and two Delta snipers lost their lives defending the wreck. When rescuers arrived, there was no sign of anyone at the wreckage. Durant was later released after a time spent in captivity. (Courtesy of Tom Daze.)

Below Right: Pieces of wreckage from Super 64 on display at the Airborne and Special Operations Museum in Fayetteville, NC. (Author's photo.)

Left and Below: Artifacts from the October 3, 1993 Battle of Mogadishu on display at th Airborne and Special Operations Museum in Fayetteville, NC. Above is damaged door panel from a HMMWV used by LTC McKnight's ground convoy element to transport apprehended Somalis captured during the raid; below is depiction of a Technical used by Aideed's gunmen during the battle. (Author's photos.)

VII

JTF Somalia

Left: Ambassador Oakley was once again chosen to serve as the Special Representative to Somalia. He is shown here during Operation Restore Hope. Ambassador Oakley had also been the former ambassador to Somalia. (DOD.)

Right: The aftermath of the October 3rd Battle of Mogadishu. The scenes of wrecked helicopters, dead and wounded Rangers, and bodies of U.S. servicemen dragged through the streets led to a public outcry in the United States to depart. Shown here is the rotor hub from Super 64; on the wall in the left rear are artifacts from Super 61, the two downed Black Hawks, on display with other artifacts at the Airborne and Special Operations Museum in Fayetteville, NC. A robust Joint Task Force with reinforcements was deployed to Somalia in late October to protect the force until the 31 March 1994 withdrawal date. (Author's photo.)

Left: As the SOF component, SOCCENT deployed a JSOTF with COL Richard Stimer (USAF) commanding, with four AC-130s (plus their KC-135 refuelers) and additional Special Forces "A" teams. Shown here is the JSOTF-Somalia headquarters building on the left in photo, and the construction of personnel tents. Laborers are erecting sandbag protective barriers. The 10th Mountain Division (LI) headquarters is in the far background. (Courtesy of Paul Holthaus.)

Right: Four Spectre AC-130s from the 16th SOW were stationed at Mombasa, Kenya. The AC-130s provided armed reconnaissance nightly sorties over Mogadishu. (Courtesy of Al Glover.)

Left: The 10th Mountain Division (Light) headquarters building in November 1993. This was previously the old commissary on the Embassy compound. COL Lawrence E. Casper led the Falcon Brigade in Somalia from September of 1993 to March 1994. (Courtesy of Paul Holthaus.)

Right: U.S. Army reinforcements after October 3rd included armor, artillery, and mechanized infantry assets (24th Mech Division). An M2 Bradley stands guard near the airport. (Courtesy of Al Glover.)

Below Left and Below Right: From October 1993 to March 1994, NAVCENT deployed carrier Task Groups and Amphibious Readiness Groups, with on-board SEALs and MEU (SOC)s. The CVBGs USS *Abraham Lincoln* and USS *America* provided the JTF with air support. Marines from the 11, 13th, and 24th MEUs (SOC) and SEALS from ST-5 and ST-8 served as snipers ashore, attached to the Special Forces company. (Courtesy of Bill Robinson.)

Right: UNOSOM II forces maintained the responsibility for the security of the Mogadishu airport, the New Port Area, and the environs of Mogadishu. Army Special Forces teams provided Coalition Warfare Teams to the Pakistanis, Egyptians, and Malaysians. (Courtesy of Al Glover.)

Left: A UH-60 Black Hawk from the Ravens, Falcon Brigade, 10th Mountain Division (LI) lands at Jaybird helo pad adjacent to the JTF and the U.S. Embassy building. From late October onward, movement from the Mogadishu airport and New Port areas was conducted by helicopter or using the five-mile bypass road to limit exposure to warring clans. (Courtesy of Al Glover.)

Right: The camouflaged water tower inside the Embassy compound. The attempt to cover it with netting to deny its use as a spotting landmark to Somali mortar fires ended after a few weeks of wind. In background, EOD destroys excess ammunition prior to the withdrawal (always seemingly unannounced, and raising anxieties among the JTF of a mortar attack). (Courtesy of Paul Holthaus.)

Left: Aerial view of SWORD base, located in the western-most area of Mogadishu. (Courtesy of Moe Elmore.)

Above Left: LTC Moe Elmore served as the JSOTF Deputy Commander and was used as the Liaison Officer to UNOSOM II headquarters. He was the team leader responsible for the JTF's Force Protection Assessment of all facilities. Shown here, HUNTER Logistics base on left.

Above Right: LTC Elmore atop the Austrailian's building, reacting to Somali small arms fire one day during his assessment tour. (Courtesy of Moe Elmore.)

Below: A portion of the five-mile bypass road heading out west near the abandoned refinery. (Courtesy of SOCCENT staff).

Above left: The JSOTF SOCOORD to JTF-Somalia. MAJ Hawk Holloway, team leader.

Right: CPT Tim Chyma and SFC Shane Irwin. CPT Gaylen Cross was also on the team. (Courtesy of Hawk Holloway.)

Coalition Activities

Left: An Italian strongpoint manned with Armored Personnel carriers in northeast Mogadishu. (Courtesy of Al Glover.)

Right: Peacekeepers from Botswana. (Courtesy of Chaplain Wylie.)

Left: Pakistani peacekeeper atop building K7 observation post. (Courtesy of Al Glover.)

ODB590 Activities

Right: MAJ Bill Robinson commanded ODB590 with five Special Forces teams and attached snipers from the Marines, SEALs, and the 5th Group sniper cell (led by SFC Harvey Sills). His company provided a demonstration of SOF equipment and capabilities during the visit of the CJCS, General John Shalikashvili. Just to the left of him and a little in the background is the JSOTF-Somalia commander, COL Rich Stimer. (Courtesy of Bill Robinson.)

Left: SF sniper in position at the Paki Stadium. (Courtesy of Bill Robinson.)

Right: DARPA conducted an initiative to line the HMMWVs with Kevlar blankets in response to ambushes on coalition forces. ODA592 receives instruction from the DARPA team while at the Pakistani camp. (Courtesy of Bill Robinson.)

Top Left: ODB590 operations center and headquarters with team DMVs. On top the roof is a sand-bagged machine gun position and observation post with a TOW sight. (Courtesy of Bill Robinson.)

Middle Left: The K7 sniper and observation post across from the U.S. Embassy compound, on Afgooye Road. (Courtesy of Bill Faistenhammer.)

Below Left: At left is the water tower sniper post located along 21 October Road, on the northern fringe of Mogadishu, across the street from HUNTER base. SF and Navy SEALs used this position. (Courtesy of Al Glover.)

Below Right: At right, an SF team from ODB590 conducts training on the Barrett .50-caliber sniper rifle for a UNOSOM II contingent; UAE forces, Egyptians, and Malaysians were all trained. (Courtesy of Bill Robinson.)

ODB590 Activities

Right: ODA592 prepares for daily recon patrol out in Pakistani sector of Mogadishu. MSG Gentz, Team Sergeant, secures his weapon. (Courtesy of Dan Weber.)

Below Left and Below Right: C-3/5th SFG(A) passes out donated school supplies and children's clothing on a Humanitarian Assistance and Civil Affairs mission; team medics treat local villagers.

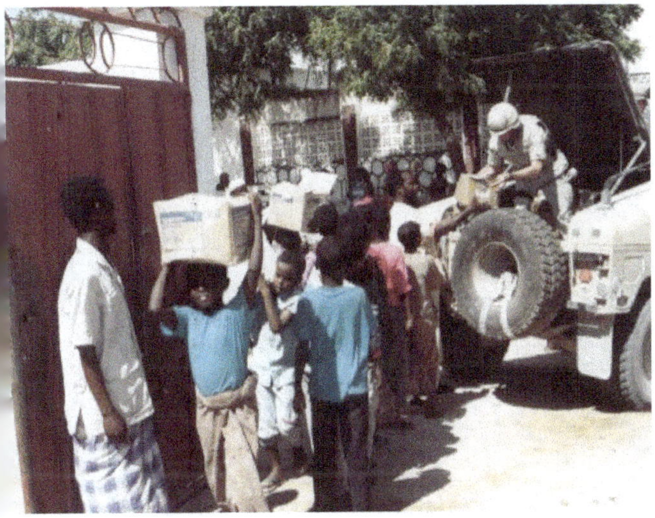

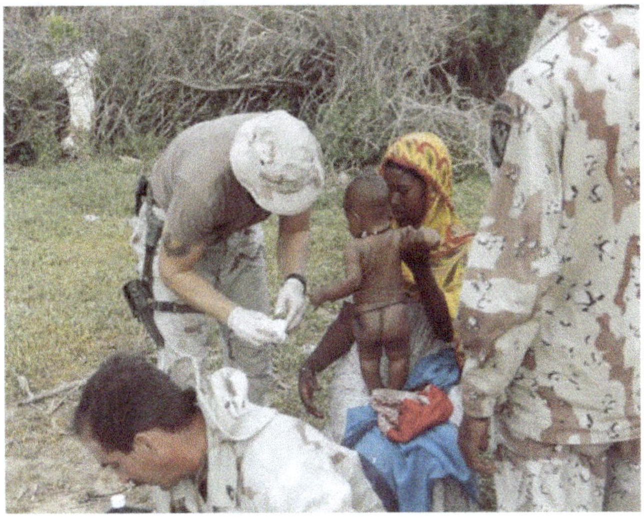

Left: ODA592 mounted team prepares for recon to Kismayo region to support a UN assessment of the airfield for the Indian contingent. The team was transported by C-130. (Note distinctive vehicular camouflage and switch to woodland pattern uniforms to operate near Kismayo and blend into local terrain.) (Photo courtesy of Bill Robinson.)

Left: A large crowd of Somalis gather near the Pasta Factory on 21 October Road to hear anti-UN and anti-U.S. rhetoric. (DOD.)

Right: Egyptian M113 guards along 21 October Road near HUNTER base. View is looking west. Note attempts by Somalis to throw obstacles down on the road. See sniper water tower position in far rear. (Courtesy of Moe Elmore.)

Left: On 10 November, SEAL Team 8, DELTA Platoon snipers attached to the SF company engaged Somalia gunmen at the K4 traffic circle near a suspected weapons cache site. It was the largest SEAL engagement since Vietnam. Three Somalis were killed and five were wounded. Gunmen fired automatic small arms weapons and RPGs at the SEALs positioned in the Soap Factory. (Courtesy of USSOCOM History Office.)

Right: In the final phase of the withdrawal, American forces moved their units and commands to the Mogadishu airfield. The sand berm hill on its southern side soon cropped a village of tents and antennas. On right, the SOCCENT JSOTF's final camp on the sand berm. (Courtesy of SOCCENT staff.)

Left: In late October 1993, MAJ James Realini was selected from the XVIII Airborne SOCOORD to serve with JTF Somalia. Major Realini was one of three Special Forces graduates of the Advanced School of Military Studies. He was assigned to the J-5 of the UNOSOM II staff and was instrumental in the planning for withdrawing all U.S. forces from Somalia. (Courtesy of Jim Realini.)

Right: Tragically, the final death in the SOF experience in Somalia occurred when the Air Commando gunship *Jockey 14* caught an engine fire and crashed in fifteen feet of water just off the coast of Kenya. One Air Force Air Commando was lost (missing after three of the crew parachuted into the water; three of ten inside survived and were rescued). (Courtesy of 16th SOS.)

VIII

Operation UNITED SHIELD

Above Left and Above Right: Prior to the UN final withdrawal from Somalia, USCENTCOM prepared for a noncombatant evacuation of American citizens. During the summer of 1994, SOCCENT was tasked to form a NEO force to cover a gap created with the absence of the USS *Peleliu ARG*. Mombasa was chosen as the Safe Haven, with the airfield at Bale Dogle serving as the Forward Support Base for the operation (shown above left). The Kenyan government provided access to the airstrip at Malindi, used previously by the Marines, for an emergency and refueling airstrip (right).

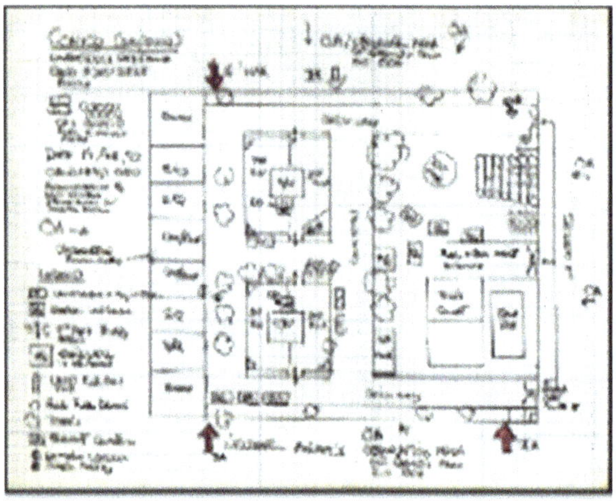

Bottom Left: In September 1994, with a NEO not needed, the United States withdrew its last diplomatic outpost from Somalia. Ambassador Simpson, U.S. State Department workers, and around fifty Marines from the FAST evacuated the USLO from the leased-CONOCO compound in downtown Mogadishu. (Force Protection sketch and photograph of USLO compound reprinted from personal briefing by LtCol John Allison – "Force Protection During Urban Operations" RAND publication – "Capital Preservation" by Russell W. Glenn, Apx E, 2001.)

Right: Cargo from the USLO compound being collected to load aboard contracted Russian cargo aircraft during the November 1994 withdrawal. (Courtesy of Al Glover.)

Left: Contracted Russian cargo aircraft were used to withdraw USLO materials and USMC equipment and armored HMMWVs during November 1994. (Courtesy of Al Glover.)

Right: Donated M-16 rifles for the much-plagued (and failed) UN Somalia Police training program await inventory and removal during the withdrawal period of the USLO in 1994. (Courtesy of Al Glover.)

Above: By November 1994, the remaining UNOSOM II forces in Mogadishu had withdrawn to positions on the Mogadishu airfield in preparation for their final evacuation during Operation United Shield. Above, a UN M113 APC returns from the Old Port area to the Mogadishu Airport, via the bypass road. It was at this location Pakistanis conducted a passage-of-lines thru the Marine security check point. (Courtesy of SOCCENT staff.)

Right: Lieutenant General Anthony C. Zinni, I MEF commander, was chosen to be the task force commander for Operation United Shield. (USMC Photo.)

Left: A Pakistani M60A1 tank from the Paki Lancer Brigade occupies its final position or the Mogadishu airfield just prior to link-up with USMC forces and SF coalition support teams. (Courtesy of Steve Cain.)

Left: ODAs from ODB520, 1/5th SFG(A), prepare for amphibious insertion into Mogadishu during Operation United Shield. Shown here inside the USLO-Kenya warehouse on the Mombasa airfield. (Courtesy of SFC Lance Caffrey, ODA524.)

Right: SF Mounted Teams from B Company, 1/5 SFG(A) ready to board the USS *Belleau Wood* (LHA-3) at the port of Mombasa, Kenya, en-route to Mogadishu. (Courtesy of SFC Rickie Young.)

Above and Right: MAJ Bryan Whitmann commanded Company B, 1/5th (SFG) during Operation United Shield. He is shown here while afloat during steaming from Mombasa Kenya to GREEN Beach at Mogadishu. LCACs from the USS *Belleau Wood* (LHA-3) landed on GREEN Beach transporting SF mounted teams on 8 February 1995, south of the Mogadishu airfield. The teams also extracted by this method once the operation ended. (Photos courtesy of Steve Cain and Bryan Whitmann.)

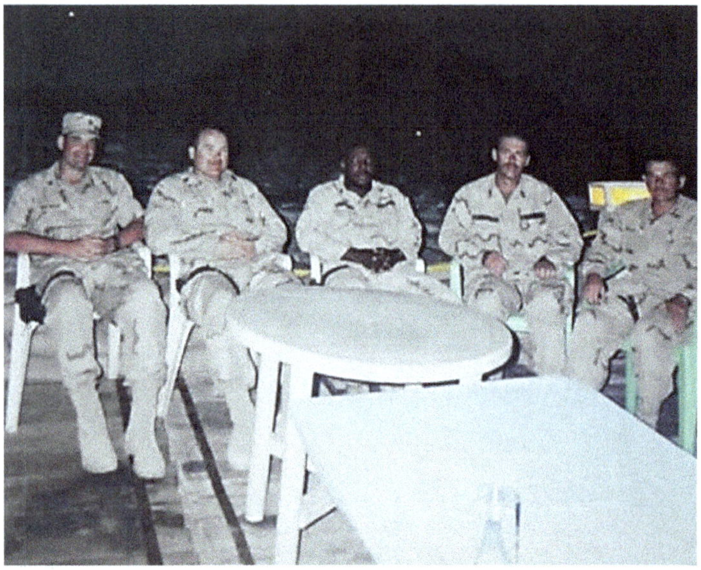

Left: (Left to right) MAJ Bryan Whitman, SGM Hank Gallahan, SGT Bobby Brown (attached), SFC Dave Asher, and an unknown Sergeant. (Courtesy of Dave Asher.)

Right: ODA operators, support personnel, and B-team staff at the ODB520 headquarters on Mogadishu airfield. Note United Shield CTF naval support in background. (Courtesy of Dave Asher.)

Left: UNOSOM II and JTF headquarters final location on sand berms south of Mogadishu airfield during Operation United Shield. (Courtesy of Dave Asher.)

Right: ODA526, commanded by CPT Steve Cain, poses with Commander and CSM of Pakistani 5th Frontier Force. CPT Cain is to the Pakistani commander's immediate right and Ricky Young is to his immediate left (wearing a load bearing vest). (Courtesy of Ricky Young.)

Above Left and Above Right: SFC Jose Bailey, ODA526 team sergeant, poses next to the team's Desert Mobility Vehicle (DMV). Members of ODA526 man their vehicle position along the airport perimeter. (Courtesy of Jose Bailey.)

Left: Pakistani M113 Armored Personnel Carrier from the 4th Punjab Battalion (supported by SF CST-1, ODA522) provides security at the southern tip of the Mogadishu airfield. (Courtesy of Steve Cain.)

Left: ODA526 assisted the 5th Frontier Force by conducting joint patrols, taking turns manning sniper and observation post positions, and establishing a vehicular fire support position around the Mogadishu airport. During their deployment, these positions were under sporadic gunfire from Somali positions outside the airport perimeter. They were augmented with STS member Doug Bauer. ODA526 team members, SSG James Korenowski and SSG Rick Young, return from patrol of the south perimeter. (Courtesy of SFC Rickie Young.)

Right: The airport field location for the 11th Pakistani Lancer Battalion (with M60A1 tanks). A ground mobility vehicle (GMV) from ODA524 is parked alongside the SF team tents. (Courtesy of Lance Caffrey.)

Right: Defensive position of the 11th Pakistani Lancer Battalion at the southern tip of the Mogadishu airfield. (Courtesy of Lance Caffrey.)

Right: ODA524 (Courtesy of Lance Caffrey.)

Left: SSG Patrick Mitchell (ODA522, CST-1 for the 4th Punjab Battalion, commanded by CPT Joe Bovy) observes for technicals which have been sniping at ODA526 main gate position and at other CST positions. (Courtesy of Lance Caffrey.)

Right: SFC Lance Caffrey, ODA524, at the 11th Lancers Battalion. He is shown here with the Executive Officer of the battalion. (Courtesy of Lance Caffrey.)

Left: Marines from the 13th MEU-SOC move their LAV into position to support the passage-of-lines of the 11th Lancer Battalion from their positions at the Mogadishu airfield for movement to the New Port area to load aboard shipping and return to Pakistan. (Courtesy of Steve Cain.)

Right: ODA524 (in the lead with their GMV and trailer) escorts the 11th Lancer Battalion's M60A1s, with their barrels to the rear, down the main Mogadishu airport runway for the passage of lines through the Marines and on to New Port. (Courtesy of Steve Cain.)

Left: A fitting, final picture seen by forces from Operation United Shield during the final withdrawal. A Somali Technical outside the wire firing random shots, while crowds of Somalis (". . . like a swarm of ants") stormed the abandoned site to loot the facilities. (Courtesy of Steve Cain.)

A

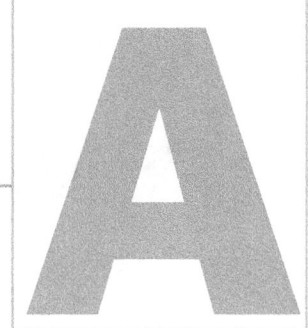

Appendix A: Clan and Political Party Affiliations

Me and Somalia against the world,
Me and my clan against Somalia,
Me and my family against my clan,
Me and my brother against my family,
Me against my brother.

— Somali proverb

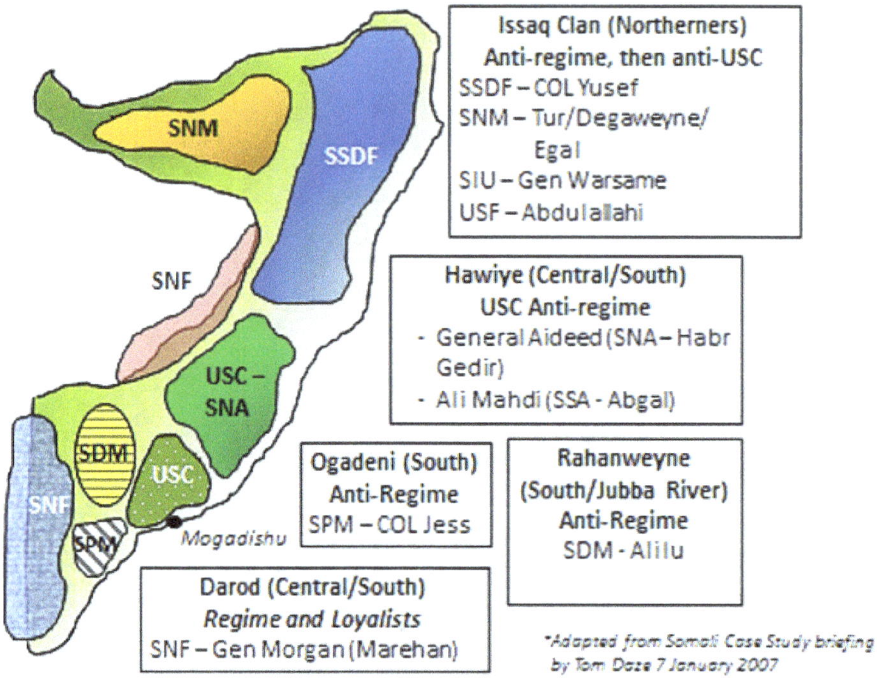

Major clan party affiliations and areas of influence.

Main Opposition Movements

There were four main opposition political movements, all aimed at the dictator, Siad Barre, and his regime. (Barre was from the Marehan sub-clan of the Darod.) Opposition movements were clan-based:

1. The Somali Salvation Democratic Front (SSDF). The first opposition party arose within the Darod clan, in the Majerteen sub-clan. The Majerteens, along with the Issaq and Hawiye clans were somewhat prominent in government and in state bureaucracies prior to Siad Barre's takeover after the 1969 military coup. Barre sensed the other Majerteen officials were not supportive of his policies and reform efforts; he was very vocal against them and denounced their dissatisfaction with his policies, along with replacing Issaq and Hawiye officials in important positions with Majerteens. However, in 1978 Majerteen officers and soldiers attempted a coup against Barre in disfavor of his performance during the failed Ogaden War. About 500 of the rebelling troops, along with sixteen Majerteen officers, were killed or executed as a result of the failed coup. COL Yusuf Ahmed escaped Barre's dragnet and fled abroad to form the Somali Salvation Front (SSF) with the objective of overthrowing the Siad Barre regime. The SSF grew to become the Somali Salvation Democratic Front in 1981. The movement's first armed elements began operating in the Mudug region, with support from Ethiopia. Colonel Abdullahi Yusef led the Galcaio branch of the SSDF.[1]

2. ***Somali National Movement*** **(*SNM*)**. The disaffected Issaq clans formed the Somali National Movement, when Issaq dissidents met in London in 1981 to challenge the Barre regime. The Somali National Movement announced itself as a nationwide opposition movement, and appealed to other clans in Somalia. The SNM operated in northwestern Somalia with armed elements operating out of Ethiopia. The SNM was both a political opposition party and a military movement. Its most famous strike against the Barre regime was a guerrilla attack against the Mandera prison near Berbera in 1983, freeing many dissident political prisoners. The SNM launched a large military campaign in May of 1988; in return, the Somali National Army and Air Force conducted brutal reprisals against the Issaq clan in northern Somalia. Led by Abdulrahman Ali Tur, the SNM declared former British Somaliland independent and formed the Somaliland Republic. (The SNM opposed the USC.)[2]

3. ***Somali Patriotic Movement*** **(*SPM*)**. The Somali Patriotic Movement was formed in 1985 as a split off of the SSDF, when Ogadenis became dissatisfied with Siad Barre's failure to address pan-Somali issues over the Ogaden region. The poor performance of the regime during the failed Ogaden War resulted in the desertion of numerous Ogaden officers. The SPM benefited greatly from its armed elements trained and experienced in the Somali Army and was able to take over territory and government positions in the south, particularly around Kismayo. Colonel Ahmed Omar Jess led the Aideed-allied SPM in that location; the other faction of the SPM was led by Colonel Aden Gabiyu, aligned with the General Morgan of the SNF. (The Ogadenis are a sub-clan of the Darod clan family.)

4. ***The United Somali Congress*** **(*USC*)**. As Barre increased his repression of the dissident clans, the Hawiye clan (who had worked with the SNM) formed their own opposition group in 1989, the United Somali Congress, and became the largest opposition party in Somalia. They were also apprehensive of Issaq claims over territory liberated from the regime. It was the armed militias within the USC who pushed Barre's military into the environs of Mogadishu and eventually took over the city and caused Barre to flee to the south. (Armed elements of the Hawiye sub-clans were the Abgal, led by Ali Mahdi under the banner of the Somali Salvation Alliance [SSA] and the Habr Gedir in the Somali National Alliance [SNA] led by General Mohamed Farah Hasan Aideed. Aideed later changed his SNA military faction to the Somali Liberation Army – the SLA.)

Regime Military-Political Party

The Somali National Front **(*SNF*)**. The Somali National Front was established in March of 1991 by General Mohamed Said Hersi Morgan, the former Minister of Defense in the Barre regime (and his brother-in-law). It was a predominantly Marehan sub-clan organization (of the Darod clan family) and composed of Barre loyalists attempting to restore the Dictator and his regime back into power. The SNF operated mainly in the Kenya-Somalia border region, from around Kismayo in the south reaching north to Baidoa and Bardera. Other leaders included Ghanni and Hashi.

Minor Political-Military Clan Opposition Movements[3]

***Somali Democratic Alliance* (*SDA*).** Formed in 1990 by the Gadabursi sub-clan (Dir clan-family).

***Somali Democratic Movement* (*SDM*).** Formed by Rahanweynes, this opposition group operated west of Mogadishu in the areas of Baidoa and Bardera, with its headquarters established in Bardera. This movement was led by Alilu and Zoppo and was loosely allied with Aideed's faction.

***Somali Independent Union* (*SIU*).** This independent organization was led by General Warsame, operating around the Bossaso region.

***Somali Salvation Front* (*SSF*).** This was an Ethiopian-based organization formed by the Majerteen sub-clan.

***Southern Somali National Movement* (*SSNM*).** The SSNM operated around Kismayo, formed by the Biyamaal sub-clan clan of the Dir clan family.

Islamic Groups[4]

Al-Itihad al-Islamiya (*AIAI*). The Islamic Front was a fundamentalist group operating in small elements throughout southern Somalia, with its center located at Luuq.

(The Muslim Brotherhood, the *Akhwan al-Muslimeen*, was another political organization and operated throughout Somalia.)

Endnotes

1. Metz, Country Handbook, 164–166.
2. Harned, Glenn. *Stability Operations in Somalia 1992 – 1995: A Case Study*. Carlisle Barracks, PA: Personal Monograph Series PKSOI (19950612 009) United States Army War College PKSOI: War College Press, July 2016, 5.
3. Ibid, 8.
4. Ibid.

B

Appendix B: Chronology of Somalia Crisis and Introduction of U.S. Army Special Forces

July 1, 1960	British Somaliland and Italian Somaliland unify to establish the independent Somali Republic.
October 15, 1969	President Abdirashid Ali Shermaarke is assassinated, apparently over a clan grievance.
October 21, 1969	Major General Mohammed Siad Barre seizes power in a bloodless coup and assumes leadership as dictator of the country, which he renames as the Somali Democratic Republic.
July 13, 1977 – March 15, 1978	Somalia's unsuccessful invasion of the disputed Ogaden region of Ethiopia.

1982 – 1989	The 5th Special Forces Group (Airborne) conducts Security Assistance and Foreign Internal Defense missions into Somalia. The Group deployed mobile training teams (MTTs) and Technical Assistance Fielding Teams (TAFTs). These missions included weapons, TOW missile training, and parachute operations with the Somali Commando Brigade.
May 23, 1986	General Barre is injured in an automobile accident, after which rivals within his government and insurgent groups in northern Somalia enter into open rebellion.
December 30, 1990	The United Somali Congress (USC) forces enters the streets and begin taking on the Somali government forces to gain overall control of the city.
January 1, 1991	Ambassador formally requests from the Department of State a Non-combatant Evacuation (NEO) of his remaining staff. The next day, USCENTCOM was tasked to undertake the NEO, code-named Operation EASTERN EXIT.
January 5, 1991	U.S. Central Command (USCENTCOM) conducts Operation EASTERN EXIT to evacuate American Embassy personnel from the Somali capital city of Mogadishu.
January 27, 1991	Dictator Siad Barre flees Mogadishu as various rebel factions, primarily the United Somalia Congress (USC) led by the powerful warlord General Mohamed Farrah Aideed, Chairman of the USC, take control of the Somali capital.
August 14, 1992	President George H.W. Bush announces U.S. participation to assist the UN World Food Program (WFP) and the ICRC food relief operations in Somalia.
August 15, 1992	Operation PROVIDE RELIEF begins and U.S. Army Special Forces participation lasts until 9 December 1992. The mission to provide humanitarian food flight security is tasked to the 5th SFG(A); subsequently, 2/5th receives the mission and Co A, 2/5th deploys to Mombasa, Kenya, with five SF teams.
August 16, 1992	The company commander A-2/5th, deploys with CENTCOM HAST to Mombasa, Kenya.
August 20, 1992	2/5th commander deploys with a small contingent of his battalion staff (S3, S4, S2, C&E officer, etc. approximately six personnel). Along with the SF company, the total personnel deployed are 70+ pax and six DMVs, arriving to Mombasa on August 22, 1992.
August 28, 1992	SF teams begin airfield assessments inside Somalia, followed by the ASAT mission once C-130s food relief flights are cleared to land.
September 16, 1992	Mission is adjusted and downsized; the battalion commander and his staff, along with select members from the ODA, redeploy back to Ft. Campbell.
November 17, 1992	A-2/5th SFG(A) is replaced with C-2/5th SFG(A), and then downsized to only two SF teams required.

December 3, 1992	As clan violence continues to impede UN humanitarian relief operations, UN Resolution 794 authorizes "all means necessary to establish a secure humanitarian assistance environment."
December 8, 1992	Operation RESTORE HOPE begins. This is a U.S.-led joint and multi-national force to answer the call to use military means to enforce the UN mandate. The multi-national military forces are designated as UNITAF (United Task Forces).
December 27, 1992	On 21 Dec 1992, the 2/5th SFG(A) (-) receives a warning order to end Operation PROVIDE RELIEF and prepare to move to Mogadishu in support of the UNITAF operations – RESTORE HOPE. On 27 Dec 1992, the ADVON for Company C departs from Mombasa to Mogadishu, followed by the remainder of the company, arriving in Belet Weyne, their new AOB site, on the 30th. All SOF would come under the command and control of a SOCCENT JSOTF, headquartered in the Embassy compound. FOB52(-) arrives on 12 January 1993, with two additional 2/5th Group SF teams. Additionally, to support coalition operations, the 1st SFG(A) deploys two SF teams (one for Canada and one for Australia), the 3rd SFG(A) deploys two SF teams (Moroccan and Botswana CSTs), and the 10th SFG(A) with three SF teams (French, Italian, and Belgian CSTs).
Mid-January 1993	SOCCENT downsizes the CST requirement as they are no longer needed when UNITAF secures the HRSs earlier than anticipated. (Only 2/5th CSTs remain in-country.)
January 15, 1993	Additional teams and DMVs deploy from Ft. Campbell to support 2/5th's operations in Somalia.
Late January 1993	SOCCENT redeploys back to MacDill AFB, leaving the 2/5th battalion commander as the senior officer for remaining SF teams.
December 1992 – April 1993	SF Coalition Support Teams provide support to all Humanitarian Support Sectors. SF teams conduct area assessments, route reconnaissance, inspections of suspected weapons caches, convoy security escort, weapons collection and destruction, essential meetings with village elders, and meetings with various warlords outside of Mogadishu.
January 29, 1993	ODB560 becomes a SOCCE(-) and, with ODA561 and ODA564, deploys to Kismayo in support of the 10th Mountain (LI) effort in that HRS.
February 6, 1993	ODA becomes involved in a confrontation with IAIA fundamentalists at the village of BeledXaaho (near Luuq). No gunfire is exchanged when fundamentalists challenge the team.

March 3, 1993	SFC Robert Deeks is killed by a landmine emplaced in the road outside of Balenbale, HRS Belet Weyne. SFC Deeks becomes the only active SF operator from the SF Groups killed in Somalia. Four other members of ODA562 are wounded in the incident.
March 27, 1993	B Company (ODB520), 1/5th SFG(A), deploys to Mogadishu with three SF teams.
April 8, 1993	The 2/5th (-) SFG(A) receives orders to redeploy with the transition of Operation RESTORE HOPE to Operation CONTINUE HOPE. C Company withdraws to Mogadishu for the redeployment. C Company and its detachments closed Ft. Campbell around 8 April 1993. A 5th Group SOCOORD was deployed to the 10th Mountain to liaise with UNOSOM II Headquarters.
April 23, 1993	ODA525 skirmishes with Islamic Fundamentalists near vicinity of Luuq. There were no friendly casualties. The SF team is credited with one Somali KIA, some small WIA, and a destroyed vehicle, along with disabling its .50-caliber machine gun.
May 4, 1993	Operation CONTINUE HOPE begins as UNOSOM II, with the withdrawal of U.S. forces and other UNITAF contingents. The U.S. continues support to UNOSOM II with logistics, a brigade from the 10th Mountain Division (LI), and an SF ODB supporting QRF operations. All U.S. forces are stationed in Mogadishu.
May 6 – 16, 1993	SOCCE 520 participates in a combined contingency operation with the QRF and Belgians and conducts special reconnaissance missions in and around Kismayo. Mopping up operations lasts one week, and scores of Somali gunmen, hiding in the bush, are captured along with their weapons.
May 1993	ODB520 and its teams conduct short duration site surveys outside of Mogadishu. ODB520 is lead agent for planning operations to seize Radio Mogadishu.
June 9, 1993	SOCCENT deploys a lightly staffed JSOTF to Mogadishu to command and control AC-130s, stationed in Djibouti.
June 10, 1993	ODA525 deploys with the 10th Mountain QRF companies to recon a bunker complex outside of Mogadishu. SF snipers with night capability provide protective fires during the destruction of equipment in AWSSs #A1 and #A4 when Somalis fire at the QRF. Five armed Somalis are killed.
June 12, 1993	AC-130s attack pre-approved targets in Mogadishu (lasting until 3 July). 1/22nd Infantry conducts raids on weapons storage sites with the assistance of ODA engineers to help destroy heavy weapons. ODB520 members were involved in the combat that day, with company snipers claiming several hits.

June 13, 1993	10th Mountain Warrior Brigade conducts sweeping attacks on suspected enemy positions vicinity of Afgoye Road. ODA523 assists as the CST to various coalition elements.
June 17, 1993	Special Forces teams and advisors participate in Operation CASABLANCA (the sweep vic. of Afgoye Road), supported with pre-strikes from AC-130s. A 5-hour firefight ensues; the SF ODA returns fire with .50-caliber and MK-19 weapons, along with their small arms. SF snipers on K7 also engage targets during the fight.
June 18, 1993	ODB520 is tasked to develop an operational plan to locate and capture Aideed. "Eyes Over Mogadishu," an airborne sniper program, begins as the reconnaissance component of the concept.
June 24, 1993	ODB520 provides support to Ambassador Glaspie's trip to meet with COL Yusef's SSDF faction in Bossaso.
July 1993	ODB520 conducts limited area reconnaissance mission, inspection of weapons storage sites, and humanitarian activities. ODA523 conducts limited FID mission to train Pakistani contingents on their newly equipped M113 APCs.
July 12, 1993	10th Mountain takes down the Abdi house; several civilians are killed and wounded, along with 14 dead Somali irregulars. July 17, 1993 SOCCENT redeploys to Tampa, MacDill AFB.
Late July 1993	A representative of the SOCOORD leads a program to conduct a physical security assessment of U.S. and coalition bases. On the night of the 25th, the team along with a security element from ODA525, receive mortar and automatic weapons fire at SWORD Base. The ODA directs Cobra fire on suspected enemy locations. Throughout the remainder of the month, SF sniper teams support U.S. bases with force protection.
July 24, 1993	A two-vehicle convoy from ODB520 is ambushed en-route to the Mogadishu airport. Two SF operators are wounded; five Somalis are killed by the return fires. The aerial SF sniper teams continue the "Eyes Over Mogadishu" mission.
August 1993	SF teams conduct FID with the Malaysian sniper units.
August 10, 1993	SF aerial sniper team engages Somali truck which was firing on the United Nations base. Two Somali militia are killed and the truck disabled. The helicopter suffers damage from small-arms fire.
August 18, 1993	ODA525 conducts area assessment, a civil affairs project, and a small MEDCAP in Bossaso.
September 8, 1993	ODB590 replaces ODB520 with three SF teams and a 5th Group Sniper Cell, becoming operational on 8 September.

September 12, 1993	ODB590 conducts SR mission to Hobyo to verify presence of SA-7s. They are inserted, with vehicles, into the area using a UN MI-26 helicopter. The contingent extracts to Mogadishu on 15 September, without results. The 5th Group Sniper Cell begins supporting the "Eyes Over Mogadishu" mission with aerial snipers. Ground sniper outposts are also established throughout Mogadishu.
September 10 – 13, 1993	ODB520 redeploys to Ft. Campbell, KY.
September 20, 1993 – October 5, 1993	ODA593 conducts area assessment of Bossaso in support of UN. ODAs and the ODB begin coordination and provide support to TF Ranger. One is the guarding and handover of Osman Atto from his custody location, to the Task Force.
October 3 – 4, 1993	The Battle of Mogadishu ensues between TF Ranger and Aideed's militia as a result of RPG fire downing two Task Force helicopters. No SF operators from ODB590 are involved in the battle.
October 7, 1993	President Clinton announces the termination of U.S. military involvement in Somalia, with end date as 31 March 1994.
October 1993	SF continue to expand the sniper program, conduct HA and MEDCAPs, and react to harassing mortar and small arms fires on American compounds. (One SF sergeant is wounded by shrapnel from a mortar round during these skirmishes.)
October 18, 1993	JTF-Somalia is operational with MG Ernst as commander. SOCCENT establishes the JSOTF-Somalia and assumes command of SF contingents and AC-130 gunships. ODB590 expands its sniper operations.
October 29, 1993	Thirteen USMC snipers from the MEU-ARG are CHOP'ed to ODB590 to assist with the sniper program. Also, in late October snipers from ST-8 (SEALs) platoons were added to the program. By early November, nine sniper teams were in positions throughout Mogadishu. AC-130s begin armed reconnaissance missions over Mogadishu, nightly.
November 10, 1993	SEAL snipers engage Somalis off-loading weapons near the K4 traffic circle. A heavy firefight ensues, ended when Cobra helicopters intervene.
November 25, 1993	A contingent of SEAL snipers from ST-5 begins operations with ODB590, and remains deployed until the American withdrawal from Somalia, January through March 1994. By mid-December ODB590 and its attached snipers are involved in over 29 engagements on heavily armed militia.
December 1, 1993	Two additional ODAs arrive for ODB590.
December 14, 1993	ODA592, and later ODA554, assists with transition of the Indian contingent into Kismayo. Other activities consist of coalition support to UN contingents on the Mogadishu airport.

January 6, 1994	ODB550 with two SF teams deploys to Mogadishu to replace ODB590, and prepare for U.S. forces withdrawal. The JSOTF-Somalia begins preliminary planning and reconnaissance for any potential NEO. All ODB550, with its SEAL snipers and attached STS and ANGLICO, move to the Mogadishu airport. Sniper operations continue throughout the city.
February 1994 – March 15, 1994	ODA551 conducts FID with Malaysian Rangers. Other SF teams continue their coalition support mission. Minor skirmishing with Somali militias continues. Reconnaissance for the potential NEO is conducted.
March 19, 1994	SOCCENT and ODB550 depart Somalia and return to home stations.
April 1994	SOCCENT is tasked by CENTCOM to be prepared to perform a NEO operation between the time period April through September. SOCCENT designates the 5th SF Group commander as the ARSOTF, with 3/5th SFG(A) and a contingent from the 75th Rangers assigned to the operation. A Civil Affairs and PSYOP detachment also is assigned. The 160th SOAR (-) and an AFSOD from the 16th SOW become the JSOACC
May 12, 1994	SOCCENT deploys a 4-man MSALT to Mogadishu to coordinate the NEO mission with the State Department, and remaining Embassy Marine security force.
June 8 – 9, 1994	A full-blown NEO rehearsal (Exercise INTREPID WARRIOR) is conducted at Ft. Campbell, KY, with success. (The mission tasking ends in the summer of 1994, with no NEO required.)
August 15, 1994	CENTCOM issues warning order and Course of Action preparation to withdraw all UNOSOM II forces from Somali, during the period 1 January 1994 – 31 March 1995. 1st MEF is given overall lead for the now-named operation, UNITED SHIELD.
August 1994	5th SF Group tasks 1/5th SFG(A) to support the operation with one ODB. The four SF teams would be utilized as coalition support teams at the Mogadishu airport during the withdrawal. The battalion commander of the 1/5th SFG(A) becomes the SOCOORD afloat for the operation.
January 29, 1995	ODB520 arrives to Mogadishu airport, after steaming aboard the USS *Belleau Wood* (LHA) from the port of Mombasa. USMC LCACs are used to transport the SF contingent from the ship to GREEN Beach.
February 1995 – March 1995	ODAs serve as coalition support teams (CSTs) to facilitate the passage of lines of their assigned coalition country and to provide access to CAS through their attached STS.
February 20, 1995 – March 1, 1995	Passage of lines and relief-in-place of coalition UN forces with USMC forces begins, with SF CSTs escorting their contingents.

March 1, 1995	ODB520 departs Somali with USMC elements, returning to the USS *Belleau Wood*. American involvement in Somalia UN humanitarian operations ends on 2 March, 1995. By 15 March, all U.S. participating elements return to home station.
January 19, 2007	African Union adopts mission to intervene in Somalia, named the African Union Mission in Somalia (AMISOM); first troops from Uganda begin arriving 6 March 2007.
January 7, 2007	The United States begins campaign of drone, cruise missile, and air strikes against al-Shabaab and Al Qaeda targets in Somalia.
2014	U.S. Army Special Forces begin training of the elite Somali Danab Brigade at Baledogle Airfield.
January 2021	President Trump orders 600 American military troops out of Somalia.
May 2022	President Biden orders return of 500 military personnel to Somalia counterterror operation.

Glossary

AIAI: Al-Itihad al-Islamiya, a Somali separatist Muslim Organization considered as extremist

AFRICOM: The United States Africa Command

ALO: Air Liaison Officer

AMCIT: American Citizen

AMEMB: American Embassy

AMISOM: African Union Mission in Somalia

AQ: Al Qaeda

AQAP: Al Qaeda in the Arabian Peninsula

ARFOR: Army Forces

ARSOF: U.S. Army Special Operations Forces

ARPCT: Alliance for Restoration of Peace and Counterterrorism

ASAT: Airborne Security Assault Team

AU: African Union

AWSS: Authorized Weapons Storage Sites; these sites were used as cantonments for weapons and ammunition used by the clans, as part of their agreement with the UN in exchange for relief and humanitarian operations

CA: Civil Affairs

CAS: Close Air Support

CCT: Combat Control Team (normally USAF ground personnel who coordinate aircraft for combat)

CENTCOM: Central Command, located in Tampa Florida at MacDill AFB, and responsible for U.S. military operations in the Middle East, East Africa, and Central Asia regions

CIA: Central Intelligence Agency

CJCS: Chairman of the Joint Chiefs of Staff

CMOC: Civil Affairs Operations Command

CONTINUE HOPE: Continuing UN military mission in Somalia after withdrawal of U.S. forces in March of 1994; UN forces continued the reduced mission from March 1994 until their withdrawal in March of 1995

CSAR: Combat Search and Rescue

CST/CWT: SF Coalition Support Team/Coalition Warfare Team

CWO: Chief Warrant Officer

DA: A SOF mission for Direct Action

DAP: Direct Action Penetrator, a variant of the MH-60 SOAR helicopter

DMV: Desert Mobility Vehicle, the Special Forces variant of the HMMWV, highly modified to meet SOF needs

EAP: Emergency Action Plan

"Eyes Over Mogadishu": A program started in the summer of 1993 to fly helicopters with snipers over the city of Mogadishu at night to suppress clan militia activities

E&E: Escape and Evasion

F-77: Roster List of American Employees working for a U.S. Embassy, Consulate, or Delegation used as a contact reference in the event of emergencies or noncombatant evacuation operations

FAC: Forward Air Controller

FAST: Forward Air Support Team

FID: Foreign Internal Defense (a Special Operations Forces mission to train foreign and indigenous troops)

FSB: Forward Support Base

GOTHIC SERPENT: The operation to capture Aideed and his infrastructure

GWOT: Global War on Terrorism

HAST: Humanitarian Assistance Survey Team

HMMWV: The "Humvee", High Multipurpose Military Wheeled Vehicle, which replaced the older Army ¼ Ton jeep

HRS: Humanitarian Relief Sector

HVT: High Value Target (such as Aideed, Osman Atto, etc.)

HUMINT: Human Intelligence

Hunter Base: The location of U.S. Army mechanized and armored forces stationed in Mogadishu October 1993 thru March 1994

ISB: Interim Support Base

ISOFAC: Isolation Facility

JAYBIRD: Main helicopter landing pad near U.S. Embassy

JCS: Joint Chiefs of Staff

JOC: Joint Operations Center

JPOTF: Joint Psychological Operations Task Force

JSOACC: Joint Special Operations Air Component Commander

JSOAT: Joint Special Operations Assessment Team

JSOC: Joint Special Operations Command (Tier 1 units)

JSOTF: Joint Special Operations Task Force

JSSA: Joint Search and Survival Agency

JTF: Joint Task Force

K4: The well-known traffic circle in Mogadishu located just north of the Mogadishu airfield

Khat: An addictive stimulant derived from chewing the leaves of the Khat bush

KUSLO: Kenya United States Liaison Office

LAV: Light Armored Vehicle (USMC)

LCAC: Landing Cushion Assault Craft (Marine)

LHA: Landing Helicopter Amphibious Assault Ship (for example, the LHA-3 USS *Belleau Wood* used by SF teams in UNITED SHIELD)

LHD: Landing Helicopter Dock (USS *Essex*)

LSD: Landing Ship Dock (USS *Fort Fisher*)

"Littlebirds": The nickname for the 160th SOAR AH-6J helicopters

LZ: Landing Zone

MAGTF: Marine Air Ground Task Force

MARFOR: Marine Forces

MEDCAP: A military Medical Capabilities humanitarian operation; if for dental purpose, then known as a DENCAP

MEF: Marine Expeditionary Force

MEU: U.S. Marine Expeditionary Unit

MEU-ARG: Marine Expeditionary Amphibious Ready Group

MEU-SOC: A MEU with Special Operations Capability

MSALT: Marine Survey and Liaison Team

MSR: A Main Supply Route used in Somalia to deliver humanitarian relief supplies

MTT: Military Training Team

NAVFOR: Naval Forces

NEO: Non-Combatant Evacuation Operation

NGO: Non-Governmental Organization

NODS: Night Observation Devices

NSW: Naval Special Warfare

NSWTU: Naval Special Warfare Task Unit

ODA: Operational Detachment "A" – the Army Special Forces Team comprised of 12 Green Beret qualified personnel

ODB: Operational Detachment "B"; in Army Special Forces this is a company command element

ODC: Operational Detachment "C"; in Army Special Forces this is a battalion command element

OPCON: Operational Control

PROVIDE RELIEF: Named U.S. military operation to assist UN in airlifting food to Somalia, conducted August 1992 through February 1993

Peacekeeping: Military operations undertaken with the consent of all major parties to a dispute, designed to monitor and facilitate implementation of an agreement, such as truces or ceasefires, while supporting diplomatic efforts to reach a long-term political settlement

Peacemaking: The process of diplomacy, mediation, negotiation, or other forms of peaceful settlements that arranges an end to a dispute and resolves issues that led to it; may be backed up with military force

PSYOP: Psychological Operations

QRF: Quick Reaction Force

Radio Mogadishu: The propaganda radio broadcast sponsored by Aideed

"Rajo": A weekly newspaper published by PSYOPs, both in the Somali and the English language

RESTORE HOPE: The UN military operation to conduct peacemaking and provide security and relief supplies to Humanitarian Relief Sectors, under Title VII of the UN Charter, throughout southern Somalia; UN forces assigned to this mission were designated as the United Task Force, including U.S. military forces

ROE: Rules of Engagement

RPG: Rocket Propelled Grenade (such as the Soviet-made RPG-7 used in Somalia by irregulars)

R&R: Rest and Relaxation tour; Mombasa, Kenya was the most oft-used place to visit

"Sammies": Derogatory term used by troops in Somalia to refer to a Somali; also named "Skinnies"

SAR: Search and Rescue

SATCOM: Satellite Communication radio

SDA: Somali Democratic Alliance

SDM: Somali Democratic Movement

SEAL: Sea, Air, Land naval special warfare personnel

SF: U.S. Army Special Forces (Green Berets)

SFG(A): U.S. Army Special Forces Group (Airborne), a Brigade-sized organization

SFOD: Special Forces Operational Detachment – Delta

Shir: A village or town respected elder's meeting

SMU: Special Mission Unit; a moniker used when referring to Delta Force, Seal Team 6, and intelligence units associated with Tier 1 assets

SNA: Somali National Alliance

SNDU: Somali National Democratic Union

SNF: Somali National Front

SNM: Somali National Movement

SNU: Somali National Union

SP: Strong Point

SOAR: Special Operations Aviation Regiment, the 160th

SOCCE: Special Operations Command and Control Element

SOCCENT: Special Operations Command Central, a sub-unified command to employ and control SOF in the CENTCOM region

SOF: Special Operations Forces (Army, Navy, Marines, Air Force)

SOS: USAF Special Operations Squadron

SOT: Special Operations Team (Electronic Intercept)

SOW: USAF Special Operations Wing

Spectre: The name for USAF AC-130 gunships

SPM: Somali Patriotic Movement

SR: A SOF mission for Special Reconnaissance

SSDF: Somali Democratic Salvation Front

SSE: Sensitive Site Exploitation

SSNM: Southern Somali National Movement

STS: Special Tactics Squadron

Sword Base: The main logistical base of the U.S. Army support to UNOSOM II and U.S. Army elements of UNITAF; located on the northern edge of Mogadishu

TACON: Tactical Control

Technical: A term for any clan, irregular gang, or hire who mounted guns, anti-aircraft guns, or RPGs and recoilless rifles on the bed of commercial vehicles and trucks; these were used by humanitarian organizations who hired them to protect their operations. Since it was forbidden to expend funds on military style equipment, these assets were listed as a technical service in vouchers—thus the term "technicals."

TF: Task Force

TFG: Transitional Federal Government (Somalia)

TF-Ranger: Elements of the 75th Ranger Regiment participating in Operation GOTHIC SERPENT (also seen as TFR)

TSOC: Theater Special Operations Command

UN: United Nations

UNITAF: Unified Task Force (United Nations international forces), those units, including American, participating in Operation RESTORE HOPE

UNITED SHIELD: The U.S.-led military operation to withdraw all UN forces from Somali, January through March 1995, ending UN humanitarian relief operations in Somalia

UNOSOM: United Nations Operation in Somalia

USAID: United States Agency for International Development

USC: United Somali Congress

USFORSOM: United States Forces Somalia

USLO: United States Liaison Office

USSOCOM: United States Special Operations Command

WFP: World Food Program

Bibliography

Allard, Kenneth. *Somalia Operation: Lessons Learned*. Washington DC: National Defense University Press, 1995.

Baumann, Robert F. and Lawrence A. Yates. *My Clan Against the World: U.S. and Coalition Forces in Somalia, 1992 – 1994*. Fort Leavenworth, KS: Combat Studies Institute, 2004.

Bowden, Mark. *Black Hawk Down*. New York, NY: Signet, 2001.

Breen, Bob. *A Little Bit of Hope: Australian Force Somalia 1993*. Victoria, Australia: Barrallier Books Pty Ltd published under Echo Press, 2018.

Brocades Zaalberg, T. *Soldiers and Civil Power: Supporting or Substituting Civil Authority in Peace Operations During the 1990s*. Amsterdam: 2005.

Casper, Lawrence E. *Falcon Brigade: Combat and Command in Somalia and Haiti.* Boulder, CO: Lynne Riener Publisher, 2001.

Chun, Clayton K.S. *Gothic Serpent: Black Hawk Down Mogadishu 1993*. Oxford: Osprey, 2012.

Clarke, Walter S. *Somalia. Background Information for Operation RESTORE HOPE 1992 – 93.* Carlisle Barracks, PA: U.S. Army War College, December 1992.

Clarke, Walter and Jeffrey Herbst (Editors). *Learning from Somalia: The Lessons of Armed Humanitarian Intervention*. Boulder, CO: Westview Press, 1997.

Dawson, Grant. *"Here is Hell": Canada's Engagement in Somalia*. Vancouver, BC: UBC Press, 2007.

DeLong, Kent and Steven Tuckey. *Mogadishu! Heroism and Tragedy*. Connecticut: Praeger Publishers, 1994.

Drysdale, John. *Whatever Happened to Somalia?: A Tale of Tragic Blunders*. London, UK: Haan Publishing, 1994.

Durant, Michael J. *In the Company of Heroes*. New York: G. P. Putnam's Sons, 2003.

Eversmann, Matt and Dan Schilling. *The Battle of Mogadishu: Firsthand Accounts from the Men of Task Force Ranger*. New York: Ballantine Books, 2004.

Fogarassy, Helen. *Mission Improbable: The World Community on a UN Compound in Somalia.* Lanham, Maryland: Lexington Books, 1999.

Hansen, Stig Jarle. *Al-Shabab in Somalia*. New York: Oxford University Press, 2013.

Hashim, Alice B. *The Fallen State: Dissonance, Dictatorship and Death in Somalia.* Boston, MD: University Press of America, 1997.

Hirsch, John L. and Robert B. Oakley. *Somalia and Operation Restore Hope: Reflections on Peacemaking and Peacekeeping.* Washington, DC: Institute of Peace Press, 1995.

Horan, Mike. *Eyes Over Mogadishu*. USA: Xlibris Corporation (www.Xlibris.com), 2003.

Kapteijns, Lidwien. *Clan Cleansing in Somalia: The Ruinous Legacy of 1991*. Philadelphia, PA: University of Pennsylvania Press, 2013.

Katz, Samuel M. *Operation Restore Hope and UNOSOM: The International Military Mission of Mercy in Somalia.* Hong Kong: Concord Publications, 1993.

Kelly, Michael J. *Peace Operations: Tackling the Military Legal and Policy Challenges.* Canberra, Australia: Australian Government Publishing Service, 1997.

Knigge, Timothy M., Major. *Operation CASABLANCA: Nine Hours in Hell*. Chapel Hill, NC: Professional Press, 1995.

Loomis, Dan. *The Somalia Affair*. Ottawa, Canada: DGL Publications, 1996.

Matteson, Wallace E., Lieutenant Colonel (Director). *Somalia: Operations Other Than War No. 93-1*. Ft. Leavenworth, KS: Center for Army Lessons Learned, Jan 1993.

McKnight, Danny R. (COL Ret.). *Streets of Mogadishu*. Chester MD: LEADING FOR FREEDOM Publishing, 2011.

Metz, Helen C. *Somalia: A Country Study* (4th Edition). Washington, D.C.: Federal Research Division, Library of Congress, 1993.

Mroczkowski, Colonel Dennis P. (USMC, Retired). *Restoring Hope: In Somalia with the Unified Task Force, 1992–1993: U.S. Marines in Humanitarian Operations.* Washington, D.C.: History Division, U.S. Marine Corps, 2005.

Neville, Leigh. *Day of the Rangers: The Battle of Mogadishu 25 Years On*. Oxford, UK: Osprey, 2018.

Oloya, Opiyo. *Black Hawks Rising: The Story of AMISOM's Successful War Against Somali Insurgents, 2007 – 2014*. UK: Helion and Company, 2016.

Peterson, Scott. *Me Against My Brother*. New York: Routledge, 2002.

Rutherford, Kenneth R. *Humanitarianism Under Fire: The U.S. and UN Intervention in Somalia.* Sterling, VA: Kumarian Press, 2008.

Sahnoun, Mohamed. *Somalia: The Missed Opportunities*. Washington, D.C.: United States Institute of Peace Press, 1994.

Samatar, Ahmed I. (Editor). *The Somali Challenge: From Catastrophe to Renewal*. Boulder, CO: Lynn Rienner Publishers, 1994.

Schultz, Richard H. Jr. and Andrea J. Dew. *Insurgents, Terrorists, and Militias: The Warriors of Contemporary Combat*. New York: Columbia University Press, 2006.

Stanton, Martin. *Somalia on $5 a Day: A Soldier's Story*. New York: Ballantine Books, 2001.

Stevenson, Jonathan. *Losing Mogadishu: Testing U.S. Policy in Somalia.* Annapolis, MD: Naval Institute Press, 1995.

Stewart, Richard W. *The United States Army in Somalia 1992 – 1994*. Washington, DC: Center of Military History, 2002.

Stevens, Charles "C. B." *Honor Bound: A Special Forces Detachment in Somalia*. Houston, Texas: Personal Draft, 1996.

Tucker, David and Ambassador Robert B. Oakley. *Two Perspectives on Interventions and Humanitarian Operations*, Carlisle, Pennsylvania: Strategic Studies Institute, July 1, 1997.

Turney-High, Harry Holbert. *Primitive War: Its Practices and Concepts.* Columbia, SC: University of South Carolina Press, 1991.

Wheeler, Ed (Brigadier General, USA, retired) and Lieutenant Craig Roberts (USAR). *Doorway to Hell. Disaster in Somalia*. Tulsa, OK: Consolidated Press International, 2002.

Whetstone, Michael (LTC, USA, Retired). *Madness in Mogadishu.* Mechanicsburg, PA: Stackpole, 2015.

Articles

Antal, Major John F. (USA), and Captain Robert L. Dunaway (USA), "Peacemaking in Somalia: A Background Brief," *Marine Corps Gazette*, Vol. 77, No. 2 (February 1993), 38–43.

Atkinson, Rick. "The Raid That Went Wrong," *The Washington Post*, January 30, 1994.

____________. "Firefight in Mogadishu, The Last Mission of Task Force Ranger." *The Washington Post*, January 30 and January 31, 1994.

Bartlett, Tom, "Guns of Many Nations: The Disarming of Somalia," *Leatherneck*, Vol. 76, No. 4 (April 1993), 12–15.

Borchini, Charles P., Lieutenant Colonel and Mari Borstelmann. "PSYOP in Somalia: The Voice of Hope," Ft. Bragg, NC: *Special Warfare*, October 1994.

Celeski, Joseph D. "A History of SF Operations in Somalia: 1992 – 1993." Fort Bragg, NC, Department of the Army, U.S. Army John F. Kennedy Special Warfare Center and School, *Special Warfare* Magazine, Vol. 15, No. 2: June 2002.

Celeski, Joseph D. "Special Forces in Somalia 1992 – 1995." Published for the Special Forces Association for the Golden Jubilee (50 years), The United States Army Special Forces 1952 – 2002 by Fairmount LLC, in the magazine *Special Forces: The First Fifty Years*, 2002.

Cooling, Norman L. "Operation Restore Hope in Somalia: A Tactical Action Turned Strategic Defeat."

Marine Corps Gazette, September 2001.

Crocker, Chester A., "The Lessons of Somalia: Not Everything Went Wrong," *Foreign Affairs*, Vol. 74, No. 3 (May/June 1995), 2–8.

Ecklund, Marshall V., Major. "Analysis of Operation Gothic Serpent: TF Ranger in Somalia." Fort Bragg, NC: *Special Warfare Magazine*, May 2004.

Ecklund, Marshall V., Major. "Task Force Ranger vs. Urban Somali Guerrillas in Mogadishu: An Analysis of Guerrilla and Counterguerrilla Tactics and Techniques used during Operation GOTHIC SERPENT." *Small Wars and Insurgencies*, Winter 2004, Volume 15, Number 3.

Farrell, Theo. "United States Marine Corps Operations in Somalia: A Model for the Future," *Amphibious Operations: A Collection of Papers*, The Occasional, No. 31., edited by Geoffrey Till, Mark J. Grove, and Theo Farrell. London: Strategic and Combat Studies Institute, October 1997.

Hoffman, Lieutenant Colonel Frank G. (USMCR (Retired)), "One Decade Later – Debacle in Somalia," U.S. Naval Institute *Proceedings*, Vol. 130, No. 1 (January 2004), 66–71.

Hooker, Richard D., Jr., "Hard Day's Night: A Retrospective on the American Intervention in Somalia," *Joint Force Quarterly*, Issue 54 (Third Quarter 2009).

Moore, Molley. "Deep in the Desert with a Somali Militia." *The Washington Post*, Feb 23, 1993.

Murphy, John R., "Memories of Somalia," *Marine Corps Gazette*, Vol. 82, No. 4, April 1998.

Oakley, Ambassador Robert B., "An Envoy's Perspective," *Joint Force Quarterly*, Issue 2 (Autumn 1993), 44–55.

Pelton, Robert Young. "Somalia – Technicals." 26 November 2006, http://www.comebqackalive.com/df/dplaces/somalia/dthing7.htm.

Piasecki, Eugene. "If You Loved Beirut, You'll Love Somalia." Fort Bragg, NC: *Veritas*, USASOC History Office Vol. 3, No. 2, 2007. Schilling, Dan (LTC, USAF, retired). "Operation Gothic Serpent: Two Decades On, A Reflection." Fort Walton, FL: *Air Commando Journal*, Vol.5, Issue 3, January 2017.

Siegel, Adam B., "Mogadishu One: The NEO Prelude, Eastern Exit Set Stage for Restore Hope," *Seapower*, Vol. 36, No. 3, March 1993.

Wu, Wei. "Why the Bush Administration Decided to Intervene in Somalia." 26 November 2006, http://www.empereur.com/somaliaus.html.

Case Studies, Papers, and Monographs

Borchini, Charles P., (LTC, USA). "Psychological Operations Support for Operation Restore Hope 9 December 1992 – 4 May 1993," *Personal Perspective Monograph*, 1 June 1994, USASOC History Office Files, Ft. Bragg, NC.

Coon, Robert C. Somalia. *UNOSOM, UNITAF, UNOSOM II, Case Study*, Campaign Analysis Course, Carlisle Barracks, PA: U.S. Army War College, 2003.

Day, Clifford E. (MAJ, USAF). "Critical Analysis on the Defeat of Task Force Ranger." Maxwell AFB,

AL: Air Command and Staff College (AU/ACSC/0314/97-03), March 1997.

Daze, Thomas J. *Centers of Gravity of United Nations Operation Somalia II.* Thesis for Master of Military Art and Science, Ft. Leavenworth, KS: U.S. Army Command and General Staff College, 1995.

Di Tomasso, Thomas. *The Battle of the Black Sea: Rangers in Mogadishu, Somalia.* Fort Benning, GA, Infantry Officer's Advance Course Paper, April, 1994: Made in the USA, Columbia, SC, 8 August 2020.

Dworken, Jonathan T., *Operation Restore Hope: Preparing and Planning the Transition to U.N. Operations* (CRM 93-148), Alexandria, Virginia: Center for Naval Analyses, March 1994.

Hall, Gideon S. (MAJ, USAF). *Warlords of Somali Civil War (1988 – 1995).* Maxwell Air Force Base, AL: Air Command and Staff College, Air University, April 2015.

Harned, Glenn. *Stability Operations in Somalia 1992 – 1995: A Case Study.* Carlisle Barracks, PA: Personal Monograph Series PKSOI (19950612 009) United States Army War College PKSOI: War College Press, July 2016.

Klein, Gregory F., Major. *Operation UNITED SHEILD: A Case Study*, CSC 1999, 23 March 2006, http://192/168/2/226/isysquery/irl1ff5/28/doc.

McGrady, Katherine A. W. "The Joint Task Force in Operation Restore Hope." Alexandria, VA: Case Study CRM93-114, The Center for Naval Analysis, March 1994.

Mellor, W.J.A. (COL). "Operation Restore Hope – The Australian Experience." Personal Monograph Series (19950612 009), U.S. Army War College, Carlisle Barracks, PA: 1995.

Ohls, Gary J. *Somalia. . . From the Sea.* Newport Papers #34. New Port, RI: U.S. Naval War College, July 2009.

Piasecki, Eugene. *The History of U.S. Army Special Operations Forces in Somalia.* Fort Bragg, NC: USASOC History Office Research Paper (Draft), August 2007.

Ratliff, Leslie L., Lieutenant Colonel. "Joint Task Force Somalia: A Case Study." Newport, RI: Naval War College, March 1995.

Turbeville, Graham and Josh Meservey and James Forest. *Countering the al-Shabaab Insurgency in Somalia: Lessons for U.S. Special Operations Forces.* MacDill AFB, Tampa, FL: the Joint Special Operations University, JSOU Report 14-1, February 2014.

Ucko, David H. and Thomas A. Marks. "Crafting Strategy for Irregular Warfare: A Framework for Analysis and Action." Washington, DC: National Defense University Press, July 2020.

Government Publications

After Action Report Summary, Subject: *U.S. Army Forces Somalia*, undated, 10th Mountain Division (LI), USASOC History Offfice Files, Fort Bragg, NC.

ARCENT Rear CAC. "Operation Restore Hope," *G-5 Civil Affairs Smart Book*, December 1992, USASOC History Office Files, Fort Bragg, NC.

Biller, Robert E., Major. Company C, 96th Civil Affairs Battalion. *After Action Report, USCINCCENT operation Restore Hope*, 8 March 1994, USASOC History Office, Fort Bragg, NC.

Brown, John S., BG, Chief of Military History, *United States Forces, Somalia After Action Report and Historical Overview: The United States Army in Somalia, 1992 – 1994*. Washington DC: Center for Military History United States Army, 2003.

Cahill, Dennis J., Captain. *After Action Report, Subject: After Action Report for Operation Restore Hope*, dated 1 July 1993, USASOC History Office, Ft. Bragg, NC.

Canadian Land Forces Command, National Defence. *In the Line of Duty: Canadian Joint Forces Somalia 1992–1993*. Quebec, Canada: National Defence, 1994.

Csrnko, Thomas R., Major General, CG USASFC, *Operation United Shield Briefing Charts*, 5 December 2006, USASOC History Office Files, Fort Bragg, NC.

Department of Defense. *SOMALIA. Country Handbook*. United States: Department of Defense Intelligence Production Program, 2001.

Department of Defense American Forces Information Service, "*Somalia: Operation RESTORE HOPE,*" Current News Special Edition, 16 December 1992, USASOC History Office Files, Fort Bragg, NC.

Department of Public Information, United Nations. "Somalia – UNOSOM 1," 21 March 1997, http://www.un.org/Depts/DPKO/Missions/unosomi.htm.

Department of Public Information, United Nations, New York. "The United Nations and the Situation in Somalia," *United Nations Reference Paper,* New York: United Nations Reproduction Section, 30 April 1993.

Department of Public Information, United Nations. "Somalia - UNOSOM I," 21 March 1997, http://www.un.org/Depts/DPKO/Missions/unosomi.htm.

Garrison, William F., Major General. Classified After Action Report, *Subject: Enclosure 3 (Operations) to After Action Report of Task Force Ranger in Support of UNOSOM II,* 28 October 1993, USASOC History Office Classified Files, Fort Bragg, NC.

Headquarters Department of the Army. Field Manual 3-05, Army Special Forces, September 2006.

Home Office. "Somalia: Majority clans and minority groups in south and central Somalia." London, UK: Country Policy and Information Note, Independent Advisory Group on Country Information, Independent Chief Inspector of Borders and Immigration, Version 3.0, January 2019.

Joint Chiefs of Staff, Joint Pub 3.0, *Doctrine for Special Operations*, Washington DC: U.S. Government Printing Office, September 9, 1993.

Joint Staff, Joint Publication 3-0, *Doctrine for Joint Operations*, September 9, 1993.

Lofland, Valerie J., "Somalia: U.S. Intervention and Operation Restore Hope," http://www.au.af.mil/au/awc/awcgate/navy/pmi/somalia1.pdf, accessed on October 26, 2015.

McGrady, Katherine A.W., *The Joint Task Force in Operation Restore Hope* (CRM 93-114), Alexandria, Virginia: Center for Naval Analyses, March 1994.

Mateer, Shawn M., Major. 9th Psychological Operations Battalion, *After Action Report – Operation*

CONTINUE HOPE, Somalia, 3 January 1994, USASOC History Office Files, Fort Bragg, NC.

Poole, Walter S. *The Effort to Save Somalia August 1992 – March 1994*. Joint History Office, Office of the Chairman of the Joint Chiefs of Staff, Superintendent of Documents, Washington DC: U.S. Government Printing Office, 2005.

Steigman, Dave, LCDR, and Linda Herlocker, CDR. *Naval Special Warfare Forces in Somalia 1992–1995*. Tampa, FL: U.S. Special Operations Command History and Research Office, May 2001.

Thompkins, Kevin L., Major. Company B, 9th Psychological Operations Battalion, *Lessons Learned for Tactical Psychological Operations in Somalia* (After Action Review), 10 March 1994, USASOC History Office Files, Fort Bragg, NC.

Tucker, David and Ambassador Robert B. Oakley, *Two Perspectives on Interventions and Humanitarian Operations*, Carlisle, Pennsylvania: Strategic Studies Institute, July 1, 1997. p. 105.

Unified Task Force Somalia. "Psychological Operations in Support of Operation RESTORE HOPE," 4 May 1993, USASOC History Office Files, Fort Bragg, NC.

United Nations. *Annex to UNOSOM II Campaign Plan for Somalia Special Operations* (DRAFT). UNOSOM II, Mogadishu, Somalia, 4 June 1993.

______________. Map No. 3690 Rev. 6 "SOMALIA," Department of Peacekeeping Operations, Cartographic Section, July 2004.

______________. "The Blue Helmets: A Review of United Nations Peace-Keeping." Third Edition. New York: United Nations Department of Public Information, 1996.

______________. *The United Nations and Somalia, 1992 – 1996*. UN Blue Book Series, Volume 3, New York: United Nations Department of Public Information, 1996.

______________. "United Nations Operations in Somalia." New York, NY, Department of Public Information, United Nations: 21 March 1997.

U.S. Army, Department of the Army Pamphlet 550-86, *Somalia: A Country Study*, 1993.

U.S. Army Center of Military History. *Somalia After Action Report and Historical Overview: The United States Army in Somalia 1992 – 1994*. Washington, DC: Center of Military History, 2003.

______________. "Task Force Ranger Operations in Somalia 3 – 4 October 1993," U.S. Special Operations Command and U.S. Army Special Operations Command History Office, 1 June 1994.

______________. *The United States Army in Somalia, 1992 – 1994*, http://www.history.army.mil/brochures/somalia/somalia.htm, accessed on October 26, 2015.

______________. *United States Forces, Somalia After Action Report*, http://www.history.army.mil/html/documents/somalia/SomaliaAAR.pdf, accessed on October 26, 2015.

U.S. Naval War College, Operations Department, *White Paper: An Analysis of the Application of the Principles of Military Operations Other Than War (MOOTW) in Somalia* (NWC 2243), Langley Air Force Base, Virginia: Army-Air Force Center for Low Intensity Conflict, February 1994.

Wikipedia, the free encyclopedia. "Technical (fighting vehicle)," 26 November 2006, http://en.wikipedia.org/wiki/Technical_ (fighting_vehicle).

______________. "United Nations Operations in Somalia." New York, NY, Department of Public Information, United Nations: 21 March 1997.

Other References

Barth, Major Fritz J. (USMCR), "A System of Contradiction," *Marine Corps Gazette*, Vol. 82, No. 4 (April 1998), 26–29.

Bartlett, Tom, "Guns of Many Nations: The Disarming of Somalia," *Leatherneck*, Vol. 76, No. 4 (April 1993), 12–15.

Cooling, Major Norman L. (USMC), "Operation Restore Hope in Somalia: A Tactical Action Turned Strategic Defeat," *Marine Corps Gazette*, Vol. 85, No. 9 (September 2001), 92–106.

Crocker, Chester A., "The Lessons of Somalia: Not Everything Went Wrong," *Foreign Affairs*, Vol. 74, No. 3, May/June 1995.

United Nations. "United Nations Peacekeeping," http://www.un.org/en/peacekeeping/operations/ peace.shtml, accessed on November 24, 2015.

Army SF in Somalia Interviews

Operation Provide Relief

Interview with CSM Jose Bailey conducted by COL (Ret.) Joseph D. Celeski 8 Feb 06 at Ft. Bragg, NC.

Conversation and interview between DCO of USASFC(A) and SFC Kent Barriger, spring of 2002, held at USASFC(A) Headquarters, Ft. Bragg, NC.

Conversation between COL Joseph D. Celeski and CAPT Randy Goodman, conducted at JSOU, Hurlburt Field, FL, summer of 2005 during JSOTF Training Conference.

Discussion with MG (Ret.) Kenneth R. Bowra in November 2006 on particulars concerning the deployment of one FOB Light to augment the SF company. Information provided by Wendell Greene during interview with COL (Ret.) Celeski, 2 July 2005, 1530 hrs, in Nashville, TN. Subject: Operation PROVIDE RELIEF. Transcribed from tape recording.

Information and pictures provided to the author during the month of October 2006 by BG (Ret.) Mark Hamilton (USAF) who served on the CJTF staff during Operation PROVIDE RELIEF.

Interview between MAJ Kent Listoe, CO A, 2/5th SFG and de-briefer from 44th Military History Detachment, conducted at Ft. Campbell, KY 13 May 1993, pp. 1–6.

Interview with LTC Steve Moniz by COL (Ret.) Joseph Celeski, conducted on 29 Nov 06 in Raipur, India.

Interview with CPT Steven P. Moniz, ODA542 Detachment Commander and MSGT Daniel J. Kaiser, ODA542 Team Sergeant on the roles and missions of ODA542 in Operation PROVIDE RELIEF.

Interview with CPT Steven P. Moniz, ODA542 Detachment Commander and MSGT Daniel J. Kaiser, ODA542 Team Sergeant on the roles and missions of ODA542 in Operation PROVIDE RELIEF.

Interview conducted by the 44th Military History Detachment on 10 May 1993 in the 2/5th SFG(A) Classroom at Ft. Campbell, KY.

Interview between COL Joseph D. Celeski and SGM Sloniger 1/3rd SFG(A) during OEF, at Kabul Military training Facility, late spring of 2003.

Interview between SSG Glenn Wharton and 44th Military History Detachment in the 2/5th SFG(A) classroom on 10 May 1993, at Ft. Campbell, KY.

Operation Restore Hope

Oral history interview with ODA561, Co C, 2/5th SFG(A), comments by SFC Breed, team sergeant, conducted by the 44th Military History Detachment in the 2/5th SFG(A) classroom, 13 May 1993, pp. 10 – 11. Extracted from notes of CWO3 Jon Concheff's personal log provided to COL(Ret.) Joseph D. Celeski in March 2003, pp. 1–5. CWO3 Concheff's team consisted of six personnel from the 10th SFG(A) formed as a composite Coalition Warfare team.

Somalia interview between LTC Kevin Murphy and COL (Ret.) Joseph D. Celeski on 30 May 2007, 1500 – 1630 hrs. Interview conducted in the USASOC Historian's Conference Room, Ft. Bragg, NC.

Personal note from Gary Ramsey to COL (Ret.) Joseph D. Celeski describing his activities in Somalia operations in 1992 and 1993.

Interview with Charles B. Smith on his team activities in HRS *Belet Weyne* during Opertion Resstore Hope. after reviewing the initial draft of his autobiography.

Daily patrol log recorded by CPT Tim Williams, detachment commander for ODA526, on operations of the team Dec 1992 through March 1993. A copy of the patrol log was personally provided to the author by LTC (P) Williams at Hurlburt Field, FL during the JSOU SOF pre-command course in 2005.

Oral history interview with ODB560, Co C, 2/5th SFG(A), conducted by the 44th Military History Detachment in the 2/5th SFG(A) classroom, 12 May 1993, p. 31.

Oral history interview with ODA561, Co C, 2/5th SFG(A), conducted by the 44th Military History Detachment in the 2/5th SFG(A) classroom, 13 May 1993.

Oral interview of ODA563 Co C, 2nd Bn, 5th SFG by the 44th Military History Detachment on 13 May 93 in the 2/5th SFG classroom at Ft. Campbell, Ky.

Oral Interview of ODA565 Co C, 2nd Bn, 5th SFG by the 44th Military History Detachment on 12 May 93 in the 2/5th SFG classroom at Ft. Campbell, Ky.

Oral interview of ODA565 Co C, 2nd Bn, 5th SFG by the 44th Military History Detachment on 12 May 93 in the 2/5th SFG classroom at Ft. Campbell, Ky.

Mission highlights briefing files from the five ODAs of Co C, 2/5th who participated in Operation Restore Hope. These files contain annotated calendars and sketch maps of key operations performed by each ODA.

Operation Continue Hope

E-mail between COL Joseph D. Celeski and CSM Patrick Ballog, Monday Feb 4, 2002, 5:15 p.m. Subject: Somalia.

E-mail between COL Joseph D. Celeski and CSM Patrick Ballog, Monday Feb 5, 2002, 10:21 am.

E-mail from LTC (Ret.) Moe Elmore on the subject of the ambush from his perspective, 17 Mar 2002, 11:51 hrs.

Notes and discussion from Mr. Al Glover, one of the JSOTF staff members who accompanied Ambassador Glaspie on this trip. Provided to COL (Ret.) Joseph D. Celeski in June 2005.

Detachment daily journal from Mike Hurst provided to COL (Ret.) Joseph D. Celeski in 2005. Also notes from executive summary and AAR in the Memorandum dtd 18 Sep 1993, Subject: AAR, SF ODB520. Deployment to Somalia, 24 Mar – 19 Sep 1993, signed by MAJ David G. Jesmer.

Witness statement provided by then MAJ Timothy M. Knigge to the SOCCE as supporting document for Silver Star awards submission for SF CST team who accompanied him that day as the Moroccan LNO. MAJ Knigge described the fight and the actions of the SF team members in order to provide a record of the events for later submission of awards. Copy of the 5-page statement provided to COL (Ret.) Joseph D. Celeski by Mike Hurst in late November of 2004.

Team log notes of CPT Mike Hurst. Provided to COL (Ret.) Joseph D. Celeski in late November 2004.

JTF Somalia

Notes from NSWTU-Alpha post operations report provided by CDR (Retired) Tom Bunce to COL (Ret.) Joseph D. Celeski on 5 Mar 2002. CDR Bunce was the commander of NSWTU-A during this period.

-mail response to Somalia questionnaire prepared by COL (Ret.) Joseph D. Celeski, dtd 29 Jan 2005.

E-mail from Dwight Comer to COL (ret.) Joseph D. Celeski about his sniper experiences in Somalia, dtd 30 Sep 2006.

Personal log of Al Glover covering the events of his tour in Somalia, entry dtd 5 January 1994.

Interview between Brigadier General (Ret.) Russ Howard and COL (Ret.) Joseph D. Celeski conducted at Tufts University, Jebsen Center 27 – 28 Feb 2008, 1430 hrs.

E-mail from Wylie W. Johnson to COL (Ret.) Joseph D. Celeski dtd 29 January 2005.

E-mail from Moe Elmore to COL Joseph D. Celeski, 14 Mar 2002, 4:22 pm, describing the force protection assessment mission.

E-mail from David B. Plummer to COL (Ret.) Joseph D. Celeski on his tenure as JSOTF commander during January–March 1994 period, dtd 30 January 2002.

Letter provided by Bill Robinson to COL (Ret.) Joseph D. Celeski outlining the role of ODB590 during his command, 8 Feb 2002.

Company C, 3/5th SFG(A) Unit History Report 1993, Executive Summary Fiscal Year 1993, dtd 24 May 04. This summary contained the unit's activities in Somalia for 4th QTR CY1993.

Somalia NEO

Interview with retired Colonel Chip Paxton in Alexandria, VA 2007.

Interview with COL Pat Higgins conducted on 29 May, 2007 at Fort Bragg, NC.

Operation United Shield

Discussion and e-mail between Dave Asher and COL (Ret.) Joseph D. Celeski 5 Feb 2006. The interview was conducted over the phone between Joe Celeski at home in Buford, GA on 5 Feb 2006, 0930 hrs, and Dave Asher from his location in Iraq near the Syrian border.

Interview between CSM Jose Bailey and COL (Ret.) Joseph D. Celeski at Ft. Bragg, NC, 8 Feb 2006.

Interview with Lance Caffrey on 25 Oct 2006, 0900 hrs, conducted by COL (Ret.) Joseph D. Celeski.

Interview conducted by COL (Ret.) Joseph D. Celeski with LTC Steve Cain conducted at Ft. Bragg, NC on 21 Sep 2004.

Phone discussion between BG Csrnko and COL (Ret.) Joseph D. Celeski conducted at Ft. Bragg, NC, 3 Aug 2006.

E-mail between CSM (Ret.) Frank McFadden and COL (Ret.) Joseph D. Celeski, 13 Oct 2006.

In memory of SFC Robert H. "Bob" Deeks, Jr., killed by a landmine during the conduct of a mounted patrol, in the province of Belet Weyne, 3 March 1993 during Operation RESTORE HOPE.

About the Author

COL Joseph D. Celeski (1954–2023) retired from a thirty-year career with the U.S. Army in September, 2004 after successful completion of commanding the 3rd Special Forces Group (Airborne), Fort Bragg, North Carolina. He assumed command of the Group in May 2002 in Afghanistan where he also served as the commander of the Combined and Joint Special Operations Task Force (CJSOTF) for two tours in Operation ENDURING FREEDOM.

After his initial career in the Armor branch, Colonel Celeski volunteered for Special Forces in 1984. He served in all command positions from a Special Forces A-team to a Special Forces Group Commander. He was an armor advisor in the 1st Royal Jordanian Armor battalion during 1989. Late in his career, he served in staff positions as the G3, Chief of Staff, and Deputy Commander of the United States Special Forces Command (Airborne).

While serving as the SOCCENT J3 Ground Operations Officer, he deployed to Somalia in late 1993 to serve as the JSOTF-Somalia J3. Later, he was one of the primary plans officers for the Somalia NEO plan in 1994 and the UN withdrawal operation, United Shield, in 1995.

His published works include articles on Special Forces in Somalia, the use of Special Forces in joint urban combat, and SOF Strategic Application. While serving as a senior fellow for the Joint Special Operations University he published three monographs on the application of SOF in counterinsurgency. His two recent published books include *The Green Berets in the Land of a Million Elephants* (Casemate, 2019) and *Special Air Warfare and the Secret War in Laos* (Air University Press, 2019), both revealing the role of special operators in Laos from 1959 to 1975.

Colonel Celeski was a graduate of the U.S. Army Command and Staff College and the Army War College. He earned a Master's Degree in Public Administration from Shippensburg University, Pennsylvania and a Master's Degree in Strategic Studies from the Army War College, Carlisle, PA.

Colonel Celeski was also a veteran of Desert Storm (Ceasefire Phase), Bosnia, and two tours in Afghanistan. He was a joint specialty officer and also qualified in the Arabic language.

Colonel Celeski was among one of the first recipients of the St. Philip Neri awards (Bronze) for his active service in Special Forces. Colonel Celeski passed away during the final production of this book. He is remembered not only for his strategic acumen and published scholarship, but also for his enduring commitment to the Special Forces community. He resided in Buford, Georgia, with his beloved wife, Judy.

www.ingramcontent.com/pod-product-compliance
Lightning Source LLC
Chambersburg PA
CBHW060302010526
44108CB00042B/2610
* 9 7 8 1 9 5 9 2 0 3 1 2 4 *